W9-CSP-540

Selected Papers, with Commentary, of Tony Hilton Royle Skyrme

World Scientific Series in 20th Century Physics

Published

Vol. 1 Gauge Theories – Path and Future
edited by R. Akhoury, B. de Wit and P. van Nieuwenhuizen

Vol. 2 Scientific Highlights in Memory of Léon van Hove
edited by F. Nicodemi

Forthcoming

Vol. 3 Selected Papers, with Commentary, of T. H. R. Skyrme
edited by G. E. Brown

Vol. 5 Selected Papers of Abdus Salam (with Commentary)
edited by A. Ali, C. Isham, T. Kibble and Riazuddin

Vol. 6 Research on Particle Detectors
by G. Charpak

Vol. 7 A Career in Theoretical Physics
by P. W. Anderson

World Scientific Series in 20th Century Physics – Vol. 3

Selected Papers, with Commentary, of Tony Hilton Royle Skyrme

Editor

Gerald E. Brown

Institute of Theoretical Physics
State University of New York
Stony Brook, New York, USA

World Scientific
Singapore • New Jersey • London • Hong Kong

Published by

World Scientific Publishing Co. Pte. Ltd.

P O Box 128, Farrer Road, Singapore 9128

USA office: Suite 1B, 1060 Main Street, River Edge, NJ 07661

UK office: 73 Lynton Mead, Totteridge, London N20 8DH

The editor and publisher would like to thank the authors and the following publishers of the various journals and books for their assistance and permission to reproduce the selected reprints found in this volume:

The American Institute of Physics (*J. Math. Phys.*); Elsevier Science Publishers B. V. (*Nucl. Phys. B, Phys. Lett. B, Phys. Rep.*); Taylor & Francis (*Phil. Mag.*); The Royal Society (*Proc. Roy. Soc. Lond.*).

While every effort has been made to contact the publishers of reprinted papers prior to publication, we have not been successful in some cases. Where we could not contact the publishers, we have acknowledged the source of the material. Proper credit will be accorded to these publishers in future editions of this work after permission is granted.

We are grateful to Professor R. H. Dalitz and Dr. D. Skyrme for providing the photographs of Professor T. H. R. Skyrme found in this volume except that taken in 1946, courtesy of Trinity College, Cambridge.

SELECTED PAPERS, WITH COMMENTARY, OF T. H. R. SKYRME

Copyright © 1994 by World Scientific Publishing Co. Pte. Ltd.

All rights reserved. This book, or parts thereof, may not be reproduced in any form or by any means, electronic or mechanical, including photocopying, recording or any information storage and retrieval system now known or to be invented, without written permission from the Publisher.

For photocopying of material in this volume, please pay a copying fee through the Copyright Clearance Center, Inc., 27 Congress Street, Salem, MA 01970, USA.

ISBN 981-02-1646-7

Printed in Singapore by Utopia Press.

This book is dedicated to Professor R. E. Peierls,
Tony's teacher and mine

Tony Skyrme
Former Fellow, Trinity College, Cambridge. Reprinted with permission
© 1946 Trinity College, Cambridge.

Taken in 1948 at Cornwall by Dr. Dorothy Skyrme.

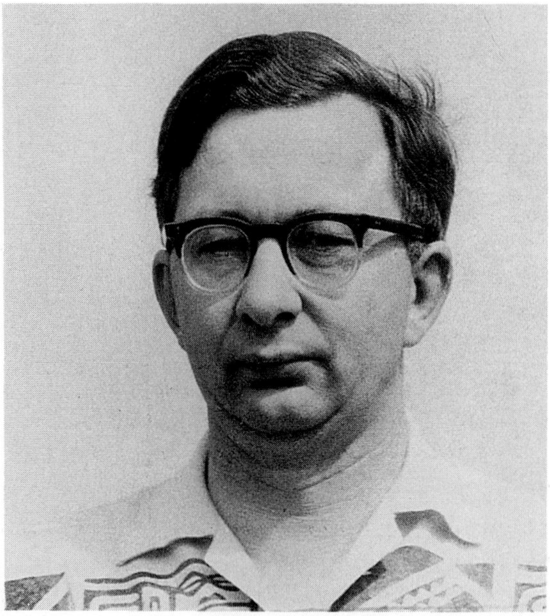

Taken in 1958 by Dr. Dorothy Skyrme.

Preface

Some time ago, Dr. K. K. Phua, Chairman of World Scientific Publishing, asked me to edit a collection of Tony Skyrme's papers. Dr. Phua had carried out his Ph.D. work in Tony's department in Birmingham, England. I had been in Birmingham from 1950–1960, working with Professor R. E. Peierls, who also was Tony's research supervisor for many years. University people in the industrial city of Birmingham were thrown inwards upon themselves, and a great sense of comraderie developed. Certainly I felt that with the Peierls and, through them, with Tony Skyrme. Even though Dr. Phua was in Birmingham at a very different time, he and I have a friendship founded upon red brick. "Red brick" denotes, in England, the discomfort — and strength and vitality — of a university like Birmingham, as opposed to the stuffy grandeur of Oxford and Cambridge.

In the spring of 1951, while I was a boarder in the Peierls' house, Tony Skyrme phoned and said that he and his wife Dorothy were coming to Birmingham for a visit. The Peierls were away, but knowing what good friends Tony and Prof.* Peierls were, I invited the Skyrmes to stay in the Peierls' house. Little did I realize that Mrs. Peierls had dragged Prof. Peierls away on an impromptu spring vacation, so that she would be prepared to entertain 350 participants in the coming British Association for Science meeting at a buffet supper in their house. At least I, and Tony, if he was still staying there, dutifully buttered and laid the thousand or so sandwiches Mrs. Peierls prepared. Certainly we Birmingham products were all united through the strength of Mrs. Peierls.

Of course, I visited Harwell when Brian Flowers, now Lord Flowers, was director and Tony the head of theoretical physics, and participated in "skyrmishes".

I had thought, in agreeing to edit this volume, to give some biographical material on Tony Skyrme, but each time I started, I found that Dick Dalitz had done it better, so his "An outline of the life and work of Tony Hilton Royle Skyrme" is reprinted in entirety in this book.

In the summer of 1960, I left Birmingham, taking up a professorship in NORDITA, the Nordic Institute for Theoretical Atomic Physics, in Copenhagen, at the invitation of Niels Bohr. In the autumn of 1960, Tony Skyrme gave a seminar

*Prof. was, and is, a term of endearment, not formality, with Rudi Peierls.

in Birmingham on skyrmions. Had I still been in Birmingham, I would certainly have attended that seminar. Although the concepts were so far ahead of their time, probably something would have stuck in my mind, because 20 years later, when Mannque Rho and I were in a dilemma in our construction of the chiral bag model of a nucleon, the fundamental concept in skyrmions was needed to rescue us. It took us more than a year in 1980 to resurrect it, but I want to tell that story later in my commentary on skyrmions.

The reader can ask how, aside from my common association with Birmingham, am I suited to make a selection of Tony's papers and to edit this volume? On the formal aspects of skyrmions, I am no great expert. On the other hand, there is an excellent article by Valery Sanyuk, which is reproduced in this volume. I was a "user" and I needed skyrmions, not because they were the first three-dimensional soliton, nor because of their formal beauty, but because they solved a real problem for Mannque Rho and me, as I shall describe in my commentary on skyrmions.

On the other hand, Tony did a lot of other things in nuclear physics. He proved the completeness of the Kapur–Peierls dispersion theory, although this proof was never published, as Dalitz relates. I used the Kapur–Peierls theory to construct in nuclei what is an analogue of one of the Anderson models in condensed matter physics. It was comforting to know that the theory was complete. Doubts about the completeness had been expressed.

I have included the Perring and Skyrme work on the alpha-particle and shell models of the nucleus. This is to show the modern reader that nuclear physics can be elegant, and that elegance is useful. These are often forgotten in the number-crunching age when huge shell model calculations devour input parameters and turn out wave functions, often untouched by the human mind.

Skyrme's work of most use to nuclear physicists is the nuclear Skyrme model. It is highly amusing to me that skyrmions were motivated, as he tells it in his "The origins of skyrmions", by the "fluid drop model of the nucleus", outlined in his papers "A new model for nuclear matter" and "Meson theory and nuclear matter", published here. By the mesonic fluid, Tony meant the π-meson field, which permeated the nucleus. In his picture, the nucleonic sources, or singularities of this field, were confined by the strong interaction between field of the fluid and sources. Two decades later, Mannque Rho and I used the strong pion field to compress the quarks in the nucleon to make the nonperturbative chiral bag, and in the process, rediscovered Skyrme's work.

From the nonlinear σ-model, invented by Tony, it has proved possible to construct the scalar interaction, composed of correlated pairs of pions coupled to angular momentum zero, the exchange of which holds the nucleus together. As noted later in my commentary on the Skyrme interaction, hard cores are present in the interaction between nucleons, and these come about from the exchange of the vector ω-mesons. These are easily introduced into the theory, coupled to the conserved isoscalar nucleon current. Tony doesn't initially give any physical significance in terms of meson exchange that provides the fourth-order term in the Skyrme equation. "It is conjectured that in the quantized theory the value of ϵ will be determinate, possibly unity, for reasons similar to those which made $\omega Q^2 = 1/\pi$ in the model of I" from "A non-linear field theory". It is now known (see the review of Zahed and Brown, the final paper in this volume) that the fourth-order term derives from ρ-meson exchange, in zero range approximation, where the ρ-meson is the partner of the ω in a $U(2)$ symmetry.

Using the mesons discussed above, J. D. Walecka has, in many papers, constructed the theory of Nuclear Hadrodynamics, which is very much used. Skyrme was probably influenced by the earlier arguments of Bohr and Mottelson, based on the liquid drop model of the nucleus. In any case, Skyrme's fluid drop model was very useful, in that it led him to the idea of skyrmions.

In my commentary on the Skyrme interaction, I note, however, that it is useful more in the context of the nuclear shell model, more in the nature of the Landau–Fermi liquid theory, which has, as a central idea, the concept of quasiparticles.

Of course, Tony's time in Harwell was a time of great reconciliation between the various theories. J. P. Elliott, in the $SU(3)$ formalism, showed how, with the proper interactions, the shell model could reproduce the Bohr–Mottelson collective model. Perring and Skyrme, in an article reproduced here, show how the alpha-particle model relates to the shell model. As Weisskopf has explained, a model "exaggerates the particular features" one wishes to deal with, and this was true, also, of Skyrme's alpha-particle model.

It must have been incredibly frustrating for Tony that almost no one understood or appreciated his theory of skyrmions until about two decades after he invented it. It took great courage and perseverance to keep on developing the theory. I heard Morris Pryce, then professor at Oxford, and sponsor of "A new model for nuclear matter", for the Royal Society, say that "Tony was too clever for his own good".

Dick Dalitz notes about Tony that in 1985, after he was awarded the very prestigious Hughes Medal by the Royal Society, "As Tony came back to his seat, grasping his medal and with a grin all over his face, he muttered with justice, 'They none of them could understand any of it at the time.'"

I am reminded of Henry Thoreau's rejoinder, upon being asked by Ralph Waldo Emerson, when Thoreau was in jail for failure to pay taxes because of his opposition to the Mexican–American war, "Why are you here?" Thoreau replied, "Why are you not here?"

Tony was never elected to the Royal Society, although the Hughes Medal was somewhat of an apology. He gave us the most conceptually profound work from England since the war. Perhaps, looking at the Fellows of the Royal Society, he could, with reason, have reversed the Emerson–Thoreau exchange and asked "Why are you there?"

G. E. B.

Contents

Preface ix

1. A. S. Goldhaber, "Obituary of T. H. R. Skyrme", *Nucl. Phys.* **A487** (1988). 1

2. R. H. Dalitz, "An outline of the life and work of Tony Hilton Royle Skyrme (1922–1987)", *Int. J. Mod. Phys.* **A3** (1988) 2719–2744. 6

I. THE ALPHA-PARTICLE AND SHELL MODELS OF THE NUCLEUS 33

1. J. K. Perring and T. H. R. Skyrme, "The alpha-particle and shell models of the nucleus", *Proc. Phys. Soc. Lond.* **A69** (1956) 600–609. 37

II. THE SKYRME INTERACTION 47

1. T. H. R. Skyrme, "The effective nuclear potential", *Nucl. Phys.* **9** (1959) 615–634. 51

2. J. S. Bell and T. H. R. Skyrme, "The nuclear spin–orbit coupling", *Phil. Mag.* **1** (1956) 1055–1068. 71

3. T. H. R. Skyrme, The spin–orbit interaction in nuclei", *Nucl. Phys.* **9** (1958/59) 635–640. 85

4. T. H. R. Skyrme, "A new model for nuclear matter", *Proc. Roy. Soc. Lond.* **A226** (1954) 521–530. 91

5. T. H. R. Skyrme, "Meson theory and nuclear matter", *Proc. Roy. Soc. Lond.* **A230** (1955) 277–286. 101

III. SKYRMIONS 111

1. T. H. R. Skyrme, "The origins of skyrmions", from *A Breadth of Physics, Proc. of the Peierls 80th Birthday Symposium*, Oxford Univ., 115

June 1987, World Scientific Pub., eds. R. H. Dalitz and R. B. Stinchcombe, pp. 192–202; also published in *Int. J. Mod. Phys.* **A3** (1988) 2745–2751.

2. Valery I. Sanyuk, "Genesis and evolution of the Skyrme model from 1954 to the present", *Int. J. Mod. Phys.* **A7** (1992) 1–40. 126

3. T. H. R. Skyrme, "A non-linear theory of strong interactions", *Proc. Roy. Soc. Lond.* **A247** (1958) 260–278. 166

4. T. H. R. Skyrme, "A unified model of K- and π-mesons", *Proc. Roy. Soc. Lond.* **A252** (1959) 236–245. 185

5. T. H. R. Skyrme, "A non-linear field theory", *Proc. Roy. Soc. Lond.* **A260** (1961) 127–138. 195

6. T. H. R. Skyrme, "Particle states of a quantized meson field", *Proc. Roy. Soc. Lond.* **A262** (1961) 237–245. 207

7. J. K. Perring and T. H. R. Skyrme, "A model unified field equation", *Nucl. Phys.* **31** (1962) 550–555. 216

8. T. H. R. Skyrme, "Kinks and the Dirac equation", *J. Math. Phys.* **12** (1971) 1735–1743. 222

IV. THE REDISCOVERY OF SKYRMIONS 231

1. E. Witten, "Baryons in the $1/N$ expansion", *Nucl. Phys.* **B160** (1979) 57–115. 235

2. E. Witten, "Global aspects of current algebra", *Nucl. Phys.* **B223** (1983) 422–432. 294

3. E. Witten, "Current algebra, baryons, and quark confinement", *Nucl. Phys.* **B223** (1983) 433–444. 305

4. G. S. Adkins, C. R. Nappi, and E. Witten, "Static properties of nucleons in the Skyrme model", *Nucl. Phys.* **B228** (1983) 552–566. 317

5. G. S. Adkins and C. R. Nappi, "Stabilization of chiral solitons via vector mesons", *Phys. Lett.* **B137** (1984) 251–256. 332

6. I. Zahed and G. E. Brown, "The Skyrme model", *Phys. Rep.* **142** (1986) 2–102. 338

Selected Papers, with Commentary, of Tony Hilton Royle Skyrme

Obituary of T. H. R. Skyrme

December 5, 1922 — June 25, 1987

T. H. R. Skyrme

(December 5, 1922 — June 25, 1987)

Tony Hilton Royle Skyrme died unexpectedly on 25 June 1987, at the age of 64. In 1985 he received the Hughes Medal, which throughout this century has been the rarest honor conferred on British physicists by the Royal Society (election to Fellowship in the Society is common by comparison, so that Skyrme's status as a Medalist but not a Fellow must be nearly unique).

The list of Skyrme's publications presented at the end of this article contains 23 items. Of these, no fewer than 9 are related to the Skyrmion [2,3,15-21]). Among his other notable works are several on many-body theory, particularly focused on applications in nuclear physics. The short-range "Skyrme forces" provided a physically sensible and computationally practical description of the residual interaction which must be added to the shell model to give a more accurate picture of nuclear structure [12-14]). This formulation is still used today, nearly 30 years after its introduction. A systematic method of isolating the degrees of freedom corresponding to center-of-mass motion of a many-body system was formulated [5]). An analysis of nuclear surface dynamics was presented [6]). These too are often cited today.

Skyrme's thinking about what is now called the Skyrmion evolved from an early paper in which he suggested that nuclear dynamics might be described in terms of pion degrees of freedom [2]). The attractive notion that nucleons might be made up of the much lighter pions was barely more than implicit in this work. Others may well have shared this idea, but only Skyrme appears to have pursued it, encountering and overcoming formidable obstacles to make a viable model. From the modern point of view, there are four principal questions which must be asked about such a model:

(1) Why is the skyrmion stable against decay, and why does it carry an integral quantum number which can be identified with baryon number?

(2) How can an entity constructed from spinless mesons exhibit half integral spin and Fermi-Dirac statistics?

(3) Can such a simple model reproduce accurately the results of many different experimental observations?

(4) Given that high-energy physics points to quarks and gluons as the fundamental constituents of strongly interacting particles, is this very different model compatible with and connectable to quantum chromodynamics?

In two key papers [17,20]) Skyrme addressed and solved the first two questions, and did significant initial work on the third, which has only been much advanced during the revival of interest in his work over the past 5 years. Identification of connections with quantum chromodynamics is also a recent development, reinforcing the continuing phenomenological advances. Skyrme's answer to the first question was to introduce a dynamics which enforce the continuity of a map from real space including the point at infinity to the parameter space of the group SU(2), which is simply the surface of a sphere in 4 dimensions, i.e., S_3. This map is characterized by a winding number which is automatically quantized and conserved. To carry out computations, Skyrme introduced an ansatz with a peculiar rotational symmetry mixing space and isospin coordinates. This device has been used many times since in related problems. He computed a lower bound on the energy of the skyrmion, and his result was later generalized to apply for arbitrary configurations and for larger groups.

Perhaps the most puzzling and subtle issue was the appearance of half integral spin. Here, Skyrme's point was that the manifold of states which go into each other under rotation is intrinsically different for any finite collection of spin-0 particles from what it is for a classical pion field configuration corresponding to an indefinite number of particles. Consequently, in the latter case, spin-half representations become allowed though not required. Indeed, the results of more recent work are that the spin of such an object will be half-integral or integral according as the number of colors in the fundamental chromodynamic group is odd or even. Meanwhile, work by others showed that the same complexity of the manifold would allow but not require an extra minus sign in the phase of a quantum mechanical wave function for interchange of two skyrmions.

As part of the effort to understand the nature of such objects, Perring and Skyrme [19]) studied the behavior of analogous objects in one space dimension. Their result, indicating that scattering occurs without radiation, provided evidence that the phenomenon now known by the name of soliton is rather general, and not confined to the solutions of one non-linear differential equation only. A further indication of the profundity of Skyrme's vision is the fact that in order to assure the topological stability of a pion field configuration he found himself compelled to introduce the non-linear sigma model, which was independently developed by others for totally different reasons, and which has turned out to be phenomenologically exceedingly successful.

In short, Skyrme's work was brilliant, deep, and right: brilliant, in that he provided an original solution to the problem of constructing fermions from bosons; deep, in that it has far reaching implications in a number of directions; and right, in that his two-parameter model yields a remarkably accurate description of low-energy hadron phenomena. Late in 1983 there occurred a "phase transition" in the rate of citations to the key paper [17]), from order of magnitude 10 to order of magnitude 100. Since then the rate has continued to increase steadily, a powerful indication of ever-widening recognition for one of the small number of truly original ideas in mathematical physics since quantum theory was invented. However belated, this recognition came in time for Skyrme to enjoy it, and for that, as well as the work itself, we can be grateful.

Alfred S. Goldhaber

T. H. R. Skyrme

Selected References

Statistics of Kinks

D. Finkelstein and J. Rubinstein, J. Math. Phys. **9** (1968) 1762-79

Soliton Physics

M.J. Ablowitz, D.J. Kaup, A.C. Newell and H. Segur, Studies in Appl. Math. **53** (1974) 249-315

Topological Stability and Energy Bounds

L.D. Faddeev, Letts. Math. Phys. **1** (1976) 289-93

Quantum Chromodynamics

E. Witten, Nucl. Phys. **B223** (1983) 433-44

Publications

1) The theory of the double Compton effect (with F. Mandl), Proc. Roy. Soc. **A215** (1952) 497-507
2) A new model for nuclear matter, Proc. Roy. Soc. **A226** (1954) 521-30
3) Meson theory and nuclear matter, Proc. Roy. Soc. **A230** (1955) 277-86
4) Quantum field theory, Proc. Roy. Soc. **A231** (1955) 321-35
5) Centre-of-mass effects in the nuclear shell-model (with J.P. Elliott), Proc. Roy. Soc. **A232** (1955) 561-6
6) The nuclear surface, Phil. Mag. **1** (1956) 1043-54
7) Re nuclear spin-orbit coupling (with J.S. Bell), Phil. Mag. **1** (1956) 1055-68
8) Re alpha-particle and shell-models of the nucleus (with J.K. Perring), Proc. Phys. Soc. **A69** (1956) 600-9
9) Nuclear moments of inertia, Proc. Phys. Soc. **A70** (1956) 433-44
10) Collective motion in quantum mechanics, Proc. Roy. Soc. **A239** (1957) 399-412
11) Fusion induced by mu-mesons, Phil. Mag. **2** (1957) 910-16
12) The effective nuclear potential, Nucl. Phys. **9** (1959) 615-34
13) The spin-orbit interaction in nuclei, Nucl. Phys. **9** (1959) 635-40
14) Some distortion effects in the nuclear p-shell, Nucl. Phys. **9** (1959) 641-9
15) A non-linear theory of strong interactions, Proc. Roy. Soc. **A247** (1958) 260-78
16) A unified model of K- and π-mesons, Proc. Roy. Soc. **A252** (1959) 236-45
17) A non-linear field theory, Proc. Roy. Soc. **A260** (1961) 127-38
18) Particle states of a quantized meson field, Proc. Roy. Soc. **A262** (1961) 237-45
19) A model unified field equation (with J.K. Perring), Nucl. Phys. **31** (1962) 550-5
20) A unified field theory of mesons and baryons, Nucl. Phys. **31** (1962) 556-69
21) Kinks and the Dirac equation, J. Math. Phys. **12** (1971) 1735-43
22) On an identity relating to partitions and repetitions of parts (with M.S. Kirdar), Can. J. Math. **34** (1982) 194-5
23) The maximum of a random walk whose mean path has a maximum (with H.E. Daniels), Adv. in Appl. Probability **17** (1985) 85-99

International Journal of Modern Physics A, Vol. 3, No. 12 (1988) 2719 – 2744
© World Scientific Publishing Company

AN OUTLINE OF THE LIFE AND WORK OF TONY HILTON ROYLE SKYRME (1922–1987)

R. H. DALITZ
Department of Theoretical Physics, Oxford University, Oxford OX1 3NP, UK

Received 24 September 1988

1922–43: Youth and Education

Tony Hilton Royle Skyrme was born on 5 December 1922 at 7 Blessington Road, Lewisham (Kent), London, the family house occupied by his maternal grandparents. His parents were John (sometimes Jack) Hilton Royle Skyrme, a bank clerk, and Muriel May née Roberts, who had been married at St. Margaret's Church in the parish of St. Margaret's and Eastney, in Portsmouth (Hants.), on 25 March 1922. Tony's paternal grandparents were James Henry Rowland Skyrme and Minnie née Hilton, the former being a schoolmaster at Combwitch, near Bridgewater (Somerset), when Tony's father was born in 1896. Tony's maternal grandfather was Herbert William Thomson Roberts, a tidal computer for the Admiralty by profession. The inclusion of Lord Kelvin's baptismal name (William Thomson) among his forenames reflects the professional contact which Tony's great-grandfather had with Lord Kelvin and the high regard in which he held the latter.

This great-grandfather of Tony's on the maternal side was Edward Roberts, born on 1 May 1845 at Deptford Broadway, a son of John Roberts of Greenwich.[1] From 1860, he worked as an assistant at the Royal Observatory, Greenwich, but was transferred to the Nautical Almanac Office in 1864. In 1868, he was appointed Secretary to the Tidal Committee of the British Association for the Advancement of Science, being made responsible later for the construction of the first Tidal Predicter, which had been designed by Lord Kelvin for this Committee,[2] and which was constructed during the winter months of 1872/3 and exhibited first at the Bradford Meeting of the British Association in 1873. Later, he played a large part in the design and construction of the Universal Tide-predicting Machines used by the Indian and Colonial Government and by the Admiralty Hydrographic Office, work for which he received the Imperial Service Order in 1907, when he retired from his post as Chief Assistant to the Superintendent of the Nautical Almanac Office. Edward Roberts was also a Fellow of the Royal Astronomical Society,[3] having been elected in 1872. After his retirement, he and his wife Louisa (née Webb) moved to Park Lodge, Eltham, and later to Broadstairs (Kent), leaving his house at 7 Blessington Road for the use of his son and other family members. It was this house which held the Tidal Predicter, the first model of the machine, which made such a strong impression on the young Tony and influenced so greatly the development of his later ideas, as Tony himself recounted in a lecture given at a Workshop on Skyrmions in 1984, a lecture which has been reconstructed and reprinted in this journal.[4] Edward Roberts lived to be 88 years old, his death being at Woolwich on 4 August 1933, so Tony knew and remembered him as well as his tide-predicting machine.

Tony's early education was at Belmont House School, a boarding school at 40 Lee Terrace, Blackheath (now known as Lewisham), which had been operated by Helen and Jessie Barff of 5 Blessington Rd. (next door to the Edward Roberts house) until their deaths in 1928 and 1930, respectively, so that it was well-known to Tony's family. Tony started at this School in 1930, just after it came under the management of Frank H. Ashley, who had purchased it from the Barff estate and operated it until World War II. This preparatory school had quite a high reputation and a good many of its pupils went on to major public schools, as did Tony, whose unusual ability with mathematical calculations had already become clear at this school.

In 1936, Tony took the entrance examination for Eton College, gaining fifth place and the award of a scholarship on the Foundation provided by King Henry VI when he established the College. As one of the 70 King's Scholars, he lived in the King's House at Eton. During his four years at Eton, he was consistently outstanding in Mathematics, gaining the Hawtrey Prize in 1938, the Tomline Prize in 1939 and the Russell Prize in 1940, for example. He played in the Mixed Wall, a school team made up of both Collegers (= Scholars) and Oppidans (= Commoners) for the traditional Eton Wall Game, which played from time to time against Wall teams made up from former Etonians, and he was twelfth man in College Wall, the team of Scholars. The Eton College Chronicle for 30 November 1939, describing the annual St. Andrew's day Wall game between College and Oppidans, noted that 'Steward and Skyrme were very efficient and tough'. In 1940, Tony took the entrance examination for Cambridge University. He was awarded the Reynolds Scholarship at King's College, but chose instead to become a Scholar of Trinity College.

At Cambridge, Tony became an outstanding mathematician, gaining 1st Class Honours in Part II of the Mathematical Tripos in 1942, proceeding to Part III on a Senior Scholarship in the following year, and so gaining his B. A. degree, as was the custom in those days. During his undergraduate years Tony was particularly active in the Archimedeans, the Mathematical Society of Trinity College, being its Secretary during Lent term of 1942, and its President during the Michaelmas term of 1942 and the Lent term of 1943.

1943–46: The War Years

After Tony's final examinations at Cambridge in June 1943, the Manpower Board assigned him to work with the Tube Alloys Directorate,[5] a body which came under the Department of Scientific and Industrial Research (D. S. I. R) and had responsibility for U. K. research and development work towards an atomic bomb. The centre of its theoretical work was at Birmingham University where Professor Rudolf Ernst Peierls (later Sir Rudolf) headed a group, funded by D. S. I. R., which was investigating a wide range of questions posed by the design and construction of such a bomb, beginning with the problems of uranium isotope separation, necessary in order to obtain the fissile isotope ^{235}U in sufficient quantity for its construction, and Tony was sent to work under Peierls. Among others, this group included A. H. Wilson, K. Fuchs, B. Davison and some Birmingham University staff, such as G. J. Kynch.

Tony soon made his presence felt in the Peierls group, taking part in a wide variety of its activities. The first two problems which he worked on had come from the nuclear experimenters and were concerned with instrumental questions, both being about corrections needed for objects of finite geometry and the effects of secondary scattering of neutrons in these objects. Such work, the practical resolution of experimenters' problems, became characteristic of Tony in later life. His door was always open to such enquirers, whether experimenters or theoreticians, and this trait soon became well-known, well-used and much appreciated. Most of this "consulting work" did not reach the pages of scientific journals, except perhaps to their 'Acknowledgements' sections, but his work on these two problems did become widely known and were much used by experimenters. The first problem [1][a] was the detailed calculation of the reduction in the intensity of a neutron atmosphere, caused by placing an absorbing disc within it, in terms of the size and shape of the disc and the total cross sections for neutrons incident on the nuclei of the material composing the disc. These calculations are of particular significance for the neutrons within a nuclear reactor or an atomic bomb. The declassification[6] of this report in 1947 specifies its date to be September 1944, although no date appears on the paper itself. Its number MS. 91 confirms that it originated from the Peierls group at Birmingham, since they had adopted the prefix MS in numbering their reports. There was so much demand for this report after the war that it was re-issued in 1961, after some revision and the incorporation of an appendix previously numbered separately as MS. 91A, as a U. K. A. E. A. Research Group document, also available in microfilm from H. M. S. O. The second problem [2] concerned the Wall Corrections for a finite ionization chamber of cylindrical shape, exposed to a uniform neutron beam parallel to its axis. The calculation determined the distribution of energy loss for the energy generated by a neutron interaction in the chamber, resulting from the escape of recoil protons through the cylindrical walls, taking into account secondary scattering. According to the declassification list,[6] this report was produced in October 1944, although the document itself bears no date. The second paper was put to experimental test much later in two experimental studies of neutron counters [10, 16], both publications bearing Tony Skyrme's name as co-author. These later papers cover much of the work reported in the document MS. 94, but include also unpublished work by Skyrme. This further work on wall corrections has survived in manuscript form, that for proportional counters [U2] in manuscript WC, undated but of vintage about 1952, and that for ^{3}He counters [U6] in manuscript 2WC, undated but probably written in 1955.

For some time before Tony's arrival at Birmingham, the Peierls group had been working on the equation of state for air appropriate to the wide range of temperatures and pressures which would occur in consequence of an atomic bomb explosion. This equation of state was needed as input into the equations of motion governing the course of the outgoing blast wave resulting from the initial energy release generated by the nuclear fission processes ultimately responsible for this explosion. With A. H. Wilson, Tony worked to develop an analytic expression for this equation of state, sufficiently valid over this range of temperatures and pressures, an important aim because an approximate, but qualitatively correct, analytic form would allow the numerical work to move ahead much more quickly.

[a] Numbers in square brackets refer to items in Skyrme's publication list, printed at the end of this article.

As a result, he became directly involved in the numerical solution of these equations of motion, which was already going on when he joined the group, and this led him to devise a new and more convenient form for them by changing the variables to a set more appropriate to the boundary conditions natural for this kind of motion. However, his formalism was not made use of, because the numerical calculations had already reached such an advanced state that it was not considered reasonable to go back and start the calculations again using the new formalism.

Tony also worked carefully through memorandum MS. 12A, the Fuchs-Peierls theoretical discussion of the separation of isotopes by a multi-stage gaseous diffusion plant, and became expert in his understanding of the processes and plant involved. After the Quebec Agreement between Churchill and Roosevelt in August 1943, when British-American cooperation in atomic bomb development was resumed, a British Mission including Peierls visited U. S. A. at the end of that month, charged with deciding how the British scientists could best help speed up the bomb development. At this time, the design and construction of the gaseous-diffusion ^{235}U separation plant at Oak Ridge by the Kellex Corporation was at a crucial stage, and they requested that Peierls should join their project as a consultant as soon as possible, together with a team of British engineers, since his experience and independent opinions were invaluable to the Kellex engineers. In the event, Peierls and his family left Birmingham in early December and travelled by ship to America, landing at Newport News (Virginia) and settling at New York, where the Kellex design offices were located, just before the end of 1943. However, as a result of several visits Peierls made to the Los Alamos Laboratory in the subsequent months, Oppenheimer soon came to feel that the British experience with the hydrodynamics of the blast wave initiated by an atomic bomb explosion was much needed at Los Alamos, especially as the British approach to solving the differential equations involved was quite different in spirit and in detail to that of the Los Alamos group. Already early in 1944, he was urging Peierls to move to Los Alamos. By March 1944, the Kellex engineers were left with only one serious problem, the control of the purge cascade (i.e. the counter-flow of uranium hexafluoride gas depleted in ^{235}U) in their plant. Peierls therefore suggested that Tony Skyrme should be brought over to take his place with the British Mission in New York, since Tony had quite sufficient experience and understanding to deal with this problem, thereby releasing Peierls for Los Alamos work. Thus, Tony joined the British Mission in New York, arriving by air about 9 March 1944. He was by far the youngest scientist in the British Mission, but with a reputation which made this fact irrelevant. The Kellex problem was settled by Tony in the course of several months, after which he moved on to work in the Theoretical Physics Division of the Los Alamos Laboratory.

As Peierls has related,[8a] Tony had a notable adventure in New York City, amusing to recall but alarming and inconvenient at the time. One hot Saturday evening, while he was out walking in Central Park, a patrol searching for draft evaders stopped him. Since he had no papers with him to identify himself, this patrol handed him into a police station. The police considered him to be a suspicious character, since he had 'a guttural accent', in their view, and he was held at the police station overnight. On the Sunday, federal agents took him to a federal prison, where he was fingerprinted, dressed in prison clothes and put in a cell. He was, of course, released at once on the Monday, when he came before the magistrate at a local court and had his account of himself verified by the British Supply Mission.

When Tony arrived at Los Alamos, in July 1944, the most important question in the air was whether a plutonium bomb could be constructed, for it had become realized only in that month that the assembly of a critical mass of plutonium would have to be carried out far more rapidly than was possible by firing two subcritical masses together with a gun, the method satisfactory for a ^{235}U bomb. The great advantage of plutonium was that it could be separated efficiently from ^{238}U by well-known chemical methods and did not require the arduous isotope separation process necessary to separate out ^{235}U. The difficulty was that one plutonium isotope had been shown to have a high decay rate for spontaneous fission, in consequence of which the masses of plutonium being brought together to exceed the critical mass would begin to fizzle long before the critical mass could be assembled, unless this could be done exceedingly rapidly. The solution proposed was that a subcritical sphere of plutonium should be surrounded by a thick shell of conventional explosives, the whole being contained within a strong outer shell made of some inert, massive material. Simultaneous detonation of the conventional explosives at a number of points on their outer surface would then result in a spherical detonation wave travelling inwards towards the center of the plutonium sphere. The high central pressure developed in this way would compress the plutonium sphere until it reached a critical density at which the nuclear explosion would begin and quickly take over. There were many difficult questions posed by this scheme, especially about the stability of the spherical blast wave between the conventional and the nuclear detonations. However, above all, it was vital that the implosion process following the conventional detonation should be fully understood, and quantitatively so. Numerical calculations on the implosion process, using IBM business card machines, were already underway, as has been related recently by Feynman,[9] who did much to improve the efficiency and rapidity of these calculations. Graham McCauley[10] recalls some remarks made by Tony about those times, in response to a comment by a third person that a BBC microcomputer was too short of programme space for any significant applications. Tony's response was that ambitions always soar above what is already possible, irrespective of what computer is available, and that he would have been very glad to have had that microcomputer's memory in their work at Los Alamos. As it was, he said, only ingenuity got them past their difficulties there:

"For the implosion calculations, we had available only IBM business machines using cards. The differential equations to be solved had to satisfy nasty conditions on moving boundaries, the latter to be determined by the calculations. Rather than trying to implement these boundary conditions as implicit conditions on the differential equations, which would have taken much more programme space than was available, these boundary conditions had to be set by hand as the IBM card machines went along, so that we had a frantic time picking up cards from the region already covered and inserting them into the region ahead of the machine calculation. We would have been very glad of the BBC microcomputer's capacity, but we got there, all the same!"

The equations of motion for the development of the implosion were of course of the same general type as those investigated by Tony in his formulation of the out-going shock-wave problem at Birmingham, so that it was quite natural for him to carry out the same procedures for the implosion problem. In the end, by a sound understanding of the physics involved, using the method of characteristics, and by the judicious use of approximate expressions for the behavior of the quantities of most importance, some of these approximations being based on general features of the results already obtained in the IBM machine calculations but justifiable *a posteriori,* he had been able to develop a mathematical procedure powerful enough that one man with a Marchant hand calculator could obtain in one day results comparable in accuracy with those obtained in weeks, even with all that team work just mentioned, using the largest array of IBM business machines available at the time. Although Tony's achievement was recognized as quite remarkable at the time, his procedure was not used at Los Alamos because the practical needs for plutonium bomb design had already been met by the IBM machine calculations, supplemented by enlightenment from some special cases and by some results from experiment. It had already been decided what array and strength of conventional explosives would be used to generate the implosion. No complete record of Tony's treatment of the implosion appears to have survived. His report [3] is probably relevant to this work but it is not available for us to determined the precise connection since it is not yet declassified.

In 1946, Tony submitted a dissertation [4] to Trinity College, Cambridge, in application for a Research Fellowship there. This was based on classified work which he had done at Los Alamos. The examiners' report on his dissertation states:

'The work is an excellent example of mathematical reasoning applied to physical picture. The mathematical method used is original and implies a mature ability of extricating the relevant features from the complicated mathematical situation which has been described. It has made an important contribution to the understanding of the phenomena. It is regrettable that, for the moment, it cannot be made generally available in published form'.

Besides Peierls, G. I. Taylor (a Fellow of Trinity College) and W. G. Penney, both of whom had also worked at Los Alamos, wrote reports on Tony's dissertation and the Research Fellowship was awarded to Tony on the basis of these three opinions of this work. The dissertation was a classified document, so that no copy could be retained for the Trinity College Library at that time, but the College Council had discretionary power (noted in a Minute dated 26 July 1946) to exempt from deposit any dissertation to which security considerations applied. According to a letter from General Groves's office, dated 18 November 1946,[11] Tony's dissertation was declassified in that month. This letter also gives us the title of the dissertation, namely 'Hydrodynamical Theory of the Reaction Zone in High Explosives' [5]. Two copies of it were mailed from the British Supply Mission at Washington to Peierls at Birmingham on November 20th.[12] Peierls sent one copy to J. H. Awbery of the Department of Atomic Energy, Ministry of Supply, London, on November 28th,[13] asking that it be sent to K. Fuchs after due note was taken of it; Awbery[14] acknowledged its receipt on November 29, saying that he had sent it on to Fuchs. The second copy was held by Peierls for some time, for others to see, and then sent to Awbery, who acknowledged its receipt on 19 Februrary 1947.[15] Although a copy of the declassified dissertation could then have been deposited in the Trinity College Library, this was not done. Apparently Tony himself did not retain a copy and, despite considerable enquiry, it is not known today where (and whether) any copy is extant.

We believe[16] that Tony's dissertation consisted of a fundamental discussion of the thickness and detailed structure of a detonation wave front. From work of Lord Rayleigh and of G. I. Taylor in 1908, it has been long known that the thickness of a shock wave is of the order of a mean free path for a molecule in the medium, but this result does not necessarily hold for a detonation wave, where energy is being fed into the wave as it progresses, from the exploding medium in the neighbourhood of the wave front. From letters dated 1946, we learn that the dissertation was based on chapters of the Los Alamos Technical Series which had been written by Tony[17] but that two sections of the latter were held back by the classification authorities, one giving a comparison with experimental data obtained at Los Alamos and the other dealing with the propagation of a detonation along a curved surface. The bulk on Tony's work on this topic was described[18] as being "of a relatively unclassified nature", the part held back being so held[19] 'not because of its content but because it reveals the fact that Los Alamos was interested in this particular subject'. Unfortunately, it has not yet been possible to identify this particular Los Alamos Technical Report.

We know little further about Tony's work at Los Alamos. He worked in the Theoretical Physics Division which was headed by H. A. Bethe. He was a member of its Section T-1 which took responsibility for the Hydrodynamics of Implosion and was headed by R. E. Peierls. He may have worked with R. P. Feynman at some point, as McCauley's story (told above) suggests, a plausible conclusion since they were both interested in numerical predictions about the implosion process; Feynman was head of the T-4 Section, charged with the solution of Diffusion Problems. Feynman and Skyrme did know each other from those times, as we can recalled from one meeting between them at A. E. R. E. (Harwell) many years later.

1946–50: The Fellowship Years

Tony Skyrme was admitted a Research Fellow at Trinity College (Cambridge) on 8 October 1946. He was granted permission to carry out his research at Birmingham University initially. Such permission is not usual, but throughout his four-year tenure of this Fellowship, he was not ever resident in Cambridge. In his curriculum vitae, he described it as a "Non-resident Fellowship"; although it may have been that in fact, this is not a classification recognized by the Trinity College Statutes and By-laws. However, what this meant in practice is that Tony received a dividend from Trinity College only during his first Fellowship year, and that he would have held a Fellow's rights during any period he may have spent at Cambridge during the three following years 1947–50.

For 1946–48, he held a University research fellowship in the Department of Mathematical Physics of Birmingham University, where Peierls had begun to form a powerful research group in the field of fundamental nuclear and particle physics. Tony had to learn modern quantum mechanics, including quantum field theory, the latter being an area which had not received much attention from physicists during the war years. He soon became known as a most useful member of this Department, a man to be consulted by experimenter or theoretician on any difficult point in their work, a man with thoroughly sound opinions. One topic which drew his attention was the formalism developed by Kapur and Peierls[20] in 1938 for the description of nuclear reaction processes in terms of resonant nuclear states. These states are defined to have only out-going waves, a condition which requires them to be eigenstates of the Hamiltonian of the system with a complex boundary condition at infinity. The completeness of the set of eigenstates so obtained was not guaranteed by the usual theorems of quantum mechanics, although it was considered to be plausible, at least in the regime of narrow resonance levels, and heuristic arguments leading to this conclusion were published by Peierls[21] in 1948. Since the Kapur-Peierls formalism was of great importance for nuclear physics, Tony provided a rigorous demonstration [5] that these eigenstates do form a complete orthogonal set, which he wrote up for publication, submitting his paper to the London Mathematical Society. His paper was accepted for publication, in principle, but the referee requested that one section (three lines in length) should be expanded because the remarks made did not prove the conclusion needed there, although the conclusion was correct. This amplification was never made and the paper remained unchanged, lying in some bottom drawer[b], unseen and forgotten. Peierls referred to Tony's forthcoming paper both in his 1948 paper and later, and so did other authors who used the Kapur-Peierls formalism.

An international conference on nuclear and particle physics was held at Birmingham University in September 1948, at which Tony gave a discussion paper on the status of the theory of beta-decay. No volume of conference proceedings was published for this meeting, but Tony's lecture was written up in 1949, in an extended form, and published [7] in the review journal Progress of Nuclear Physics.

In Birmingham, Tony soon met Dorothy Mildred Millest, a lecturer in experimental nuclear physics in the Physics Department of Birmingham University, and found that they had many interests in common. She was a daughter of Frank Charles Millest, a commercial traveller, and had taken her first degree in physics at Royal Holloway College, Egham (Surrey), in 1940. She had stayed there as a member of staff during the years 1941–44, carrying out research under Prof. Frank Horton, and taking her Ph. D. degree in 1944, following which she was appointed to her post at Birmingham. On 26 August 1949, near the end of their first period at Birmingham University, they were married in St. George's church at Edgbaston, a suburb of Birmingham. It was quite a formal affair, morning suits and all, well remembered by those who attended it.[8b] Tony's best man was Max Krook, a South African then in Peierls's department and later at the Harvard Observatory.

In September 1948, Tony took up a research associateship at M. I. T., in Viki Weisskopf's theoretical nuclear physics group there. In this stimulating and sophisticated physics atmosphere, many of his later ideas about nuclei began to take form. For the academic year 1949–50, he was a member of the Institute for Advanced Study at Princeton (New Jersey), where his central interest was in the new developments going on in quantum field theory.

[b] This paper has now been found and is held in Prof. Skyrme's Archive in the University of Birmingham Library. It is of interest to note that a rigorous proof of the completeness required by Kapur and Peierls was published in 1961 by McLeod.[22] It is curious that McLeod's paper was not drawn to the attention of Peierls, nor to that of any nuclear physicist known to us, at that time. Subsequently, rather general theorems have been established by other mathematicians, concerning the completeness of the eigenfunctions of complex differential operators, which cover the Kapur-Peierls question as a special case.

1950–61: A. E. R. E. (Harwell)

In the early summer of 1950, Tony accepted an appointment as Senior Principal Scientific Officer in the Division of Theoretical Physics at A. E. R. E. Professor M. H. L. Pryce, the Wykeham Professor of Physics at Oxford University, was Acting Head of this Division at the time, several months after Klaus Fuchs had ceased to be its Head. Tony took up this post on 4 September 1950, his wife Dorothy taking up a research post in the Nuclear Physics Division. Tony's experimental colleagues soon found their way to his door and he carried out a number of calculations related with the experimental programme at A. E. R. E., on the photo-disintegration process $\gamma^{12}C \to 3\alpha$ [8], on the double Compton effect $\gamma e \to e\gamma\gamma$ [11], on the spectrum of neutrons resulting from the interaction of 171 MeV protons from the A. E. R. E. synchrocyclotron on a range of nuclei [9], and on the total cross sections for such neutrons, with energies from 60 to 153 MeV, incident on aluminium and lead targets [12]. These papers reached publication quite promptly, no doubt because Tony had a co-author in each case. Other papers were not so fortunate and it became common knowledge that Tony had a desk drawer full of manuscripts awaiting completion and submission for publication.

When Brian Flowers returned to A. E. R. E. from Birmingham in 1952, to become Head of the Theoretical Physics Division, he learned of this drawer-full of unpublished papers, and made it one of his priorities to move as many of these papers as possible out and into the scientific journals. Tony's response to Flowers was: "I know that you are right. I just don't like to see my name in print". Flowers's success in this endeavour may be measured by noting the rapid rise in Tony's publication rate, which amounted to ten papers in the one year 1956. Nevertheless, not all of Tony's papers emerged, as is shown by the list of unpublished manuscripts given below, in Appendix U; see, for example, his calculations of the total cross sections [U4] and the stripping cross section [U5] for high energy deuterons incident on nuclei, which appear to have been written in the year 1953–54. However, the most serious omission was Tony's early and novel work on the three-body problem at low energies [U3]. In this paper, his aim was to set the problem up as far as possible in terms of the shape-independent parameters a and r_{eff} defined, for each two-body interaction, by the expansion

$$k \cot \delta_2 = -\frac{1}{a} + \frac{1}{2} r_{\text{eff}} k^2 + \frac{1}{6} P k^4 , \tag{1}$$

where δ_2 denotes the phase-shift for the two-body elastic scattering considered, as a function of their c.m. momentum k. He also allowed the possibility that a small number of additional parameters might be necessary to characterize the wave function in the central region where all three particles are close together. In this aim, Tony's paper has some similarity in spirit with the later (1956) work of Skornyakov and ter-Martirosyan[23], except that the latter did not introduce any additional parameters and even took $r_{\text{eff}} = 0$, a situation which Tony asserted to be singular and to lead to infinite binding energy for the three-body system[c]. Indeed, Tony found his three-body equations to be nonsingular only if a nonzero P were included in (1). He also required as input an additional phase δ_3, an unknown function of energy, to characterize the three-body interactions in the central region. There is no date on Tony's manuscript, but there is a brief reference in the text to work[24] 'published by M. Verde (H. P. A., 1951)'. I remember that Tony presented this work at a seminar at Birmingham in the summer of 1952, adding some ideas about the method by which the final integral equations might be solved, so that it seems reasonable to suppose that this three-body paper was prepared in 1951–52. I believe that Tony's method was never put to the test of numerical calculation; certainly,[25] from time to time, he endeavoured to interest some of his younger colleagues to carry out this work, but there were no takers. Tony's paper was labelled 'Part I', but 'Part II', the numerical part, did not ever get started. When asked about this paper, Tony always replied that his paper was not worth publishing until there were numerical calculations done to show that his three-body method could be carried through in practice, and this was probably a sound decision. His work was done far ahead of Faddeev's time (1960), but it appears incomplete, in that it makes approximations on intuitive grounds; for example, he assumed the space wavefunction to be purely symmetric in the central region. No doubt it is reasonable to make such approximations in a first approach, calculating the necessary corrections in the next approximation or justifying them *a posteriori,* but this was never done.

It was at this point that Tony's new thoughts on field theory and nuclei began to emerge. He developed his own approach to quantum field theory [15] by constructing first a classical functional analogous to the scattering matrix of quantum theory, and only then introducing the postulate of quantization. This allowed him to reach all the usual formulae of quantum field theory, including the Edwards-Peierls expression for the one-particle propagator. The main new feature of this paper was the derivation of a field-theoretic variational principle which is a relativistic analogue of that well-known nonrelativistically for the Schrodinger equation. He used this variational principle in work much later with J. S. Bell [32], providing a nonperturbative estimate for the magnetic moments of the nucleons.

As for nuclei, Tony introduced in 1954 a new view [13] concerning their structure, placing the emphasis on their content of a 'mesic fluid', neutral in the mean but undergoing spin and charge fluctuations locally, envisaging that the nucleonic sources (i. e. the nucleons of the nucleus) were coupled strongly to the mesic fluid rather than to each other, a situation leading to some understanding of the shell-model aspects of the nucleus. In a related paper [14], he went on to derive a self-consistent mesonic field, acting on the nucleons but also generated by the nucleons, for the case of a symmetric pseudoscalar meson theory, the idea being that the nucleon mass should stem from its coupling with the fluctuating mesonic field. The calculation assumed that the mesonic field was determined by the mean nucleonic density in the nucleus, in an essentially classical way, the quantum-mechanical fluctuations then being added as a perturbation and leading to the residual spin- and isospin-interactions between the nucleons in a nucleus. He had expressed these ideas quite early on, and he thanks Weisskoff, which whom he had spent the academic year 1948–9, for discussions concerning them. As he described in a recent lecture [44], he was groping his way towards a continuum theory of nuclei, based on semi-

[c]The work of Ref. 23 was directed towards the calculation of the neutron-deuteron scattering length at zero kinetic energy and did not discuss bound states for the three-nucleon system, so that it does not necessarily conflict with Tony's statement.

classical ideas, nucleons being objects emergent from theory rather than being point-like objects imposed from without. The whole panorama of his thoughts is well expressed in a summary document printed here as Appendix T. Unfortunately, this document was not dated, but it seems a reasonable presumption that it was written down about five years later, than these papers [13, 14], when he had reached the point of identifying nucleons with the solitons of a nonlinear meson theory, which are now known as Skyrmions.

The next five years 1955–59 represent the peak of Tony's research activity. He was then Head of the Nuclear Theory Group within the Theoretical Physics Division under B. H. Flowers. J. S. Bell, J. P. Elliott, A. M. Lane, F. Mandl, J. Perring and R. J. N. Phillips were all members of his group, each for at least a large part of this period. Tony kept in touch with them through a weekly meeting in his office, during the early afternoon, for which one or two members of the group had been assigned to report on some new papers or some recent article in the journals. Discussion was vigorous and incisive, to the extent that these meetings became known as the "Skyrmishes", consisting essentially of a dialogue between Tony and the speaker, although others were not reluctant to join in, especially not John Bell. Those who were there all express admiration of Tony's "deep understanding of almost any topic which came up", "his quickness of understanding, whether you were right or wrong" and describe him as "critical but never unpleasant or discouraging". The "Skyrmishes" were the most lively, stimulating and constructive of the seminars any of them have known anywhere. At the Division seminars, I am told, the front row regularly consisted of Skyrme, Bell and Walter Marshall, the most intimidating front row that any speaker could ever expect to face.

This period was also the most lively as regards Tony's research. It will be convenient to describe this work under a series of sub-headings:

(a) *Nuclear Shell Model*

As a result of some unbelievable results obtained by Phil Elliott with the shell model, Tony and he came to realize that the use of a potential fixed in space for the generation of one-particle orbitals for the nucleons had one serious defect, quite outstandingly so for light nuclei, namely that the center-of-mass of the nuclear system described by the shell-model wavefunction was not at rest. They pointed out that the shell-model wavefunctions then in use had too many quantum numbers, three of them being (N, L, M_L) for the description of the motion of center-of-mass, whereas the physical states of interest were only those with $(L, M_L) = (0, 0)$ and $N = 1$ (the ground state of the c.m. motion), the others all being spurious [17]. They discussed explicit examples, illustrating the errors made by the inclusion of such spurious states on the calculated properties of nuclear states, and showed how to remove them from the shell-model wavefunctions. One of these illustrations was the calculation afresh of the magnetic moments for light nuclei [18]. These papers were the first to point out the quantitative importance of removing these spurious components of the usual shell-model wavefunctions and the first to show how to remove them in a systematic way. This was a significant step forward in establishing the nuclear shell model on a quantitative basis and all nuclear physicists had to become familiar with the details of this paper [17].

In that era, nuclear physicists were much confounded by the success of nuclear models which appeared quite different, indeed even contradictory, in their underlying motivations and assumptions. Even apart from the rotating droplet model of Bohr and Mottelson, which emphasized the collective aspects of nuclear excitation, there were the shell-model, which assumed the nucleons to move independently in definite orbital states, and the alpha-particle model for $A = 4n$ nuclei, which pictured the nucleons to be clustered in groups of four, each cluster having zero isospin and being tightly bound into an alpha-particle form, the low-lying excited states of the nucleus being associated with the rotations and the vibrations of this systems of n alpha particles. For example, for ^{16}O, where $n = 4$, the lowest excited levels are known to be $(0+)^*$ at 6.06 MeV and $(3-)$ at 6.13 MeV, and these have natural interpretations in the alpha-particle model, the first being due to radial pulsation of the 4α system, the second being the lowest mode of rotational excitation permitted for a tetrahedral structure by the Bose character of alpha-particles. For ^{8}Be, ^{12}C and ^{16}O, Perring and Skyrme [25] wrote down wavefunctions of n alpha-particles which interact through harmonic interactions, the internal alpha-particle wavefunctions being of Gaussian form, and antisymmetrized these wave-functions, with respect to interchange of any two nucleons, so they become permissible wavefunctions for $4n$ nucleons expressible as a sum of shell-model configurations. They then calculated the energies to be associated with these states. When this was done for ^{16}O, the first four excited states were calculated to be $(3-)$, $(0+)^*$ $(2+)$ and $(1-)$, the same as those observed. The main discrepancy was that the $(3-)$ level lay lowest, at energy about 3 MeV, which means that the calculated moment of inertia is too large by a factor of 2. This shows that there are shell-model wavefunctions which describe 'alpha-particle-like' states, with reasonable energy values. The important part of Perring and Skyrme's calculations was to examine the structure of these states in terms of the usual shell-model basis. This was done in detail for the $(0+)^*$ state, with the conclusion that its shell-model configuration was 50% $(1p)^{-1}(2p)$, 30% $(1s)^{-1}(2s)$ and 20% $(1p)^{-2}(1d)^2$, not a surprising result, except that shell-model calculations at that time did not include two-nucleon excitation terms. On the other hand, the calculated $(3-)$ and $(1-)$ states were found to have wavefunctions with large admixtures even of three-nucleon excitation terms. This calculation was immensely instructive, showing that each successful nuclear model could illuminate the validity and use of other (successful) nuclear models. This paper also became well-known, as the first major step in the coordinating and reconciliation of different nuclear models.

(b) *Brueckner Theory for Nuclear Matter*

This work was naturally attractive to Tony, because it emphasized the mean, coherent properties of the nuclear medium in which each nucleon moved, rather than its particulate structure. He quickly became an expert with the Brueckner model and used it to investigate a number of questions.

His first application was a discussion of the surface of a finite nucleus. He set the problem up generally but then specialized the discussion to very large A, such that the problem became one-dimensional, the case of a semi-infinite slab of nuclear matter with a plane surface. It was in this work that Tony first made use of a pseudo-potential for the nucleon-nucleon interaction within nuclear matter (see below). His estimation of the surface thickness and the surface energy involved the use of a density-dependent effective mass and of a Thomas-Fermi estimate for the relation between the local potential energy and the density. He concluded that his calculation could account for the observed values of both the surface energy and the surface thickness, while fitting also the optical model potential for bulk nuclear matter, and predicting a greater spatial extension for the nuclear potential than for the charge distribution, beyond the mean nuclear surface [20].

With J. S. Bell, Tony also applied the Brueckner model to the calculation of the spin-orbit coupling between a nucleon and a nucleus, since the nucleon-nucleon spin-orbit interaction had recently been determined from the analysis of nucleon-nucleon scattering data. These calculations were carried through for both light and heavy nuclei, with a view for use in estimating the energy separation between nucleon orbitals $j = l \pm \frac{1}{2}$. They also carried these calculations through for nucleon-nucleus scattering for energies of order 100 MeV and found an appreciable increase (of order 20%) in the effective strength of the spin-orbit potential, relative to that for a bound nucleon [22].

Tony also calculated nucleon magnetic moments within nuclei [29] with J. S. Bell and R. J. Eden, treating the nuclear environment by the methods of Brueckner.

(c) *Collective Motion*

It was very typical of Tony to develop a new way of looking at problems already much discussed in the literature, and then to go on to see how this new idea appeared when used for other problems. He introduced collective co-ordinates to describe in a systematic way a band of states, all having essentially the same internal structure but associated phenomenologically with some collective mode of excitation. The co-ordinates he introduced were not redundant but replaced combinations of the usual co-ordinates; their most efficient definition was determined by a variational principle. He applied this procedure to discuss the shell-model, the alpha-particle model and the Bohr-Mottelson model, in the domain of nuclear physics, but then went on to the discussion of plasma oscillations and of excited nucleonic states [21]. No new results were obtained, nor was the method deep; it stands as a practical example of how to approach new problems where the way is not clear.

At the Physical Society Conference on Nuclear Physics held on 30–31 March 1960 at Liverpool,[26] Tony gave an invited paper on 'A Collective Model for ^{16}O', in which he discussed and compared the various collective models which may be used to understand its spectrum of excited states, as a means of illustrating quantitatively the collective motions possible in nuclei more generally [U7].

(d) *Muon-induced Fusion*

At the beginning of January 1957, Alvarez *et al.*[27] reported that they had observed the reaction

$$\mu^- + p + d \longrightarrow {}^3\text{He} + \mu^- + 5.4\ \text{MeV}\ , \qquad (2)$$

for μ^- mesons stopping in a hydrogen bubble chamber containing a small admixture of deuterium. As soon as the news reached A. E. R. E., Tony began calculations on his blackboard to account quantitatively for the complex sequence of processes which lead to this end result. This involved consideration of the atomic processes leading to the formation of the 'molecular' system $(pd\mu^-)$, of the reaction rate for the nuclear fusion process $(p + d) \to (^3\mathrm{He} + '\gamma')$ taking place in this state, and of the internal conversion process by which the 'photon' energy is taken up by the μ^- meson. Within a few days, Tony showed [27] that his estimates for these rates were reasonably consistent with the rate observed for the overall process (2) and then gave a seminar talk at A. E. R. E. about his conclusions. Similar work was published at about the same time by Jackson[28] and by Sakharov and Zeldovich.[29]

(e) *The Skyrme Potential*

At the Rehovoth Conference on Nuclear Structure in 1957, Tony proposed [31] the use of an NN effective potential for the form,

$$T = \sum_{i<j}\sum t_{ij} + \sum_{i<j<k}\sum\sum t_{ijk} \ , \tag{3}$$

where the functions t_{ij} and t_{ijk} took the momentum dependent forms

$$\begin{aligned} t_{ij} = \Big\{ & t_0 (1 + x_0 P^\sigma_{ij})\,\delta(\mathbf{r}_{ij}) + \frac{1}{2} t_1 (1 + x_1 P^\sigma_{ij})(\delta(\mathbf{r}_{ij})\,k^2 + \text{conj.}) \\ & + t_2 (1 + x_2 (P^\sigma_{ij} - \frac{4}{5}))\,\mathbf{k}' \cdot \delta(\mathbf{r}_{ij})\,\mathbf{k} \\ & + \frac{1}{2} T(\delta(\mathbf{r}_{ij})(\boldsymbol{\sigma}_i \cdot \mathbf{k}\,\boldsymbol{\sigma}_j \cdot \mathbf{k} - \frac{1}{3}\boldsymbol{\sigma}_i \cdot \boldsymbol{\sigma}_j k^2) + \text{conj.}) \\ & + \frac{1}{2} U(\boldsymbol{\sigma}_i \cdot \mathbf{k}'\,\delta(\mathbf{r}_{ij})\,\boldsymbol{\sigma}_j \cdot \mathbf{k} - \frac{1}{3}\boldsymbol{\sigma}_i \cdot \boldsymbol{\sigma}_j\,\mathbf{k}' \cdot \delta(\mathbf{r}_{ij})\mathbf{k} + \text{conj.}) \\ & + iV(\boldsymbol{\sigma}_i + \boldsymbol{\sigma}_j) \cdot \mathbf{k}'\,\delta(\mathbf{r}_{ij}) \times \mathbf{k} \Big\} \end{aligned} \tag{4a}$$

where $P^\sigma_{ij} = (1 + \boldsymbol{\sigma}_i \cdot \boldsymbol{\sigma}_j)/2$ and 'conj' means "reverse the order of the terms k', $\delta(r_{ij})$ and k, and then interchange k and k'". The only three-body term used was simply

$$t_{ijk} = t_3\,\delta(\mathbf{r}_{ij})\,\delta(\mathbf{r}_{ik}) \ . \tag{4b}$$

These potentials are of zero range, but momentum dependent. The terms t_0, t_1, t_2, t_3 are central, T and U are of tensor character, while V is a spin-orbit interaction. Including the coefficients x_0, x_1, and x_2, and three-body term t_3, there are ten parameters. The three-body term (4b) mimics the effect of a density dependence for the mean potential. Tony advocated that nuclear shell-model calculations of bulk nuclear properties (as opposed to effects depending critically on nucleon-nucleon correlations) should be made using this effective nucleon-nucleon potential, with parameters which were to be chosen to fit empirical data on nuclei, such as the mean density and binding energy per particle for nuclear matter and more detailed properties (e.g. binding energy and energy level structure) for individual light nuclei. Tony then went on to determine from the current data, estimates [33] for the central parameters t_0, t_1, and t_3, showing that it was vital to include a nonzero t_3, to express the result of many body effects. His discussion [34] of the spin-orbit interaction did not work out so well, for the data available did not really appear consistent with a single choice of V. He used this potential also for a discussion of configuration mixing in nuclei with $A = 6$ and 14, the main aim being to account for the slow rate observed for ^{14}C beta decay [35].

Potentials of the above form have proved very convenient for nuclear physicists to use in shell-model and Hartree-Fock calculations for nuclear structure. Earlier calculations used a variety of representations of the nucleon-nucleon potentials for these calculations and comparison between different calculations was then difficult, since two chosen potentials could differ in so many ways. Simply to have all such calculations done for the same effective potential was already an advantage, and to have such a simplified effective potential made the calculations so much easier to do. This potential has become very widely used in nuclear structure theory, where it is known today as the *Skyrme Potential.*

(f) *Nonlinear meson theories and Skyrmions*

Tony was at the peak of his originality in the period 1958–61, and published a series of papers on nonlinear meson theories and their application to nuclear physics in the most fundamental sense. These papers express a number of the ideas he wrote down, which we have included here as Appendix T. They arise, in a sense, out of two earlier papers [13,14] about nuclear matter and its saturation. Indeed, he referred to the papers [30, 36, 38, 39] by the notation I to IV, as the core of his ideas, but the further papers [37, 40, 41], which we might number V–VII respectively, either exemplify the ideas of these papers by concrete calculations [37, 41], or apply them to questions about the real world (e.g. the existence of 'strangeness' as a conserved quantum number [40]), or demonstrate that the entities derived do have the properties attributed to them in some conventional theories.

The essential point arises already in Paper I. Tony discussed a four-dimensional isospin space for a spinless meson fluid ϕ_α for $\alpha = 1$ to 4, having a constant magnitude Q

$$\sum_{\alpha=1}^{4} \phi_\alpha^2 = Q^2 \tag{5}$$

and notes that the new component ϕ_4 may be written in terms of the constant Q and a position-dependent angle variable $\theta(x)$. This led him to consider the nonlinear differential equation now known as the Sine-Gordon equation, which takes the following form in two space-time dimensions

$$\frac{d^2\theta}{dt^2} - \frac{d^2\theta}{dx^2} + \frac{1}{4}\kappa^2 \sin 4\theta = 0 \tag{6}$$

and which reduces to the Klein-Gordon equation for small θ. The question arises as to the boundary equations at $x = \pm\infty$, for which the natural supposition is $\sin 2\theta = 0$, in both limits. Equation (6) then allows the static solution

$$\tan(\theta - \frac{1}{2}n\pi) = \exp(\pm\ \kappa\, x/x_0) \tag{7}$$

such that

$$\theta(+\infty) - \theta(-\infty) = \pm\frac{\pi}{2} \ . \tag{8}$$

This solution (7) is localized at $x = x_0$, being what has now become termed a *soliton,* which may represent a particle. Moving solitons are obtained by replacing x in (7) by $(x \pm vt)$. There are two types of soliton, corresponding to the ± sign in expressions (7) and (8), and they may be considered as representing particle and antiparticle, respectively. When there are a number of x_0 values, sufficiently far apart, the net function $\theta(x)$ may be well approximated by

$$\theta(x) = \sum_i \theta^+(x - x_0^i) + \sum_j \theta^-(x - x_0^j) \ , \tag{9}$$

where the $\theta^\pm$ are the solutions (7) for the sign ±. Each term of (9) contributes a kink or anti-kink to $\theta(x)$, and the net sum, which is

$$\theta(+\infty) - \theta(-\infty) = (\sum_i n_i^+ - \sum_j n_j^-)\pi/2 \tag{10}$$

will remain constant throughout the motion, even though the solitons overlap. This equation (10) is identical with the statement of a conservation law for the number of particles, minus antiparticles which is reminiscent of the conservation laws already known to govern fermion field theories. Indeed, Tony's original Lagrangian had included both boson and fermion fields, but he then noticed that his theory appeared to have two objects which followed the conservation law appropriate to fermions. At the end of paper I, he remarks that 'the Boson has been fundamental' in this work, expressing his aim to get 'a unified description of particles and fields . . . in terms of the nonlinear meson field equations alone' (by 'particles' and 'fields', he means fermions and bosons, respectively), and that 'this point of view is completely opposite to that of Heisenberg,[30] who regarded the spinor field as fundamental'.

It is not appropriate to expound here other aspects of Tony's approach to a fundamental theory of elementary particles. Some of them were usually imaginative, but are excluded by our knowledge of the facts today. For example, Paper I in 1958 referred to the possibility that isospin might be a discrete symmetry, and proposed that strangeness might play a role analogous to that of isospin but in a different three-dimensional space, corresponding to a symmetry $SU(2)_T \times SU(2)_S$, while parity conservation would be violated in some strong interactions and charge might be conserved only mod(4); Tony recognized that there was no empirical evidence for any of these last four conclusions, of course. Paper II in 1959 pushed these ideas further, relating the four real fields ϕ_α directly with the K meson and $\bar{K}$ meson isospin doublets and introducing three angles $\theta(x)$, $\phi(x)$ and $\psi(x)$. The physical idea was that K mesons and the π mesons might represent two different kinds of wave motion in the space of the variable labelled by α. In the Conclusion of Paper II, Tony draws attention to the fact that, since θ, ϕ and ψ are angles, the state functions will necessarily be multivalued, having separate branches associated with the multiple values of these angles at one x value. He ended with the conjecture 'that the Fermion fields form an equivalent way of describing this multi-valuedness', coming closer to the recognition of topological quantum numbers associated with fermion objects. Paper

III in 1961 considers the semi-classical aspects of a meson field theory in (3 + 1) space-time dimensions. Tony emphasized that the essential feature of his nonlinear field theory is the representation of the fundamental field quantities in terms of angular variables rather than linear variables. The state functions are then multivalued, each branch lying on a different sheet, and singularities arise when two different sheets cross. To specify the form of the field configuration at a singularity, Tony was much attracted to the form used by Pauli,[31] which occurs in the strong-coupling theory of symmetrical pseudoscalar pion fields and is specified by the orthogonal matrix.

$$e_{i\alpha} = \sigma_i \tau_\alpha \ . \tag{11}$$

This measures the relative orientation of the spin and isospin co-ordinate frames and involves both the source and the field in its vicinity. Just as for the (1 + 1) dimensional case of Paper I, there are particle-like and anti-particle-like solutions and their total number is rigorously conserved (since this number is now linked with the topology to the state). Paper IV of 1961 discussed the quantization of Tony's nonlinear theory. He was able to find a creation operator U for a 'particle', which is neutrino-like (massless) and obeys anti-commutation relations, but whose interpretation was not completely clear then. In the treatment given, these 'particle' states exist only if an energy-scale factor ϵ satisfies the condition $\epsilon = \hbar c/2$.

This series of papers were followed by two further papers [37, 40], both submitted for publication in 1961 just a few days before Tony left A. E. R. E. The first, joint with J. K. Perring, returned to the Sine-Gordon equation in (1 + 1) space, considered in Paper I, and calculated the behavior of 'particles' interacting with 'particles', and of 'particles' interacting with 'anti-particles', including bound states for the latter. This was done directly by numerical integration of the Sine-Gordon equation, carried out by A. L. Leggett while he was a vacation student working in Tony's group. These interactions were found to be simple elastic scattering processes, and it was possible to deduce a 'particle-anti-particle' potential energy, with a straightforward physical interpretation. In the second of these two papers, the point was being made very clearly: 'The objective is the construction of a theory of self-interacting meson fields, which will admit states that have the phenomenological properties of particles, interacting with mesons'. Again we note explicity that by 'meson field' Skyrme means a '*boson*' and by 'particle', a '*fermion*'. The nonperturbative solution found for N ≠ O is

$$\text{for } \alpha = 1,2,3 \qquad \phi_\alpha = \left(\textstyle\sum_i e_{i\alpha}\, x_i/r\right) \sin\theta(r) \ , \tag{12a}$$

$$\text{and for } \alpha = 4 \qquad \phi_4 = \cos\theta(r) \ . \tag{12b}$$

This solution is centered on the point $x_{0i} = 0$, and $(e_{i\alpha})$ here denotes an arbitrary constant orthogonal matrix. Choosing $\theta(\infty) = 0$, by convention, it was found that there is just one solution of the Sine-Gordon equation in three-dimensions for each possible value $\theta(0) = N\pi$. The number N has the same interpretation as in the (1 + 1) case of Paper I [30], namely as the (fermion) baryon number. Two such 'baryons' were shown to have a repulsion at all distances in this theory. A representation of a meson state was also found, so that the nature of meson-meson and meson-particle interactions could also be discussed; for example, it was found that the $J = T = 1$ $\pi\pi$ interaction is always attractive, whereas the p wave πN interaction is repulsive, on the average. However, this theory still needed quantization, which must be nonperturbative for a system with $|N| \geqslant 1$, and must therefore have a different starting point. His conjecture was that this would need the use of singular operators involving branch-points and 'that these operators may have many of the properties of fermion field operators in conventional theories'.

During the period just covered, Tony's base was A. E. R. E., at Harwell. He had become thoroughly established there, having been promoted to a B-band appointment, namely to the rank of D. C. S. (= Deputy Chief Scientist) effective from 1 July 1957. This meant that he was free to carry out research as he saw fit, irrespective of the mission of the Harwell Laboratory. His research group was outstanding, by any standards, and he was highly regarded by the A. E. R. E. community. Dorothy and he lived in a prefabricated bungalow on the A. E. R. E. housing site just north of the Laboratory. One strong interest which Dorothy and he had in common was gardening and in those years they achieved a garden which was both beautiful and interesting, as many people have remarked. Tony believed that serious brain work should be confined to working time and came back home to the garden or other pursuits at the end of the day. He was a very methodical man and kept much of his knowledge in a series of note books. His day kept pace with the clock. He was skilful with his hands and built his own television and hi-fi systems. When they bought their long-base Land Rover in 1961, they both learned how to strip it completely and how to put it all together again. This was a precaution against the possibility of breakdowns in remote mid-East and far-East places, and they carried a large stock of spare parts, ready for almost all eventualities.

Tony attended a considerable number of conferences on nuclear and particle physics in this period. Some of those to which he made a leading contribution have been mentioned above, such as the Birmingham Conference in 1948 and the Liverpool Nuclear Physics Conference in 1960. He reported to Nature [19] about the 1956 Nuclear Physics Conference at Harwell, and spoke about both his mesic fluid [23] and his variational principal in quantum field theory [24] at the 1956 Conference on Nuclear Physics and Elementary Particles at Pisa. He put forward his nuclear pseudo-potential [31] for the first time at the Rehovoth Conference on Nuclear Structure in 1957, and travelled a good deal with Dorothy in Israel and the Middle East after it was over. I recall that he attended the High Energy Physics Conference at Rochester (New York) in 1957, and doubtless he attended many other conferences besides these.

In September 1958, Tony took a year's unpaid leave from A. E. R. E. first of all to work with Professor K. A. Brueckner at the University of Pennsylvania at Philadelphia as a visiting professor for their Fall semester, where he lectured and completed Paper II [36], submitted for publication from Philadelphia. In February 1959, Tony and Dorothy began a long return journey to England, travelling west across U.S.A., but taking in many notable National Parks, such as Big Bend National Park, and with a stop at Los Alamos. They then crossed the Pacific Ocean to Sydney, Australia, where they purchased a Land Rover for their Australian journeys. They spend about a month at the Australian National University in Canberra, visiting Professor K. J. Le Couteur and his Department of Theoretical Physics in its Research School of Physical Sciences. Tony gave a number of seminar talks there, until they left for Darwin about the end of May. To reach this take-off point for Indonesia and Malaysia, they travelled across much of Australia's large and lonely interior, by way of Camooweal in Western Queensland. Their next stop was Kuala Lumpur, Malaysia, where they visited relatives. As gardeners, they were much impressed by the lush growth and beautiful flowers of that tropical place, which gave them a flavor of the attractions which life could offer there. Its joys lingered long in their memories and must have influenced them greatly in their decisions later on. They bought a Hillman car there and then drove on west, across Burma, India and Pakistan. In Iran, they suffered an alarming experience when they both fell seriously ill of salmonella poisoning in a barren and little-populated region. Since Dorothy was much less affected, she had to leave Tony, making him as comfortable as possible, in order to seek help. Fortunately, they knew that there was a group from A. E. R. E. working at the Nuclear Training Centre at Teheran, set up by the Central Treaty Organisation (CENTO) in April 1959. She was able to contact them by telephone quite quickly, to obtain their help urgently and to secure proper medical treatment for Tony, until he recovered sufficiently for them to continue their journey home.

After their return to A. E. R. E., Tony took up work again on 12 October 1959. Much was changing at A. E. R. E.. Brian Flowers had moved on to a professorship at Manchester University and W. M. Lomer had become Tony's Division Head. The emphasis on basic research was being replaced by an increasing internal pressure for the Harwell scientists to take up research problems from outside, and so to act as paid consultants to Industry or to other Government bodies. Tony was free from any direct pressure of this kind, in view of his D. C. S. appointment, but it did change the atmosphere in the Division. In 1960, John Bell left for C. E. R. N., Phil Elliott left for the University of Sussex and Walter Marshall became Division Head. Tony came to feel that ten years in one place was enough, perhaps thinking back to the beauty of Malaya, and that it was time for a change. He therefore approached Professor C. J. Eliezer, the Head of the Mathematics Department of the University of Malaya, who was at Cambridge on sabbatical leave at that time. Eliezer was much pleased by the prospect of having Tony Skyrme as a member of his department and soon arranged an appointment for him as Senior Lecturer. Tony then resigned his A. E. R. E. post, effective 30 September 1961.

1962–64: The University Of Malaya

Tony and Dorothy had made careful plans for their overland journey to Kuala Lumpur, where the University was located. They had bought a new long-based Range-Rover, and had made themselves thoroughly familiar with its innermost details, as mentioned above. They had a very complete kit of spare parts, to cover any eventuality, and knew how to replace them. They reached Kuala Lumpur early in 1962, just as the University of Malaya (Kuala Lumpur Division) was gaining independent status as the University of Malaya, its Singapore Division becoming the University of Singapore. On arrivial there, Tony found the Mathematics Department much smaller in active staff than he had expected. The practice was to take the brightest graduates each year on to the teaching staff, and to have them work on a Master's thesis. When such a postgraduate student gained his M. Sc. degree, he was given three years leave of absence, and the finance necessary, to go abroad to a major center of research to gain further experience and to work for a doctorate, following which he would return to his teaching post in the University of Malaya. Consequently, a large fraction of the teaching staff listed in the University Calender were actually working abroad, each on leave of absence until he gained his Ph. D. degree. The teaching load for the staff at Kuala Lumpur was correspondingly heavy, especially since the course programme in Mathematics was being extended, as a result of the new status of the University. Tony worked very hard, but quite happily, planning out the new lecture courses and preparing detailed lecture notes and problem sheets for them. For the academic year 1963 – 4, Tony was made Acting Head of the Mathematics Department while Eliezer was again on sabbatical leave at Cambridge. More generally, Tony took an active part in the mathematical life of Malaysia, not only through its University lectures but also in the Malaysian Mathematical Society, to whom he gave a talk on 'Symmetry' at one of their meetings in 1964.

During his year as Acting Head of the Mathematics Department, Tony was beset by a number of administrative problems, arising partly from his own high standards and his integrity, which blunted the joys of his everyday life at Kuala Lumpur through that period. Also there was increasing pressure for the University teaching to be in the Malayan language rather than in English. Nevertheless, the beauty of Malaysia was still there, and the fascination of the East. The Skyrmes managed to travel widely over Malaysia and beyond, on weekends and vacation periods, and came to know the country thoroughly and to enjoy its great floral displays and its striking scenery.

Tony's contract with the University of Malaya was for three years. However, in 1963, Professor R. E. Peierls moved from Birmingham to take up the Wykeham Chair of Theoretical Physics at Oxford University and the question came up of his replacement in the Department of Mathematical Physics at Birmingham. Tony was a logical candidate for this Chair, a first-class mathematician highly skilled in its applications to Nuclear Physics and Elementary Particles. Fortunately, the University of Malaya made no difficulty about the breaking of his contract with them and he accepted the Chair, taking it up for the academic year 1964–5, after another long overland drive back to England.

1964–87: The University of Birmingham

In Peierls's time at Birmingham, the Department of Mathematical Physics was primarily a research department, filled with research students and postdoctoral visitors from all parts of the world. There were few undergraduates at Birmingham who specialized in mathematical physics, but the permanent staff of the department provided lectures for other departments, especially for the Mathematics and Physics departments and for many of the Engineering departments. In the cold winds that began to blow through British universities by 1970, the student/staff ratio became a figure of merit by which to compare different departments, whether or not it was really relevant in the particular circumstance considered. Tony's department had a second professor (D. Thouless), and, although it continued to be quite well supplied with research students and visitors, the student/staff ratio stood out and it became apparent that Tony's department would have to seek shelter in a larger department. The Physics department also suffered from another internal figure of merit, namely, the cost per student, which had always and everywhere been higher for

Physics departments than for other departments simply because physics required advanced and expensive equipment, more or less, depending on the department's speciality, but more in the case of Birmingham which was strong in nuclear and high-energy physics. The amalgamation which was most logical, for the Mathematical Physics Department to merge with the Physics department, was not feasible, because it would increase the Physics department's problems, through the student/staff ratio. The amalgamation which finally took place about five years ago was of the Mathematics and the Mathematical Physics departments, Tony becoming the Professor of Applied Mathematics in the combined Mathematics department. The new set-up probably suited Tony quite well. He did not like the task of administration and his interests had tended to drift back towards mathematics, the subject in which he had taken his first degree.

He was not very active in research in this last period of his life. He published one further paper (VII) in the series I, II . . . coming under the heading (f) for the period 1950–61, a demonstration [41] that his mesonic field model of a fermion would still go through in the case of relativistic velocities and that its state function then had the structure expected for a spin-½ fermion, namely that of a solution to the Dirac equation. In this period he did not attend conferences, nor give seminar talks on other campuses, in general. However, he had two successful research students supported by S. E. R. C. studentships at Birmingham, Drs Jeffery G. Williams, who gained his doctorate in 1969 and has published work on kinks in nonlinear field theories and on topological questions arising out of them, and Dr. Graham Ringwood who gained his doctorate in1972 and has published papers on the properties of monopoles.

As always, Tony's door was open to the enquirer seeking enlightenment. Many engineers and statisticians sought his advice and were given detailed assistance with their problems. He generally dismissed any suggestion that his name might appear on the paper resulting from his advice and work. We know of only one exception to this pattern, a paper [43] on a question of random-walk statistics published jointly with Professor H. E. Daniels. Tony's vital contribution was the explicit solution in 1976 of an integral equation which had arisen in Daniels's work on random walk theory,[32] but Tony did not consider this solution worthy of publication. When Daniels found that this same integral equation came up again in a number of other problems, he felt it right that Tony should have the credit for its solution. He therefore included Tony's solution as an appendix to a 1985 paper which was dependent on it and added Tony's name as a co-author of the paper, without receiving any response from Tony to his letter requesting approval of this action.

In the late 1970's, when emphasis began to focus on nonlinear theories, both classical and quantum, the significance of Tony's kinks and their relationship with fermion states, generated within a purely mesonic theory, began to be apparent; they became a topic of widespread interest and there was an explosion of literature concerning them, their significance and their implications.[33] The question arose as to whether they might not lead to satisfactory theories of baryons and baryonic resonances, without need for quarks. These hopes seem to have been dashed, but Skyrmion theory has continued to shed increasing light on the structure and implications of field theories. There have been many nuclear or particle physics conferences in which there was a session devoted to Skyrmions and related topics. Although invited to speak at, or even just to attend, a number of such conferences, Tony would not accept, until he was finally persuaded to take part in a small 'Workshop on "Skyrmions" ', held on 17–18 November 1984 at Cosener's House, Abingdon (U.K.), and organized by the Theoretical Physics group of the Rutherford Appleton Laboratory, and to give the opening talk, entitled "Historical Introduction". Unfortunately, there was no Proceedings volume prepared from this Workshop nor was any record made of Tony's contribution. So, when Sir Rudolf Peierls reached his 80th Birthday, and a symposium was organized to mark this occasion, in his honour, we asked Tony Skyrme to attend and to repeat the talk which he had given at Cosener's House in 1984, under the title "The Origins of Skyrmions". Tony agreed to do this, but died two days before he was to speak. A reconstruction of his Cosener's House talk [44] has been printed in the volume "A Breadth of Physics", the proceedings of the Peierls 80th Birthday Symposium, and also in this issue of the International Journal of Modern Physics A.

In 1985, the Royal Society awarded Tony its Hughes Medal, which recognizes "an original discovery in the physical sciences, particularly electricity and magnetism or their applications". It was a happy choice, uniformly supported from all sides—what discovery could be more original than that of how to construct fermions from spinless boson fields. It is reported by those near by that, as Tony came back to his seat, grasping his medal and with a grin all over his face, he muttered, with justice, "They none of them could understand any of it at the time". I have heard it said that Tony's great trouble was that he was completely unable to explain his ideas to anybody else. However, one must reflect and remember just how unconventional and revolutionary those ideas were. The central point emerged slowly through that sequence of papers I, II, . . . etc. Although the essential idea was already present in paper I, it is stated only as a conjecture in Paper II, and not demonstrated until Paper III, where a special case is worked through. Even later, more papers by Tony, running up to Paper VII, were necessary to clarify this idea and its relationship with real physical particles. The last paper logically necessary, for the completion of this part of his programme (as specified here in Appendix T) was not published until 1971.

Tony was a theoretical physicist of unusual imagination, but having both a deep sensitivity and an abnormally strong degree of self-criticism. His mathematical perceptions were very deep, and he could see layers deeper than anyone else in his locale. In 1950, one referee wrote that 'there is hardly a problem in which he is not willing to take an intelligent interest'. Another wrote in 1957, that 'he has the highest intellectual abilities of any one at Harwell'. He was unusually quick in understanding, on any topic. Professionally, although mathematical in style, he was in close touch with physical significance in his work. An assessment in 1945 was that 'he handles these complicated problems very well, and has shown a remarkably good sense in picking out the important factors and in making sensible approximations'. He was immensely strong in setting out work for numerical calculation and was generally able to develop procedures superior to those already in use, although he was not himself an enthusiastic calculator. On the mathematical side, he was an excellent problem solver, but not only that, for his own approach to functional integration was both original and elegant. At the time of his death, he was engaged in writing a book on the path integral approach to quantum mechanics, including quantum field theory.

Acknowledgements

In conclusion, I want to say thanks to many people for discussion and correspondence about Tony's work and about various periods of his life. Above all to Professor Sir Rudolf Peierls, but also to Lord Flowers, J. K. Perring, J. S. Bell, Phil Elliott, J. E. Bowcock, A. M. Lane, G. McCauley, R. J. N. Phillips, A. T. C. Ferguson, C. J. Eliezer and a number of others, too numerous to list here. I also wish to acknowledge the helpful assistance of Mrs. P. Hatfield, the Eton College Archivist, and the Library staff of Trinity College, Cambridge, concerning Tony Skyrme's student years.

References

1. Who's Who for 1929 (Adam & Charles Black, London) p. 2596.
2. Lord Kelvin's Mathematical and Physical Papers (ed. J. Larmor, C. U. P., 1911), Vol. VI, No. 271: "The tide gauge, tidal harmonic analyser and tide predicter."
3. Monthly Notices of the *Roy. Astron. Soc.* **94** (1933/4) 284.
4. T. H. R. Skyrme, *Int. J. Mod. Phys. A,* this issue, p. 2745.
5. M. Gowing, *Britain and Atomic Energy 1939–1945* (Macmillan, London, 1964).
6. See *Nature,* issue of 22 March 1947, p. 411.
7. R. E. Peierls and K. Fuchs, "Separation of isotopes", meno. MS. 12A (undated, but probably written late in 1941). A revised version of this paper was declassified in February 1947 and issued as report BDDA-97.
8. R. E. Peierls, *Bird of Passage* (Princeton University Press, Princeton, New Jersey, 1985) pp. 194-5.
9. R. P. Feynman, *Surely You are Joking, Mr. Feynman* (Norton, New York and London, 1985) pp. 125–132.
10. Graham McCauley, private communication (January, 1988).
11. Ms. Eng. Misc. b. 214 C-286. Letter dated 18.11.1946, from O. G. Hayward to G. A. McMillan.
12. *Ibid.* Letter dated 20.11.1946, from G. A. McMillan to R. E. Peierls.
13. *Ibid.* Letter dated 28.11.1946, from R. E. Peierls to J. H. Awbery.
14. *Ibid.* Letter dated 29.11.1946, from J. H. Awbery to R. E. Peierls.
15. *Ibid.* Letter dated 19.2.1947, from J. W. Awbery to R. E. Peierls.
16. R. E. Peierls, provide communication (1988).
17. M. S. Eng. Misc. b. 214 C-286. Letter dated 26.9.1946 from R. E. Peierls to A. H. Wilson.
18. *Ibid.* Letter dated 1.6.1946, from R. E. Peierls to T. H. R. Skyrme.
19. *Ibid.* Letter dated 2.10.1946, from R. E. Peierls to G. I. Taylor.
20. P. L. Kapur and R. E. Peierls, *Proc. R. Soc.* **A166** (1938) 277.
21. R. E. Peierls, *Proc. Camb. Phil. Soc.* **44** (1948) 242.
22. J. B. McLeod, *Quart. J. Math. Oxford* (2), **12** (1961) 291.
23. G. V. Skornyakov and K. A. ter-Martirosyan, *Dokl. Akad. Nauk. SSR* **106** (1956) 425.
24. Presumably the following paper: A. Troesch and M. Verde, *Helv. Phys. Acta.* **24** (1951) 39. This is the only paper by Verde published in 1951 in this journal.
25. R. J. N. Phillips, private communication (1968).
26. Year Book of the Physical Society (London, 1960) p. 91.
27. L. W. Alvarez *et al., Phys. Rev.* **105** (1957) 1127.
28. J. D. Jackson, *Phys. Rev.* **106** (1957) 330.
29. A. D. Sakharov and Ya. B. Zeldovich, *Sov. Phys. J. E. T. P.* **5** (1957) 775.
30. W. Heisenberg, *Rev. Mod. Phys.* **29** (1957) 269.
31. W. Pauli, *Meson Theory of Nuclear Forces* (New York: Interscience, 1946).
32. H. E. Daniels, *Adv. Appl. Prob.* **6** (1974) 607.

33. I. J. R. Aitchison, *Surveys in High Energy Physics* (ed. J. Charap) (1988), in press.
34. *A Breadth of Physics,* Proceedings of the Peierls 80th Birthday Symposium, eds. R. H. Dalitz and R. B. Stinchcombe, (World Scientific, Singapore, 1988).

List of Publications of Prof. T. H. R. SKYRME

1. "Reduction in neutron density caused by an absorbing disc." Memorandum MS. 91. (September, 1943), Revised edition (including Memo. MS. 91A) reprinted in June, 1961. 18 pp. Also issued by H. M. S. O. as microfilm BDDA 38.
2. "Escape of energy from a cylindrical ionization chamber" (including appendix). Memorandum MS. 94 (October, 1943).
3. "Treatment of discontinuities in Lagrangian integration of symmetrical hydrodynamical problems", a report in the Los Alamos Technical Series (1944).
4. "Hydrodynamical theory of the reaction zone in high explosives," Trinity College Fellowship dissertation, declassified November 1946, pp. 24.
5. "A system of complex orthogonal functions" (1948, unpublished).
6. "Tank model for magnetic problems of axial symmetry" (with R. E. Peierls), *Phil. Mag.* **40** (1949) 269.
7. "Theory of beta-decay", *Prog. Nucl. Phys.* **1** (1950) 115.
8. "Shell model calculation of the photodisintegration of ^{12}C into three α-particles" (with M. J. Brinkworth) AERE rept. T/R 802 (1951).
9. "Analysis of high-energy neutron production" (with F. Mandl), *Proc. Phys. Soc. A.* **65** (1952) 101.
10. "A proportional counter for neutron flux measurement in the energy range 0.1 to 1 MeV." (with P. R. Tunnicliffe and A. G. Ward), *Rev. Sci. Instrum.* **23** (1952) 204.
11. "The theory of the double Compton effect", (with F. Mandl), *Proc. R. Soc. A.* **215** (1952) 497.
12. "Energy dependence of neutron total cross sections", (with F. Mandl), *Phil. Mag.* **44** (1953) 1028.
13. "A new model for nuclear matter", *Proc. R. Soc. A.* **226** (1954) 521.
14. "Meson theory and nuclear matter", *Proc. R. Soc. A.* **230** (1955) 277.
15. "Quantum field theory," *Proc. R. Soc. A.* **231** (1955) 321.
16. "Helium-3 filled proportional counter for neutron spectroscopy" (with R. Batchelor and R. Aves), *Rev. Sci. Instrum.* **26** (1955) 1037.
17. "Centre-of-mass effects in the nuclear shell model" (with J. P. Elliott), *Proc. R. Soc. A.* **232** (1956) 561.
18. "The effect of centre-of-mass motion on nuclear moments" (with J. P. Elliott), *Nuovo Cimento* **4** (1956) 115.
19. "High-energy nuclear physics" (with T. G. Pickavance and G. H. Stafford), *Nature* **78** (1956) 115.
20. "The nuclear surface", *Phil. Mag.* **1** (1956) 1043.
21. "Collective motion in quantum mechanics", *Proc. R. Soc. A.* **239** (1956) 399.
22. "The nuclear spin-orbit coupling" (with J. S. Bell), *Phil. Mag.* **1** (1956) 1055.
23. "The concept of mesic fluid in relation to pseudoscalar meson theory", *Suppl. Nuovo Cimento* **4** (1956) 749.
24. "A variational method in relativistic quantum field theory", *ibid*, p. 753.
25. "The alpha-particle and shell-models of the nucleus" (with J. K. Perring), *Proc. Phys. Soc. A.* **69** (1956) 600.
26. "Nuclear moments of inertia", *Proc. Phys. Soc. A.* **70** (1956) 433.
27. "Fusion induced mu-mesons", *Phil. Mag.* **2** (1957) 910.
28. "Parity nonconservation in weak interactions, Part II", *Prog. Nucl. Phys.* **6** (1957) 274.

29. "Magnetic moments of nuclei and the nuclear many-body problem" (with J. S. Bell and R. J. Eden), *Nucl. Phys.* **2** (1957) 586.
30. "A nonlinear theory of strong interactions", *Proc. R. Soc. A.* **247** (1958) 260.
31. "A nuclear pseudo-potential", *Proceedings of the Rehovoth Conference on Nuclear Structure*, 8–14 September 1957, ed. H. J. Lipkin (North-Holland, Amsterdam, 1958), p. 20.
32. "The anomalous moments of the nucleons" (with J. S. Bell), *Proc. R. Soc.* **242** (1958) 129.
33. "Effective nucleon potential", *Nucl. Phys.* **9** (1959) 615.
34. "The spin-orbit interaction in nuclei", *Nucl. Phys.* **9** (1959) 635.
35. "Some distortion effects in the nuclear *p*-shell", *Nucl. Phys.* **9** (1959) 641.
36. "A unified model of *K*- and *π*- meson", *Proc. R. Soc. A.* **252** (1959) 236.
37. "A unified field equation", (with K. J. Perring), *Nucl. Phys.* **31** (1961) 550.
38. "A nonlinear field theory", *Proc. R. Soc. A.* **260** (1961) 127.
39. "Particle states of a quantized meson field", *Proc. R. Soc. A.* **262** (1961) 237.
40. "A unified field theory of mesons and baryons", *Nucl. Phys.* **31** (1962) 556.
41. "Kinks and Dirac equation", *J. Math. Phys.* **12** (1971) 1735.
42. "On an identity relating to partitions of parts", (with M. S. Kirdar), *Can. J. Maths.* **34** (1982) 194.
43. "The maximum of a random walk whose mean path has a maximum" (with H. E. Daniels), *Adv. Appl. Prob.* **17** (1985) 85.
44. "The origins of Skyrmions", in *A Breadth of Phyiscs,* Proceedings of the Peierls 80th Birthday Symposium (eds. R. H. Dalitz and R. B. Stinchcombe, 1988), p. 193.

Appendix U

Unpublished research manuscripts held in the T. H. R. Skyrme Archive at Birmingham University.

U1 "Wall corrections for a monoenergetic isotropic reaction" (1952?), Memo. gw. 2 pp.
U2 "Wall corrections in proportional counters" (1952?), Memo. WC., 8 pp.
U3 "Three body problem at low energies Part I" (1952?).
U4 "Deuteron cross-sections of heavy nuclei at high energies" (1953?), 13 pp.
U5 "The stripping of fast deutrons" (1953–4?), 10 pp.
U6 "Wall corrections for ^{3}He counter" (1955), Memo. 2WC, 6 pp.
U7 "A collective model for 0^{16}", an invited paper to the Nuclear Physics Conference at Liverpool on March 30–31, 1960, manuscript BC (incomplete), 11 pp.
U8 "Symmetry", talk given to the Malaysian Mathematical Society on 22 January 1964, 7 pp.

Appendix T

Characteristics of a theory *T. H. R. Skyrme*

1. "Electromagnetic waves are the propagation of irrotational motions through a fluid,? a 'sea' of ± electrons in occupied states.
2. *Electrons, protons*, are vortex singularities in *this fluid.*
3. The fluid occupies all space except that occupied by nuclear and mesic matter.
4. *Nucleons* are disturbances central within drops of *mesic fluid*.
5. Electromagnetic properties of nucleons and nuclei are determined by *surface conditions* on the boundary between *mesic* and *electric* fluids.
6. Nuclei in their ground and low states consist of a number of quantized singularities or disturbances within a drop of *mesic fluid. Within* this fluid forces 'between nucleons' are small, but an incident particle has to exchange energy through its drop of fluid.

7. Mesic fluid has a considerable surface tension,? viscous.
8. When an incident nucleon hits a nucleus, the fluids amalgamate and thereafter the disturbances mingle freely within the compound nucleus.
9. At higher energies the fluid drops may pass through one another without intermingling.
10. If mesic fluid is viscous, nucleonic disturbances are quantized and cannot lose any further energy to the fluid - cf. electron in Bohr orbit.
11. A droplet of mesic fluid cannot easily separate from the mass of a nucleus unless it contains a 'spinning part' i. e. a nucleon.
12. Mesic fluid is not conserved.
13. Beta-decay of proton involves stripping 'distribution of charge' off nucleon and onto a positron disturbance. This seems a difficult process!

I. THE ALPHA-PARTICLE AND SHELL MODELS OF THE NUCLEUS

J. K. Perring and Tony Skyrme wrote this paper (page 37 of this volume) in order to reconcile the apparent differences between various nuclear models, in this case the shell model and the α-particle model. Dalitz gives an excellent discussion of this work in the first part of his outline in the first article of this volume.

Most interesting in the Perring–Skyrme paper is the case of ^{16}O. Here, the first excited state is a 0+ state at 6.06 MeV. In any simple shell model description this would be an odd parity state. Perring and Skyrme point out that in the α-particle model this is the symmetrical "breathing mode" of vibration. One can think of displacing the α-particles slightly, as done earlier by D. M. Dennison (*Phys. Rev.* **96** (1954) 378), with this mode representing the "breathing" of the resulting tetrahedron.

As Dalitz notes, the important part of Perring and Skyrme's calculations was to examine the structure of the α-particle states in the shell model basis. In the case of the 0+ breathing mode, the shell model configuration was 50% $(1p^{-1})(2p)$, 30% $(1s^{-1})(2s)$ and 20% $(1p^{-2})(1d)^2$. Shell model calculations at that time did not include two-nucleon excitation terms.

It is amusing, at least for me, to trace the further development of this 0+ state. If the state had the large $(1p^{-1})(2p)$ and $(1s^{-1})(2s)$ components, which were in phase for the pair (e^+e^-)-decay, then the probability of this decay would be large, as remarked in the Perring–Skyrme article. This seemed to agree, at least in general magnitude, with the then measured matrix element (S. Devons, G. Goldring, and G. R. Lindsey, *Proc. Phys. Soc.* **A67** (1954) 134). Later, the ratio of experimental to theoretical matrix element, calculated with the above configurations, turned out to be very small.

It was later understood that the 6.06 MeV 0+ state was best described as a four-particle, four-hole state in a highly deformed basis (G. E. Brown and A. M. Green, *Nucl. Phys.* **75** (1966) 401). Roughly speaking, the intrinsic state, out of which the 6.06 MeV state is projected, has the shape of a bathtub, with a ratio of 2 to 1 for major to minor axes. It was the first identified superdeformed state. In terms of spherical shell model wave functions, Haxton and Johnson (W. C. Haxton and Calvin Johnson, *Phys. Rev. Lett.* **65** (1990) 1325) were able to reproduce the Brown and Green results with $\sim$ 86,000 states.

The 2+ state at 6.92 MeV is the first excited member of a rotational band built on the 6.06 MeV 0+ state, members of which have been measured up through the 8+ state. The second 2+ state at 9.84 MeV is the lowest member of a $K = 2$ band. The breathing mode (single-particle excitations) of Perring and Skyrme is not to be seen until higher energy.

The simple connection between collective modes, such as the breathing mode, has, however, been very useful in describing the Roper Resonance, the breathing mode of the nucleon at 1440 MeV, about 500 MeV above the nucleon ground state. A very satisfactory description of the Roper Resonance is as a breathing mode of the skyrmion, to be discussed later.

Although deformation greatly modifies the picture, the α-particle description is very instructive for the 1− states. The lower state of Perring and Skyrme comes quite close to the empirical 1− state at 7.12 MeV. This lies just below the ^{4}He + ^{12}C threshold. The next 1− state is at 9.58 MeV; it is known to have a large α-width, essentially the single-particle width. In Perring and Skyrme, the second 1− state comes at 13 MeV. Haxton and Johnson, in their large shell model calculation, do not find the 9.58 MeV state at all. The reason can be found in unpublished ^{12}C(Li^6,d) ^{16}O experiments (Terry Fortune, private communication). These show the second excited 1− state to result from excitation of one of the particles in the four-particle, four-hole excited 0+ state at 6.06 MeV, up from the $2s, 1d$ shell up into the $2p, 1f$ shell. The state constructed by Perring and Skyrme must lie higher, probably somewhat lower than 13 MeV because of the effects of deformation. There are, therefore, three 1− states, which can be characterized as an α-particle in a chiefly $3P$, $4P$ and $5P$ state, relative to the ^{12}C core. Because of the large deformation in the $5P$ state, a 1− excitation built on the four-particle, four-hole 0+ mode of ^{16}O, the $5P$ comes lower than the $4P$, the latter being the Perring–Skyrme 13 MeV state.

Through two-particle, two-hole excitations, the $3P$ state will mix with the $4P$ one, the $4P$ with the $5P$, giving a 3×3 matrix as found by Brown and Green for the 0+ states. The 1− state at 7.12 MeV is a coherent mixture of $3P$ and $4P$ states. Of course, a small piece of $5P$ will also be mixed in. Although the 7.12 MeV state is a coherent mixture of $3P$ and $4P$, there is one more node in the latter, so the two have opposite signs at large distances where the α-widths are determined, choosing phases so that the signs are the same at short distances. Thus, the α-capture on ^{12}C to the 1− state at 7.12 MeV is small, about 0.1 single particle units (L. Buchmann *et al.*, *Phys. Rev. Lett.* **70** (1993) 726).

The elegant considerations of Perring and Skyrme give a simple framework in which to view problems like the α capture on ^{12}C. Although, in the end, we cannot calculate the capture width to the 7.12 MeV 1− state quantitatively, we can understand why it is small and why it cannot be accurately calculated. These days one would simply put the problem into a large shell model program that mixed millions of states. If one were lucky, one would come out with an answer not far from that which one would guess on the basis of simple considerations, with the correct physical picture.

The Alpha-Particle and Shell Models of the Nucleus

By J. K. PERRING AND T. H. R. SKYRME
Atomic Energy Research Establishment, Harwell, Berks.

Communicated by B. H. Flowers ; MS. received 23rd February 1956

Abstract. It is shown that it is possible to write down α-particle wave functions for the ground states of ^{8}Be, ^{12}C and ^{16}O, which become, when antisymmetrized, identical with shell-model wave functions. The α-particle functions are used to obtain potentials which can then be used to derive wave functions and energies of excited states. Most of the low-lying states of ^{16}O are obtained in this way, qualitative agreement with experiment being found. The shell structure of the 0^+ level at 6·06 MeV is analysed, and is found to consist largely of single-particle excitations. The lifetime for pair-production is calculated, and found to be comparable with the experimental value. The validity of the method is discussed, and comparison made with shell-model calculations.

§ 1. Introduction

It has been known for some time that the shell-model wave function for a light 4n-type nucleus will automatically give something of an α-particle structure. This is purely a consequence of the exclusion principle and the symmetry properties of the individual orbitals, and holds quite independently of inter-nucleon forces. A similar result holds in atomic structure; for example, it is well known that the sp^3 configuration of carbon can be described by tetrahedrally directed orbitals. The shell-model in its simplest form does not allow for correlation between nucleons of differing spin and isotopic spin, and we might expect an α-particle model, which emphasizes the positive correlations between nucleons, to be a better representation of the structure of the nucleus.

The α-particle model does, indeed, seem to work fairly well in certain ways, for instance in accounting for the binding energies of the 4n nuclei; and Dennison (1954) has recently been able to interpret most of the low-lying levels of ^{16}O on this basis. An advantage of such a model is that it requires us to solve a problem with only one quarter of the number of degrees of freedom of the individual-particle model, using, in general, methods that are well known from the theory of molecular spectra. If such a simplification can be achieved it will be very desirable, since the algebraic and numerical manipulations of the shell model become complex very quickly.

In this paper we shall use a method which will enable us to predict energy levels for α-particle nuclei, and which will also allow us to derive shell-model wave functions for excited states. Our aim is to compare our conclusions with those of the shell-model, and to see to what extent these are independent of particular assumptions about inter-nucleon forces; then, perhaps, we may be able to use the method as a guide to the structure of more complex systems. Our approach is based on a simple but surprising mathematical identity. This enables us to write down a simple and reasonable α-particle wave function which turns out to be identical with a shell-model function.

§ 2. General Method

The transformations we shall make depend on the simple and convenient properties of harmonic-oscillator wave functions. In the individual-particle model we suppose that we can take the nucleon wave functions to be the solutions of the equation

$$\sum_{i=1}^{N} [-(\hbar^2/2m)\nabla_i^2 + \tfrac{1}{2}k(\mathbf{r}_i - \mathbf{S})^2]\Psi = E\Psi \qquad \ldots\ldots(1)$$

where $\mathbf{S}$ is the centre-of-mass coordinate.

If we measure lengths in units of $(\hbar^2/km)^{1/4}$ and energies in units of $(k\hbar^2/m)^{1/2}$ this becomes

$$\tfrac{1}{2}\sum_{i=1}^{N} [-\nabla_i^2 + (\mathbf{r}_i - \mathbf{S})^2]\Psi = E\Psi. \qquad \ldots\ldots(2)$$

The wave functions of the individual particles are then of the form (polynomial) $\times \exp\{-(\mathbf{r}-\mathbf{S})^2/2\}$ and their energy levels are spaced by single units. Many-particle wave functions are built up of products of these, and therefore contain the factor $\exp[-\frac{1}{2}\sum(\mathbf{r}_i - \mathbf{S})^2]$. A general shell-model wave function can thus be written

$$\sum \pm P(\mathbf{r}_i - \mathbf{S}, \ldots, \mathbf{r}_N - \mathbf{S}) \exp[-\tfrac{1}{2}\sum(\mathbf{r}_i - \mathbf{S})^2] F(1, \ldots, N), \qquad \ldots\ldots(3)$$

where P is a polynomial in the $3N$ variables $(\mathbf{r}_i - \mathbf{S})$, $F(1,\ldots,N)$ is a spin–charge function, and the product is antisymmetrized with respect to all the variables.

Now let us consider an α-particle nucleus consisting of $4N$ nucleons. We may split the coordinates $\mathbf{r}_i$ into groups of four, with centres $\mathbf{R}_j$, say, to correspond to α-particles,

$$\mathbf{R}_j = \tfrac{1}{4}\sum_{i=4j-3}^{4j} \mathbf{r}_i, \qquad \ldots\ldots(4)$$

and it is easily shown that

$$\sum_{i=1}^{4N} (\mathbf{r}_i - \mathbf{S})^2 = \sum_{j=1}^{N}\sum_{i=4j-3}^{4j} (\mathbf{r}_i - \mathbf{R}_j)^2 + 4\sum_{j=1}^{N} (\mathbf{R}_j - \mathbf{S})^2. \qquad \ldots\ldots(5)$$

In the expression (3) the exponential factor is completely symmetric and can therefore be taken outside the antisymmetrization. Then for certain types of polynomial P it is possible to express the anti-symmetrized expression $\Sigma \pm PF$ as the result of antisymmetrizing an expression

$$Q(\mathbf{R}_1, \ldots, \mathbf{R}_N) \prod_{j=1}^{N} F_\alpha(4j-3, \ldots, 4j) \qquad \ldots\ldots(6)$$

where Q is a symmetric polynomial and F_α is the α-particle spin charge function. When this condition holds we may re-write the wave function (3) as

$$\sum \pm Q(\mathbf{R}_1, \ldots, \mathbf{R}_N) \exp\{-2\sum(\mathbf{R}_j - \mathbf{S})^2\} \prod_{j=1}^{N} [\exp\{-\tfrac{1}{2}\sum(\mathbf{r}_i - \mathbf{R}_j)^2\} F_\alpha(4j-3, \ldots, 4j)] \qquad \ldots\ldots(7)$$

each term of which may be thought of as the product of an α-particle wave function with the internal wave functions of its constitutents.

These ideas allow us to construct a shell-model wave function, corresponding to any α-particle wave function of the form

$$\psi_\alpha = Q(\mathbf{R}_1, \ldots \mathbf{R}_N) \exp\{-2\sum(\mathbf{R} - \mathbf{S})^2\}. \qquad \ldots\ldots(8)$$

The converse is not true in general. To any such wave function (8) there is a corresponding α-particle potential $U(\mathbf{R}_1, \ldots \mathbf{R}_N)$, such that

$$\{(-\tfrac{1}{8})\sum_j \nabla_j^2 + U(\mathbf{R}_1, \ldots \mathbf{R}_N)\}\psi_\alpha = E\psi_\alpha. \qquad \ldots\ldots(9)$$

We can take this potential to represent the combined effects of the central potential and the exclusion principle. The forms of ψ_α and U derived are not unique, and it is necessary to impose a further condition on them. If we start from a shell-model wave function (3) representing the ground state of the system it is reasonable to demand that ψ_α and U shall be such that ψ_α is the lowest eigenfunction of U; this excludes some choices, but still leaves a range of functions U.

Once we have fixed U we can consider other states of the system. There is no guarantee that this effective potential U is the same for excited states as for the ground state, but in order to make progress we shall assume that it is. With this assumption we can solve the Schrodinger equation (9) to obtain wave functions for the excited states, which we can interpret in terms of the rotations and vibrations of the system of α-particles. From each new function we can construct the corresponding shell-model function, by antisymmetrizing, and thus find the fractions of the various configurations which it comprises.

This procedure is a way of choosing particular configurations so as to keep together the α-particle groups. Physically, this is effected by predominantly Majorana inter-nucleon forces. The relation between these and our effective potential U is rather obscure; the origin of the excitation energy seems to be entirely different in the two models.

We shall now consider in more detail the three simplest α-particle nuclei: ^{8}Be, ^{12}C, ^{16}O. We have been most interested in the last because of the possibility of comparing our results with the shell-model calculations recently made by Elliott (to be published).

§ 3. Beryllium 8

We may expect the ground state of ^{8}Be on the shell model to be composed principally of the most symmetric state of the configuration s^4p^4. This we may write in the form (3) with

$$\left.\begin{aligned} P &= (\mathbf{r}_5\cdot\mathbf{r}_6)(\mathbf{r}_7\cdot\mathbf{r}_8)+(\mathbf{r}_5\cdot\mathbf{r}_7)(\mathbf{r}_6\cdot\mathbf{r}_8)+(\mathbf{r}_5\cdot\mathbf{r}_8)(\mathbf{r}_6\cdot\mathbf{r}_7) \\ F &= F_\alpha(1,\ldots,4)F_\alpha(5,\ldots,8) \end{aligned}\right\} \qquad \ldots\ldots.(10)$$

The polynomial P is equivalent under antisymmetrization to the α-particle expression $Q=(\mathbf{R}_1-\mathbf{R}_2)^4$. This may be verified directly by antisymmetrizing each expression in the pairs 1 and 5, 2 and 6, 3 and 7, 4 and 8, when each gives

$$[(\mathbf{r}_1-\mathbf{r}_5)\cdot(\mathbf{r}_2-\mathbf{r}_6)]\,[(\mathbf{r}_3-\mathbf{r}_7)\cdot(\mathbf{r}_4-\mathbf{r}_8)]+[(\mathbf{r}_1-\mathbf{r}_5)\cdot(\mathbf{r}_3-\mathbf{r}_7)]\,[(\mathbf{r}_2-\mathbf{r}_6)\cdot(\mathbf{r}_4-\mathbf{r}_8)]$$
$$+[(\mathbf{r}_1-\mathbf{r}_5)\cdot(\mathbf{r}_4-\mathbf{r}_8)]\,[(\mathbf{r}_2-\mathbf{r}_6)\cdot(\mathbf{r}_3-\mathbf{r}_7)].$$

It seems to be true in general that it is only necessary to antisymmetrize with respect to particles with the same spin–charge function, which is a fortunate simplification. We can now write ψ_α as $\psi_\alpha = R^4\exp(-R^2)$ where

$$\mathbf{R}=\mathbf{R}_1-\mathbf{R}_2. \qquad \ldots\ldots(11)$$

This ψ_α satisfies the Schrödinger equation with $U(\mathbf{R})=R^2+5/R^2$ in which we may regard the second term as a manifestation of the exclusion principle. We could have taken for Q any polynomial in $\mathbf{R}$ of the fourth degree and obtained the same equivalent shell-model wave function (terms of lower degree than the fourth disappear on antisymmetrization). We should then have obtained a different U, and ψ_α might no longer have been the ground state of U. The present choice is, however, the simplest.

The first excited state of this potential has angular momentum $J=2$. (Only even angular momenta and even parity states are allowed, since the α-particles obey Bose statistics.) The wave function is

$$R^n \exp(-R^2) Y_2(\theta, \phi) \quad \text{with} \quad n = \tfrac{1}{2}(\sqrt{105}-1) \simeq 4{\cdot}62 \qquad \ldots\ldots(12)$$

and the energy of excitation is 0·62 unit. It can be seen that there is a certain amount of rotation-vibration interaction; a pure rotational wave functon would have $n=4$ in (12).

In order to obtain the shell-model wave functions we must put (12) into a rational form. The minimum of the potential $U(R)$ is at $R=5^{1/4}$. Hence we write $R^{4\cdot62} = R^4\{5^{1/2}+(R^2-5^{1/2})\}^{0\cdot31}$ and expand by the binomial theorem. The leading term gives just the ^{1}D state of the configuration $1s^4 1p^4$; further terms are equivalent to higher configurations.

The first pure vibrational state is $(R^2-11/4)R^4 \exp(-R^2)$ with an energy of excitation of 2 units. The resulting shell-model configurations are those in which one particle is doubly excited, e.g. $1s^3 1p^4 2s$. No configurations occur for which two particles are singly excited (e.g. $1s^2 1p^6$). The reason for this is that the present form of the α-particle model implies the exclusive use of configurations of the maximum spatial symmetry [44]. There are five such states involving single excitation of two particles, but all of these are spurious states which describe motion of the centre of mass (Elliott and Skyrme 1955; we are indebted to Dr. Elliott for discussions on this point).

§ 4. Carbon 12

If we take the ground-state of ^{12}C to be that complementary to ^{8}Be, so that the four holes have P given by (10), we can carry out a similar analysis. The α-particle wave function may be taken as

$$\psi_\alpha = |N|^4 \exp\{-2\sum(\mathbf{r}_i - \mathbf{S})^2\} \qquad \ldots\ldots(13)$$

where

$$N = (\mathbf{R}_1 \times \mathbf{R}_2 + \mathbf{R}_2 \times \mathbf{R}_3 + \mathbf{R}_3 \times \mathbf{R}_1) \qquad \ldots\ldots(14)$$

which is twice the vector area of the triangle formed by the three α-particles. The corresponding potential is

$$U(\mathbf{R}_1, \mathbf{R}_2, \mathbf{R}_3) = 2\sum_i (\mathbf{R}_i - \mathbf{S})^2 + 2\sum_i (\mathbf{R}_i - \mathbf{S})^2/N^2 \qquad \ldots\ldots(15)$$

which has a minimum when $\mathbf{R}_1$, $\mathbf{R}_2$, $\mathbf{R}_3$ lie at the vertices of an equilateral triangle of side $(4/3)^{1/4}$.

Excited states may be dealt with in much the same way as for ^{8}Be, except that it is not always possible to write down exact wave functions. The analysis may be simplified somewhat by eliminating the centre-of-mass motion. Introducing the internal coordinates

$$\left.\begin{aligned} \mathbf{G} &= \tfrac{1}{3}(\mathbf{R}_1 + \omega\mathbf{R}_2 + \omega^2\mathbf{R}_3) \\ \mathbf{H} &= \tfrac{1}{3}(\mathbf{R}_1 + \omega^2\mathbf{R}_2 + \omega\mathbf{R}_3) \\ \boldsymbol{\omega} &= \exp 2\pi i/3 \end{aligned}\right\} \qquad \ldots\ldots(16)$$

(13) then becomes

$$\psi_\alpha = 3^6 |\mathbf{G} \times \mathbf{H}|^4 \exp(-12\mathbf{G}.\mathbf{H}) \qquad \ldots\ldots(17)$$

with the corresponding Schrödinger equation

$$\left\{-\frac{1}{12}\nabla_G\cdot\nabla_H+12\mathbf{G}\cdot\mathbf{H}-\frac{4}{9}\frac{\mathbf{G}\cdot\mathbf{H}}{|\mathbf{G}\times\mathbf{H}|^2}\right\}\psi_\alpha=E\psi_\alpha. \quad \ldots\ldots(18)$$

Although the coordinates are complex, all quantities appearing in (17) and (18) are real.

§ 5. Oxygen 16

We shall treat the oxygen 16 nucleus in rather more detail. The ground state is the closed shell $1s^4 1p^{12}$, and the corresponding α-particle wave function is

$$\psi_\alpha = V^4 \exp -2\sum(\mathbf{R}_i-\mathbf{S})^2 \quad \ldots\ldots(19)$$

where V is six times the volume of the tetrahedron formed by $\mathbf{R}_1, \mathbf{R}_2, \mathbf{R}_3, \mathbf{R}_4$. Analytically, V is given by the determinant whose rows are $(1, X_i, Y_i, Z_i)$ where X_i, Y_i, Z_i, are the components of the vector $\mathbf{R}_i$.

The corresponding potential is

$$U(\mathbf{R}_1,\ldots\mathbf{R}_4)=2\sum_i(\mathbf{R}_i-\mathbf{S})^2+\frac{3}{2}\sum_i|N_i|^2/V^2 \quad \ldots\ldots(20)$$

in which the $\mathbf{N}_i$ are twice the vector areas of the faces of the tetrahedron, and are given by expressions similar to (14). U has a minimum for a regular tetrahedral configuration of side $(108)^{1/4} = 3{\cdot}2$.

The analysis can be considerably simplified by eliminating the centre-of-mass motion, introducing internal coordinates analogous to (11) and (16):

$$\left.\begin{aligned}\mathbf{F}&=\tfrac{1}{4}(\mathbf{R}_1+\mathbf{R}_2-\mathbf{R}_3-\mathbf{R}_4)\\ \mathbf{G}&=\tfrac{1}{4}(\mathbf{R}_1-\mathbf{R}_2+\mathbf{R}_3-\mathbf{R}_4)\\ \mathbf{H}&=\tfrac{1}{4}(\mathbf{R}_1-\mathbf{R}_2-\mathbf{R}_3+\mathbf{R}_4)\end{aligned}\right\} \quad \ldots\ldots(21)$$

Then V is proportional to the triple scalar product $\mathbf{F}\cdot(\mathbf{G}\times\mathbf{H})=(\mathbf{FGH})$, and $\Sigma_i(\mathbf{R}_i-\mathbf{S})^2=4(F^2+G^2+H^2)$; so that $\psi_\alpha=(\mathbf{FGH})^4\exp\{-8(F^2+G^2+H^2)\}$ and

$$\{-\tfrac{1}{32}(\nabla_F{}^2+\nabla_G{}^2+\nabla_H{}^2)+8(F^2+G^2+H^2)+\tfrac{3}{8}[(\mathbf{F}\times\mathbf{G})^2+(\mathbf{G}\times\mathbf{H})^2 +(\mathbf{H}\times\mathbf{F})^2]/(\mathbf{FGH})^2\}\psi_\alpha=E\psi_\alpha. \quad \ldots\ldots(22)$$

We can now look for excited states, by analogy with the case of ^{8}Be, with wave functions of the form $\psi=P\psi_0$ where $P(\mathbf{F},\mathbf{G},\mathbf{H})$ must satisfy

$$\sum_{F,G,H}[\nabla_F{}^2+8\{(\mathbf{G}\times\mathbf{H})(\mathbf{FGH})^{-1}-4\mathbf{F}\}.\nabla_F]P=-32(E-E_0)P,$$

which has elementary solutions $(F^2-11/32)$, $\mathbf{F}.\mathbf{G}$,$(F_iF_j-\frac{1}{3}F^2\delta_{ij})$, etc. with excitation energy of two units. Now α-particles obey Bose statistics, so we must take combinations of these solutions with the correct symmetry. The effect of a permutation of α-particles is to permute $\mathbf{F}$, $\mathbf{G}$ and $\mathbf{H}$, either with no change of sign, or with the signs of two coordinates changed simultaneously. Thus allowed forms of P are

$$(F^2+G^2+H^2-33/32) \quad \ldots\ldots(23)$$

$$\sum_{F,G,H}(F_iF_j-\tfrac{1}{3}F^2\delta_{ij}). \quad \ldots\ldots(24)$$

The first expression (23) is the wave function for the symmetrical 'breathing' mode of vibration (ω_1 in the language of molecular spectroscopy), and (24) is a combined vibrational–rotational mode with angular momentum $J=2$ and even parity. The other vibrational modes, which would, if allowed, be degenerate with ω_1, are ω_2: (F^2-G^2), $(2H^2-F^2-G^2)$; ω_3: $\mathbf{F}.\mathbf{G}$, $\mathbf{G}.\mathbf{H}$, $\mathbf{H}.\mathbf{F}$.

We may identify (23) with the 0^+ pair-emitting level at 6·06 MeV and (24) with the 2^+ level at 6·91 MeV. The remaining low-lying levels known are a 3^- at 6·14 MeV a 1^- at 7·12 MeV (Ajzenberg and Lauritsen 1955) and a 2^- at 8·85 MeV (Wilkinson 1955). It does not seem possible to obtain exact solutions of (22) with odd parity, and we have been forced to use approximate forms, and to seek the aid of the variation principle. Expressions for P with the correct symmetry requirements are

$$1^-: \sum_{F,G,H} (\mathbf{F}.\mathbf{G})(\mathbf{F}\times\mathbf{G})(\mathbf{FGH})^{-1}, \sum_{F,G,H} (\mathbf{F}.\mathbf{G})\mathbf{H} \quad \ldots\ldots(25)$$

$$2^-: \sum_{F,G,H} (F^2G_iG_j - G^2F_iF_j)(\mathbf{FGH})^{-1} \quad \ldots\ldots(26)$$

$$\left.\begin{aligned} 3^-: &\sum_{F,G,H} (F_iG_j - \tfrac{1}{5}\mathbf{F}.\mathbf{G}\delta_{ij})(\mathbf{F}\times\mathbf{G})_k(\mathbf{FGH})^{-1} \\ &\sum_{F,G,H} [F_iG_jH_k - \tfrac{1}{5}(F_i\delta_{jk} + F_j\delta_{ki} + F_k\delta_{ij})\mathbf{G}.\mathbf{H}] \end{aligned}\right\} \ldots\ldots(27)$$

It should be noted that whereas there are two simple tensors of 1^- and 3^- symmetry, there is only one of 2^-.

If we now work out $(\psi|H|\psi)/(\psi|\psi)$ for each of these expressions, we shall obtain estimates of the energy. Better estimates can be obtained for the 1^- and 3^- levels by allowing mixtures of the two types of expressions in (25) and (27). The levels obtained, after some complicated and laborious algebra, are shown in table 1.

It so happens that the two functions taken for the 3^- state overlap very considerably, the integral of the normalized functions being 0·92; this explains the very high energy obtained for the higher level. If we take our unit of energy to be about 3 MeV, to give reasonable agreement for the 0^+ and 2^+ states, we find the 1^- level in good agreement with experiment, but the 2^- and 3^- levels completely wrong (see table 1). It should be noted that the second form of P for the 1^- and 3^- states is the more important, the coefficients being in the ratio 0.63 : 1 for the 1^- state and 0.18 : 1 for the 3^-. This implies that the main constituents of these states should correspond to triple rather than single, shell-model excitations, i.e. to configurations such as $p^{-3}\,d^3$ rather than $p^{-1}\,d$.

It is possible to obtain some exact results for the states in which P is a quartic expression. These have excitation energies of 4 units. Three 0^+ states are obtained, for which the P's are orthogonal combinations of

$$\left.\begin{aligned} &(F^2+G^2+H^2)^2 - (35/16)(F^2+G^2+H^2) + 1155/1024 \\ &(F^2G^2+G^2H^2+H^2F^2) - (11/16)(F^2+G^2+H^2) + 363/1024 \\ &[(\mathbf{F}.\mathbf{G})^2 + (\mathbf{G}.\mathbf{H})^2 + (\mathbf{H}.\mathbf{F})^2] - (1/16)(F^2+G^2+H^2) + 33/1024 \end{aligned}\right\} \ldots\ldots(28)$$

together with one 2^+ state

$$\sum (F_iF_j - \tfrac{1}{3}F^2\delta_{ij})(F^2+G^2+H^2-37/32) \quad \ldots\ldots(29)$$

and one 4^+ state

$$\sum(F_iF_j - \tfrac{1}{3}F^2\delta_{ij})\{\sum(F_iF_j - \tfrac{1}{3}F^2\delta_{ij}) - (4/35)(F^2+G^2+H^2)\}, \quad \ldots\ldots(30)$$

It seems likely that there are also neighbouring 2^+ and 4^+ states, the wave functions of which can be approximated by such expressions as

$$P = \sum \mathbf{F}.\mathbf{G}(F_iG_j + F_jG_i - \tfrac{2}{3}\mathbf{F}.\mathbf{G}\delta_{ij}) \qquad (2^+),$$

$$P = \sum F^2(G_iG_j - \tfrac{1}{3}G^2\delta_{ij}) \qquad (2^+),$$

$$P = \sum (F_iG_j + F_jG_i - \tfrac{2}{3}\mathbf{F}.\mathbf{G}\delta_{ij})^2 \qquad (4^+), \text{ etc.}$$

Energies for these functions could be estimated by the variation principle. This has not been done, as the algebra involved would be extremely complex.

Energies for the states given in (28), (29) and (30), are shown together with the others in table 1. It is apparent that no precise correspondence with experiment can be established. The experimental level scheme becomes complicated, and is not definitely complete, above 11 MeV excitation.

Table 1.

State	Experimental Energy (MeV)	Calculated Energy (units of $(k\hbar^2/m)^{1/2}$)	$((k\hbar^2/m)^{1/2}=3$ MeV)
0^+	6·06	2	6·0
2^+	6·91	2	6·0
1^-	7·12	2·22	6·7
1^-	9·58	4·33	13·0
2^-	8·85	4·32	13·0
3^-	6·14	0·97	2·9
3^-	11·62	14·9	45
0^+ (3 states)	11·25, etc.	4	12·0
2^+	9·84	4	12·0
4^+	10·36	4	12·0

Experimental data, with one exception, are taken from the compilation of Ajzenberg and Lauritsen (1955). The value for the energy of the 2^- state is due to Wilkinson (1955).

§ 6. Shell-Model Wave Functions

We shall now reverse our initial procedure, and find to what shell-model wave functions the 0^+ excited-state wave function (23) corresponds. The method is simplified by using instead of the individual-particle coordinate $\mathbf{r}_i$, certain combinations of them

$$\boldsymbol{\rho}_i = \tfrac{1}{4}(\mathbf{r}_i + \mathbf{r}_{i+4} - \mathbf{r}_{i+8} - \mathbf{r}_{i+12}),$$
$$\boldsymbol{\sigma}_i = \tfrac{1}{4}(\mathbf{r}_i - \mathbf{r}_{i+4} + \mathbf{r}_{i+8} - \mathbf{r}_{i+12}),$$
$$\boldsymbol{\tau}_i = \tfrac{1}{4}(\mathbf{r}_i - \mathbf{r}_{i+4} - \mathbf{r}_{i+8} + \mathbf{r}_{i+12}),$$

so that $\mathbf{F} = \frac{1}{4}\Sigma\boldsymbol{\rho}_i$, $\mathbf{G} = \frac{1}{4}\Sigma\boldsymbol{\sigma}_i$, $\mathbf{H} = \frac{1}{4}\Sigma\boldsymbol{\tau}_i$. Then P of (23) is proportional to

$$\sum_i(\boldsymbol{\rho}_i^2+\boldsymbol{\sigma}_i^2+\boldsymbol{\tau}_i^2)+2\sum_{i>j}(\boldsymbol{\rho}_i\cdot\boldsymbol{\rho}_j+\boldsymbol{\sigma}_i\cdot\boldsymbol{\sigma}_j+\boldsymbol{\tau}_i\cdot\boldsymbol{\tau}_j)-33/2. \qquad \ldots\ldots(31)$$

We take the spin–charge functions in the form

$$(\alpha_1\beta_2\gamma_3\delta_4)\,(\alpha_5\beta_6\gamma_7\delta_8)\ldots$$

Then we need to antisymmetrize within the four sets (1, 5, 9, 13), (2, 6, 10, 14), ... Under a permutation of one of these sets the variables ρ_i, σ_i, τ_i are permuted among themselves, with either no sign changes, or two simultaneous changes. In particular the triple product $(\rho_i\sigma_i\tau_i)$ is completely antisymmetric, and is the expression of lowest degree which is so. Hence the factor $(FGH)^4$ must reduce to a multiple of

$$(\boldsymbol{\rho}_1\boldsymbol{\sigma}_1\boldsymbol{\tau}_1)(\boldsymbol{\rho}_2\boldsymbol{\sigma}_2\boldsymbol{\tau}_2)(\boldsymbol{\rho}_3\boldsymbol{\sigma}_3\boldsymbol{\tau}_3)(\boldsymbol{\rho}_4\boldsymbol{\sigma}_4\boldsymbol{\tau}_4). \qquad \ldots\ldots(32)$$

The remaining terms in (31) contribute two steps of excitation to the wave function; either a single particle can be doubly excited, or a pair of particles can each be singly excited. For the moment we need consider only single or double excitation of sets. Double excitation of a set gives a contribution.

$$(\rho_i^2+\sigma_i^2+\tau_i^2)(\boldsymbol{\rho}_i\boldsymbol{\sigma}_i\boldsymbol{\tau}_i) \qquad \ldots\ldots(33)$$

to the wave function, and single excitation expressions of the type

$$Q(a,b;\ c,d)=\begin{vmatrix} \rho_a\rho_b & \rho_c\rho_d & 1 \\ \sigma_a\sigma_b & \sigma_c\sigma_d & 1 \\ \tau_a\tau_b & \tau_c\tau_d & 1 \end{vmatrix} \qquad \ldots\ldots(34)$$

where a, b, c, d, are coordinate indices. The extraction of the correct constants to multiply the expressions (32), (33), and (34) is extremely laborious and will not be given in full here.

If we antisymmetrize with respect to the sets 3 and 4, keeping only the parts corresponding to excitations of sets 1 and 2, the wave function reduces to a multiple of

$$[(\boldsymbol{\rho}_1+\boldsymbol{\rho}_2)^2+(\boldsymbol{\sigma}_1+\boldsymbol{\sigma}_2)^2+(\boldsymbol{\tau}_1+\boldsymbol{\tau}_2)^2-K](\boldsymbol{\rho}_1+\boldsymbol{\rho}_2, \boldsymbol{\sigma}_1+\boldsymbol{\sigma}_2, \boldsymbol{\tau}_1+\boldsymbol{\tau}_2)^2(\boldsymbol{\rho}_3\boldsymbol{\sigma}_3\boldsymbol{\tau}_3)(\boldsymbol{\rho}_4\boldsymbol{\sigma}_4\boldsymbol{\tau}_4) \qquad \ldots\ldots(35)$$

where $K=495/56$. Antisymmetrizing the part of this which depends on sets 1 and 2, we find that

$$(\boldsymbol{\rho}_1+\boldsymbol{\rho}_2, \boldsymbol{\sigma}_1+\boldsymbol{\sigma}_2, \boldsymbol{\tau}_1+\boldsymbol{\tau}_2)^2 \to 4(\boldsymbol{\rho}_1\boldsymbol{\sigma}_1\boldsymbol{\tau}_1)(\boldsymbol{\rho}_2\boldsymbol{\sigma}_2\boldsymbol{\tau}_2)$$

which must be multiplied by

$$\sum_{i=1,2}(\rho_i^2+\sigma_i^2+\tau_i^2)-K$$

to give the contribution to (35), and

$$(\boldsymbol{\rho}_1\cdot\boldsymbol{\rho}_2+\boldsymbol{\sigma}_1\cdot\boldsymbol{\sigma}_2+\boldsymbol{\tau}_1\cdot\boldsymbol{\tau}_2)(\boldsymbol{\rho}_1+\boldsymbol{\rho}_2, \boldsymbol{\sigma}_1+\boldsymbol{\sigma}_2, \boldsymbol{\tau}_1+\boldsymbol{\tau}_2)^2$$
$$\to \sum_{i=1,2}(\rho_i^2+\sigma_i^2+\tau_i^2)(\boldsymbol{\rho}_1\boldsymbol{\sigma}_1\boldsymbol{\tau}_1)(\boldsymbol{\rho}_2\boldsymbol{\sigma}_2\boldsymbol{\tau}_2)-\tfrac{1}{3}\sum\epsilon_{abc}\epsilon_{uvw}Q_1(d,a;b,v)Q_2(d,u;c,w)$$

in which the second summation is over all repeated indices. In this way the total wave function is reduced to

$$[6\sum_{i=1}^{4}(\rho_i^2+\sigma_i^2+\tau_i^2)-4K](\boldsymbol{\rho}_1\boldsymbol{\sigma}_1\boldsymbol{\tau}_1)(\boldsymbol{\rho}_2\boldsymbol{\sigma}_2\boldsymbol{\tau}_2)(\boldsymbol{\rho}_3\boldsymbol{\sigma}_3\boldsymbol{\tau}_3)(\boldsymbol{\rho}_4\boldsymbol{\sigma}_4\boldsymbol{\tau}_4)$$
$$-2\sum\sum\epsilon_{abc}\epsilon_{uvw}Q_1(d,a;\ b,v)Q_j(d,u;\ c,w)(\rho_k\sigma_k\tau_k)(\rho_l\sigma_l\tau_l).$$

The summations in the second term are over $i<j<k<l$, and repeated indices.

It is now necessary to transform back to the original coordinates. Using the relations

$$(\boldsymbol{\rho}_1\boldsymbol{\sigma}_1\boldsymbol{\tau}_1)=-\frac{1}{16}\begin{vmatrix} 1 & x_1 & y_1 & z_1 \\ 1 & x_5 & y_5 & z_5 \\ 1 & x_9 & y_9 & z_9 \\ 1 & x_{13} & y_{13} & z_{13} \end{vmatrix}$$

$$Q(a,b;\ c,d)=-\frac{1}{64}[(1,\boldsymbol{r}_a,\boldsymbol{r}_c,\boldsymbol{r}_b\boldsymbol{r}_d)+(1,\boldsymbol{r}_a,\boldsymbol{r}_d,\boldsymbol{r}_b\boldsymbol{r}_c)+(1,\boldsymbol{r}_b,\boldsymbol{r}_c,\boldsymbol{r}_a\boldsymbol{r}_d)$$
$$+(1,\boldsymbol{r}_b,\boldsymbol{r}_d,\boldsymbol{r}_a\boldsymbol{r}_c)]$$

this is readily accomplished, and it only remains to transform the elements of the determinants into the orthogonal polynomials of the single-particle wave functions.

When this is done, and identical terms are collected, the intensities (amplitudes squared) of the different configurations found in the wave functions are as shown in table 2.

Table 2.

Configuration	$1s^{-1}\,2s$	$1p^{-1}\,2p$	$1p^{-2}1d^2$	$1p^{-2}1d2s$	$1p^{-2}\,2s^2$
Intensity (%)	29	52	17	1·9	0·15

Strictly, we should include the ground state $1s^4 1p^{12}$, which would amount to about 20%. This arises because the method is not self-consistent. However, it would not be possible to attach any firm significance to the resulting wave function, and we have therefore omitted the ground state.

A noteworthy feature of these figures is the large proportion of single-particle excitation, which is in agreement with the lifetime for pair emission. The matrix element for this transition is

$$\int \psi_0^* \sum r_p^2 \psi_{ex}\, dT,$$

the mean square radius of the proton distribution, to which only single-particle excitations contribute. If we take the excited state to be at 6 MeV, the unit of length is $3{\cdot}7 \times 10^{-13}$ cm. The matrix element is then $11 \times 10^{-26}\,\mathrm{cm}^2$, which is comparable with the experimental value of $3{\cdot}8 \times 10^{-26}\,\mathrm{cm}^2$ (Devons, Goldring and Lindsey 1954). Alternatively, we can take this result to define our unit of length, and hence find our unit of energy to be 8·4 MeV. The energy of the 3^- state then becomes 8·2 MeV, which is too high; however, the change is in the right direction.

This brings out a point of interest: a different unit of length seems to be required for properties which depend on the nuclear size—the energy of the pure rotational state and the value of the matrix element above—from that required for the vibrations, which involve the inter-particle forces. The latter makes the nucleus far too large.

§ 7. Conclusions

Considering the crudity of the approximations we have made, the agreement, though qualitative, is surprisingly good. Starting from a shell-model wave function, we have been able to deduce, qualitatively, the ^{16}O spectrum. In particular, the order of the lowest odd-parity levels, 3^-, 1^-, 2^-, is that found experimentally and the ratio of their separations is of the right order. However, Elliott (to be published), in a shell-model calculation, has been able to fit these levels considering only single excitations, whereas our model predicts large amounts of triple excitation in the 1^- and 3^- states.

The quantitative agreement of the 0^+, 2^+, and 1^- levels with experiment seems to be largely coincidental. If we are to obtain quantitative agreement for all the lower levels, we must be able to find vibration and rotation constants of the order of Dennison's empirical values. In our model the three vibration frequencies are equal, whereas Dennison's are $\omega_1 = 6{\cdot}05$, $\omega_2 = 6{\cdot}77$, $\omega_3 = 4{\cdot}7$ MeV, for his scheme (*a*), and our rotation parameter $R = \hbar^2/I$ is far too small. Now according to Dennison's scheme (*a*) $E(0^+) = \omega_1$, $E(1^-) = E(2^+) = \omega_3 + 9R/4$; since we have found $\omega_1 = \omega_3$ and R to be small, we do find that the three levels lie close together. However, we cannot fit the remaining low-lying levels satisfactorily. Similar reasoning applies to Dennison's scheme (*b*).

There is a further experimental feature which finds no place in either our model or Dennison's. This is that the threshold for the dissociation $^{16}O \rightarrow {}^{12}C + \alpha$ lies at an excitation of 7·15 MeV (Ajzenberg and Lauritsen 1955) just above the first excited states. We should therefore expect the potential for any motion which separates one α-particle from the other three to differ considerably from the harmonic approximation and to be much more like a Morse function with a rather shallow well. The vibration ω_3 is such a motion; we therefore cannot expect to find any good agreement between a harmonic-oscillator analysis and experiment for the 1^- and 2^+ levels. Similar considerations will also hold for the interpretation of higher levels.

It seems difficult to devise any satisfactory way of improving the present model. To begin with, one has no guide to the validity of the fundamental approximation, in which the potential is assumed to be independent of the state considered. In order to obtain any better agreement with experiment it would be necessary to alter this potential, which would certainly complicate the solution of the wave equation. A further problem which would have to be solved is that of finding the shell-model wave functions. The method used in § 6 is extremely complex, and it would be impracticable to apply it to a large number of states, or to larger nuclei, such as ^{20}Ne or ^{24}Mg.

References

Ajzenberg, F., and Lauritsen, T., 1955, *Rev. Mod. Phys.*, **27**, 77.

Dennison, D. M., 1954, *Phys. Rev.*, **96**, 378.

Devons, S., Goldring, G., and Lindsey, G. R., 1954, *Proc. Phys. Soc.* A, **67**, 134.

Elliott, J. P., and Skyrme, T. H. R., 1955, *Proc. Roy. Soc.* A, **232**, 561.

Wilkinson, D. H., 1955, *Phys. Rev.*, **99**, 631.

II. THE SKYRME INTERACTION

In the late 1040s, the idea of a very strong short-range interaction between nucleons, often employed as a hard-core potential, gained a following in the nuclear physics community because it appeared to easily account for saturation. With this strong repulsion, nucleons simply could not be pushed together, beyond a certain density. (Now we know that the short-range repulsion is only one of several important elements in nuclear saturation, and that nuclei saturate at a substantially lower density than would be given by this mechanism alone.)

The problem with hard-core potentials is that they are very strong, infinite over a certain range of r, so that the conventional methods could not handle them. Keith Brueckner brought forward a theory, now called after him, in which the effects of the hard core were summed to all orders, giving a well-behaved effective potential, the so-called t-matrix. This t-matrix was simply related to the nucleon–nucleon scattering amplitude.

Dick Dalitz writes in his outline of the life and work of Tony Skyrme (first paper in this volume) that the Brueckner theory for nuclear matter was naturally attractive for Tony, because it emphasized the mean, coherent properties of the nuclear medium in which each nucleon moved. Yet the Brueckner theory is close to the Hartree–Fock theory, of the type that had been used for atoms, only the nuclei didn't have a central potential like the Coulomb field from the nucleus in an atom. Brueckner's contribution was his way of regularizing the infinite hard-core repulsion, in some sense summing its effects to all orders, which meant replacing it by the t-matrix. Further, the Brueckner theory looked very much like the Hartree–Fock theory used for atoms.

In "A new model for nuclear matter" (page 91 of this volume) "the meson field is supposed to condense into an incompressible fluid and the nucleonic sources are confined to its interior by a strong interaction between the sources and the fluid as a whole." The collective model of the nucleus influenced this stronger-interaction fluid way of looking at it.

Whereas this meson fluid picture of the nucleus was not directly useful, Tony's belief in this picture strongly influenced his invention of the skyrmion. "It seemed to me that a type of fluid drop picture of the nucleus was simpler and more appropriate, the individual nucleons being some kind of local twists in the field, rather than independently interacting particles. Trying to understand the nature of the solutions, I gradually came to look more at the ideal of a single nucleon as a twist in some ether-like fluid — such as would be described now by the preferred direction in a theory with a spontaneously broken internal symmetry." (from "The origins of skyrmions" reprinted on page 115 of this volume).

Still, Tony Skyrme worked with the Brueckner theory, in several papers, applying Brueckner's idea that the nucleon–nucleon scattering amplitude should, with medium corrections, be the effective interaction in nuclei. This idea is used, for example, in "The nuclear spin–orbit coupling", by J. S. Bell and T. H. R. Skyrme (page 71 of this volume) and "The spin–orbit interaction in nuclei", by T. H. R. Skyrme (page 85). The spin–orbit interaction was very much needed in the nucleon–nucleon scattering experiments in order to explain the observed polarization; these experiments were going on about Tony in Harwell. This spin–orbit interaction between nucleons had been a key element in the shell model of the nucleus, formulated by Maria Mayer and J. H. D. Jensen.

At the *Rehovoth Conference on Nuclear Structure* in 1957, Tony proposed the effective nucleon–nucleon potential to be used in nuclei of the form given by Eq. (3) in Dalitz's introductory outline for this volume. In the paper, accompanying his article with John Bell on the spin–orbit interaction, "The effective nuclear potential", reproduced here, Skyrme laid out what is now called the Skyrme potential. In this, the two-body potential is expressed as

$$t_{12} = \delta(\mathbf{r}_1 - \mathbf{r}_2)t(k', k)$$

where $t(k', k)$ is the scattering matrix. Medium corrections are expressed in a three-body term

$$t_{123} = \delta(\mathbf{r}_1 - \mathbf{r}_2)\delta(\mathbf{r}_3 - \mathbf{r}_2)t_3 \, .$$

Because of the zero-range nature of these potentials, they are immensely easy to use. Generations of nuclear theorists have now been brought up on them. They work extremely well, and capture nearly all of the important elements of the *in medium* nucleon–nucleon interaction. The long-range pion exchange part of the interaction is not well handled by the zero-range approximations, in general. The spin–orbit interaction can be put in by hand, but its strength cannot be simply connected with that of the nucleon–nucleon interaction, and this is why Skyrme had difficulty in getting this strength right in his papers. In the Walecka relativistic mean field theory, which followed Skyrme's theory, an additional factor m_n/m_n^* comes in front of the spin–orbit interaction, and this makes it large enough to form the pronounced j-subshells observed in nuclei. But this large magnitude comes at the expense of an effective mass m_n^* which is small, $\sim 0.55 - 0.6\ m_n$ for nuclear matter density, much smaller than the 0.8 m_n usually employed with the Skyrme nonrelativistic theory. Such a small effective mass naturally results in the saturation mechanism of the Walecka theory, I believe, and have written in several papers, that the large effective mass generally found by investigators using the Skyrme theory is correct, and that the large spin–orbit interaction results from new effects; namely, density-dependent meson masses. But this is not the place to expound on these.

correct, and that the large spin-orbit interaction results from new effects, namely

1.C:
1.D.1

Nuclear Physics **9** (1959) 615—634; © *North-Holland Publishing Co., Amsterdam*
Not to be reproduced by photoprint or microfilm without written permission from the publisher

THE EFFECTIVE NUCLEAR POTENTIAL

T. H. R. SKYRME

Atomic Energy Research Establishment, Harwell, Didcot, Berks.

Received 18 October 1958

Abstract: An empirical analysis is made of the mean effective internucleon potential required in the shell-model description of nuclei, allowing for the presence of many-body effects as suggested by current theory. A consistent description is found in which the effective two-body interaction acts almost entirely in even states, and the many-body effects are simulated by a repulsive three-body contact interaction. The strength of the two-body interaction is consistent with that expressed by the free scattering matrix of the two-nucleon system, and that of the three-body interaction with the 'rearrangement energy' calculated in the many-body theory.

1. Introduction

In the decade that has elapsed since the publication of Rosenfeld's study of nuclear forces [1]) there have been several developments of great significance for the problem of nuclear forces in relation to the structure of nuclei. The shell model [2]) of nuclei has acquired a firm empirical standing, based on the multiplicity of facts that can be explained with the assumption that nucleons move in independent orbits inside a nucleus, generalised, in the more complex cases, to transmigration between a relatively small number of orbits. The character of the single-particle potential needed to define these orbits is known to be that of a rounded square well with a strong attractive spin-orbit coupling; the occupancy of nearly degenerate states is determined by residual interactions, whose form and strength are similar to those effective in the deuteron.

The detailed nature of this interaction is unimportant for many results; usually a simple radial shape has been chosen for the potential, gaussian or Yukawa, and an exchange character has been taken either to fit the facts as well as possible in the region considered or such as to describe the low-energy properties of the two-nucleon system; it has often been assumed that the interaction must have the 'symmetrical' character, suggested by second-order meson theory, to ensure nuclear saturation. It is now clear that such an interaction is very different from the real potential acting between two (free) nucleons. The form of this real potential is still uncertain, but recent analyses [3,4]) of nucleon-nucleon scattering in terms of a potential indicate the presence of strong non-central forces and of a 'repulsive core' at small distances.

This last feature especially has stimulated a critical examination of the theory of many-body systems and of the foundations of the empirical shell-model structure; a review has been made by Eden [5]). At an early stage in this investigation it was suggested by Brueckner [6]) that the effective potential used in shell-model calculations could be approximated by the scattering matrix of the two-nucleon system, generalised off the energy shell when necessary. This simple identification is not now believed to give a reasonable first approximation; it is necessary to allow for the average effects of the presence of other nucleons on the scattering of any one pair, through the mean potential they provide and through the effects of the exclusion principle. If it is possible then to construct an effective interaction that can be handled in the simple manner of shell-model calculations it will not necessarily possess also the attributes of a 'realistic' two-body interaction; it will not necessarily describe the deuteron system adequately, nor will nuclear saturation be related only to the exchange character.

The purpose of this paper † is to re-examine the empirical evidence about the potential effective in complex nuclei in the light of the current discussions of the many-body system; a preliminary account of this work has been given elsewhere [7]). It is now customary to speak of the shell-model system, described by a wave-function Φ, as providing a *model* of the real system described by the wave-function Ψ; their relation is described symbolically by

$$\Psi = F\Phi \tag{1}$$

where F introduces all the local correlations characteristic of the actual system. If a correspondence is established between the real states Ψ_n and an orthonormal set Φ_n, the effective Hamiltonian may be defined as that which has the eigenfunctions Φ_n with the eigenvalues E_n equal to those of the real system.

If the effective Hamiltonian is defined in this formal way it is evident that the usual variational principle is applicable, and may be used to determine the ground-state model wave-function in the Hartree-Fock manner. For practical use, however, we must make the additional assumption that the effective Hamiltonian can be expressed in a comparatively simple way in terms of internucleon potentials; the empirical test of this hypothesis is the main purpose of the present paper.

Two kinds of assumption have to be made in this analysis, the form of the Φ_n and the form of the effective interaction. The 'model' for infinite nuclear matter is taken, as usual, to be a Fermi gas; for closed shell nuclei, and those adjacent, nuclear theory suggests a wave-function describable by a single Slater determinant, that is unique apart from the unknown shape of the

† The present paper, planned as the first of a series, is confined to discussion of the ***mean*** potential.

single-particle orbital functions. But for nuclei further removed from closed shells the model should probably include long-range correlation effects, such as intrinsic distortion or surface vibrations.

The effective interaction potential will *not* be the same as that defined in the self-consistent many-body theory (to which the variational principle is not applicable). The potential used in our analysis must contain three-body, and generally many-body, terms which describe the way in which interaction between two particles is influenced by the presence of others; the two-body terms alone should be related closely to the scattering between free nucleons.

These considerations have led to the following ansatz for the form of the effective potential:

$$T = \sum_{i<j}\sum t_{ij} + \sum_{i<j<k}\sum\sum t_{ijk} \tag{2}$$

in which the many-body effects have been simulated by three-body terms alone, for the sake of simplicity of calculation. From the view-point of many-body theory there is no reason why the two-body terms should have the form of a static potential though with our philosophy they should be functions of the *relative* coordinates of the nucleons; for the sake of a simple comparison with the scattering matrix we have expressed them as

$$t_{12} = \delta(\mathbf{r}_1 - \mathbf{r}_2) t(\mathbf{k}', \mathbf{k}) \tag{3}$$

where $\mathbf{k}$ is the operator corresponding to the relative wave-number,

$$\mathbf{k} = \tfrac{1}{2} i (\nabla_1 - \nabla_2); \tag{4}$$

placed on the *right* of the delta-function $\mathbf{k}'$ denotes the same operator placed on the *left*; this form was used in an earlier discussion [8]) of the spin-orbit potential. An ordinary static (Wigner) interaction would then be described by a t dependent only on $\mathbf{k}' - \mathbf{k}$.

It is generally believed that the most important part of the two-body interaction can be represented by a contact potential, i.e. by constant $t(\mathbf{k}', \mathbf{k})$; this suggests an expansion in powers of $\mathbf{k}'$ and $\mathbf{k}$. If this expansion is stopped at the quadratic terms only a small number of undetermined coefficients occur, and an attempt can be made to determine these by comparison with experimental energies. However, this can only be done in conjunction with a similar determination of the parameters of the three-body interaction, and to reduce the problem to manageable proportions only the simplest possible form for the latter has been considered, the contact interaction

$$t_{123} = \delta(\mathbf{r}_1 - \mathbf{r}_2)\delta(\mathbf{r}_3 - \mathbf{r}_1) t_3 \tag{5}$$

with one undetermined constant t_3. Averaged over one of the particles this gives a contribution to the two-body contact interaction proportional to the local density. Some discussion of this assumption has been given by Squires [9])

This form of potential is given in more detail in the following section. In the approximation which has been most extensively investigated there are four undetermined constants involved in the 'mean' potential (i.e. the average involved in nuclei with equal numbers of protons and neutrons, each with their spins equally in both directions). These have been fixed, approximately, by considering the properties of infinite nuclear matter (section 3), including surface effects, and of light nuclei near closed shells using oscillator wave-functions for those of the single-particles (section 4), with the results given in section 5, which are the same as those previously reported [7]). The spin-orbit part of the interaction cannot be adequately described by the extreme short-range approximation and will be discussed separately in a following paper [10]). The remaining parts of the interaction, i.e. the tensor components and spin-dependent parts of the central potentials, are much more difficult to determine unambiguously. In principle they may be inferred from the level structure of nuclei more distant from closed shells and earlier estimates were based upon a study of the configuration ($s^4 p^n$) for nuclei between He and O; we now believe however, that even in this case, in the p-shell, distortion may introduce configuration interaction effects that should more naturally be included in the model wave-function than in the effective potential [11]). For some estimates made in this paper the spin-dependence of the central interaction has been represented by a single term.

We have also considered briefly some simple generalisations of the form assumed for the mean potential (sections 6 and 7); they do not appear to alter the qualitative features previously found. Finally, in sections 8 and 9, we compare the strength-constants of our potential with other available evidence.

2. Assumed Potential

To complete the specification of the potential given by equations (2) to (5) above we must introduce an explicit form for $t(\mathbf{k}', \mathbf{k})$. A polynomial expansion in powers of $\mathbf{k}$ and $\mathbf{k}'$ has been chosen on account of the comparatively simple algebraic expressions for the matrix elements; on the other hand this form is unrealistic for large momentum transfers, so that it is not suitable for the discussion of second-order effects, unless some momentum cut-off is introduced. If the polynomial is limited to terms quadratic in the momenta (analogous to the effective mass approximation), the most general form is

$$\begin{aligned} t(\mathbf{k}', \mathbf{k}) = {} & t_0(1+x_0 P^\sigma)+\tfrac{1}{2}t_1(1+x_1 P^\sigma)(\mathbf{k}'^2+\mathbf{k}^2) \\ & + t_2[1+x_2(P^\sigma-\tfrac{4}{5})]\mathbf{k}'\cdot\mathbf{k} \\ & + \tfrac{1}{2}T[\boldsymbol{\sigma}_1\cdot\mathbf{k}\boldsymbol{\sigma}_2\cdot\mathbf{k}-\tfrac{1}{3}\boldsymbol{\sigma}_1\cdot\boldsymbol{\sigma}_2\mathbf{k}^2+\text{conj.}] \\ & + \tfrac{1}{2}U[\boldsymbol{\sigma}_1\cdot\mathbf{k}'\boldsymbol{\sigma}_2\cdot\mathbf{k}-\tfrac{1}{3}\boldsymbol{\sigma}_1\cdot\boldsymbol{\sigma}_2\mathbf{k}'\cdot\mathbf{k}+\text{conj.}] \\ & + V[i(\boldsymbol{\sigma}_1+\boldsymbol{\sigma}_2)\cdot\mathbf{k}'\times\mathbf{k}], \end{aligned} \tag{6}$$

where P^σ is the spin-exchange operator (in this form it is unnecessary also to introduce isobaric spin-dependence). The constants t_0, t_1, t_2 are a measure of the *mean* central potential, with an exchange character specified by x_0, x_1, x_2; T and U characterise a tensor potential in even and odd states respectively, and V describes a short range two-body spin-orbit potential.

The validity of this approximation to $t(\mathbf{k}', \mathbf{k})$ is considered only from the empirical point of view in our analysis. It has appeared from our study [10]) of the spin-orbit coupling that the extreme short-range V-term is quite inadequate to explain the observed variations of spin-orbit coupling among nuclei, and this may also be true of the tensor interactions; however, this does not affect the conclusions of the present paper, which is concerned primarily with the mean potential.

It is convenient to describe here some additional terms that are considered briefly in sections 6 and 7. The first is suggested by the presence of considerable D-waves in the nucleon-nucleon interaction at energies around 100 MeV, and is taken in the form

$$t_D[\mathbf{k}^2\mathbf{k}'^2-(\mathbf{k}\cdot\mathbf{k}')^2]. \tag{7}$$

The second is a four-body interaction in T, which is again assumed to have zero range:

$$t_4\delta(\mathbf{r}_1-\mathbf{r}_2)\delta(\mathbf{r}_1-\mathbf{r}_3)\delta(\mathbf{r}_1-\mathbf{r}_4). \tag{8}$$

The expectation value of the two-body interaction (6) has to be calculated for an antisymmetrised wave-function; when a *product* wave-function is used t must be multiplied by the antisymmetrising factor $(1-P^xP^\sigma P^\tau)$, so that the effective interaction becomes

$$t(\mathbf{k}', \mathbf{k})-t(\mathbf{k}'-\mathbf{k})P^\sigma P^\tau. \tag{9}$$

The *mean* potential considered in this paper is the average of (9) over spins and isobaric spins, which reduces to

$$\tfrac{3}{4}t_0+\tfrac{3}{8}t_1(\mathbf{k}^2+\mathbf{k}'^2)+\tfrac{5}{4}t_2\mathbf{k}'\cdot\mathbf{k}. \tag{10}$$

Similarly in (7), t is replaced by $\frac{3}{4}t$; the same averaging applied to the three- or four-body interactions (5) and (8), results in the multiplication of t_3 by the factor $\frac{3}{8}$, and t_4 by $\frac{3}{32}$.

This mean potential is also appropriate for systems differing by one particle from symmetrical ones, apart from an additional contribution from the spin-orbit term.

3. Infinite Nuclear Matter

We consider first idealised infinite nuclear matter, described by a Fermi gas with spin and isobaric spin states equally occupied. This is a characterised by a density ρ_0 and a mean binding energy per particle B; their values have been taken to be those given by Cameron [12]) in an analysis of the semi-empirical mass-formula †. In terms of a radius constant r_0, and a convenient unit of energy $E_0 = 1/(2Mr_0^2)$,

$$r_0 = 1.112 \times 10^{-13}\ \text{cm}, \qquad E_0 = 16.80\ \text{MeV},$$

one has

$$\rho_0 = (\tfrac{4}{3}\pi r_0^3)^{-1}, \qquad B/E_0 = 1.014. \tag{11}$$

The expectation value of the mean effective Hamiltonian for such a Fermi gas gives

$$-B/E_0 = 1.38(1+\lambda_1+\lambda_2)+\lambda_0+\lambda_3+\lambda_4 \tag{12}$$

where the dimensionless parameters λ_n are defined by

$$\begin{aligned} \lambda_0 &= \tfrac{1}{2}(\rho_0/E_0)\tfrac{3}{4}t_0 \\ \lambda_1 &= \tfrac{1}{4}(\rho_0/E_0r_0^2)\tfrac{3}{4}t_1 \\ \lambda_2 &= \tfrac{1}{4}(\rho_0/E_0r_0^2)\tfrac{5}{4}t_2 \\ \lambda_3 &= \tfrac{1}{6}(\rho_0^2/E_0)\tfrac{3}{8}t_3 \\ \lambda_4 &= \tfrac{1}{24}(\rho_0^3/E_0)\tfrac{3}{32}t_4. \end{aligned} \tag{13}$$

The condition that B be stationary for variations of ρ_0 gives

$$0 = 1.38\ (\tfrac{2}{3}+\tfrac{5}{3}\ (\lambda_1+\lambda_2))+\lambda_0+2\lambda_3+3\lambda_4. \tag{14}$$

Throughout the following analysis these relations are used to eliminate two of the parameters λ_n; it is convenient to eliminate the most important parts of the potential, which are in fact those specified by λ_0 and λ_1;

$$\begin{aligned} \lambda_0 &= -4.605+0.5\lambda_3+2.0\lambda_4, \\ \lambda_1 &= 1.60-\lambda_2-1.09(\lambda_3+2\lambda_4). \end{aligned} \tag{15}$$

The interaction given by (7) does not affect the forward or backward scattering and does not therefore contribute to the 'infinite' case.

The self-consistent single-particle potential, defined by the Hartree-Fock approximation has a quadratic momentum dependence, on account of the assumption (6), and may be described by an effective mass M^* and a well-depth V_0 (for the lowest state), where

$$\begin{aligned} M/M^* &= 1+\lambda_1+\lambda_2 = 2.60-1.09(\lambda_3+2\lambda_4), \\ V_0/E_0 &= 1.38(\lambda_1+\lambda_2)+2\lambda_0+3\lambda_3+4\lambda_4 = -7.00+2.5(\lambda_3+2\lambda_4). \end{aligned} \tag{16}$$

The energy of the topmost particle is $V_0+2.30(M/M^*)E_0 = -B$.

The surface effects in a large nucleus may be estimated in the way suggested by us in an earlier analysis [13]). In that investigation we used a simple analytic approximation to the form of the energy as a function of density; we have verified (for the values of the parameters finally selected) that this is indeed a close approximation. The density distribution, according to the

† Different values of these parameters were also considered, but consistency with light nuclei could only be obtained for values within 3 % of those quoted.

Thomas-Fermi approximation has the form †

$$\rho/\rho_0 = [\tanh(R-r)/2b]^2 \tag{17}$$

where, in the present notation,

$$b^2/r_0^2 = \tfrac{3}{4}\lambda_1-\tfrac{1}{4}\lambda_2 = 1.20-\lambda_2-0.82(\lambda_3+2\lambda_4). \tag{18}$$

The thickness measured from 10 % to 90 % of full density is $2.98b$, to be equated with an experimental value of about 2.4×10^{-13} cm.

With the same approximations the surface energy is represented by the term

$$\tfrac{8}{5}(b/r_0)E_0A^{\frac{2}{3}}. \tag{19}$$

The coefficient of $A^{\frac{2}{3}}$ given by Cameron [12]) is 25.8 MeV, while with $b = 0.80\times10^{-13}$ cm this formula gives 19.3 MeV; the discrepancy is hardly significant in view of the approximations employed.

Finally we may consider the symmetry energy that arises when the numbers of protons and neutrons are unequal, expressed in the semi-empirical mass-formula by a term

$$E_{sy}(A-2Z)^2/A. \tag{20}$$

To calculate E_{sy} it is assumed that protons and neutrons form two separate Fermi gases with densities $(Z/A)\rho_0$ and $(A-Z/A)\rho_0$ respectively. The result gives

$$\begin{aligned} E_{sy}/E_0 = 1.38(\tfrac{5}{9}-\tfrac{10}{9}\lambda_1x_1+\tfrac{8}{9}\lambda_2+\tfrac{2}{5}\lambda_2x_2) \\ -\tfrac{1}{3}\lambda_0(1+2x_0)-\lambda_3-2\lambda_4. \end{aligned} \tag{21}$$

It should be noted that this formula bears no simple relation to the effective mass; also that it depends upon the exchange character of the interactions (that is upon the x_i), but fortunately not very sensitively. An estimate based upon the apparent strengths of singlet and triplet interactions in light nuclei, suggests that the contributions of these terms to the expression (21) is about 0.25. The elimination of λ_0 by (15) then gives

$$E_{sy}/E_0 = 2.55+1.23\,\lambda_2-1.17\,\lambda_3-2.67\,\lambda_4. \tag{22}$$

The value determined empirically by Cameron [12]) is $E_s = 31.5$ MeV; this rather large value arises because Cameron includes in his formula an additional surface symmetry energy term.

Corresponding symmetry energy terms arise also in the single-particle potential. Thus we obtain for *protons* a velocity-dependent potential (exclusive of the Coulomb term) described by an effective mass M_p^* and depth V_p, where

$$M/M_p^* = 1+\lambda_1+\lambda_2+(N-Z/A)(\tfrac{1}{3}\lambda_1-\tfrac{1}{5}\lambda_2) \tag{23}$$

and

$$V_p = V_0+(N-Z/A)[-2E_{sy}+1.38\tfrac{1}{9}E_0(10+5\lambda_1+13\lambda_2)]. \tag{24}$$

For *neutrons* similar equations hold with N and Z interchanged.

† Equation (33) of ref.[13]).

The term involving E_{sy} can be related to the Coulomb potential through the stability condition (which determines Z as a function of A); this shows that the term is equal to

$$-1.2(Ze^2/R) \tag{25}$$

the factor 1.2 arising from the *average* of the Coulomb potential within the nucleus. It may easily be verified that the other terms are such that the total energy of the topmost proton

$$V_p+2.30\ E_0(M/M_p^*)[1-\tfrac{2}{3}(N-Z/A)]+1.2(Ze^2/R) = -B. \tag{26}$$

4. Light Nuclei

The expectation value of the effective Hamiltonian for light nuclei has been calculated on the assumption that the shell-model wave-functions of these nuclei can be approximated by suitable oscillator wave-functions. The scale-constant of the wave-functions is described by the parameter y such that the exponential factor is exp $[-\frac{1}{2}y^2(r/r_0)^2]$. The polynomial form assumed for $t(\mathbf{k}', \mathbf{k})$ leads then to an expression for the energy which is a polynomial in y; the value of y is determined by minimising the energy.

The contribution of the mean potential, described by the strength parameters $\lambda_0 \ldots \lambda_3$, to the total energy may be written

$$E/E_0 = Ay^2+B_0\lambda_0 4.255\, y^3+(B_1\lambda_1+B_2\lambda_2)12.766\, y^5+B_3\lambda_3 6.970\, y^6, \tag{27}$$

where the coefficients A, B_n depend upon the particular shell-model state considered; their values for a number of states (or differences of states) are given in table 1.

TABLE 1

State	A	B_0	B_1	B_2	B_3
s^3	3	$\frac{1}{2}$	$\frac{1}{2}$	0	$\frac{1}{4}$
s^4	$\frac{9}{2}$	1	1	0	1
s^4p-s^4	$\frac{5}{2}$	$\frac{1}{4}$	$\frac{1}{4}$	$\frac{1}{6}$	$\frac{1}{4}$
s^4p^{12}	$\frac{69}{2}$	$\frac{31}{4}$	$\frac{35}{4}$	4	$\frac{116}{9}$
$s^4p^{12}-s^4p^{11}$	$\frac{5}{2}$	$\frac{7}{8}$	$\frac{25}{24}$	$\frac{1}{2}$	$\frac{37}{18}$
$s^4p^{12}d-s^4p^{12}$	$\frac{7}{2}$	$\frac{9}{16}$	$\frac{35}{48}$	$\frac{13}{24}$	$\frac{38}{36}$
$s^4p^{12}2s-s^4p^{12}$	$\frac{7}{2}$	$\frac{21}{32}$	$\frac{91}{96}$	$\frac{47}{48}$	$\frac{31}{18}$
$s^4p^{12}d^{20}2s^4$	$\frac{237}{2}$	$\frac{1945}{64}$	$\frac{2625}{64}$	$\frac{1760}{64}$	$\frac{2060}{27}$

For other nuclei in the p-shell, which cannot be described by a single determinantal wave-function, we must determine the various two- (and three-)body matrix elements of the effective potential in states formed by two (or three) p-wave nucleons. These matrix elements may also be written in the form (27); the values of B_n for the contributions of the mean, spin-independent potential are given in table 2.

TABLE 2

State	B_0	B_1	B_2	B_3
$p^2[2]S$	$\frac{5}{24}$	$\frac{3}{8}$	0	$\frac{5}{18}$
$p^2[2]D$	$\frac{1}{12}$	$\frac{1}{12}$	0	$\frac{1}{9}$
$p^{-2}[2]S$	$\frac{5}{24}$	$\frac{3}{8}$	0	$\frac{25}{27}$
$p^{-2}[2]D$	$\frac{1}{12}$	$\frac{1}{12}$	0	$\frac{10}{27}$
p^2 or $p^{-2}[11]P$	0	0	$\frac{1}{15}$	0
$p^{\pm 3}[3]P$	0	0	0	$\pm\frac{7}{36}$
$p^{\pm 3}[3]F$	0	0	0	$\pm\frac{1}{18}$

The three-body contact interaction gives no contribution to states of (p^3) other than the most symmetric [3] states. The spin-dependent parts of the potential contribute proportionately; so that in a spin-triplet state B_0 should be multiplied by $(1+x_0)$, and so on.

Since the total energy should be stationary for the 'correct' value of y, it is not necessary to fix y self-consistently with great precision. In practice y has been fixed for each nucleus by choosing a particular set of λ_n, close to those finally adopted, and a particular low-lying shell-model to fix the B_n. Some of these values and the corresponding r.m.s. radii are given in table 3.

TABLE 3

Mass Number	y	Calc. r.m.s. radius 10^{-13} cm	Exp. r.m.s. radius
3	0.788	1.41	—
4	0.743	1.58	1.61
6	0.709	1.97	2.44
7	0.690	2.11	2.20
8	0.683	2.19	—
12	0.676	2.35	2.37
14	0.670	2.41	—
16	0.663	2.46	2.64
40	0.596	3.21	3.52

The experimental radius values are taken from Hofstadter [14]).

There is a tendency for our model to predict rather small radii; this might be expected on two counts: (a) oscillator wave-functions have less 'tail' than realistic ones, and (b) correlations in the real wave-function may increase the smearing of the surface region, by introducing higher single-particle states with larger r.m.s. radii.

5. Determination of Parameters

The first object of the present analysis has been to see whether a reasonable 'fit' to the experimental energies can be obtained with a mean potential depending only on the four parameters $\lambda_0 \ldots \lambda_3$. To simplify the comparison we have demanded that the conditions (15) should be satisfied exactly (with $\lambda_4 = 0$) and used these to express all calculated energies in terms of λ_2 and λ_3 only. Each comparison with experiment then determines a linear relation between these two parameters, which can conveniently be displayed graphically.

A number of these lines are shown in fig. 1. In the comparisons indicated the observed binding energies have been corrected for the Coulomb contribution, and for spin-orbit splitting, to give an 'observed' mean nuclear binding that can be compared directly with the calculations. The first point

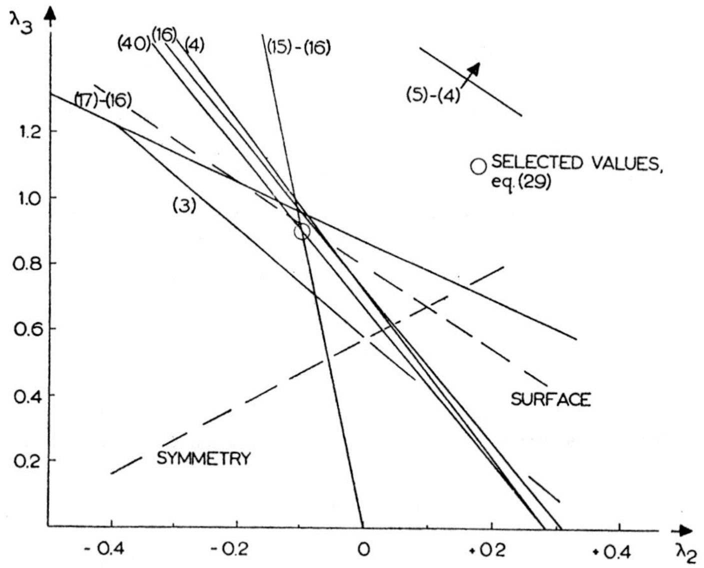

Fig. 1. Relations between the parameters λ_2 and λ_3 determined by the binding energies of some light nuclei (with mass numbers) and by the empirical surface and symmetry energies.

to note is that the total binding energies of the closed shell nuclei are consistent, and they indicate a relation between the parameters which has been taken as

$$\lambda_2 + 0.415\, \lambda_3 = 0.27. \tag{28}$$

The positions of the other lines suggest strongly that a positive value of λ_3, of the order of unity, is needed to obtain any sort of consistency among the other data. It would be pointless in this preliminary analysis to attempt to determine a 'best-possible' set of values, so a value of 0.9 was chosen for λ_3

by inspection fixing the other parameters by (16) and (28); these values are

$$\lambda_0 = -4.155 \qquad \lambda_3 = 0.9$$
$$\lambda_1 = 0.723 \qquad \lambda_2 = -0.104 \tag{29}$$

and the corresponding values of the t_n in the potential (6) are ($f = 10^{-13}$ cm)

$$t_0 = -1072\, f^3 \text{ MeV} \qquad t_3 = +8027\, f^6 \text{ MeV}$$
$$t_1 = +461\, f^5 \text{ MeV} \qquad t_2 = -40\, f^5 \text{ MeV}. \tag{30}$$

The qualitative conclusions are that there is a considerable repulsive three-body force tending to reduce the strength of interaction between two nucleons in nuclear matter compared with free space; and that this effective interaction is predominantly of 'Serber' type, being very weak in odd states. The uncertainty in the determination of λ_3 is of the order of ± 0.2, and λ_2 is scarcely significantly different from zero. (See also section 7, below.)

With these values (29) we obtain the following results. The mean single-particle potential (eq. (17)) has

$$V_0 = -80 \text{ MeV}, \qquad M/M^* = 1.62, \tag{31}$$

the quotient of which is $M^* V_0/M = -49$, close to the values commonly used in optical model analyses. The formulae (23) and (24) indicate an appreciable difference in the values for protons and neutrons in heavy nuclei, but the changes are such that the product M^*V would be expected to remain nearly constant.

The surface thickness constant b, determined by eq. (18), comes out to be $0.76\, r_0 = 0.84 \times 10^{-13}$ cm.

The symmetry energy E_s, eq. (22), is 92 MeV.

For the light nuclei the calculated radii have been given in table 3; some calculated binding energies are given in table 4, and compared with the observed values, corrected for Coulomb and spin-orbit effects.

TABLE 4

Mass Number	Calculated B.E. (MeV)	'Corrected' Experimental B.E. (MeV)
3	11	8.5
4	27	29
5 relative to 4	−4	−1
15 relative to 16	−19	−19
16	138	140
17(d) relative to 16	1	1
17(2s) relative to 16	0	3
40	410	413

The calculation of the energies of other nuclei in the p-shell requires knowledge of the spin-orbit coupling and of the exchange nature of the interactions and lies outside the scope of the present paper.

6. Four-Body Interactions

The agreement obtained in the previous section by use of the very simple effective mean potential is satisfactory, but it is also necessary to examine how far the conclusions would be modified by including additional terms in the potential. As far as the many-body parts are concerned the simple three-body contact interaction (5) could be generalised both by including momentum-dependent terms (i.e. finite range) and by adding four- (or more) body interactions. The former possibility would require the introduction of several new parameters, and determination of significant values for them would be impractical at the present stage; we consider briefly below the effects of introducing a four-body interaction in the simplest possible way.

This interaction, taken in the zero-range form given by eq. (8), introduces just one new parameter t_4, or λ_4 defined by eq. (13). This term has already been included in the formulae of section 3; for the light nuclei described by oscillator wave-functions there will be an additional term in the expression for the energy (27), which may be written analogously as

$$B_4\lambda_4\, 13.622\, y^9. \tag{32}$$

The definition is chosen so that $B_4 = 1$ for mass 4 (i.e. for the state s^4); the value for the state s^4p^{12} is found to be $\frac{5089}{256}$. We have used the total binding energy of O^{16} to fix relation between the parameters in addition to eq. (15), analogous to eq. (28); this is

$$\lambda_2+0.415\,\lambda_3+0.680\,\lambda_4 = 0.27, \tag{33}$$

and with it all energies have been expressed in terms of λ_3 and λ_4 only.

In many of the formulae of section 3 these parameters occur in the combination $\lambda_3+2\lambda_4$, which may be understood in the following way. The three- and four-body terms may be represented by an equivalent density-dependent two-body contact interaction,

$$t'_0(\rho) = t_0+\tfrac{1}{6}\rho t_3+\tfrac{1}{96}\rho^2 t_4. \tag{34}$$

TABLE 5

Energy	X
Surface Energy	2.3
Symmetry Energy	2.1
Single-Particle	2.0
s^4	2.4
s^4p-s^4	1.6
$s^4p^{12}-s^4p^{11}$	1.6
$(p^2)S$	2.5
$(p^{-2})S$	2.0

The derivative t'_0 with respect to density at the equilibrium density is proportional to $t_3+\frac{1}{8}\rho_0 t_4$, that is to $\lambda_3+2\lambda_4$.

We have found that the combinations which occur are closely similar in all cases. Writing them as $\lambda_2+X\lambda_3$, values of X are given in table 5.

The scatter of values is such that a significant determination of λ_4 is impossible. The inference suggested is that only the first derivative of an equivalent density-dependent two-body interaction is important, or that the interaction at densities far from ρ_0 is relatively unimportant.

7. D-Wave Interactions

The shortcomings of the assumed form (6) for the effective two-body interaction may be divided into incorrect *energy* dependence (i.e. upon $\mathbf{k}^2$ and $\mathbf{k}'^2$) and *angular* dependence (i.e. upon $\mathbf{k}\cdot\mathbf{k}'$). Since the range of momenta is much the same in all nuclei it is possible that the former type of error may not seriously affect matrix elements; on the other hand the small angular momenta involved in light nuclei mean that scattering at angles far from 0 or π may be important. The form of (6) corresponds to scattering only in S or P states; the analysis of nucleon-nucleon scattering at similar relative momenta indicates considerable amounts of D-wave (see section 9); it is expected therefore that an important correction to (6) might be represented by a term

$$t_D[\mathbf{k}^2\mathbf{k}'^2-(\mathbf{k}\cdot\mathbf{k}')^2]. \tag{7}$$

We define a dimensionless parameter

$$\lambda_D = \tfrac{1}{4}(\rho_0/E_0 r_0^4)\tfrac{3}{4}t_D \tag{35}$$

and a coefficient B_D such that the contribution of (7) to the energy of a state built of oscillator wave-functions is

$$B_D\lambda_D\, 12.766\, y^7. \tag{36}$$

To the surface thickness parameter (7) contributes

$$\Delta(b^2/r_0^2) = \lambda_D\, 1.8\, r_0^2\rho^{\frac{2}{3}} \approx 0.72\,\lambda_D \tag{37}$$

on average. The values of B_D for some of the states listed in table 1 and 2 are given in table 6.

As with the four-body interaction we have eliminated a parameter by fitting exactly the binding energy of O^{16}, giving the relation

$$\lambda_2+0.415\,\lambda_3-0.716\,\lambda_D = 0.27. \tag{38}$$

It turns out that when λ_0, λ_1 and λ_2 are eliminated by (15) and (38), the surface energy and the energy of p^{-1} in O^{16} are almost independent of λ_D. As these two quantities were fitted well by the values in section 5, no change in the value of λ_3, nor therefore of λ_0, is indicated.

On the other hand there does appear to be a small gain in consistency by assuming a small positive value for λ_D. The calculations that have been made are not extensive enough to justify a definite choice, but for illustration we quote the following possible values of λ_1, λ_2 and λ_D,

$$\lambda_1 = 0.62, \qquad \lambda_2 = 0, \qquad \lambda_D = 0.15. \tag{39}$$

TABLE 6

State	B_D
s^4	1
$s^4p - s^4$	$\frac{1}{4}$
s^4p^{12}	$\frac{31}{4}$
$s^4p^{12} - s^4p^{11}$	$\frac{7}{8}$
$(p^2)S$	$\frac{55}{72}$
$(p^2)D$	$-\frac{1}{36}$

8. The Three-Body Potential

The preceding analysis suggests the need for an appreciable three-body interaction in calculations using shell-model wave-functions. We must now ask how far this conclusion is compatible with the general success of such calculations in which two-body forces alone have been used.

In the p-shell intermediate coupling calculations, such as have been extensively made by Kurath [15]), depend only upon three parameters L, K and a but even these are not unambiguously given by the comparison of calculated and experimental energy levels. The principal effect of the three-body interaction would be to *decrease* the central-force integrals towards the end of the shell, (apart from any effects arising from changes in the single-partical orbits). There is indeed some indication of this, but its significance cannot be assessed so long as the single-particle spin-orbit is regarded as an arbitrary parameter. In three-particle calculations, such as those of Elliott and Flowers [16]), for mass 19, the effect of the three-body interaction will be to raise the levels of states of maximum orbital symmetry by a few tenths of a MeV; since the low-lying levels are mainly composed of such states, the additional interaction is unlikely to cause great changes in the level order predicted.

In their analysis of interactions in mass 40, Pandya and French [17]) have found some evidence for a three-body contact interaction precisely of the form that we have postulated. They find for its strength $t_3 = 28\,000\ f^6$ MeV, which is several times *larger* than our preferred value (about $8\,000\ f^6$ MeV); we would ascribe some part of this difference to the fact that Pandya and

French used an oscillator constant $r_f = 3f$, larger than the value derived from table 3, which is $r_0\sqrt{2}/y = 2.64f$; the scaling factor $(2.64/3.00)^6$ would reduce the value of t_3 needed to give the same values for the matrix elements to 13 000 f^6 MeV.

The results of Talmi and Thieberger [18]) on the binding energies of shells in the j-j coupling model appear, on the other hand, to indicate that two-body interactions alone give a good account of the variation of binding energy within a shell. However, the fact that cubic terms in the energy are small does not necessarily imply that three-body interactions are absent, because the latter may be marked by changes (e.g. in the single-particle wave-functions) which have been ignored in the empirical argument. To illustrate this point suppose that the 'core' is kept fixed, but that the orbit considered can be described by a variable scale parameter y, such for example as was used in section 4; the energy of a configuration j^n will contain terms such as

$$nA(y)+\tfrac{1}{2}n(n-1)B(y)+\tfrac{1}{6}n(n-1)(n-2)C(y) \tag{40}$$

where A, B and C are the single-particle energy, the mean binding energy between a pair, and that between three particles. For $n = 1$, y will have the value y_0 that makes A a minimum, and we can write

$$\begin{aligned} A(y) &= A_0+\tfrac{1}{2}A_2(y-y_0)^2+\ldots, \\ B(y) &= B_0+\tfrac{1}{2}B_1(y-y_0)+\ldots. \end{aligned} \tag{41}$$

We substitute these in (40) and minimise with respect to y; the terms in the energy up to order n^3 are then

$$nA_0+\tfrac{1}{2}n(n-1)[B_0-\tfrac{1}{4}B_1^2/A_2]+\tfrac{1}{6}n(n-1)(n-2)[C_0-\tfrac{3}{4}B_1^2/A_2]. \tag{42}$$

Since A_2 is positive the correction term tends to decrease the real three-body interaction measured by C_0. In the p-shell typical mean values taken from our calculations are

$$y_0^2A_2 = 79 \text{ MeV}, \qquad y_0B_1 = -3.4 \text{ MeV}, \qquad C_0 = 0.18 \text{ MeV}$$

for which $\tfrac{3}{4}B_1^2/A_2 = 0.21$ MeV, reducing considerably the apparent three-body interaction.

In another direction some confirmation of our estimate for the strength of the three-body interaction is found in the value of the 'rearrangement energy' arising in the many-body theory [19]). This is the difference between the real binding energy of the topmost particle in the nucleus (equal, for infinite nuclear matter, to B) and that given by the topmost energy level in the self-consistent single-particle potential as conventionally defined for pairwise interactions. In our analysis, with the explicit introduction of many-body potentials, these two quantities are the same, and Squires [9]) has shown that the rearrangement energy is simply related to the contribution of the many-body potentials to the binding energy. Comparison of eqs. (16) and (12) shows that this 'rearrangement energy' must be equal to

$$E_0[(2\lambda_0+3\lambda_3+4\lambda_4)-2(\lambda_0+\lambda_3+\lambda_4)] = E_0(\lambda_3+2\lambda_4) \tag{43}$$

again depending only on the combination $\lambda_3+2\lambda_4$ as we should expect.

Our estimate for this quantity is about 15 MeV (with an uncertainty of the order of 20 %), to be compared with the value of 17 MeV found in the numerical calculations of the many-body theory.

9. The Two-Body Potential

Our analysis of the effective potential has yielded

(a) A representation of the mean (i.e. average of singlet and triplet) even-state potential in the form

$$\delta(\mathbf{r}_1-\mathbf{r}_2)[t_0+\tfrac{1}{2}t_1(\mathbf{k}^2+\mathbf{k}'^2)+t_3\bar{\rho}] \tag{44}$$

where $\bar{\rho}$ is an average of the nuclear density at the point of interaction.

(b) The conclusion that the odd-state average interaction (which is 90 % the triplet-triplet interaction) is relatively weak, possibly slightly attractive. We can compare the matrix elements calculated with (44) with the corresponding averages used in shell-model calculations.

At the beginning and end of the p-shell we find the values given in table 7 for the average S and D matrix elements.

TABLE 7

Matrix Element	Energy (MeV)
$(p^2)\overline{S}$	−7.9
$(p^2)\overline{D}$	−5.0
$(p^{-2})\overline{S}$	−2.0
$(p^{-2})\overline{D}$	−2.1

The values at the beginning are in accord with those commonly assumed; taking the conventional singlet-triplet ratio of 0.6, they correspond to $L/K = 6$, $K = 1.25$ MeV. At the end of the shell however, the repulsive effects of the three-body interaction have greatly reduced their strength. The effect has probably been exaggerated by the simple forms that we have assumed, but is not necessarily in violent contradiction with observation because the level scheme is dominated by the spin-orbit coupling towards the end of the shell †.

There is very little evidence on the odd-state interaction from level structure. In the calculations of Kurath [15]) an exchange mixture was used in which the triplet-odd interaction is strongly repulsive, but the effect on the low-lying levels which are observed is generally small. In his analysis of the

† See, for example fig. 2 of ref. [7]). The disagreement for mass 13 in fig. 3 is due, in part at least, to incorrect assumptions about the tensor force.

structure of Li^6, Soper [20]) has found the 3.3 interaction to be weakly attractive, but the 1.1 to be undetermined. Huber [21]) has made a systematic analysis of the p-shell with two-body forces and also finds a weak attraction in the mean in odd states.

Beyond the p-shell so many different matrix elements are involved in the calculations that this type of comparison is not helpful.

It is also interesting to compare our results with a simple estimate based on the phase-shift analysis of nucleon-nucleon scattering. For this we assume that

(a) $t(\mathbf{k}', \mathbf{k})$ is proportional to the scattering amplitude from relative momentum $\mathbf{k}$ to $\mathbf{k}'$ with every term $e^{i\delta} \sin \delta$ replaced by δ,

(b) the phases δ are the same as for free-particle scattering,

and when only S and P waves are included in our approximation for t, also that

(c) the comparison is made only for forward and backward scattering, $\mathbf{k} = \pm\mathbf{k}'$.

These assumptions lead to the identifications

$$\begin{aligned} (\rho_0/E_0)(t_0+t_1k^2) &= -(6/kr_0)(\bar{\delta}_0+5\bar{\delta}_2+9\bar{\delta}_4+\ldots) \\ (\rho_0/E_0x_0{}^2)t_2k^2 &= -(6/kr_0)(3\bar{\delta}_1+7\bar{\delta}_3+\ldots) \end{aligned} \tag{45}$$

where the even-state averages are of 50 % triplet and 50 % singlet, and the odd state of 90 % triplet and 10 % singlet, and the average over J for the triplet states has the usual $(2J+L)$ weighting. The laboratory energy of scattering is

$$E = 67(kr_0)^2\text{MeV}. \tag{46}$$

The numerical comparison is illustrated in fig. 2. The solid lines represent the left-hand sides of (45) as a function of energy with the values of the t_n given by (30). The points marked represent values of the right-hand sides computed from the phase-shifts given by Gammel and Thaler [3]) or by Marshak and Signell [4]). The agreement is quite good in view of the limitations of the assumed form for $t(\mathbf{k}', \mathbf{k})$.

In the discussion of the symmetry energy, eq. (21), we required an estimate of the exchange dependence of the two-body potential. The odd-state exchange dependence, specified by x_2, cannot easily be determined and not at all without discussion of the non-central tensor force components of the potential. In even states x_0 and x_1 determine the relative strengths of triplet and singlet interactions; in most cases they occur in a similar combination (approximately $x_0-0.5x_1$), and if, arbitrarily, we put $x_1 = 0$ inspection of the level structure in masses 6 and 14 suggests a value of x_0 around 0.1. This can be compared with the phase-shifts by making the identification

$$(\rho_0/E_0)(x_0t_0+x_1t_1k^2) = -(6/kr_0)(\tilde{\delta}_0+5\tilde{\delta}_2+\ldots). \tag{47}$$

In this formula $\tilde{\delta}$ denotes half the difference of triplet and singlet phases. A good fit to the phase-shifts is given by $x_0 = x_1 = 0.12$ approximately.

For the D-wave contribution to the potential, defined by (7), the phase-shift identification is

$$(\rho_0/E_0)t_D k^4 = (9/kr_0)5\tilde{\delta}_2 \tag{48}$$

or equivalently

$$\lambda_D = \tfrac{27}{16}5\tilde{\delta}_2/(kr_0)^5. \tag{49}$$

The phase-shifts at 100 MeV then give a value around 0.3 for λ_D; this is larger by a factor 2 than the value previously suggested, eq. (39).

If the D-wave contribution were increased to this value the empirical analysis would then determine a compensatingly smaller value for t_1 which would make the agreement in eq. (45) less satisfactory. This however, might be compensated by momentum-dependent terms in the three-body interaction which we have ignored for the sake of simplicity.

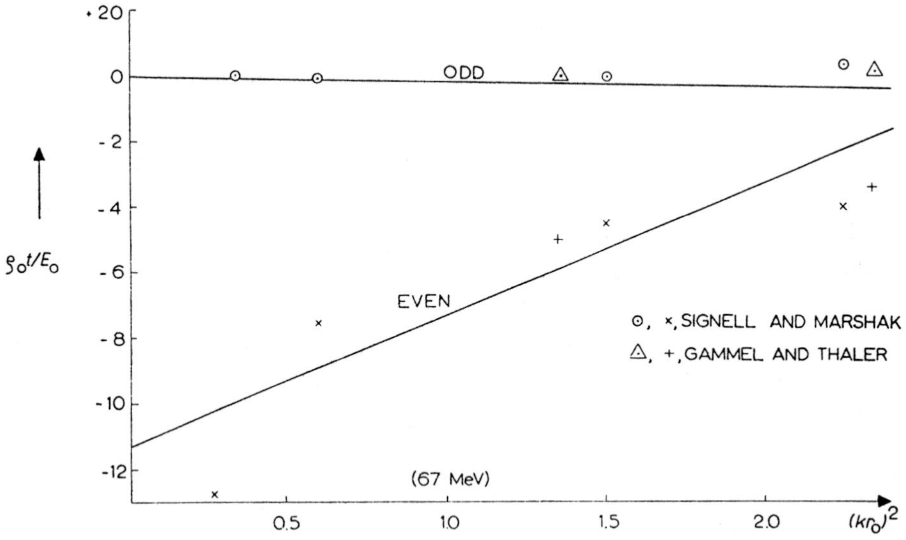

Fig. 2. The mean central potential in even and odd states compared with the estimate based on scattering phase-shifts.

10. Conclusion

We have shown that, within the limitations imposed by the simple forms assumed for ease of computation, a consistent account of nuclear binding energies requires the introduction of many-body terms, or that if the interaction is forced into a two-body form the latter must be allowed to be density dependent (at least); this conclusion is in agreement with the results of numerical calculations made on the many-body problem. Our empirical argument ultimately rests upon the fact the two- and three-body interactions contribute in different relative strengths to the *total* binding energy and to the interaction between two given particles or to the self-consistent single-particle potential; this, in effect, removes the discrepancy that has been known for a long time between the strengths of two-body interaction needed to explain level structure and to explain total energies.

We have incidentally constructed a simple model for the effective interaction, but we must emphasise that it can only reasonably be used in Born approximation where high momentum components are irrelevant. The Serber character of the interaction is in agreement with evidence provided by nucleon scattering off light nuclei.

The empirical elucidation of the spin-dependent parts of the interaction is much more difficult. This is because we must now consider nuclei removed from the closed shells, and the appropriate 'model' wave-functions are less certain on account of possible collective long-range correlations. These are present even in the p-shell, so that it is uncertain how far the empirical energy matrices used in shell-model calculations reflect the effective interaction without modifications due to the neglect of distortion in the model wave-function.

The most important problem is to relate the effective interaction to some 'real' interaction between the particles. Current formulations of the many-body problem emphasise the self-consistent determination of a two-body interaction; while this may be a suitable approach for the idealised problem of 'infinite nuclear matter', we suggest that it may be more profitable, for the application of the theory to finite nuclei, to leave explicitly the many-body effects as additional terms in the energy, so as to give a Hamiltonian that is independent of the particular state considered.

References

1) L. Rosenfeld, Nuclear Forces (North-Holland Publ. Co., Amsterdam, 1948)
2) A. M. Lane, The Nuclear Shell-Model in Handbuch der Physik (Springer-Verlag, Berlin, 1957)
3) J. L. Gammel and R. M. Thaler, Phys. Rev. **107** (1957) 1337
4) P. S. Signell and R. E. Marshak, Phys. Rev. **109** (1958) 1229
5) R. J. Eden, in Nuclear Reactions, ed. by P. J. Endt and M. Demeur (North-Holland Publishing Co., Amsterdam, 1959) Ch. I
6) K. A. Brueckner, C. A. Levenson and H. H. Mahmoud, Phys. Rev. **95** (1954) 217
7) T. H. R. Skyrme, Proc. Rehovoth Conf. Nucl. Structure (North Holland Publ. Co., Amsterdam, 1958) p. 20
8) J. S. Bell and T. H. R. Skyrme, Phil. Mag. **1** (1956) 1055
9) E. J. Squires, Nuclear Physics (1958) (in course of publication)
10) T. H. R. Skyrme, Nuclear Physics **9** (1958/59) 635
11) T. H. R. Skyrme, Nuclear Physics **9** (1958/59) 641
12) A. G. W. Cameron, Can. J. Phys. **35** (1957) 1021
13) T. H. R. Skyrme, Phil. Mag. **1** (1956) 1043

14) R. Hofstadter, Ann. Rev. Nuc. Sci. **7** (1957) 231
15) D. Kurath, Phys. Rev. **101** (1956) 216
16) J. P. Elliott and B. H. Flowers, Proc. Roy. Soc. A **229**
17) S. P. Pandya and J. P. French, Ann. Phys. **2** (1957) 166
18) I. Talmi and R. Thieberger, Phys. Rev. **103** (1956) 718
19) K. A. Brueckner, Phys. Rev. **110** (1958) 597
20) J. Soper, Phil. Mag. **2** (1957) 1219
21) R. Huper, Z. f. Naturforschung **12a** (1957) 295

CVIII. *The Nuclear Spin–Orbit Coupling*

By J. S. Bell and T. H. R. Skyrme
Atomic Energy Research Establishment, Harwell, nr. Didcot, Berks.†

[Received May 29, 1956]

Abstract

Analysis of the nucleon–nucleon scattering around 100 MeV has determined the spin–orbit coupling part of the two-body scattering matrix at that energy, and a reasonable extrapolation to lower energies is possible. This scattering amplitude has been used, in the spirit of Brueckner's nuclear model, to estimate the resultant single-body spin–orbit coupling for a single nucleon interacting with a large nucleus. This resultant potential has a radial dependence approximately proportional to $r^{-1}\, d\rho/dr$, and with a magnitude in good agreement with that required to explain the doublet splittings in nuclei and the polarization of nucleons scattered elastically off nuclei.

§ 1. Introduction

The coupling between the spin of a single nucleon and its orbit within a nucleus is manifested by the large splitting between the levels of a doublet in the nuclear excitation spectrum, interpreted in terms of a shell model, and by the strong polarization detected in the scattering of nucleons by nuclei. It is well known (Sternheimer 1955 and references given therein) that both these effects can be described approximately by an attractive potential $(\sigma\, .\, l)U(r)$ between a single nucleon and the whole nucleus, and that the strength of this potential is some twenty times greater than the relativistic Thomas term associated with the mean central potential.

Attempts have been made to explain this large value of the interaction as an amplification due to mesonic effects; in terms of radiative corrections (Chisholm and Touschek 1953) or in terms of a reduced inertial mass of the nucleon (Johnson and Teller 1955, Skyrme 1955). Our present knowledge of the mesonic fields within a nucleus is so uncertain, however, that an analysis along these lines remains speculative. A simpler, and not necessarily inconsistent, approach lies in an analysis in terms of two-body interactions. The only type of interaction which gives a spin–orbit coupling in the first approximation is the two-body spin–orbit force (Hughes and Le Couteur 1950, Elliott and Lane 1954, Blin-Stoyle 1955). The introduction of this force is however an additional assumption, and attempts have been made to explain the spin–orbit coupling in terms of higher order effects of the two-body tensor force,

† Communicated by Dr. B. H. Flowers.

which latter must also be present; the calculations of Keilson (1951) do not suggest however that this is a good explanation, but the evaluation of higher order effects is a very complicated matter.

In the model of nuclear matter that has been discussed by Brueckner (1954) and co-workers in a series of papers, the two-body interactions are introduced through the two-body scattering matrix. From their original viewpoint the scattering matrix was to be determined by the experimental information on nucleon–nucleon scattering; subsequently it was seen, that it was necessary to recalculate the scattering matrix for two particles interacting in the presence of the mean field of the other nucleons and subject to the exclusion principle. Apart from the complication introduced by this extra calculation there is the additional uncertainty due to ignorance of the potentials which give rise to the scattering and of how they may themselves be modified within nuclear matter. In this paper we have adopted the original point of view with regard to the relevant part of the scattering matrix.

An approximate determination of the scattering matrix has recently become possible as a result of experimental and theoretical developments. The angular distribution of the n–p polarization at 95 MeV has been measured by Hillman and Stafford (1956), and this has made possible a phase-shift analysis at that energy by Phillips (private communication), these phase-shifts are in close agreement with those given by Feshbach and Lomon (1956), whose results give also an estimate of the energy variation of the phases. The energies important in the present analysis lie below 100 MeV, and approximate analytical formulae have been determined to represent the spin–orbit part of the scattering amplitude over this energy range; the uncertainties in this could be reduced by the extension to lower energies of the measurements of polarization.

This scattering amplitude has been interpreted for the finite nuclear system in a manner similar to that employed by Skyrme (1956) in his discussion of the nuclear surface; as in that work the present discussion assumes a standard nucleus with equal numbers of protons and neutrons and does not allow for a possible difference between their distributions. Our analysis is closely related to that for a two-body spin–orbit potential, and may in some respects be regarded as an extension of the work of Blin-Stoyle (1955); the relation between these two approaches is discussed below.

§ 2. The Scattering Amplitude

The amplitude f that describes the scattering of two free nucleons is a scalar function of $\mathbf{k}$, $\mathbf{k}'$, the initial and final relative momenta of the nucleons, and of their spins $\boldsymbol{\sigma}_1$ and $\boldsymbol{\sigma}_2$. It is unnecessary to introduce the isotopic spin, provided that the initial and final states are properly antisymmetrized, because the p–p amplitude is part of the n–p one. The possible forms of spin-dependence of f have often been given, e.g. by Stapp (1955); those parts independent of spin are obviously irrelevant to our purpose; terms that are bilinear in the two spins also give no contribution to the resultant single-particle spin–orbit coupling, because the direct integrals vanish after averaging over the spins of one nucleon and the exchange integrals only give spin independent contributions after averaging. The only term that remains is that one linear in the spins, of the form

$$A=i(\boldsymbol{\sigma}_1+\boldsymbol{\sigma}_2)\,.\,(\mathbf{k}'\times\mathbf{k})F(\mathbf{k}^2,\,\mathbf{k}'^2,\,\mathbf{k}\,.\,\mathbf{k}') \quad . \quad . \quad . \quad (1)$$

where F is a scalar function of its arguments; on the energy shell $\mathbf{k}^2=\mathbf{k}'^2$.

For the comparison with the analysis of the experimental information the amplitude (1) must be related to the phase shifts. In terms of the scattering angle θ, and with the notation of Stapp (*loc. cit.*, eqn. (10) and table B)

$$k^2 \sin\theta F(k^2,\,k^2,\,k^2\cos\theta)=-\tfrac{1}{4}\sqrt{2}(M_{10}-M_{01}). \quad . \quad . \quad . \quad (2)$$

The matrix elements M are expressible as series of harmonics with coefficients depending upon the phase shifts; if only those terms are retained that appear to be important over the energy range considered,

$$\begin{aligned} M_{01}&=(1/4ik)(\sin\theta/\sqrt{2})[3R_{11}{}^1-3R_{11}{}^2-\sqrt{6}R^2 \\ &\quad+\cos\theta(3R_{21}{}^1+5R_{21}{}^2-8R_{21}{}^3+3\sqrt{2}R^1-4\sqrt{3}R^3)] \\ M_{10}&=(1/4ik)(\sin\theta/\sqrt{2})[-2R_{11}{}^0+2R_{11}{}^2-\sqrt{6}R^2 \\ &\quad+\cos\theta(-6R_{21}{}^1+6R_{21}{}^3+3\sqrt{2}R^1-4\sqrt{3}R^3) \end{aligned} \quad . \quad (3)$$

where the R coefficients are functions of the phase shifts δ and admixture parameters ϵ (arising from the tensor interactions) given by Stapp (eqn. (39)). In the present problem it is not however really the scattering amplitude that is wanted, but the 'reactance amplitude' corresponding to a scattered wave orthogonal to the incident wave; this is obtained from the scattering amplitude by replacing $[\exp(2i\delta)-1]$ by $[2i\tan\delta]$. Then, for $L=J$,

$$R^J_{J,\,1}=2i\tan\delta^J_J \quad . \quad . \quad . \quad . \quad . \quad . \quad . \quad . \quad (4)$$

and for the mixed triplet states with $L=J\pm1$,

$$\begin{aligned} R^J_{J+1,\,1}&=2i(\cos^2\epsilon^J\tan\delta^J_{J+1}+\sin^2\epsilon^J\tan\delta^J_{J+1}) \\ R^J&=i\sin^2\epsilon^J(\tan\delta^J_{J+1}-\tan\delta^J_{J-1}). \end{aligned} \quad . \quad . \quad . \quad . \quad . \quad (5)$$

(The quantities R^J do not appear in the expression (2).)

We want now to represent A as a simple analytic function of the momenta $\mathbf{k}$ and $\mathbf{k}'$. Comparison of (1), (2) and (3) suggests that we should take F as a linear function of $\mathbf{k}\,.\,\mathbf{k}'$ at a given energy,

$$F=A(\mathbf{k}^2,\,\mathbf{k}'^2)+C(\mathbf{k}^2,\,\mathbf{k}'^2)\mathbf{k}\,.\,\mathbf{k}'. \quad . \quad . \quad . \quad . \quad . \quad (6)$$

In principle the values of A and C on the energy shell, when $\mathbf{k}^2=\mathbf{k}'^2$, can be found from the experimental phase shifts (by using formulae (2) to (5) above), but at present information sufficient to determine the phase shifts is available at only one energy, around 95 MEV, in the important range, so that this must be supplemented by semi-empirical estimates of the energy-dependence of the phase shifts; for this we have used the work of Feshbach and Lomon (1956).

The phase shifts that occur in R can be expanded in a series of odd powers of k, so in principle A and C in eqn. (6) could be expanded, on the energy shell, in powers of k^2. As a rough approximation A can be represented by a linear expression, and this will naturally be generalized to points off the energy shell, using the condition of hermiticity, by assuming

$$A(k^2, k'^2)=A_0-\tfrac{1}{2}(A_1k^2+A_1{}^*k'^2). \quad . \quad . \quad . \quad . \quad . \quad (7)$$

The energy variation of C cannot be represented properly in this way, owing to the behaviour of the ^{3}S phase-shift $\delta_0{}^1$; this phase goes to π at zero energy and must pass through $\frac{1}{2}\pi$ somewhere around 20 MeV, at which point C will have a pole. Such low energies do not however make a significant contribution to the spin–orbit coupling so we look for a rough approximation to C over the range 40–100 MeV. In our subsequent analysis we shall first make the crude approximation

$$C(k^2, k'^2)=\text{constant} \quad . \quad . \quad . \quad . \quad . \quad . \quad . \quad (8)$$

which is very poor, but sufficient as a guide to the expected effects of the C term.

It will appear that in fact the most important part of F is the 'forward-scattering' part, and it is convenient to include in this part the contributions that will arise from exchange. Then, in a 'standard' nucleus with four nucleons in each orbital state, the amplitude concerned is

$$\begin{aligned} F_0&=F(\theta=0)+\tfrac{1}{2}F(\theta=\pi) \\ &=(3/2)A(k^2, k^2)+(1/2)C(k^2, k^2)k^2 \quad . \quad . \quad . \quad . \quad (9) \end{aligned}$$

where the factor $\frac{1}{2}$ is the expectation value of the charge exchange operator P_τ. It is possible to handle somewhat more complicated approximations to F_0; in particular a quadratic polynomial in k^2 seems to be a fair enough approximation over the important energy range, and the inadequacy of experimental or theoretical knowledge of the correct value of F does not justify more precise calculations at present. Such an approximation is used for our numerical estimates in § 4.

§ 3. Nuclear Calculation

The method of calculation is similar to that employed by Skyrme (1956); as in eqn. (12) of that reference a pseudopotential is introduced defined by

$$V_{12}=(-4\pi\hbar^2/M)A(\mathbf{k}, \mathbf{k}')\delta(\mathbf{x}_1-\mathbf{x}_2) \quad . \quad . \quad . \quad . \quad (10)$$

where A is the reactance amplitude, and the normalization is so chosen as to reproduce the scattering amplitude in Born approximation. In a matrix element of V_{12}, $\mathbf{k}$ and $\mathbf{k}'$ are to be interpreted as differential operators acting upon the states to the right and to the left respectively,

$$\left.\begin{aligned}\mathbf{k}&=\tfrac{1}{2}(\mathbf{k}_1-\mathbf{k}_2)=-\tfrac{1}{2}i(\nabla_1-\nabla_2), &&\text{on the right}\\ \mathbf{k}'&=\tfrac{1}{2}(\mathbf{k}_1'-\mathbf{k}_2')=\ \tfrac{1}{2}i(\nabla_1-\nabla_2), &&\text{on the left}\end{aligned}\right\} \quad . \quad . \quad (11)$$

In a matrix element integrated over all space, integration by parts leads to a relation between these operators, analogous to conservation of momentum,

$$\mathbf{k}_1'+\mathbf{k}_2'=\mathbf{k}_1+\mathbf{k}_2. \quad . \quad . \quad . \quad . \quad . \quad . \quad . \quad (12)$$

In the spirit of Brueckner's model of the nucleus, the contribution of the spin–orbit interaction to the total energy of the system is found from the expectation value of $\sum_{i<j} V_{ij}$ for a determinantal wave-function of single-particle states $\phi_i(\mathbf{x})$. The terms that involve a particular $\phi_0(\mathbf{x})$ are

$$\textstyle\sum_i \iint \phi_0^*(\mathbf{x}_1)\phi_i^*(\mathbf{x}_2)\, V_{12}[\phi_0(\mathbf{x}_1)\phi_i(\mathbf{x}_2)-\phi_0(\mathbf{x}_2)\phi_i(\mathbf{x}_1)]\, d\mathbf{x}_1\, d\mathbf{x}_2. \quad . \quad (13)$$

and the summation over states i may include the state ϕ_0 without error because of the antisymmetry.

When the amplitude A is expressed as a polynomial in $\mathbf{k}$ and $\mathbf{k}$, the direct integrals in (13) will lead to sums such as $\sum |\phi_i(\mathbf{x})|^2$, $\sum\phi_i^*(\mathbf{x})\nabla\phi_i(\mathbf{x})$, etc.; those that occur in our subsequent analysis can be expressed in terms of

$$\left.\begin{aligned}\rho&=\textstyle\sum |\phi_i(x)|^2\\ \tau_{\mu\nu}&=\textstyle\sum (d\phi_i^*/dx_\mu)(d\phi_i/dx_\nu), \qquad \tau=\tau_{\mu\mu}\\ \xi&=\textstyle\sum (\nabla^2\phi_i^*)(\nabla^2\phi_i)\end{aligned}\right\} \quad . \quad . \quad . \quad (14)$$

The exchange integrals involve similar sums that are outer products in spin and isotopic spin space, instead of the inner products that occur in (14). With the assumption of a standard nucleus with four particles in each state, this outer product will be a unit operator I in the spin spaces, so that

$$\textstyle\sum \phi_i(\mathbf{x})\phi_i^*(\mathbf{x}')=\tfrac{1}{4}I\sum \phi_i^*(\mathbf{x}')\phi_i(\mathbf{x}). \quad . \quad . \quad . \quad . \quad . \quad (15)$$

Finally in this perturbation calculation all these sums may be evaluated for the uncoupled states, and provided that the spin states are equally occupied these may be taken to be real in space, so that the conjugation sign may be dropped in the above definitions.

As a first step the energy (13) has been evaluated with the assumptions (7) and (8); we define the constants a, b and c so that

$$(-4\pi\hbar^2/M)A=i(\boldsymbol{\sigma}_1+\boldsymbol{\sigma}_2)\cdot(\mathbf{k}'\times\mathbf{k})[a-\tfrac{1}{2}(b\mathbf{k}^2+b^*\mathbf{k}'^2)+c\mathbf{k}\cdot\mathbf{k}']. \quad . \quad (16)$$

The calculation is straightforward; the use of the relation (15) enables the exchange integrals to be expressed as multiples of the direct ones: for the a and b terms this multiple is $\frac{1}{2}$, for the c term it is $-\frac{1}{2}$. The result may be expressed in the form

$$\int \phi_0^*(\mathbf{x})\, V(\mathbf{x})\phi_0(\mathbf{x})\, d\mathbf{x} \quad . \quad . \quad . \quad . \quad . \quad . \quad . \quad . \quad (17)$$

where then $V(x)$ is the resultant single-particle potential. Provided that the sums (14) have spherical symmetry, i.e.

$$\begin{aligned}\rho(\mathbf{x})&=\rho(r)\\ \tau_{\mu\nu}(x)&=(\tfrac{1}{3})\tau(r)\delta_{\mu\nu}+\tau_1(r)(x_\mu x_\nu-r^2\delta_{\mu\nu}/3),\end{aligned} \quad . \quad . \quad . \quad (18)$$

$V(x)$ can be expressed in the form $(\boldsymbol{\sigma}\,.\,\mathbf{l})U(r)$. The expression for U can be simplified further if it is supposed that the radial part of ϕ_0 is real; this restriction essentially means that ϕ_0 describes a stationary state.

Then we obtain from (16),

$$
\begin{aligned}
U(r)=&(\tfrac{3}{4}a)r^{-1}d\rho/dr\\
&-(3/16)(b-c/3)[(5/3)r^{-1}d\tau/dr-(2/3)rd\tau_1/dr\\
&-(10/3)\tau_1-r^{-1}d\rho/dr(\nabla^2)]\\
&+(3/32)br^{-1}d(\nabla^2\rho)/dr\\
&+(3/32)(b+2c/3)[(rd/dr+5)(r^{-1}d/drr^{-1}d\rho/dr)] \quad . \ . \quad (19)
\end{aligned}
$$

where b has been written for the real part of b, since only that occurs.

The last term can be simplified if only terms of order $1/r$ at the surface are retained: then the last two terms together give

$$(3/16)(b+c/3)r^{-1}d(\nabla^2\rho)/dr. \quad . \ . \ . \ . \ . \quad (20)$$

These last terms involve second or higher derivatives of the density, so that their average through the surface vanishes (for sufficiently large r); and as might be expected their contribution to the spin–orbit coupling is rather small; it is also rather uncertain, depending sensitively upon the density distribution and upon the shape of the wave function ϕ_0. If these are neglected the remaining terms of U can be obtained in a simple way that can conveniently be extended to more complicated expressions for the amplitude A. This method depends upon the observations that the vector product $\mathbf{k}'\times\mathbf{k}$ is equivalent with $(\mathbf{k}_2-\mathbf{k}_2')\times\mathbf{k}$, on account of the relation (12) that may be used inside the integral (13), and that the factor $(\mathbf{k}_2-\mathbf{k}_2')$ is equivalent to an operation of differentiation upon the product $\phi_i^*(x)\phi_i(x)$. It is this factor which produces the first derivatives occurring in the principal part of U; the later terms containing higher derivatives arise from the difference between $\mathbf{k}_2$ and $\mathbf{k}_2'$ in the terms of F. Thus the retention only of the leading terms of U is equivalent with putting $\mathbf{k}_2=\mathbf{k}_2'$, and so $\mathbf{k}=\mathbf{k}'$, in F, after the inclusion of the exchange contributions.

This last prescription means that then only the 'forward' part F_0 of F is required as in eqn. (9). Then the evaluation of (13) involves first the averaging of $\mathbf{k}F_0(\mathbf{k}^2)$ over $\mathbf{k}_2$ at the position $\mathbf{x}$ ignoring the gradients of density etc., and secondly a differentiation with respect to position to take care of the factor $(\mathbf{k}_2-\mathbf{k}_2')$; the result can be expressed in terms of gradients of the quantities defined in eqn. (14). It can easily be verified that the assumption $F_0=(3/2)(a-bk^2)+\tfrac{1}{2}ck^2$ indeed reproduces the first two terms in the result (19) that we found above.

As was mentioned in the previous section a fair approximation can be made to F_0 in the form

$$(-4\pi\hbar^2/M)F_0=\alpha-\beta k^2+\gamma k^4 \quad . \ . \ . \ . \ . \ . \quad (21)$$

and then this method can easily be applied to find $U(r)$. If the non-diagonal terms such as τ_1 are omitted the result is

$$\begin{aligned} U(r) = & \tfrac{1}{2}\alpha r^{-1} d\rho/dr \\ & -(1/8)\beta[(5/3)r^{-1}d\tau/dr - r^{-1}d\rho/dr(\nabla^2)] \\ & +(1/32)\gamma[(7/3)r^{-1}d\xi/dr - (14/3)r^{-1}d\tau/dr(\nabla^2) \\ & + r^{-1}d\rho/dr(\nabla^4)]. \quad (22) \end{aligned}$$

As written this expression is not properly hermitian but this does not cause any difficulty in the following applications where ∇^2 is equivalent to a real operator.

The evaluation of these formulae requires a knowledge not only of the density distribution, but also of the distribution of τ etc. For this we have employed the Thomas–Fermi approximation, as in Skyrme (1956), according to which

$$\tau = (3/5)(6{\cdot}0)\rho^{5/3}\,; \qquad \xi = (3/7)(36{\cdot}0)\rho^{7/3}. \quad (23)$$

Furthermore with the same approximation for a particle at the top of the Fermi distribution, energy $-E_0$,

$$-\nabla^2 = (6{\cdot}0)\rho^{2/3} \quad (24)$$

so that for a particle at energy level E

$$\begin{aligned} -\nabla^2 &= (6{\cdot}0)\rho^{2/3} + k^2 \\ k^2 &= (2M/\hbar^2)(E+E_0). \quad (25) \end{aligned}$$

With the help of (23) and (25), $U(r)$ defined by eqn. (22) can be written in the form

$$\begin{aligned} U(r) = & \tfrac{1}{2}r^{-1}(d\rho/dr)\{\alpha - \beta(4{\cdot}0\rho^{2/3} + \tfrac{1}{4}k^2) \\ & + \gamma[18{\cdot}0\rho^{4/3} + (10{\cdot}0\rho^{2/3})(\tfrac{1}{4}k^2) + (\tfrac{1}{4}k^2)^2]\}. \quad (26) \end{aligned}$$

Compared with eqn. (19) this neglects among other terms, those in τ_1 (defined in eqn. (18)). An estimate of the error made can be obtained by considering the momentum distribution in a Fermi gas bounded by an infinite plane wall, a model suggested by Swiatecki (1951). The anisotropy of the momentum distribution is most marked actually at the wall, where the density goes to zero ; at this point the effect would be to replace the factor (5/3) that multiplies $r^{-1}d\tau/dr$ in the second term of (22) by 1·4, and correspondingly in eqn. (26) the term $4{\cdot}0\rho^{2/3}$ that multiplies β would become $3{\cdot}6\rho^{2/3}$. Since this is only a 10% effect at its most important it is a small matter compared with other errors that can arise from our approximations.

§ 4. The Doublet Splitting

We consider in this section the doublet splitting for a nucleon at the top of the Fermi distribution, for which, in eqn. (25), $k^2=0$. Then the formula (26) gives

$$U_0(r) = \tfrac{1}{2}r^{-1}d\rho/dr[\alpha - (1{\cdot}1y^2/3)\beta + (1{\cdot}3y^{4/3})\gamma] \quad (27)$$

where we have put

$$\rho = y\rho_0\,; \qquad \rho_0 = (4\pi r_0{}^3/3)^{-1} = 0{\cdot}138 \quad (28)$$

with $r_0 = 1{\cdot}2$, and the unit of length equal to 10^{-13} cm.

If the form (21) is assumed to be a good approximation the result (27) shows by comparison that the important range of $\mathbf{k}^2$ is from 0 to 1·1, corresponding to laboratory energies of 0 to 90 MeV, with the greatest contribution near $y=\frac{1}{2}$, around 55 MeV. A more physical argument is that the important values of $\mathbf{k}^2$ are those around the average of $\frac{1}{4}(\mathbf{k}_1-\mathbf{k}_2)^2$, where $\mathbf{k}_1$ is the momentum of the particle at the top of the Fermi distribution and $\mathbf{k}_2$ of another with which it interacts ; this average has the value $(2/5)k_F{}^2=0{\cdot}63$, with $r_0=1{\cdot}2$, giving an energy of 52 MeV.

Unfortunately polarization experiments have not yet been done at this low energy, so an extrapolation downwards has been needed of the

Fig. 1

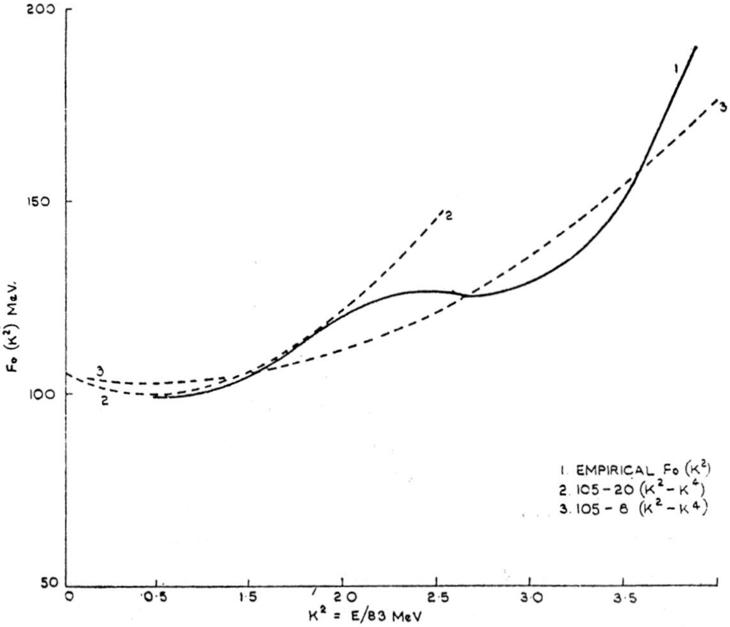

The function F_0.

phase-shift analysis available at 95 MeV (Phillips, private communication). This has been made following the estimates given by Feshbach and Lomon (1956), which are in agreement with the phases determined at 95 MeV and higher energies. The function $F_0(k^2)$, defined by eqn. (9), determined empirically in this way is shown in fig. 1. An approximation to this curve for the lower energies has been taken in the form (21) with the coefficients

$$\alpha=105, \qquad \beta=\gamma=20 \text{ MeV} \quad . \quad . \quad . \quad . \quad . \quad . \quad (29)$$

with 10^{-13} cm as the unit of length ; this approximation is also shown in the figure.

With the values given by (29), $U_0(r)$ becomes

$$U_0(r)=r^{-1}d\rho/dr[50{\cdot}2+13(y^{2/3}-0{\cdot}42)^2]. \quad . \quad . \quad . \quad . \quad (30)$$

The second term is rather small, reflecting the small dependence of F_0 on k^2 in this range ; a simple average over y, from 0 to 1, gives the value 1·3. Since the estimation of its expectation value is rather uncertain anyway, we shall adopt the simple formula

$$U_0(r)=51{\cdot}5r^{-1}d\rho/dr. \qquad (31)$$

With the perturbing potential $\boldsymbol{\sigma}\,.\,\mathbf{l}U(r)$, the splitting between levels with $j=l\pm\frac{1}{2}$ is

$$\Delta E=-(2l+1)\langle U(r)\rangle. \qquad (32)$$

For a light nucleus it may not be unreasonable to assume a parabolic density distribution

$$\rho(r)=\rho_1(1-r^2/R^2) \qquad (33)$$

for which, independently of the state of the nucleon considered, (31) and (32) give

$$\Delta E=(2l+1)(51{\cdot}5)(2\rho_1/R^2). \qquad (34)$$

If we write the standard radius $r_0A^{1/3}$ as R_0, the proper normalization of the density (33) gives the condition

$$\rho_1/\rho_0=2{\cdot}5(R/_0R)^3 \qquad (35)$$

so that ρ_1 may be eliminated from (34) to give

$$\Delta E=24{\cdot}8(R_0/R)^5(2l+1)A^{-2/3}. \qquad (36)$$

If now R be chosen so that the mean square radius is equal to that of a sphere of radius R_0, $(R/R_0)=1{\cdot}18$ and $\Delta E=10{\cdot}7\ (2l+1)A^{-2/3}$. This gives for ^{15}N and ^{17}O splittings equal to 5·3 and 8·1 MEV respectively, to be compared with the experimental values of 6·3 and 5·1 MEV respectively.

Another estimate of the density distribution in light nuclei may be found from a shell model with oscillator wave functions : the calculation of the expectation value in (32) is then equivalent to the calculation made by Lane and Elliott (1954) in the limit of zero range, as will be discussed below. The calculated splittings depend upon the scale constant r_1 of the wave functions (supposed to contain the exponential factor $\exp(-\frac{1}{2}r^2/r_1^2)$), and are for

$$^{15}\mathrm{N}:\ 23{\cdot}9(r_0/r_1)^5\ ; \qquad ^{17}\mathrm{O}:\ 33{\cdot}0(r_0/r_1)^5. \qquad (37)$$

For the same value of r_1 the ratio of these is close to the ratio found for a parabolic distribution. To reproduce the observed splittings it would be necessary to assume

$$^{15}\mathrm{N}:\ r_1=1{\cdot}56\ ; \qquad ^{17}\mathrm{O}:\ r_1=1{\cdot}74. \qquad (38)$$

As has been pointed out by Lane and Elliott (1954) such a jump in the value of r_1 going across the ^{16}O closed shell appears to be necessary to explain the Coulomb energy differences of mirror nuclei ; their unpublished calculation gives the values 1·67 and 1·90 for the two nuclei, with a ratio similar to that required in (38). The equation of the mean square radius with $(3/5)R_0^2$ for ^{16}O gives $r_1=1{\cdot}56$.

Since the spin–orbit splitting is such a sensitive function of the density distribution, exemplified by the fifth power of r_1 that occurs in (37), and since it is a rather different measure of it from the Coulomb energy difference or the mean square radius, it is hard to say whether or no the formula (31) leads to results in accord with experiment. A tentative conclusion from the analysis given would be that the constant in that formula may be too small, possibly by as much as 30%.

In the region of heavy nuclei the experimental information near ^{208}Pb has been examined by Blin-Stoyle (1955), who employed the form (31) with an adjustable numerical constant. He found that the data could be well fitted by taking for the constant (K, in his notation) the value 57 (assuming that $r_0=1\cdot2$), which is 10% greater than our calculated value.

In noting these approximate agreements we should not overlook the assumptions made in deriving the simplified formula (26); these were first the 'forward scattering' approximation used to obtain (22) and secondly the use of the Thomas–Fermi approximation to express all quantities in terms of the density distribution. These approximations are very similar and both depend for their validity on the smallness of the density gradient. This condition is hardly satisfied at the nuclear surface, but once the forward-scattering approximation has been made the remaining error will be small because of the flatness of F_0 as a function of k^2; on the other hand the separate parts A and Ck^2 of F_0, as in eqn. (9), vary more rapidly with energy and the error in the forward scattering approximation may be appreciable. It will be shown in the § 5 how an alternative evaluation of (16) may be made which indicates, as might be expected, that the approximations have tended to underestimate the magnitude of the important momenta.

§ 5. Relation with Spin–Orbit Potential

The method that we have employed is essentially the same as that generally used in shell-model calculations, namely the calculation of the expectation value of a perturbing interaction in an unperturbed individual particle state; the difference is only that we have used a singular point-type interaction instead of a potential. The singularity arises just because we have chosen to use a finite polynomial expression for the scattering amplitudes; in so far as these polynomials can be regarded as approximations to the matrix elements of some potential, equivalent results should be obtained by working with that potential.

In our problem the amplitude A, eqn. (1), resembles closely the matrix element of the two-body spin–orbit potential, and we shall now enquire how far this correspondence can be carried. We shall assume a potential of the type

$$-(1-x+xP_\tau)(\boldsymbol{\sigma}_1+\boldsymbol{\sigma}_2)\cdot(\mathbf{r}_1-\mathbf{r}_2)\times(\mathbf{p}_1-\mathbf{p}_2)V(r_{12}) \quad . \quad . \quad . \quad (39)$$

such as has been considered by Lane and Elliott (1954) and other authors. In this expression P_τ is the charge exchange operator and x measures the fraction of exchange potential. The matrix element of this potential between states of momentum **k** and **k**′ for the n–p system is easily found to be

$$i(\boldsymbol{\sigma}_1+\boldsymbol{\sigma}_2)\cdot(\mathbf{k}'\times\mathbf{k})[(1-x)W(|\mathbf{k}-\mathbf{k}'|)+xW(|\mathbf{k}+\mathbf{k}'|)] \quad . \quad . \quad (40)$$

where W is related to $V(r)$ by

$$W(q)=(8\pi/q^3)\int_0^\infty V(r)(\sin qr-qr\cos qr)r\,dr. \quad . \quad . \quad . \quad (41)$$

In particular for the form of V used by Lane and Elliott, the Yukawa form $V=V_0(\lambda/r)\exp(-r/\lambda)$, we obtain

$$W(q)=16\pi V_0\lambda^5(1+\lambda^2q^2)^{-2}. \quad . \quad . \quad . \quad . \quad . \quad . \quad (42)$$

As compared with the general form of amplitude given by (1), the expression (40) derived from a static potential in Born approximation is of a special form, and we cannot expect to be able to express the observed scattering amplitude in this way. As we mentioned in the discussion following eqn. (6) the coefficient C in that equation has a singular behaviour at low energies, and this certainly cannot be reproduced from a potential. The physical meaning behind this is that ‘ final state interactions ’ play an important role in the spin–orbit coupling between pairs of nucleons, and the use of a spin–orbit potential in Born approximation can only be regarded as a mathematical artifice. At higher energies where the phase shifts become small a direct physical interpretation as a potential might be more reasonable, but then the whole concept of a potential seems to be unsatisfactory for analysing nucleon–nucleon scattering.

It is possible however to use this idea of a potential to find the effect of using the form (19) for $U(r)$. This was derived from the form (16), which can be reproduced by expanding W to order q^2 with suitable values for the parameters V_0, λ, x. Thus to this order

$$\begin{aligned}&(1-x)W(|\mathbf{k}-\mathbf{k}'|)+W(|\mathbf{k}+\mathbf{k}'|)\\&\quad=16\pi V_0\lambda^5[1-2\lambda^2(\mathbf{k}^2+\mathbf{k}'^2)+4\lambda^2(1-2x)\mathbf{k}\cdot\mathbf{k}']. \quad . \quad . \quad . \quad (43)\end{aligned}$$

The results of Lane and Elliott may also be expanded to relative order λ^2, giving then the splitting due to the amplitude F given by (43); their results are for

$$\begin{aligned}&{}^{15}\text{N}:\ \Delta E=(108/\sqrt{2\pi})(V_0\lambda^5/r_1^5)[1-[(70+40x)/9](\lambda^2/r_1^2)],\\&{}^{17}\text{O}:\ \Delta E=(150/\sqrt{2\pi})(V_0\lambda^5/r_1^5)[1-(10+3{\cdot}2x)(\lambda^2/r_1^2)]. \quad . \quad (44)\end{aligned}$$

In terms of the constants a, b and c of eqn. (16), determined by comparison with (43), the final factors in these expressions for the splittings are proportional respectively to

$$a-(b-0{\cdot}22c)(2{\cdot}5/r_1^2)\,; \qquad a-(b-0{\cdot}14c)(2{\cdot}9/r_1^2) \quad . \quad . \quad . \quad (45)$$

which are to be compared with the estimate

$$a-(b-0{\cdot}33c)(4{\cdot}0\rho^{2/3}) \quad . \quad . \quad . \quad . \quad . \quad . \quad . \quad . \quad (46)$$

derived from eqn. (26). The comparison suggests that the coefficient of b, and therefore the nuclear momenta, have been somewhat underestimated.

As we mentioned before the form (43), equivalent to (16), is not a realistic representation of the amplitude so the analysis in terms of a potential is limited in application. Because the amplitude F_0 is a slowly varying function of energy the analysis in § 4 leads to substantially the same result as the use of a potential of zero range (when the exchange character is irrelevant). Such a potential implies that all the interaction between nucleons takes place in the ^{3}P state, but this is not really so and there is a considerable contribution from the ^{3}S–^{3}D coupled state which gives the C term in eqn. (6).

§ 6. Potential for Scattering

Another possible comparison of our results with experiment comes from the analysis of the polarization of nucleons elastically scattered from target nuclei. Our analysis has been based upon the possibility of using the form (6) for the amplitude (although this could easily be generalized) ; accordingly comparison should be limited to those energies for which it may be a fair approximation. In this section we shall refer to the calculations of Sternheimer (1955) analysing the scattering of 130 Mev protons from light nuclei ; this energy is already rather high, so that we should be prepared for some inaccuracy in our hypotheses.

The values (29) do not give a good representation of F_0 over the larger energy range that is relevant for incident nucleons of 130 Mev ; a better overall fit is obtained with the choice

$$\alpha=105, \qquad \beta=\gamma=8 \qquad (47)$$

which is illustrated in fig. 1.

In this case we must include the k^2 terms in the formula (26) for $U(r)$; k^2 was defined by eqn. (25), and in that equation the mass M should properly be the effective mass M^*. We adopt the values used by Skyrme (1956) for M^* and E_0, which give

$$\tfrac{1}{4}k^2=1{\cdot}75(1+0{\cdot}5y)^{-1} \qquad (48)$$

where as before y is the density in units of ρ_0 ; we also use the same density distribution, so that in the ‘ large nucleus ’ approximation

$$-\tfrac{1}{2}r^{-1}d\rho/dr=0{\cdot}0767A^{-1/3}y^{1/2}(1-y)$$
$$y=\rho/\rho_0=[\tanh\,(R-r/(1{\cdot}5))]^2. \qquad (49)$$

The formula (26) thus gives

$$-A^{1/3}U(r)=K(1-y)y^{1/2} \qquad (50)$$

where the coefficient K varies from 8·85 to 10·25 as y goes from 0 to 1. For comparison we note that eqn. (31) gives $K=7{\cdot}9$, so that there is a 20% increase in the strength of the potentials compared with that felt by a particle at the top of the nuclear well.

Before we compare our potential with that used by Sternheimer, we should make allowance for the mass variation. In the approximate expression for the phase shifts used by Sternheimer the perturbing potential is multiplied by the factor $(\hbar k/T)$, where T is the kinetic energy of the nucleon and k its wave number. This factor is equal to $(2M^*/(E-V))^{1/2}$, where E is the energy of the incident nucleon and V

is the nuclear potential, and when the effective mass changes appreciably through the surface the potential V is increased so that the variation of this factor is not negligible. An effective potential has accordingly been estimated by multiplying (50) by this factor, expressed as a function of y with the same hypotheses. The resulting modified coefficient K' then varies from 8·4 to 6·8 ; this potential is shown in fig. 2 as a function of distance from the outer edge of the nucleus, and it is not significantly different from the potential (31).

The maximum value of the potential is about $3A^{-1/3}$ MeV and its width is very similar to that assumed by Sternheimer, whose potential is also shown in the figure, adjusted to have a coincident maximum ;

Fig. 2

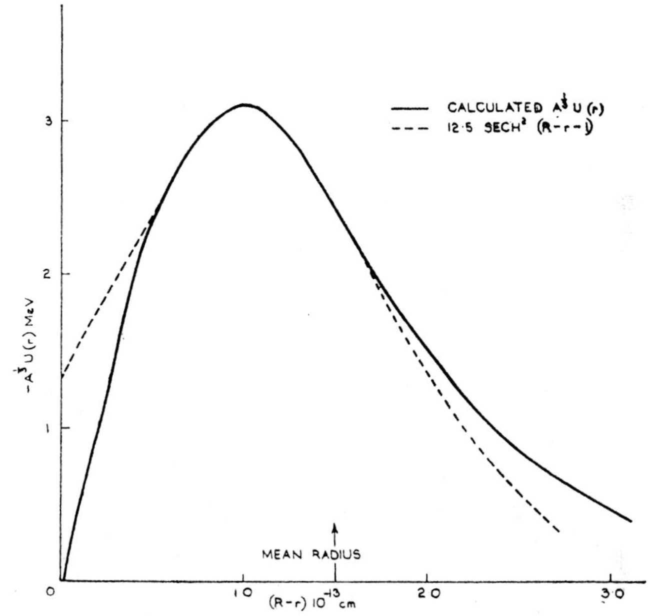

The potential $U(r)$.

the maximum value is close to that found necessary by Sternheimer to explain the experimental results. The main difference is that the maximum of our potential occurs $0{\cdot}5\times10^{-13}$ cm outside the mean radius $r=r_0A^{1/3}$ instead of at that point (always speaking of the potential in a large nucleus, ignoring the r^{-1} factor outside). It is not possible to say without detailed calculation whether this and other small differences would help to remove the difficulties that Sternheimer found in fitting properly the angular distributions.

Summary

The general shape of the phase-shift analysis of the nucleon–nucleon scattering for energies up to 100 MeV now seems well established and it has been possible to fix the spin–orbit part of the scattering amplitude with reasonable certainty.

This empirically determined amplitude has been applied in the spirit of Brueckner's model of the nucleus to find the effective single-particle spin–orbit coupling. The essential assumptions are that

(i) The energy of the nuclear system may be determined by using a 'model' wave function which is approximately that of a set of independent particles moving in a central scalar potential.

(ii) The energy change due to a perturbing interaction is approximately the sum of changes due to two-body interactions and can be evaluated by using the reactance amplitudes and model wave function.

(iii) The scattering associated with these two-body interactions is approximately the same as that between free nucleons, so far as the spin–orbit amplitude is concerned.

The numerical result that we obtain from our analysis appears to be in agreement with experimental evidence, within the limitations arising from ignorance of the nuclear wave functions. However this agreement really involves only one quantity, the strength of the coupling, so that it would be premature to claim that the method of calculation was thereby justified.

Acknowledgments

The authors are grateful to Dr. R. J. N. Phillips for his help with the scattering analysis, in particular for providing formula (3), to Dr. J. P. Elliott for many discussions and for providing the results in eqn. (44), and to Dr. R. J. Blin-Stoyle for discussion of his work.

References

Blin-Stoyle, R. J., 1955, *Phil. Mag.*, **46**, 973.
Brueckner, K. A., 1954, *Phys. Rev.*, **96**, 908; *Ibid.*, **97**, 1353.
Chisholm, J. C. R., and Touschek, B. F. X., 1953, *Phys. Rev.*, **90**, 763.
Elliott, J. P., and Lane, A. M., 1954, *Phys. Rev.*, **96**, 1160.
Feshbach, H., and Lomon, E., 1956, *Phys. Rev.*, **102**, 891.
Hillman, P., and Stafford, G. H., 1956, *Nuovo Cimento*, **3**, No. 3, 663.
Hughes, J., and Le Couteur, K. J., 1950, *Proc. Phys. Soc.* A, **63**, 1219.
Keilson, J., 1951, *Phys. Rev.*, **82**, 759.
Johnson, M. H., and Teller, E., 1955, *Phys. Rev.*, **98**, 783.
Skyrme, T. H. R., 1955, *Proc. Roy. Soc.* A, **230**, 277.
Skyrme, T. H. R., 1956, *Phil. Mag.*, *preceding paper.*
Stapp, H. P., 1955, *Thesis*, University of California, UCRL-3098.
Sternheimer, R. M., 1955, *Phys. Rev.*, **100**, 886.
Swiatecki, W. J., 1951, *Proc. Phys. Soc.* A, **64**, 226.

1.C:
1.D.1

Nuclear Physics 9 (1958/59) 635—640; © *North-Holland Publishing Co., Amsterdam*
Not to be reproduced by photoprint or microfilm without written permission from the publisher

THE SPIN-ORBIT INTERACTION IN NUCLEI

T. H. R. SKYRME

Atomic Energy Research Establishment, Harwell, Didcot, Berks.

Received 18 October 1958

Abstract: The analysis previously made of the average nuclear potential has been extended to consideration of the spin-orbit interactions. It has not been possible to find a satisfactory two-body interaction consistent with all the data; that suggested by the phase-shift analysis of nucleon-nucleon scattering is just within the region of possible forms.

1. Introduction

In a previous paper [1]) we have examined empirically the nature of the *average* interaction between nucleons which is effective in the shell-model description of a nucleus. The next logical step in the determination of the effective interaction is a discussion of the spin-orbit coupling terms, since these are already manifested for a single particle outside a closed shell. This paper contains an account of our analysis of the data made in a way similar to that used for the mean potential, and incidentally brings up to date an earlier investigation of the problem [2]).

The problem of spin-orbit coupling is distinguished from that of the average potential by the comparative paucity of data, and the sensitive dependence of their interpretation upon details of the wave-functions assumed. In our first investigations we had assumed that the spin-orbit interaction might be described by a two-body interaction of very short range, written in the notation of ref. [1]) as

$$\delta(\mathbf{r}_1-\mathbf{r}_2)i(\boldsymbol{\sigma}_1+\boldsymbol{\sigma}_2)\cdot\mathbf{k}'\times\mathbf{k}V \tag{1}$$

with some constant V.

This simple hypothesis cannot be made to fit the facts adequately. It fails

(a) to describe the strong variation of coupling through the p-shell,

(b) to account for the different couplings in masses 15 and 17,

(c) to explain the large amount of coupling indicated by the calculations on the order of single-particle levels.

It can be modified either by allowing V to be dependent upon the momenta $\mathbf{k}'$ and $\mathbf{k}$, or by allowing it to depend upon density through the introduction of similar three- (or many-)body terms. The possible forms of three-body interaction are numerous and have not been explored, but as trial calculations have not indicated any obvious gain in consistency through their inclusion they will not be discussed further in this paper.

We have considered the three-parameter form of V that was introduced in ref.[2]), allowing linear dependence upon $\mathbf{k}^2$ and upon $\mathbf{k}\cdot\mathbf{k}'$. The former dependence does not appear to be significant and cannot be determined empirically; on the other hand the introduction of 'angular' dependence in V leads to greater consistency.

The formulae needed for this discussion are collected in the following two sections. The possible determination of the parameters is discussed in section 4, together with an estimate based upon free-nucleon phase shifts as in ref.[2]); those numerical values were founded on the phase-shift analysis of Feshbach and Lomon [3]), and have been replaced by values derived from the analyses of Gammel and Thaler [4]) and of Marshak and Signell [5]).

2. The Two-Body Interaction

The discussion of $V(\mathbf{k}', \mathbf{k})$ will be based on the three-parameter form

$$V = v_0+\tfrac{1}{2}v_1(\mathbf{k}^2+\mathbf{k}'^2)+v_2\mathbf{k}'\cdot\mathbf{k}. \tag{2}$$

It is convenient also to introduce the dimensionless parameters

$$\begin{aligned}\mu_0 &= (\rho_0/E_0 r_0^{\,2})\tfrac{3}{2}v_0\\ \mu_1 &= (\rho_0/E_0 r_0^{\,4})\tfrac{3}{2}v_1\\ \mu_2 &= (\rho_0/E_0 r_0^{\,4})\tfrac{1}{2}v_2;\end{aligned} \tag{3}$$

$\rho_0 = \frac{4}{3}\pi r_0^{\,3}$ is the density of nuclear matter, and $E_0 = 1/(2Mr_0^{\,2})$ as in ref.[1]).

In the p-shell two distinct matrix elements only of the spin-orbit interaction are needed. In the notation of Elliott [6]) they have the following values for oscillator wave-functions (characterised by an exponential factor $\exp[-\frac{1}{2}(yr/r_0)^2]$):

$$\begin{aligned}D = E &= -0.177\, y^5(\mu_0+2.5\, y^2\mu_1)E_0,\\ G &= -0.532\, y^5\mu_2 E_0.\end{aligned} \tag{4}$$

The single-particle spin-orbit coupling is usually written as $a\,\mathbf{l}\cdot\mathbf{s}$. In terms of the matrix elements E and G, its values at the beginning and end of the p-shell are

$$\begin{aligned}&\mathrm{p}; && a = 6E\\ &\mathrm{p}^{-1}: && a = -9E-5G.\end{aligned} \tag{5}$$

In mass 6, the splitting of the three ^{3}D levels (which is the most obvious manifestation of the spin-orbit coupling (Soper [7])) is determined by

$$a_{\text{eff}} = 6E+2G \tag{6}$$

apart from effects of the tensor forces.

According to eq. (5) the value for mass 15 is

$$a = 1.60\, y^5(\mu_0+2.5\, y^2\mu_1+1.67\, y^2\mu_2) \tag{7}$$

while in mass 17 a straightforward calculation gives for the d-level

$$a = 1.33\, y^5(\mu_0+2.9\, y^2\mu_1+1.2\, y^2\mu_2). \tag{8}$$

3. The One-Body Interaction

The effective single-body spin-orbit interaction corresponding to the form (2) for $V(\mathbf{k}', \mathbf{k})$ was worked out by Bell and Skyrme [2]) for the case of a shell-model wave-function with four nucleons in each orbital state and a spherically symmetric density distribution. The result could be written in the form of an interaction

$$A(r)\mathbf{l}\cdot\mathbf{s} \tag{9}$$

where

$$\begin{aligned} A(r)/E_0 = {} & \mu_0(r_0^2/\rho_0)r^{-1}\mathrm{d}\rho/\mathrm{d}r \\ & + \tfrac{1}{4}(\mu_2-\mu_1)(r_0^4/\rho_0)r^{-1}(\mathrm{d}/\mathrm{d}r)(\nabla^2\rho) \\ & + \tfrac{1}{4}(\mu_2+\mu_1)(r_0^4/\rho_0)(\tfrac{5}{3}r^{-1}\mathrm{d}\tau/\mathrm{d}r-r^{-1}(\mathrm{d}\rho/\mathrm{d}r)\nabla^2). \end{aligned} \tag{10}$$

In this expression ρ is the nucleon density, τ is the density of (momentum) [2]) and an anisotropic term in the momentum density has been dropped.

In a large nucleus the contribution of the terms involving $r^{-1}(\mathrm{d}/\mathrm{d}r)(\nabla^2\rho)$ will be assumed to be small †. In the last group of terms we make an estimate by using the Thomas-Fermi approximation; for a particle at the top of the Fermi distribution $\nabla^2\rho = -\mathrm{d}\tau/\mathrm{d}\rho$, so the significant terms in (10) can be written in the form

$$(r_0^2/\rho_0)(r^{-1}\mathrm{d}\rho/\mathrm{d}r)[\mu_0+(\mu_1+\mu_2)\tfrac{2}{3}(\mathrm{d}\tau/\mathrm{d}\rho)r_0^2]. \tag{11}$$

From the T-F approximation $\frac{2}{3}r_0^2(\mathrm{d}\tau/\mathrm{d}\rho) = 1.5(\rho/\rho_0)^{\frac{2}{3}}$; with the surface distribution used in ref.[1]), the average value of $(\rho/\rho_0)^{\frac{2}{3}}$ is $\frac{3}{4}$. Finally therefore the combination of parameters that enters into the magnitude of the single-particle coupling in a large 'average' nucleus is

$$\bar{\mu} = \mu_0+1.12(\mu_1+\mu_2). \tag{12}$$

In comparing this result with the experimentally observed coupling in heavy nuclei it must be remembered that

(i) This formula can only be expected to be approximately correct for the average of neutron and proton spin-orbit couplings.

(ii) There may be other contributions to the single-body coupling arising from central and tensor forces, because the single-particle states are already in *j-j* coupling.

† This is hard to justify, but is supported by the manner in which the lines of the figure approach the 'infinite limit' as the nuclei become more tightly bound.

The term involving $(\mathrm{d}/\mathrm{d}r)(\nabla^2\rho)$ may be interpreted in terms of a finite range of the interaction, by writing

$$\begin{aligned}\bar{\mu}\rho(\mathbf{r})+\tfrac{1}{4}(\mu_2-\mu_1)r_0{}^2\nabla^2\rho \\ = \text{Average}\ \rho(\mathbf{r}+\mathbf{s}),\end{aligned} \tag{13}$$

in which the r.m.s. value of the range is given by

$$\text{Average}\ \mathbf{s}^2 = \tfrac{3}{2}r_0{}^2(\mu_2-\mu_1)/\bar{\mu}. \tag{14}$$

4. Determination of Parameters

From the light nuclei of masses 5, 6, 15 and 17 four relations between the parameters μ_n are obtained, given by eqs. (4) to (8) above when equated to the experimental splittings. In all of these the ratio of the coefficients of μ_1 and $\bar{\mu}_0$ varies only between 1.1 and 1.25, and the same is true of the large nucleus combination $\bar{\mu}$ given by eq. (12). It is not therefore possible to determine μ_0 and μ_1 separately, but only a combination which is approximately

$$\mu_0+1.2\,\mu_1 = \bar{\mu}_0; \tag{15}$$

this is equivalent to saying that the only important relative momenta are those near $(kr_0)^2 = 1.2$. In using the formulae we have accordingly replaced μ_0 by μ_0, and left out the μ_1 terms.

The evidence concerning $\bar{\mu}$, the effective coupling strength in heavy nuclei, is confusing. The only direct evidence available on doublet splittings is around Pb^{208}. These data have been analysed in terms of a short range interaction by Blin-Stoyle [8]), who found that they were consistent with a strength $\bar{\mu} = 0.44$, in the present notation.

On the other hand a number of calculations have been made of single-particle levels in a realistic well with spin-orbit coupling, and these indicate that a much larger value is needed to obtain the right level sequence. We have, in particular, examined the calculations of Ross, Lawson and Mark [9]); these authors took

$$A(r) = -\tfrac{1}{2}\lambda(\hbar/Mc)^2r^{-1}\mathrm{d}V/\mathrm{d}r$$

with

$$V(r) = -V_0/[1+\exp\alpha(r-\mathrm{a})] \tag{16}$$

as the central potential; λ was an adjustable parameter. Calculations were made with and without an effective mass correction, and we have interpolated between them to the value $M/M^* = 1.62$, favoured in ref. [1]).

We have estimated the relation between $\mathrm{d}V/\mathrm{d}r$ and $\mathrm{d}\rho/\mathrm{d}r$ in two ways: on the basis of the Thomas-Fermi approximation and by direct numerical comparison with the surface shape derived in ref. [1]). They agree in giving, approximately

$$\mathrm{d}V/\mathrm{d}r = -0.8V_0\rho_0^{-1}\mathrm{d}\rho/\mathrm{d}r \tag{17}$$

which then leads to the identification

$$\bar{\mu} = \tfrac{1}{2}0.8(\hbar/Mcr_0)^2\lambda V_0/E_0 = 0.0144\lambda V_0/E_0. \tag{18}$$

The interpolated numerical value is then $\bar{\mu} = 1.8$, much larger than Blin-Stoyle's estimate.

Part at least of this discrepancy appears to be due to the fact that the calculations of Ross, Mark and Lawson give excessively large doublet splittings at the top of the well, e.g. the 1i splitting is some three times that estimated (from experiment) by Blin-Stoyle, while part may be due to the approximations used by the latter.

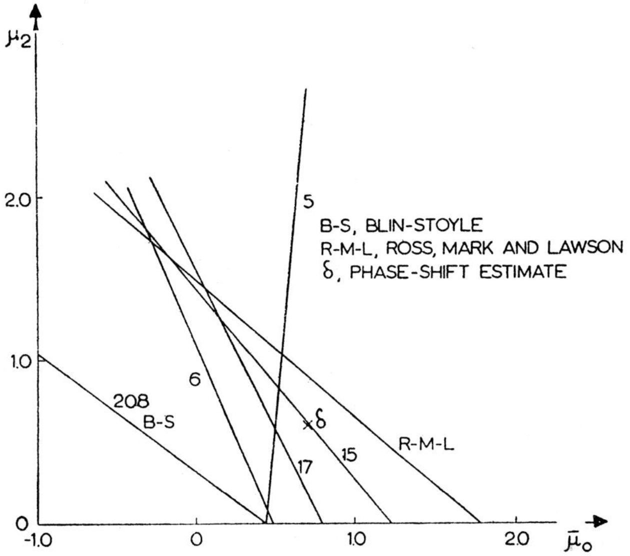

Fig. 1. Relations between $\bar{\mu}_0$ and μ_2 determined by doublet splittings in the mass numbers indicated, by single-particle level order and by the phase-shift estimate.

The various relations obtained between $\bar{\mu}_0$ and μ_2 are shown in the figure; the lines for masses 5 and 6 should be allowed considerable uncertainty on account of the greater inaccuracy of the assumed wave-functions for these lightly bound systems. There is no well-defined best pair of values. At the one extreme (disregarding mass 5 and Blin-Stoyle's value) a small value of $\bar{\mu}_0$ and large μ_2 are possible, with the interaction occurring principally in even states ($T = 0$); on the other hand the data may not be impossible to reconcile with the point marked δ which is that estimated on the basis of the nucleon-nucleon scattering phases.

This estimate has been made in the way described in refs. [1,2]), using δ not $\tan\delta$ as a measure of the energy shift; the formulae relevant here are

$$\bar{\mu}_0 = \tfrac{9}{8}(kr_0)^{-3}(5\delta_1{}^2-2\delta_1{}^0-3\delta_1{}^1),$$
$$\mu_2 = \tfrac{3}{8}(kr_0)^{-5}(14\delta_2{}^3-9\delta_2{}^1-5\delta_2{}^2). \qquad (19)$$

For $(kr_0)^2 = 1.2$, the value suggested by the ratio of the coefficients of μ_1 and μ_0, we obtain from the phase-shifts given by Marshak and Signell [5])

$$\bar{\mu}_0 = 0.7, \qquad \mu_2 = 0.5.$$

5. Conclusions

The analysis made in this paper does not unfortunately lead to such a clear conclusion as our analysis of the *mean* potential. Lane and Elliott [10]) found preference for a neutral interaction over a symmetrical one, but did not explicitly consider mixtures; our results do not disagree with theirs. Abraham [11]) who analysed the Li^7 splitting and compared requirements in this case with those for mass 5 preferred the exchange character $(1-P^\tau)$, i.e. interaction only in even states. On the other hand, Huper [12]) in an analysis of structure in the p-shell prefers a $(1+P^\tau)$ dependence.

These differences, together with the discrepancies between estimates of the coupling in heavy nuclei, make the determination of the effective interaction difficult. There is a need for extensive numerical calculations with realistic wave-functions.

It is not possible yet to say whether or not the two-body interaction will provide a sufficient account of all the phenomena. Jancovici [13]) has estimated the contribution of tensor forces to the reaction matrix, and finds that they are inadequate. The 'real' spin-orbit interaction between two nucleons may then account for most of the effective interaction investigated here, or it may be that many-body effects are relatively important: in the latter case the analysis made in this paper would have no significance.

References

1) T. H. R. Skyrme, Nuclear Physics **9** (1958/59) 615
2) J. S. Bell and T. H. R. Skyrme, Phil. Mag. **1** (1956) 1055
3) H. Feshbach and E. Lomon, Phys. Rev. **102** (1956) 891
4) J. L. Gammel and R. M. Thaler, Phys. Rev. **107** (1957) 1337
5) P. S. Signell and R. E. Marshak, Phys. Rev. **109** (1958) 1229
6) J. P. Elliott, Proc. Roy. Soc. A **218** (1953) 345
7) J. M. Soper, Phil. Mag. **2** (1957) 1219
8) R. J. Blin-Stoyle, Phil. Mag. **7** (1955) 1973
9) A. A. Ross, R. D. Lawson and H. Mark, Phys. Rev. **104** (1956) 401
10) A. M. Lane and J. P. Elliott, Phys. Rev. **96** (1954) 1160
11) G. Abraham, Nuclear Physics **1** (1956) 415
12) R. Huper, Z. f. Naturforschung **12a** (1957) 295
13) B. Jancovici, Nuovo Cimento **7** (1958) 290

A new model for nuclear matter

By T. H. R. Skyrme

Atomic Energy Research Establishment, Harwell, Berks

(*Communicated by M. H. L. Pryce, F.R.S.—Received* 1 *July* 1954)

The different values obtained for nuclear radii from electromagnetic interactions as compared with specifically nuclear interactions suggest a model of nuclear matter in which the meson field is supposed to condense into an incompressible fluid and the nucleonic sources are confined to its interior by a strong interaction between the sources and the fluid as a whole. The sources are also coupled to spin and charge fluctuations in the fluid, whose exchange leads to further internucleonic forces. It is necessary to postulate that the fluid have a comparatively low density; as a result rotational levels of the fluid are high, leading to a small probability of exchange of angular momentum (and charge coupled to it) with the sources. The values of the anomalous electrical interactions of nucleons deduced are in rough agreement with the facts. The nuclear structure indicated is a shell model embedded in the mesic fluid whose oscillations, strongly coupled to the nucleons, give rise to the collective features of nuclear structure as in the theory of Bohr & Mottelson. It is suggested that this picture of the mesic field may indicate where to look for solutions of the meson field equations.

1. Introduction

The evidence now available about the radii of nuclei is, *prima facie*, rather conflicting. Although it all supports the view that the density of nuclear matter is nearly constant, so that the radius R is proportional to $A^{\frac{1}{3}}$, different experimental facts suggest different constants.

Until recently a reasonable value for the radius was commonly supposed to be

$$R = 1{\cdot}5\times 10^{-13} A^{\frac{1}{3}}\,\text{cm}. \tag{1}$$

34-2

This value, or one close to it, agrees with the evidence from penetrability factors calculated for α-decay and for reactions among light nuclei, with that from the scattering of neutrons, and with that from the difference in binding energies of mirror nuclei.

On the other hand, recent evidence on the charge distribution within the nucleus suggests a rather smaller value, around

$$R' = 1{\cdot}2 \times 10^{-13} A^{\frac{1}{3}} \text{ cm}. \tag{2}$$

This is suggested by the evidence of electron scattering (Bitter & Feshbach 1953; Hofstadter, Fechter & McIntyre 1953; Schiff 1953 and others), of the levels of μ-mesic atoms (Cooper & Henley 1953; Fitch & Rainwater 1953), and of the isotope shift (Brix & Kopfermann 1951).

If we disregard the usual interpretation of the evidence from mirror nuclei, these results mean that the charge of the nucleus lies within the radius R' but that the specifically nucleonic interactions extend out to the radius R. Now the charge of the nucleus must be quite closely confined to the neighbourhood of the protons, otherwise it would be hard to understand the successful prediction of magnetic moments on the basis of the shell model (Flowers 1952), so that they also lie within the radius R'; and if we accept the charge independence of nucleonic interactions, so *a fortiori* must the neutrons.

If the nucleonic interactions of the nucleus are then described by short-range interactions centred on its constituent nucleons, it is difficult to see how this should lead to the effective radius (1), also proportional to $A^{\frac{1}{3}}$. Now these interactions are most likely to be due to the intermediary of the π-meson fields, so that if we regard the latter as constituents of the nucleus we require some model in which the mesic fields occupy a volume given by (1), while the nucleons occupy only a smaller interior region.

The nucleon-meson interactions are probably very strong, so that in the usual picture derived from perturbation theory the nucleon is surrounded by a cloud of many mesons. There is at present no satisfactory method of analyzing this situation theoretically, and we here propose a phenomenological description suggested by the problem of nuclear radii, which provides a qualitative model with several interesting features.

2. The mesic fluid

In strong-coupling theory of charged or spinning fields it is usual to separate the meson field into a scalar amplitude and a direction of spin or charge that is closely coupled to the nucleon. In a somewhat similar way we suppose that the meson cloud surrounding one nucleon, or many adjacent nucleons, can be described by an amplitude part which is pictured classically as a fluid and by superimposed fluctuations of spin or isotopic spin.

The interactions of a nucleon with this fluid fall similarly into two parts: in the first place the virtual emission or absorption of mesons will tend to alter the volume of the mesic fluid, and in the second place there will be a coupling between the spin, or isotopic spin, of the nucleon and the fluctuations in the fluid.

The fluid is considered to be a classical electrically neutral incompressible fluid. Then the interactions of the first kind, altering the volume of fluid, will be transmitted very quickly through the whole volume and the interaction will be one between the nucleon and the fluid as a whole. The processes of virtual emission or absorption will set up volume fluctuations in the fluid, and the self-interaction of these with the source will give an energy change that we may picture as a potential energy of the nucleon within the mesic drop. As a very crude analogue we can imagine the nucleonic source to be an oscillatory pump free to move within a volume of liquid.

Inside a large mass of fluid this mechanism may give a self-energy which is part of the observed nucleon mass and of no further interest, but inside a finite drop the effect of the fluid surfaces may change this energy and it is this in which we are interested. We consider for simplicity only a spherical drop of radius R and we want the energy of the nucleon source as a function of r, its distance from the centre. The only other distances that might be involved are small ones associated with the extension of the source, so it seems plausible, on dimensional grounds, that the potential energy should be a function of (r/R) only. As the nucleon source approaches the surface the latter will be distorted, and it is only in a conventional sense that we can speak of the energy at or outside the surface; this is not important, however, because we are supposing that the nucleons are bound well within the drop, so that they have only a small probability of being near the surface. If we suppose as a convenient convention that the potential vanishes at the surface, a reasonable form to assume for it is then

$$V(r) = \begin{cases} V_0(r^2/R^2-1) & (r<R), \\ 0 & (r>R). \end{cases} \tag{3}$$

The quadratic dependence on (r/R) is chosen mainly for convenience of calculation and is not otherwise essential.

When we use this potential we shall neglect any recoil of the fluid, not because the fluid is massive, but because it is reasonable to suppose that it depends more on the average position of the nucleon than on its instantaneous position, in contrast with the perturbation theory viewpoint.

The mechanism leading to the potential (3) should not give any interaction between nucleons in the same drop of fluid. The volume fluctuations produced by one nucleon give rise only to a uniform flow of fluid past another nucleon, which we must consistently suppose to have no effect if we are using the idea of a velocity-independent potential; also these fluctuations should be linearly superposable.

The interactions, between nucleons and mesons, of the second kind will give rise to self-energy effects and to internucleonic potentials; we shall consider these later (§§4 and 5).

3. Energy of a nuclear system

We ascribe to the hypothetical mesic fluid a volume energy density ρ and a surface energy density (surface tension) σ. An estimate of the magnitude of these constants and of V_0 can be found by considering the energy of a system of many (A) nucleons within a fluid drop of radius R. The energy of the fluid is, by hypothesis,

$$E_{\text{mes.}} = (4\pi/3)\,\rho R^3 + 4\pi\sigma R^2. \tag{4}$$

Each nucleon moves in the potential well (3), which we may consider infinite without much error because the nucleons stay well within the mesic drop. The energy levels above the bottom of the well are characterized by three quantum numbers l, m, n; if $l+m+n = N$ the energy is

$$E_N = (s/R)\,V_0(2N+3), \tag{5}$$

where

$$s = \hbar/(2MV_0)^{\frac{1}{2}}, \tag{6}$$

and M is strictly the mass of the 'bare' nucleon, though we here neglect the small difference from the observed nucleonic mass.

We consider the 'standard heavy nucleus', in which we neglect differences between neutrons and protons, so that the nucleons fill up the successive levels with four particles in each. For the sake of simplicity we treat the quantum numbers as continuous, and then easily find that A nucleons will fill the levels up to an energy

$$E_{\text{max.}} = (2s/R)\,V_0(3A/2)^{\frac{1}{3}}. \tag{7}$$

The average energy in this potential is $\frac{3}{4}E_{\text{max.}}$, so that this contribution of the nucleons to the total energy of the system is

$$E_{\text{nuc.}} = AV_0[(3s/2R)\,(3A/2)^{\frac{1}{3}}-1]. \tag{8}$$

There will also be contributions from the residual internucleonic potentials which we shall provisionally assume only to introduce small corrections.

The radius of the drop will presumably adjust itself so that the total energy $E_{\text{mes.}}+E_{\text{nuc.}}$ is a minimum; if we neglect the surface-energy term, which in fact gives only a trivial correction, we find at the minimum

$$R = r_0 A^{\frac{1}{3}}, \tag{9}$$

where

$$r_0^4 = (\tfrac{3}{2})^{\frac{4}{3}}\,V_0 s/4\pi\rho. \tag{10}$$

The system we have considered may reasonably be regarded as a picture of the ground state of a 'standard heavy nucleus'; in the lighter nuclei there will be an appreciable amount of dissociation, and the energy calculated above will represent more the potential energy of the system for close approach of the particles. We therefore identify formulae (9) and (1), giving

$$r_0 = 1{\cdot}5\times 10^{-13}\,\text{cm}. \tag{11}$$

With (10) this fixes one relation between V_0 and ρ. We obtain another by considering the volume occupied by the nucleons; approximately they will extend out to the radius R', such that the potential energy at R' is equal to the energy of the topmost particles in the well; this gives

$$V_0(R'/R)^2 = E_{\text{max.}} = V_0(2{\cdot}3s/r_0). \tag{12}$$

We equate R' with the experimental value (2) and obtain

$$s = 4{\cdot}2\times 10^{-14}\,\text{cm} \quad \text{and} \quad V_0 \approx 115\,\text{MeV}, \tag{13}$$

and therefore

$$(4\pi/3)\,\rho r_0^3 \approx 18\,\text{MeV}. \tag{14}$$

As the energy has been minimized with respect to R, the small surface-energy term can be identified with that found in the semi-empirical mass formula for nuclei, which gives

$$4\pi\sigma r_0^2 \approx 15\,\text{MeV}. \tag{15}$$

If the neutrons and protons are not equal in number, there will be a correction term to $E_{\text{nuc.}}$, and a simple calculation gives for this $\frac{2}{3}E_{\text{max.}}(Z-\frac{1}{2}A)^2/A$; the coefficient value of about 50 MeV is smaller than the 'observed' value of about 80 MeV; there will, however, be contributions to this term and also to the volume term in the mass formula from the internucleonic forces which we have neglected here.

A more serious difficulty comes from the Coulomb energy, which is too great if calculated in the usual way with the small radius R', disagreeing both with the binding-energy difference of light nuclei and with the Coulomb energy of heavy nuclei as deduced from the semi-empirical mass formula. The present theory cannot explain this, but at least the situation is no worse than with a more conventional picture; the explanation might have to be sought in the properties of the mesic fluid or its fluctuations in the presence of an electric field. A simple polarization effect has been considered, but as this merely leads to a redistribution of charge it cannot explain simultaneously all the facts.

4. Electrical interactions of nucleons

This problem of the Coulomb energy may be connected with the problem of the difference between the masses of proton and neutron. One might expect that the neutron had lower energy, because of attraction between the proton and negative meson cloud, as which it spends part of its time. On the present model, as we shall see below, the probability of this dissociation is rather small, and we can neglect the effect of this on the masses. We must then ascribe here the lower energy of the proton to a lower energy of the meson cloud that surrounds it, which we might loosely describe as a polarization effect; and the effect on this of the rest of the nucleus might contribute to the Coulomb energy discussed in the previous section.

The anomalous moments of the nucleons must be due to the dissociation effect when both charge and spin are transferred to the meson cloud. For a discussion of this we shall use the model constructed above, assuming that we can use it also for a single nucleon, which is at least qualitatively reasonable. As was postulated in §2 there is an interaction between the 'bare' nucleons and mesons in which the former transfer their spin or charge to the fluid. The spin of the nucleon can flip over, transferring one unit of angular momentum to the fluid, exciting therefore a rotational level of the mesic drop.

A significant feature of the present model is the low density of the mesic fluid, given by (14), and this implies that much energy is needed to excite rotation. A non-relativistic estimate of the rotational energy, using the moment of inertia, leads to an absurd result because the peripheral velocity needed is many times that of light. A rigid sphere rotating very fast would distort into a cylinder with a peripheral velocity nearly equal to c, and an angular velocity, $\omega \sim 3^{\frac{1}{2}}c/r_0$. If the system has angular momentum $\hbar$, the energy will be nearly equal to $\hbar\omega$, or about 230 MeV for our chosen value of r_0. This argument is very crude but indicates the order of magnitude of the rotational energy. It is plausible that the interaction energy between the states is rather smaller than this, so that the amplitude of the 'spin-flip' component of the lowest state will be rather small.

The model does not predict the strength of this interaction, but it can be estimated from the experimental values of the magnetic moments, whose sum is approximately 0·88 nuclear magneton. For this we also need to know about the probability of charge transfer from nucleon to mesons; it seems necessary, and quite reasonable, to assume that spin and charge transfer are linked together for the following reasons:

(i) The high rotational energy accounts for the non-existence of low-lying spin isobars. It is difficult to see why charge isobars should not occur, unless the two phenomena are linked together.

(ii) The small value of the electron-neutron interaction also requires that charge transfer should be inhibited.

(iii) Linked spin and charge will also lead to internucleonic forces mainly of Majorana type, and these are preferred as perturbations of nuclear shell structure.

The state in which one nucleon is inside a neutral spinless mesic drop may be described as a $^2S_{\frac{1}{2}}$ state in both spin and charge (the letters refer to the spin of the mesic component). The interaction will cause transitions to the $^2P_{\frac{1}{2}}.{}^2P_{\frac{1}{2}}$ state if spin and charge are linked. Within our assumptions the ground state is a mixture of these, whose levels differ by $E_{\text{rot.}} = 230\,\text{MeV}$. If the amplitude of the upper state is α, then the contribution of the proton's intrinsic moment (assumed to be unity) to the total moment of the proton is found to be $(1-10\alpha^2/9)$ and to that of the neutron $(-2\alpha^2/9)$; the mesic contributions will be equal and opposite (assuming as we have always done that the interactions are charge-independent), so that the sum of the moments should be $(1-4\alpha^2/3)$. The experimental values indicate then that $\alpha = 0{\cdot}30$, so that the probability of the upper state is 9 %.

This value of the amplitude would result from an interaction energy between the states of about $0{\cdot}36E_{\text{rot.}}$ or 83 MeV. The ground state would then be depressed below the ^2S state by $0{\cdot}32V_{\text{int.}}$ or 27 Mev.

From these numbers we can compute the mesic contributions to the moments. In only four-ninths of the upper state does the mesic cloud have both spin and charge, so the contribution to the moments are $(4\alpha^2/9)\,g$, where g is the gyromagnetic ratio for the mesic drop. Assuming a uniform distribution within the drop $g = Mc^2/(4\pi/3\rho r_0^3)$, because the nuclear magneton has been taken as unit, equal to about 52, and this gives a total moment for the neutron $\mu_n \approx -2{\cdot}1$. This is surprisingly close to the observed value $-1{\cdot}9$, considering the crudity of the arguments.

As far as the electron-neutron interaction is concerned we want the probability of charge dissociation; this is $\frac{2}{3}\alpha^2$ or 6 %. In the dissociated state we assume that the (negative) mesic charge is spread uniformly over a sphere of radius r_0, and that the protonic charge density is proportional to $e^{-r^2/r_0 s}$, the nucleon density in the potential (3). The potential of this distribution is

$$V(r) = e\begin{cases} -1/r + (2/r\pi^{\frac{1}{2}})\displaystyle\int_0^{r/(r_0 s)^{\frac{1}{2}}} e^{-x^2}\,dx & (r \geqslant r_0), \\[2ex] r^2/2r_0^3 - \frac{3}{2}r_0 + (2/r\pi^{\frac{1}{2}})\displaystyle\int_0^{r/(r_0 s)^{\frac{1}{2}}} e^{-x^2}\,dx & (r \leqslant r_0). \end{cases} \tag{16}$$

Therefore the volume integral of the *e-n* potential for this dissociated state is $-e^2r_0^2\pi(0{\cdot}4-s/r_0)$, which is equivalent to 13 keV spread over a square well of the classical electronic radius. Multiplied by 6 %, this then gives about 800 eV for the *e-n* potential arising from the charge distribution; this is sufficiently small to agree reasonably with experiment.

5. Nuclear forces

The model that we have developed has some qualitative implications for nuclear interactions. Interaction potentials will arise from three causes:

(i) When two systems approach there will be interference between the mesic clouds.

(ii) The interactions between a 'bare' nucleon and its cloud will be modified when this cloud meets, or is immersed in another.

(iii) The residual interactions between nucleons within the fluid.

The first will probably give rise to 'ordinary' forces between nuclear systems; as the energies associated with the fluid are small, these effects may be rather weak, and show no particularly interesting features.

We have seen in the previous section that spin and charge exchange should be linked. We might express this formally by writing the residual nucleon-meson interaction in the form

$$V_{\text{int.}} = \text{const.}\,\sigma_i\,\tau_j\,Q_{ij}, \tag{17}$$

where Q_{ij} is a suitable operator describing the emission or absorption of spin fluctuations in the fluid. As in the discussion of single nucleons we suppose that perturbation theory is adequate for the discussion of these fluctuations; then the transfer of one fluctuation from nucleon to nucleon within a drop should give rise to internucleon potentials of the exchange character

$$(\boldsymbol{\sigma}_1\,.\,\boldsymbol{\sigma}_2)\,(\boldsymbol{\tau}_1\,.\,\boldsymbol{\tau}_2). \tag{18}$$

As to its strength we can compare the internucleon energy with the self-energy deduced in §4. For two protons with parallel spins, the operator (18) has the value 1; in such a state only one-ninth of the self-energy processes correspond to possible exchange processes between the protons; on the other hand, either nucleon may emit the fluctuation first, multiplying the interaction by 2. These arguments suggest a constant $(\frac{2}{9})\,27 = 6$ MeV multiplying the operator (17).

This, however, is not very significant without the specification of range, which is difficult to discuss without a more mathematically formulated theory; to be consistent, however, the range cannot be greater than the order of r_0. Then this potential would only give one-third or less of the required *n-p* interaction in the deuteron, and the rest would come from the other two kinds of potential.

In calculating the interaction (18) we have ignored any influence of the extent of the fluid on the interactions, and we should expect the fluid boundary to have some effect on the fluctuations within. These effects may appear as tensor and spin-orbit forces between particles.

Suppose two nucleons approach so that the merged drops will have an ellipsoidal shape with the major axis in the direction of their approach. In this situation if the spins are alined in the direction of the axis angular momentum may be exchanged with the fluid as a whole, leading to an energy gain, but if the spins are at right angles to the axis rotation of the whole fluid will be prevented by the bare nucleons. Such could be the mechanism for a tensor force.

Again, when a nucleon crosses the surface of another nucleon or nuclear system, the tendency to induce a local rotation of the mesic fluid about the spin axis $\boldsymbol{\sigma}$ will be accompanied by a tendency to translation in the direction $\boldsymbol{\sigma} \times \mathbf{n}$, where $\mathbf{n}$ is the outward normal to the surface, i.e. the state in which the nucleon is moving in this direction will be of lower energy than that in which it moves in the opposite direction. Applied to a sphere this is equivalent to saying that there should be an attractive spin-orbit coupling on the surface.

6. Nuclear structure

The model of §3 is, by construction, a shell model in a parabolic well with the small radius (2). There is the advantage that the mean potential has an 'external' source, and is as effective for the outermost particles as for those in the core.

It was suggested above that there should be an $\mathbf{l}.\mathbf{s}$ interaction on the surface; this would lead to an $\mathbf{l}.\mathbf{s}$ coupling for a particle bound inside the nucleus, rather smaller in magnitude because the interaction is effective only when the particle comes up to the surface. The strength of interaction needed depends rather critically upon its depth of penetration, etc., but would have to be some 20 MeV over a distance of the order of r_0, to give the observed splittings of a few MeV.

The remaining interactions between the nucleons would be principally of the type (iii) above, and have the exchange character (18); according to our previous estimate these are quite weak and can justifiably be treated as perturbations of the mean potential (3).

With the mesic fluid as an extra constituent of the nucleus, there will be extra degrees of freedom corresponding to oscillations of the fluid. It will not, however, be possible to consider these as independent motions because a deformation of the fluid drop must lead to a change in the potential (3) and to coupling with the nucleons; the present theory is not complete enough to give this change in potential, but a reasonable guess suggests that when the drop is deformed into an ellipsoid with axes R_i, the potential should be of the form

$$V = V_0(\Sigma r_i^2/R_i^2 - 1), \tag{19}$$

which for small deformations gives a $Y^{(2)}$ coupling between nucleons and surface oscillations.

This coupling is strong and will therefore lead directly to the model proposed by Bohr & Mottelson (1953), from a different limit. Whereas in the theory of these authors the energy of the surface oscillations is increased by the deformability of the nucleus, in the present model the small inertia of the mesic fluid is increased by the drag of the nucleons coupled to it. In many problems it will not then be necessary to consider explicitly the motion of the fluid; this will be sufficiently well represented by the similar oscillatory motions of the nucleons coupled to it.

7. Nuclear reactions

The classification of forces introduced in §5 suggests that at low energies, in S states, the interaction between nucleons will consist principally of an ordinary force together with a tensor force; it is well known that this provides a fair description of the low-energy data.

At higher energies the description must become much more complicated. One feature may be noted in particular: the spin-orbit coupling on the surface of the drop will of course lead to an **l.s** interaction between two nucleons; it will also lead to spin-orbit coupling for single nucleons on the surface of nuclei. This collective effect may account for the large polarization produced in the near-elastic scattering of nucleons by nuclei (Rochester Conference 1954).

Also at very high energies the interaction will take place so rapidly that there will be little time for the mesic fluid to make itself felt, and the only important interactions will be short range ones between the 'bare' nucleons. The apparent nuclear radius should then also have the small value R', which appears to be confirmed by experiments with neutrons at 1000 MeV (Rochester Conference 1954).

It might also seem that the present theory should make some definite predictions about pion-nucleon interactions. There is however the difficulty of describing the relation between the observed free mesons and our hypothetical mesic fluid with its fluctuations. Presumably the pion should be pictured as a drop of fluid, containing some excitation or fluctuation that gives it its pseudoscalar character and isotopic spin unity, and also most of its mass. When such a particle collides with a nucleus this fluctuation will then be free to move in the combined fluid drops, and can interact with the nucleons in much the same way as would a meson with nucleons in the conventional picture, with weak coupling corresponding to the difficulty with which nucleons can emit or absorb fluctuations.

In particular with pion-proton scattering we might then expect to get resonance effects associated with the formation of excited nucleon states. The scattering data can in fact be roughly interpreted in terms of a resonance around 137 MeV (centre-of-mass system) (Brueckner 1952), and this value fits in with the idea of an excited rotational level about 250 MeV above the ground state as was considered in §4. There is a difficulty however because this level should give rise to states with $T = \frac{1}{2}$ as well as $T = \frac{3}{2}$, and there is no obvious reason why the former should be excluded, as experiment seems to indicate.

8. Conclusion

Many of our conclusions have only been qualitative, because the ideas introduced have not yet been fitted into a sufficiently precise or complete methematical framework. Part of this framework can possibly be constructed phenomenologically; for example, by a more detailed examination of the fluctuations in the mesic fluid and the way they interact with nucleons; if this can be done closer estimates of nuclear interactions could be made for comparison with experiment. Further clues may be provided by the study of collective phenomena in nuclei, such as the elastic scattering of nucleons and the structure of near-magic nuclei where the present model would add distinctive features to the general theory of Bohr & Mottelson.

Further, the author believes that this model, apart from giving an approximate phenomenological description of nuclear matter, may provide a clue as to the direction in which to look for solutions of the meson field equations in the strong-coupling case. The description of forces between two nucleons on this model shows a strong resemblance to that deduced from pseudoscalar meson field theory (as far as that can be done); according to the latter we expect forces of the type of (18), and similar tensor forces, and in addition scalar forces, which are singular at the origin, but when cut off at some small radius provide together a fair description of low-energy phenomena. According to the proposed model we expect similar forces acting when the particles are not too close so that an approximate meaning can be given to their separation, together with the formation of a complex compound state at close approach.

We have considered a number of facts about nuclei and shown how these may be fitted into our framework. There are some difficulties, such as the problem of reconciling nuclear Coulomb energy with the small charge radius. Our hypotheses are too crude for us to expect to explain phenomena accurately and we wish primarily to introduce a slightly new background against which other and new experimental data can be discussed.

The author is grateful to several of his colleagues for valuable criticism of these ideas throughout their formation.

References

Bitter, F. & Feshbach, H. 1953 *Phys. Rev.* **92**, 837.
Bohr, A. & Mottelson, B. R. 1953 *Dansk. Mat. Fys. Medd.* **27**, no. 16.
Brix, P. & Kopfermann, H. 1951 *Fortschr. Akad. Wiss. Gottingen*, pp. 17–49.
Brueckner, K. A. 1952 *Phys. Rev.* **86**, 106.
Cooper, L. N. & Henley, E. M. 1953 *Phys. Rev.* **92**, 801.
Fitch, V. L. & Rainwater, J. 1953 *Phys. Rev.* **92**, 789.
Flowers, B. H. 1952 *Phil. Mag.* **43**, 1330.
Hofstadter, R., Fechter, H. R. & McIntyre, J. A. 1953 *Phys. Rev.* **92**, 978.
Schiff, L. I. 1953 *Phys. Rev.* **92**, 988.

Meson theory and nuclear matter

By T. H. R. Skyrme
Atomic Energy Research Establishment, Harwell, Berks

(*Communicated by Sir John Cockcroft, F.R.S.—Received* 9 *February* 1955)

An attempt is made to justify the use of the concept of a 'mesic fluid' in connexion with the structure of nuclear matter. A transformation is made of the usual symmetric pseudo-scalar meson theory to bring into evidence certain saturation properties, which provide a natural basis for the use of a 'self-consistent' field in the discussion of nuclear structure. Fluctuations about this semi-classical saturated state will give rise to residual interparticle forces within the nucleus, and are also briefly considered in relation to electromagnetic interactions.

1. Introduction

In an earlier paper (Skyrme 1954, referred to below as I) the author has proposed a phenomenological model of nuclear matter in which saturating meson fields are supposed to form a fluid in which the nucleons are free to move, with residual interactions that result from spin-charge fluctuations in the fluid. In the present paper it will be shown how such a model might be conceived in more mathematical terms, and how it may be related to the usual type of meson theory.

Meson theory is used in a form in which the usual renormalization techniques are inapplicable, and one is therefore forced to employ a cut-off procedure; only non-relativistic field theory is used. For these reasons much importance cannot be attached to the numerical values used in the argument, whose main purpose is to show that the phenomenological picture may reasonably represent the behaviour of the coupled particle and meson fields.

In paper I it was suggested that this model of nuclear matter might equally be applicable to the structure of an individual nucleon; the author believes that this latter problem is, at least from the present point of view, more complicated than that of a nucleon in nuclear matter and it is not analyzed here. The experimental fact that the properties of a nucleon in nuclear matter do not appear to be very different from those of free nucleons is shown not necessarily to be inconsistent with the author's picture, but one cannot at present analyze further the condensation process by which the individual particle model passes into the collective one.

2. The physical model

The bare nucleons are coupled to a symmetrical pseudo-scalar meson field, whose potentials may be resolved into an amplitude and a direction (in isotopic spin space). The self-mass of the nucleons coming from mesonic fluctuations (and this is assumed to be the origin of all the mass) may be analyzed first into fluctuations of direction and then those of amplitude; the former give a self-mass that increases as the amplitude decreases, because, roughly speaking, the energy is inversely proportional to the 'moment of inertia'. After this stage of 'partial renormalization' the nucleon's mass is a function of the meson field amplitude and has a sharp minimum for a certain field strength. The sharpness of the minimum means that fluctuations of amplitude are strongly inhibited, and we can treat this amplitude approximately as a classical field variable.

In nuclear matter then the meson field amplitude approximates to a constant saturation value in the region occupied by nucleons, and outside of it the tendency of the field to fall off produces an increase of the nucleon mass above its minimum value, roughly equivalent to a potential confining the nucleons, which move with an effective mass equal to the minimum. The total energy of the system contains, in addition to small contributions from the classical energy of the meson field and the kinetic energy of the nucleons, another important fluctuation term, arising from the inhibition of amplitude fluctuations within the nuclear volume. For an individual free nucleon this energy will appear as part of its inertia, but when the nucleon enters a nuclear system this part will be 'pushed outside' and appear as energy shared by the whole system.

The directional fluctuations that produce the self-mass similarly give rise to interparticle potentials within nuclear matter, which are predominantly of an exchange character that saturate, and can plausibly be treated as perturbations to the central field produced by the primary interaction with the field amplitude. Other perturbations will arise from deformations and oscillations of the coupled nucleon-meson fluid system.

3. Mathematical formulation

We start from the usual expression‡ for the Lagrangian density of nucleon fields interacting with a symmetrical pseudo-scalar meson field, with the special assumption that all the nucleon's mass is of mesic origin so that there is no 'inertial' mass term in the primitive expression;§ this is

$$\mathscr{L}_0 = -\tfrac{1}{2}[(\mathrm{d}\underline{\phi}/\mathrm{d}x_\mu)^2 + \kappa^2\underline{\phi}^2] - \psi^\dagger[\gamma_\mu \mathrm{d}/\mathrm{d}x_\mu + \mathrm{i}g\gamma_5\underline{\tau}\,.\,\underline{\phi}(x)]\,\psi, \tag{1}$$

in which we use the convention that underlined quantities as $\underline{\phi}$ are vectors in isotopic spin space. At each point of space and time we write the meson field vector in polar co-ordinates as

$$\underline{\phi} = \underline{u}\phi, \quad \text{where} \quad \phi = +|\,\underline{\phi}\,|, \tag{2}$$

so that $\underline{u}$ is a unit pseudo-vector and ϕ a scalar, and then make the unitary transformation,||

$$\psi \to \exp(-\mathrm{i}\pi/4\,.\,\gamma_5\underline{\tau}\,.\,\underline{u})\,\psi = 2^{-\frac{1}{2}}(1-\mathrm{i}\gamma_5\underline{\tau}\,.\,\underline{u})\,\psi, \tag{3}$$

which may be regarded as the limiting form of a 'Foldy' transformation (Foldy 1951) appropriate to zero inertial mass, or as a strong coupling transformation to diagonalize the interaction. The transformed Lagrangian is

$$\mathscr{L} = -\tfrac{1}{2}[(\mathrm{d}\underline{\phi}/\mathrm{d}x_\mu)^2 + \kappa^2\underline{\phi}^2] - \psi^\dagger[\gamma_\mu \mathrm{d}/\mathrm{d}x_\mu + g\phi + \tfrac{1}{2}(\underline{\tau}\,.\,\underline{u} + \mathrm{i}\gamma_5)\,(\gamma_\mu \mathrm{d}/\mathrm{d}x_\mu \underline{\tau}\,.\,\underline{u})]\,\psi, \tag{4}$$

wherein we shall suppose that the new ψ is more directly comparable with real nucleons.

‡ We use throughout units in which $\hbar = c = 1$.

§ The possibility of renormalizing the relativistic theory is not considered.

|| We assume that g is positive, otherwise the sign of γ_5 is to be changed in (3).

From this Lagrangian the Hamiltonian may be formed in the usual way; in a non-relativistic particle approximation the interaction terms would be

$$\mathscr{H}_{\text{int.}} = \psi^*[\beta g\phi + \tfrac{1}{2}\boldsymbol{\sigma}.\nabla\underline{\tau}.\underline{u} + \tfrac{1}{2}\phi^{-1}(\underline{\tau}\times\underline{u}.\underline{\Pi})]\psi + (1/8\phi^2)[\psi^*(\underline{\tau}\times\underline{u})\psi]^2. \quad (5)$$

The objective now is to eliminate the $\underline{u}$ in the interaction terms and reduce the system to one in which the interaction term is a function only of the scalar amplitude ϕ. In a relativistic formulation, this process may be defined without ambiguity by using, for example, the methods of functional integrals (Matthews & Salam 1954).

In the following analysis we shall, however, use only an exploratory perturbation technique. To use this we need a first approximation from which to start. It will be shown below (§5) that it is consistent from the perturbation point of view to assume that in the first approximation the interaction Hamiltonian is

$$\mathscr{H}_{\text{int.}} = \psi^*\beta W(\phi)\psi, \quad (6)$$

where the function‡

$$W(\phi) = g\phi - \tfrac{1}{2}h/\phi^2 + \tfrac{1}{4}f/\phi^4. \quad (7)$$

The coupling terms

$$\mathscr{L}_2 = -\tfrac{1}{2}\psi^\dagger[(\underline{\tau}.\underline{u} + \mathrm{i}\gamma_5)(\gamma_\mu \mathrm{d}/\mathrm{d}x_\mu \underline{\tau}.\underline{u})]\psi \quad (8)$$

will then be treated as a perturbation whose reactive effects are cancelled (approximately) by the extra terms introduced into (7).

4. The first approximation

The first problem then is to discuss the nature of the system with interaction Hamiltonian given by (6) and (7). For the reasons mentioned in §2, we shall first consider this problem in a semi-classical way. We notice that $W(\phi)$ has a minimum value W_0 for a value ϕ_0 of ϕ, where

$$g + h/\phi_0^3 = f/\phi_0^5 \quad (9)$$

and

$$W_0 = \phi_0(g - \tfrac{1}{2}h/\phi_0^3 + \tfrac{1}{4}f/\phi_0^5), \quad (10)$$

and for values of ϕ nearly equal to ϕ_0 we can write

$$W = W_0 + \tfrac{1}{2}W''(\phi - \phi_0)^2, \quad (11)$$

where

$$W'' = 5f/\phi_0^6 - 3h/\phi_0^4 = K/\phi_0. \quad (12)$$

The constant K will turn out to be a large number, of the order of 100, which makes the minimum of W a very sharp one.

Consider now a large nucleus; the mean static classical field will then be determined by the equation§

$$(\kappa^2 - \nabla^2)\underline{\phi} = (\underline{\phi}/\phi)(K/\phi_0)(\phi_0 - \phi)\rho, \quad (13)$$

where ρ is the mean nuclear density. The solutions, at least those of lowest energy, are of the form $\underline{\phi} = \underline{e}\phi$, where $\underline{e}$ is a constant vector. The density is that of the nucleons contained, which move according to the wave equation

$$[\gamma_\mu \mathrm{d}/\mathrm{d}x_\mu + W(\phi(x))]\psi = 0. \quad (14)$$

‡ It will be shown that the coefficients h and f should be positive.

§ In this equation $\mathrm{d}W/\mathrm{d}\phi$ has been replaced by $(K/\phi_0)(\phi - \phi_0)$, since the nuclear density will be small where this approximation fails.

The coupled system described by (13) and (14) provides a mathematical basis for the hypothesis used in I of a fluid drop in which the particles move under a central nucleon-meson interaction. Since the effective mass W increases rapidly as ϕ decreases from ϕ_0, the state of lowest energy of this system will be one in which $\phi \sim \phi_0$ over a certain volume, with an exponential tail, and the nucleons are confined to this volume. If we neglect the surface-energy terms arising from the tail and assume that $\phi = \phi_0$ over a sphere of radius R, then the energy of the meson field is

$$E_0 = (\tfrac{1}{2}\kappa^2\phi_0^2)(4\pi R^3/3), \tag{15}$$

and the kinetic energy of the nucleons is approximately (at 4 to a state)

$$E_k = (0\cdot 69)\, A^{\frac{5}{3}}/W_0 R^2, \tag{16}$$

and the total energy is a minimum when

$$R^5 = (0\cdot 22)\,(\kappa^2\phi_0^2 W_0)^{-1}. \tag{17}$$

For the sake of numerical illustration we shall employ hereafter the numerical values‡

$$W_0 = 4\cdot 0\kappa \quad \text{and} \quad \phi_0 = 0\cdot 26\kappa, \tag{18}$$

and then we obtain from (17)

$$RA^{-\frac{1}{3}} = (0\cdot 96_5)/\kappa = 1\cdot 3_5 \times 10^{-13}\,\text{cm}, \tag{19}$$

a reasonable value since the nucleons *confined within* this radius will have an effectively smaller radius, which may further be decreased by the effect of the residual internucleonic forces. Also with these values

$$E_0 = (\tfrac{2}{3})\, E_k = (0\cdot 13\kappa)\, A = 18\,\text{MeV per nucleon}. \tag{20}$$

The remainder of the energy of the system (i.e. about $2\cdot 3\kappa$ per nucleon) must come from the vacuum polarization energy, and to estimate this we must consider the quantized meson fields.

For the quantum-theoretic discussion of the meson fields we shall write the meson field as the sum of a classical part and a fluctuation,

$$\underline{\phi} = \underline{e}\phi_{\text{cl.}} + \underline{\chi}, \tag{21}$$

where $\phi_{\text{cl.}}$ is the meson field discussed above, and $\underline{e}$ is now an auxiliary 'strong-coupling' parameter. The ground-state wave-function will be independent of $\underline{e}$, and we shall ignore excitation energy due to rotation of $\underline{e}$ in intermediate states§; this means that we calculate with $\underline{e}$ as a fixed parameter over which we perform a final averaging. Outside the nucleus $\underline{\chi}$ will satisfy the free meson field equations, but inside the Hamiltonian will contain the additional term

$$\tfrac{1}{2}\rho W''(\delta\phi)^2. \tag{22}$$

‡ The implications of (18) in terms of coupling constant and cut-off will appear later.

§ This is certainly *not* justifiable for the individual nucleon, where the volume of mesic fluid is small.

Now for small fluctuations $$\delta\phi \simeq \underline{e}.\underline{\chi}, \tag{23}$$

so that fluctuations of the field in the direction of $\underline{e}$ have an increased effective mass $(K\delta/\phi_0)^{\frac{1}{2}}$, of the order of 10κ, and are therefore strongly damped; on the other hand, fluctuations in other directions, which we shall refer to as transverse, obey everywhere the free-particle equations.

Only the longitudinal fluctuations contribute to the vacuum polarization effect, and for these fluctuations the nucleus may approximately be regarded as a hard sphere into which they cannot penetrate, as is consistent with the original hypothesis that the field inside could be considered classically. The extra energy so arising is estimated in the appendix, with the approximate result

$$E_{\text{fl.}} \simeq (R^3/2\pi)\int k^4\,\mathrm{d}k/\omega. \tag{24}$$

With the radius given by (19) the energy per nucleon is then

$$(0{\cdot}9/2\pi\kappa^3)\int k^4\,\mathrm{d}k/\omega, \tag{25}$$

and if this is to account for the remaining energy of the system, we must cut off the integration at a momentum K, where approximately

$$K = 2{\cdot}8\kappa, \tag{26}$$

corresponding to a distance of $0{\cdot}5\times 10^{-13}$ cm. This low value of the cut-off is not unreasonable considering that the physical cut-off due to relativistic effects should be comparable with W, and it is known that non-relativistic theories tend to need rather low cut-offs in their interpretation.

5. The reaction terms

We are now in a position to estimate the reaction terms by a perturbation treatment and compare them with the hypothesis (7). When the meson field is expressed in the form (21) we can expand $\underline{u}$ in powers of χ, as

$$u_\alpha = e_\alpha + (1/\phi)\,(\delta_{\alpha\beta} - e_\alpha e_\beta)\,\chi^\beta \ldots, \tag{27}$$

and substitute this in the interaction Hamiltonian (5). In the pseudo-vector coupling term we then get the interaction operator

$$(1/2\phi)\,(\boldsymbol{\sigma}.\boldsymbol{\nabla})\,(\underline{\tau}.\underline{\chi} - \underline{\tau}.\underline{e}\,\underline{e}.\underline{\chi}) \tag{28}$$

in lowest order, to which only the *transverse* fluctuations contribute. The self-energy due to this interaction calculated in a simple non-relativistic way may then be written in the form $(-\frac{1}{2}h/\phi^2)$, where we find

$$h = (1/4\phi^2)\int k^4\,\mathrm{d}k/\omega^2. \tag{29}$$

The next term in the interaction Hamiltonian also gives a self-energy proportional to $(1/\phi^2)$, but this is cancelled by the lowest contribution of the last term (the quadratic point interaction). Both these terms give contributions proportional to $(1/\phi^4)$, of which, with a cut-off, the contribution of the last term is larger; the sum of these two terms may be written in the form $(\frac{1}{4}f/\phi^4)$ with

$$f \simeq (1/4\phi^3)\int\int k^2\,\mathrm{d}k k'^2\,\mathrm{d}k'/(\omega+\omega'), \tag{30}$$

and is positive, as is essential for the existence of the saturation effects.

We identify these calculations of self-*energy* with the *mass* terms in $W(\phi)$, because it is evident that in a proper covariant treatment of the problem, the elimination of the $\underline{u}$ can lead only to such terms.

With the cut-off given by (26), and the value $\phi_0 = 0{\cdot}26$, we then obtain

$$h/2\pi\phi_0^3 = 1{\cdot}28, \tag{31}$$

while from the equations (9) and (10) the elimination of f gives

$$g/2\pi = (0{\cdot}8)\,(W_0/2\pi\phi_0) + (0{\cdot}2)\,(h/2\pi\phi_0^3), \tag{32}$$

so that the corresponding value of g is roughly

$$g = (2\pi)\,2{\cdot}22 = 14. \tag{33}$$

For consistency we then also require that the value of f should satisfy

$$f/2\pi\phi_0^5 = 3{\cdot}5, \tag{34}$$

which differs from the expression (30) only by some 10 %, which is not really significant in view of the crude procedures that we are using.

6. Nuclear forces

In our first approximation the nucleons inside nuclear matter move under the influence of the self-consistent 'potential' described by $W(\phi(x))$, as in equation (14). In addition to a central potential of approximately square-well shape, this also gives rise to a spin-orbit force after reduction to a two-component wave equation. Since the 'potential' appears as mass rather than energy, the sign of this will be *opposite* to that of the usual Thomas term; furthermore, the smaller effective mass W of the nucleon inside the nucleus means that the potential to confine the nucleons to the same volume should be bigger, roughly speaking, by a factor (M/W) and the spin-orbit coupling will then be bigger by a factor of $(M/W)^3$ or about 5.

The fluctuations of $\underline{u}$, represented in our perturbation treatment by the 'transverse' meson fluctuations, give rise to internucleon potentials analogous to the reaction terms for single particles. The pseudo-vector coupling term thus produces the potential

$$-(1/6\phi_0^2)\,(\underline{\tau}^1.\underline{\tau}^2)\,(\boldsymbol{\sigma}^1.\boldsymbol{\nabla}_1)\,(\boldsymbol{\sigma}^2.\boldsymbol{\nabla}_2)\,(\mathrm{e}^{-\kappa r}/4\pi r), \tag{35}$$

which may as usual be split up into a tensor force, and the central potential

$$V_1 = (\kappa^3/24\pi\phi_0^2)\,(\tau^1.\tau^2)\,(\boldsymbol{\sigma}^1.\boldsymbol{\sigma}^2/3)\,(\mathrm{e}^{-\kappa r}/\kappa r). \tag{36}$$

With the value $\phi_0 = 0{\cdot}26\kappa$, the numerical value of the coefficient would be about 27 MeV.

In a similar way the other interaction terms give rise to a second-order potential of $(\underline{\tau}^1 . \underline{\tau}^2)$ character; this is a 'meson-pair' potential which may be evaluated by the methods used by Levy (1952) with the result

$$V_2 = (1/6\phi_0^4)(\underline{\tau}^1 . \underline{\tau}^2)(\kappa^2/16\pi^2 r^3) K_2(2\kappa r). \tag{37}$$

This is a highly singular potential whose significance cannot really be estimated without a better understanding of the cut-off to be used.

Between individual free nucleons the potential (35) would be reduced by a factor $\frac{2}{3}$ because the $\underline{e}$-vectors of the two systems are independent; this potential would then correspond to the use of an effective (psuedo-scalar) coupling constant $g_{\text{eff.}}$ such that $(g_{\text{eff.}}/2M) = (1/3\phi_0)$, or

$$g_{\text{eff.}} \simeq 17. \tag{38}$$

In addition, between free nucleons there will be further Wigner forces arising from interference between the field amplitude distributions associated with the two particles and between the distorted vacuum fluctuations.

7. Electrical interactions

The terms that must be added to the Lagrangian (4) to allow for the presence of an external electromagnetic field A_μ can be written down by the usual rules. The linear interaction terms are

$$eA_\mu(j_\mu^0 + j_\mu^1), \tag{39}$$

where j_μ^0 is the usual current expression,

$$j_\mu^0 = (\phi_1 \, \mathrm{d}/\mathrm{d}x_\mu \, \phi_2 - \phi_2 \, \mathrm{d}/\mathrm{d}x_\mu \, \phi_1) + \psi^\dagger(\tfrac{1}{2}\mathrm{i}\gamma_\mu(1-\tau_3))\psi, \tag{40}$$

and j_μ^1 includes the extra terms that arise from the derivative interactions,

$$j_\mu^1 = \psi^\dagger\{\tfrac{1}{2}\mathrm{i}\gamma_\mu[\tau_3 - u_3(\underline{\tau} . \underline{u}) + \gamma_5(\tau_1 u_2 - \tau_2 u_1)]\}\psi. \tag{41}$$

In the Hamiltonian the charge operator takes on the form

$$Q = \psi^* \tfrac{1}{2}(1-\tau_3)\psi + \int (\Pi_1 \phi_2 - \Pi_2 \phi_1) \, \mathrm{d}^3x, \tag{42}$$

which is related to the isotopic spin in the usual way. When we make the transformation (21) we must add to Q another term to allow for the rotations of e; we introduce a vector $\underline{f}$, formally conjugate to $\underline{e}$, and the required term is

$$Q_e = (f_1 e_2 - f_2 e_1). \tag{43}$$

In the following calculations we shall consider only the effect of the pseudo-vector coupling in the perturbation Hamiltonian, because the amplitude of the state induced by the isotopic spin coupling is small. Then if ψ_0 is the wave function of a nucleon in the central potential, the normalized wave-function that includes this perturbation is

$$\psi = (1-\alpha^2)^{\frac{1}{2}}[\psi_0 - (1/2\phi_0)(1/\omega)\,\boldsymbol{\sigma} . \nabla(\underline{\tau} . \underline{\chi} - \underline{\tau} . \underline{e}\,\underline{e} . \underline{\chi})\psi_0], \tag{44}$$

where ω is the energy of the fluctuation meson, and α is the amplitude of the second component, with

$$\alpha^2/(1-\alpha^2) = (1/4\phi_0^2)(1/2\pi^2)\int k^4 \, \mathrm{d}k/\omega^3. \tag{45}$$

Using the same numerical values as before, we obtain $\alpha^2 = 0{\cdot}32$.

With this wave-function the expectation values of the various parts of the charge operator are

$$\left.\begin{aligned} \langle \tfrac{1}{2}(1-\tau_3)\rangle &= \tfrac{1}{2}-\tfrac{1}{2}(1-4\alpha^2/3)\,\tau_3, \\ \langle Q_{\text{fl.}}\rangle &= -(\alpha^2/3)\,\tau_3, \\ \langle Q_e\rangle &= -(\alpha^2/3)\,\tau_3, \end{aligned}\right\} \tag{46}$$

so that 80 % of the proton's charge remains on the nucleon, and the remainder is distributed equally between the local fluctuation field and the body of the mesic fluid.

In considering the current we shall, consistently with the use of the wave-function (44), neglect the first part of the current $\mathbf{j}^1$; the second part of that current, together with the current of the meson field, will give rise to an anomalous magnetic moment equal and opposite for proton and neutron. The sum of their moments and orbital contributions to the magnetic moment arise from the particle current operator $\frac{1}{2}\boldsymbol{\alpha}(1-\tau_3)$.

We find in a similar way for the expectation value of this operator

$$\langle \tfrac{1}{2}\boldsymbol{\alpha}(1-\tau_3)\rangle = \{\tfrac{1}{2}(1-4\alpha^2/3)-\tfrac{1}{2}(1-8\alpha^2/9)\,\tau_3\}\,\langle\boldsymbol{\alpha}\rangle_0 \tag{47}$$

with approximate numerical values of the coefficient

$$\left.\begin{aligned} 0{\cdot}65 &\text{ for proton,} \\ -0{\cdot}07 &\text{ for neutron.} \end{aligned}\right\} \tag{48}$$

The corresponding contributions to magnetic moment will be measured in units of $eh/2Wc$, and therefore in nuclear magnetons the values (48) have to be multiplied by (M/W), restoring them to values not very different from those for free nucleons (and making the sum of the moments nearly equal to unity).

The contribution of the other terms to the intrinsic anomalous moment is not large enough to agree with experiment. The discrepancy is not quite so great as that pointed out by Sachs (1952) because the reduction of effective mass 'allows' a greater amount of the state with one meson; the contribution of this state to the anomalous moment is $(\frac{4}{9})\langle M/\omega\rangle\alpha^2$, which is only of the order of $0{\cdot}6$. Either relativistic effects increase this value, or terms may be important which have here been neglected. It does not appear that the suggestion made in I that the low density of the fluid should give a high gyro-magnetic ratio can be substantiated by the present analysis.

8. Conclusions

Apart from questions of covariant treatment, renormalization and cut-off procedures which should not greatly affect the physical picture, it is felt that the principal uncertainties in the present lie in the problem of the 'first approximation' solution suggested in § 4; is the meson field amplitude really determined by the mean nucleon density, or is it more closely linked to the instantaneous positions of the sources? Is it reasonable to consider first the quasi-classical state of lowest energy and then add to it the rather large fluctuation energy term represented by (24)?

Since it is hard to answer such questions with certainty, the point of view is taken that the suggested type of solution is sufficiently attractive as an approximate description of nuclear matter to merit more detailed examination of its consequences. As far as we have here considered them, and within the limits of the crude numerical methods, they do not show any flagrant discrepancies from experimental experience. Some of the conclusions are not very different from those derivable from the individual particle point of view, suggesting that the 'saturation effects' set in before higher order effects (from the perturbation starting point) become very effective.

The author is indebted to many people for criticisms, expecially Professor R. E. Peierls, F.R.S., and Professor V. F. Weisskopf.

Appendix

For a free meson field there is a zero-point vacuum fluctuation energy of $\frac{1}{2}\hbar\omega$ for each mode of oscillation with frequency ω, so that the density of this energy is $\int (\frac{1}{2}\omega)\,\mathrm{d}^3k/(2\pi)^3$. If the vacuum state is slightly disturbed, we can calculate the change in energy by the expectation value of the perturbing interaction; for example, if the Hamiltonian contains the extra term

$$\tfrac{1}{2}V(x)\,\phi^2(x), \tag{A1}$$

then since the expectation value of $\phi^2(x)$ at any point is

$$\langle\phi^2(x)\rangle = \frac{1}{2}\int (1/\omega)\,\mathrm{d}^3k/(2\pi)^3, \tag{A2}$$

the increase of energy due to (A1) considered as a perturbation is

$$\delta E = \int V(x)\,\mathrm{d}^3x \int (1/4\omega)\,\mathrm{d}^3k/(2\pi)^3. \tag{A3}$$

However, if V is large (i.e. compared with κ^2), as in the limit of hard-sphere scattering, this formula is meaningless.

In this case we can tackle the problem in the following way: place the whole system in a very large box, and calculate the change in the energy levels due to the term (A1), which will be inversely proportional to the dimensions of the box. Then the change in energy of the whole system, integrated throughout the box, will be the finite energy change required in the limit when the size of the box becomes infinite. Suppose that $V(x)$ is a spherically symmetric potential, then these energy changes can easily be expressed in terms of the scattering phase shifts δ_l, and our desired result can be expressed in the form

$$\delta E = \int (1/4\omega)\,F(k)\,\mathrm{d}^3k/(2\pi)^3, \tag{A4}$$

where

$$F(k) = -(4\pi/k)\sum_{l=0}^{\infty}(2l+1)\,\delta_l(k). \tag{A5}$$

If Born approximation is used for the scattering, then (A 5) can be evaluated and we get back to the result (A 3). For hard-sphere scattering we estimate that on the average the phase shifts δ_l will be equal to $(-kR)$ for l-values less than kR, and when this is substituted into (A 4 and A 5) we get the formula (24).

References

Foldy, L. 1951 *Phys. Rev.* **84**, 168.
Levy, P. 1952 *Phys. Rev.* **88**, 738.
Matthews, P. T. & Salam, Abdus 1954 *Nvovo Cim.* **12**, 563.
Sachs, R. G. 1952 *Phys. Rev.* **87**, 1100.
Skyrme, T. H. R. 1954 *Proc. Roy. Soc.* A, **226**, 521.

III. SKYRMIONS

Unquestionably, Tony's invention of what are now called "skyrmions" was Tony Skyrme's most important work, "unconventional and revolutionary" as Dalitz says. Tony referred to the four papers following this commentary as I to IV; they were the core of his ideas. The further three papers, according to Dalitz, might be numbered V to VII. Preceding this commentary are the two papers Tony referred to as the genesis of many of his ideas. Following this commentary is Tony Skyrme's "The origin of skyrmions", which gives a beautiful explanation of why he was trying to do what he did. Following this I have added Valery Sanyuk's "Genesis and evolution of the Skyrme model from 1954 to the present", in which the author shows how Skyrme's model and ideas were brought into contemporary physics, especially discussing the formal aspects.

The most remarkable of these papers is the third one, "A non-linear field theory", in which the ideas crystallized.

I have little to add to the Skyrme and Sanyuk papers, which show the development of Tony Skyrme's thoughts clearly. I came upon the scene two decades later, in 1979, when I wanted to make a little bag model of the nucleon, the bag being composed of three quarks, which had by then been brought into the picture. In 1978, I gave a seminar in Caltech about work with Hans Bethe on stellar collapse. The morning following the seminar, I went to see Dick Feynman. I told him that the MIT bag model of quarks was simply too large; it had a radius R of $\sim$ 1 fm. With such a large radius, the nucleons would be like grapefruit in a bowl. It would be difficult to see how they could perform the independent motion that they exhibit in the shell model.

Feynman responded by a number of objections and penetrating questions, but he was obviously intrigued. This was a great stimulation to me, since many, if not most, particle physicists despised nuclear physics. (In fact, the only criticism that I have of Sanyuk's article is that it tries to convert Tony Skyrme retroactively into a particle theorist.) Feynman asked me how I wanted to compress the MIT bag. I told him that the pion cloud would compress the quarks. Only later, I discovered that in the Skyrme model, the pion cloud compressed the quarks to a point, the point source of the baryon number.

Upon returning to Stony Brook, I asked Mannque Rho how to formulate this quark compression. He had read a paper by Callan, Dashen, and Gross (C. G. Callan, R. F. Dashen, and D. J. Gross, *Phys. Lett.* **B78** (1978) 307), in which the external pion cloud was coupled to the interior quarks so as to make the axial vector current continuous across the boundary; in other words, to preserve chiral invariance at the joining boundary. The external axial vector current carried by the pions was, thus, added to the quark axial current in order to conserve this current, not unlike Maxwell's addition of the displacement current in order to conserve the electric current. The only trouble was that for Callan–Dashen–Gross, the pion cloud expanded the bag. But we found that they had left out a surface coupling term, and when calculated more properly, it did compress it. In retrospect, we should have looked at Chodos and Thorn (A. Chodos and C. B. Thorn, *Phys. Rev.* **D12** (1975) 2733), who formulated the chiral bag model.

In 1980, we (V. Vento, M. Rho, E. M. Nyman, J. H. Jun, and G. E. Brown, *Nucl. Phys.* **A345** (1980) 413) were able to obtain the classical hedgehog solution to the chiral bag equations. In order to solve these, we were led to the chiral angle θ, and the most convenient chiral angle was the "magic angle" $\theta = \pi/2$. At this angle, the bag radius $R = R(\theta)$ was ~ 0.5 fm, which fitted the phenomenology of the nucleon–nucleon interactions well.

We were very happy with this chiral angle until we discovered that the baryon number in the quarks at that chiral angle was $N_B = \frac{1}{2}$, and we didn't know where the other half was. After months of questioning colleagues, a visitor from Korea, Dr. Hyun Ku Lee of Han Yan University, pointed out Skyrme's work to us, where it was immediately clear that the other half was in the pion cloud. We probably would have been spared a lot of trouble, had I but stayed in Birmingham a few months longer in 1960 and heard Skyrme's lecture; I like to think that something of his model would have remained in my mind. But maybe not, because his ideas were so revolutionary to people at that time.

Colleagues had the greatest difficulty seeing how three pion fields could make up a nucleon. How can one get from bosons to a fermion? Of course, we now have Tony's paper on the origin of skyrmions.

Perhaps it's worth discussing this point now that we know about quantum chromodynamics (QCD). I like to draw a picture, based on the paper by J. Goldstone and F. Wilczek (*Phys. Rev. Lett.* **47** (1981) 986). Although the authors work in a one plus one dimensional space, their argument is instructive. Here a quark loop is coupled below the horizontal line to three pion fields. Above the line, the baryon current $J_\mu = (\bar{\psi}\gamma_\mu\psi)$ is measured. When the three classical π-fields below the line are arranged to be in the Skyrme configuration for baryon number B, then that change, as measured by the integral over J_0, made up out of the quark fields, is B. In other words, when the classical pion fields are arranged in the proper topology, then they polarize the vacuum so that the net baryon number in quarks is that of the B in the Skyrme expression in terms of pion fields. There are other examples

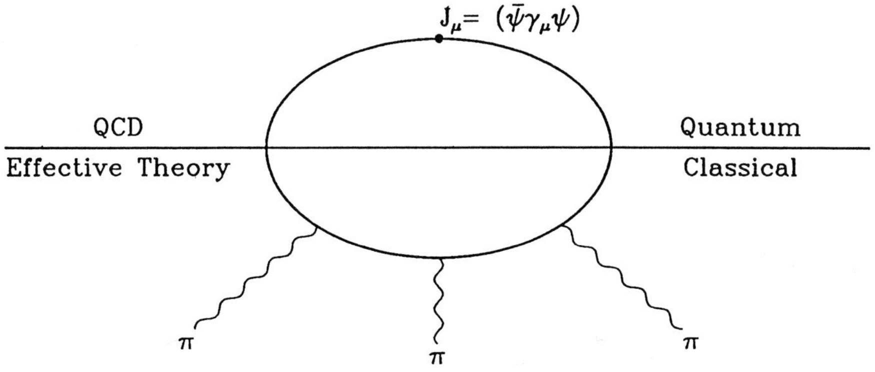

of new quantum numbers appearing in the effective sector, here that of pion fields, when it possesses a topology.

In other words, in the case of Skyrme's nucleon, made up out of three classical pion fields, the three fields polarize the vacuum so that enough quarks collect to form an object of $B = 1$, the antiquarks being sent off to infinity. Topology enforces that B be exactly unity for the nucleon.

In the spring of 1985 I made, with my wife, a sentimental journey back to Birmingham, England. The mathematical physics department of Prof. Peierls had been merged with the mathematics department, so I gave a seminar in the physics department, lecturing about "skyrmions" with Tony Skyrme in the front row of the audience. Having been greatly impressed by Witten's derivation of the Skyrme expression for the baryon number density, given in Witten's "Global aspects of current algebra", reproduced here, I paused briefly for the elegance of the derivation to sink in. Tony's response was, "Isn't that obvious?" Indeed, it had been to him for one and a half decades. It had taken us a long time to catch up.

Professor Tony Skyrme
Photograph taken about 1958 by Dr. Dorothy Skyrme

THE ORIGINS OF SKYRMIONS*†

T.H.R. Skyrme
formerly Professor of Applied Mathematics
The University of Birmingham
Birmingham, England

I am very grateful to the organizers for inviting me to give a brief talk at this meeting, but I accepted only with some reluctance, as I have really very little new to say that would be of interest to a Workshop such as this. They explained that they wanted me to say a little, quite informally, about the history and philosophy of the origins of the model that lies behind the meeting. I will do this briefly, leaving recent developments to those who have been more closely associated with them than I have been. I shall finish by mentioning briefly some related ideas in which I have recently been interested.

I had three main motives for trying to make a model of the kind we are discussing : unification, the renormalisation problem, and what I call the "fermion problem". The first of these is fairly obvious. Unification of one kind or another has always been a goal of theoretical physicists. I know that some of our colleagues are suspicious of attempts to bring in grand ideas of symmetry and unification for their own sake, unless they are clearly required by the experimental evidence. However, I believe that most progress in

---

*This talk was given as a historical introduction to the Workshop on "Skyrmions" held at Cosener's House, Abingdon, on November 17-18, 1984. It has been re-constructed by Dr. Ian Aitchison from sketchy notes left by Tony Skyrme, and from his own notes taken at that talk, aided by notes taken by several others who were present.

†Also published in Int. J. Mod. Phys. A3, 2745-2751 (1988).

physics has been due to ideas of this kind - though it is equally true that many of the ideas have been wild speculation. In the present context, this philosophy led me to think that instead of having two fundamental types of particle - bosons and fermions - it would be nice if we only had one. Heisenberg had proposed his non-linear spinor theory in 1958, in which everything was made from one fundamental self-interacting fermion field. For reasons which I will explain in a moment, I didn't like fermions much, and thought it would be fun to see if I could get everything out of a self interacting boson field theory instead!

The other two "problems" are of course very modern ones, and specific to quantum field theory; but in thinking about them I have noticed that they have appeared earlier in the history of physical ideas, in rather different guises. In particular, I have found it interesting to go back to the views of that eminent Victorian, Sir William Thomson (later Lord Kelvin (1824-1907)). Kelvin was deeply concerned with the problems of the atomic structure of matter, and its bearing on the theory of gases, and so on, just as today we are concerned with the elementary particle structure of nuclear matter.

Kelvin was very reluctant to accept the idea of infinitely rigid point-like atoms. In one of his lectures[1)] he spoke of

> "the monstrous assumption of infinitely strong and infinitely rigid pieces of matter, the existence of which is asserted as a probable hypothesis by some of the greatest modern chemists in their rashly-worded introductory statements ..."

He seems to have felt intuitively that there was something deeply wrong with the idea - though I am not sure precisely why. Anyway, I have always found the idea of any sort of elementary particle as a point-like object unreasonable - and of course we have good reason for uneasiness, because such a theory has no natural cut off and infinite renormalisation seems inevitable. Certainly, renormalisation theory has been built, through the efforts of a number of distinguished physicists, into a beautiful and ingenious form; but I still feel that it is just a very good and useful way of enabling us to live with our ignorance of what really goes on at short distances. Such accommodation should only be provisional. Indeed, these problems would not arise in the first place if the fundamental particles were actually extended objects.

The infinite divisibility of matter seemed absurd to the Greek philosophers of the Epicurean School, and to Lucretius who gives such a fine account of their ideas. This was one of their reasons for believing in atoms. Another was their feeling that there had to be some kind of atomic structure to conserve specific qualities. As Kelvin put it in the same lecture already quoted:

> "For the only pretext seeming to justify the monstrous assumption of ... their rashly-worded introductory statements, is that urged by Lucretius and adopted by Newton; that it seems necessary to account for the unalterable distinguishing qualities of different kinds of matter."

This was the point seized on by Kelvin, and given a quite different explanation in terms of "vortex atoms", inspired by Helmholtz's work in fluid motion. Kelvin continued his lecture thus:

> "But Helmholtz has proved an absolutely unalterable quality in the motion of any portion of a perfect liquid, in which the peculiar motion which he calls "Wirbelbewegung" has been once created. Thus, any portion of a perfect liquid which has "Wirbelbewegung" has one recommendation of Lucretius' atoms - infinitely perennial specific quality. To generate or destroy "Wirbelbewegung" in a perfect fluid can only be an act of creative power. Lucretius' atom does not explain any of the properties of matter without attributing them to the atom itself. Thus the "clash of atoms", as it has been well called, has been invoked by his modern followers to account for the elasticity of gases. Every other property of matter has similarly required an assumption of specific forces pertaining to the atom. It is as easy (and as improbable, if not more so) to assume whatever specific forces may be required in any portion of matter which possesses the "Wirbelbewegung" as in a solid indivisible piece of matter, and hence the Lucretius atom has no prima facie advantage

over the Helmholtz atom. A magnificent display of smoke-rings, which he recently had the pleasure of witnessing in Professor Tait's lecture room, diminished by one the number of assumptions required to explain the properties of matter, on the hypothesis that all bodies are composed of vortex-atoms in a perfect homogeneous liquid. Two smoke-rings were frequently seen to bound obliquely from one another, shaking violently from the effects of the shock."

I shall come back to this "vortex" idea in a moment, but before that I turn to the third problem - the "fermion" one. I know that the idea of an intrinsic fermion is perfectly acceptable to many - perhaps most - people, but I have always felt an unease about quantum-mechanical concepts that do not have clear classical analogues. Now that quantum mechanics can be well understood as an averaging over classical configurations, it seems even more slightly anomalous. Fundamental fermions are awkward to handle in the path integral formalism. Admittedly, we can incorporate them by using Grassmann variables, but this seems an unnatural, purely mathematical, construction. I would like to think that the fermion concept was just a good way of talking about the behaviour of some semi-classical construction, and that it was no more fundamental than renormalisation.

Here we come back to Kelvin. Almost 100 years ago, he gave a series of lectures at Johns Hopkins University - for which he was offered expenses and $1000, quite a substantial sum for those days. He subsequently wrote the lectures up, and published them as the Baltimore Lectures[2)]. Kelvin was reluctant, for much of his life, to accept unreservedly the ideas of Maxwell, his contemporary. In one of his Baltimore lectures[3)] he tried to explain why:

"I can never satisfy myself until I can make a mechanical model of a thing. If I can make a mechanical model I can understand it. As long as I cannot make a mechanical model all the way through I cannot understand; and that is why I cannot get [this is probably the reporter's Americanism for the word "accept"] the electromagnetic theory. I firmly believe in an electromagnetic theory of light, and that when

> we understand electricity and magnetism and light we shall see them all together as parts of the whole. But I want to understand light as well as I can, without introducing things that we understand even less of. That is why I take plain dynamics. I can get a model in plain dynamics; I cannot in electromagnetics. But so soon as we have rotators to take the part of magnets, and something imponderable to take the part of magnetism, and realise by experiment Maxwell's beautiful ideas of electric displacements and so on, then we shall see electricity, magnetism, and light closely united and grounded in the same system."

For Kelvin, then, an "understanding" meant a mechanical model. He and Tait spent a long time developing "smoke ring" or "vortex" models of the atom (based on Helmhotz's work), and also various mechanical models of the luminiferous ether.

Later in life Kelvin came to accept Maxwell's theory; and his own grand design of vortex atoms had to be abandoned. Sadly, he felt that his life had been a failure, despite his many brilliant achievements.

The analogies with the subject we are discussing today are fairly evident, thought I hesitate to draw any moral from history. Certainly I can remember being greatly impressed by a machine he and Tait had caused to be built in 1873 for the Tidal Committee of the British Association, for predicting the tides, world-wide[4]. This machine was in my grandfather's house, and the ingenuity of its mechanism, whereby it could produce this complicated pattern of tides, had considerable influence upon me. Anyway, I wanted a physical model which would reproduce the curious behaviour of fermions. I had been vaguely aware of Kelvin's picture of vortex atoms, and I had always been attracted by the type of structure that was possible in non-linear theories such as general relativity, or the Born-Infeld theory of electromagnetism. I liked the idea, for example, that the "sources" of gravitation might themselves be produced by the field equations - as, presumably, some kind of singularities in the fields - instead of having to be put in by hand. There was here an obvious hope that somehow fermionic sources of "strong charge" or baryon number might emerge as singularities of some non-linear classical meson field theory.

My first interest in more concrete possibilities really arose in the context of nuclear physics. I felt that since nuclear matter was so homogeneous we should not have to use a discrete (i.e. shell-model) picture to describe it. It seemed to me that a type of fluid drop model of the nucleus was simpler and more appropriate, the individual nucleons being then some kind of local twists in the fluid, rather than independent interacting particles. Trying to understand the nature of the solutions, I gradually came to look more at the idea of a single nucleon as a twist in some ether-like fluid - such as would be described now by the preferred direction in a theory with a spontaneously broken internal symmetry.

To simplify the problem I looked at the analogous problem in one space (and one time) dimension - the problem now know as the Sine-Gordon equation. This employs a single angle-type field variable $\alpha(x,t)$, in one space dimension, and the equation of motion is

$$\partial_t^2\alpha - \partial_x^2\alpha = -m^2\sin\alpha .$$

A vacuum state is such that $\cos\alpha = 1$, but of course we can have a situation in which a time independent $\alpha(x)$ interpolates from one vacuum at $x = -\infty$ (e.g. $\alpha(-\infty) = 0$) to another at $x = +\infty$ (e.g. $\alpha(+\infty) = 2\pi$), as shown in the Figure 1.

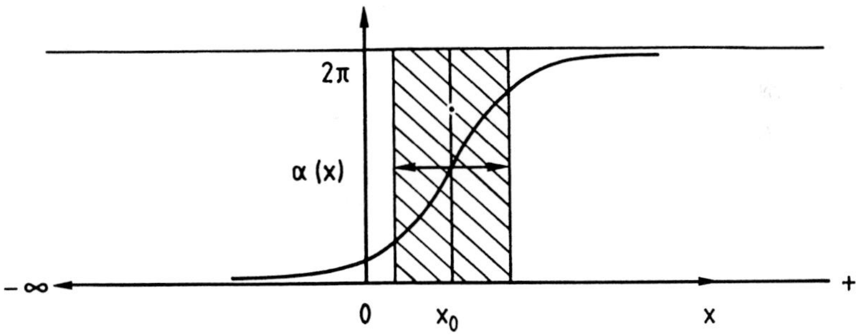

Fig. 1. A simple kink centred on $x = x_0$. The arrows show its width, the x-range over which most of the variation of $\alpha(x)$ occurs.

The "kink" or singularity, propagates according to the one-dimensional "neutrino-like" equation

$$(\partial_t - \partial_x)\psi(x,t) = 0$$

and ψ obeys anticommutation relations. If all physically observable quantities depend only on α mod(2π) then (given the boundary conditions that α tends to a multiple of 2π - i.e. a vacuum configuration - at $x = \pm\infty$), the real line is effectively compactified into a circle, S_1. α is itself defined on an S_1, so that $\alpha(x)$ provides a mapping from the S_1 of real space to the S_1 of field space, the number of times the α circle is covered being the winding number of the mapping (see Fig. 2) In three dimensions this generalises naturally to a mapping of an S_3 into an S_3, characterised again by a winding number, which is conserved. In this way I was led almost inevitably to the form of the model that does now seem to have some relevance to physical structures.

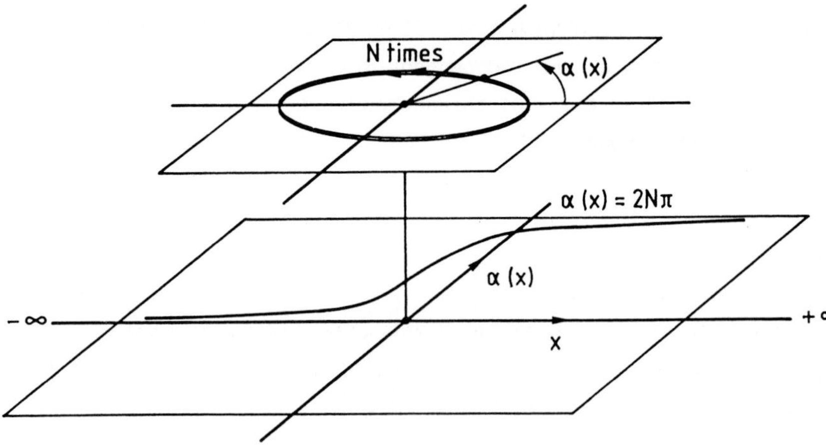

Fig. 2. The mapping from real space to field space provided by $\alpha(x)$. On the lower plane, the function $\alpha(x)$ is plotted vs. x, for the case $\alpha(+\infty) - \alpha(-\infty) = 2N\pi$. On the upper plane $\alpha(x)$ is plotted as the angle of a point on a unit circle, which winds arouned N times as x runs from $-\infty$ to $+\infty$. Such an $\alpha(x)$ has winding number N.

I identified the topologically conserved winding number with the physical baryon number. In order to get dynamical stability, I introduced into the Lagrangian (in the 3+1 case) a term of fourth order in the derivatives of the meson field[5]; its form was chosen so that it could be quantized straightforwardly. I don't have anything new to say about that term now.

I ought to add, by the way, that another very important influence leading in the same direction was the work of Pauli on the symmetric pseudoscalar meson theory of nuclear forces in the mid-forties, brought together in his well-known book[6]. The form of solution that I found useful in my work was partly inevitable, but was also strongly suggested by Pauli's semiclassical picture of spin and isospin locked together so as to minimize the energy of the system.

I still entertain the hope that some type of non-linear theory will yield an explanation of elementary particles that can be visualised in a semiclassical way, and that the quarks or leptons introduced as sources in most theories will be seen to be mathematical constructs helpful in its understanding, rather than fundamental constituents, just as the idea of vortices in a fluid is an indispensible way of talking about certain types of fluid motion.

I would like to finish by mentioning ideas along the lines which are not at all fully worked out. My model of the nucleon was of course based on a non-linear scalar field theory - but these days our fundamental theories all seem to be gauge theories, so that one can again wonder whether perhaps it is possible to have topologically interesting field configurations in such theories, and if so how the constraints of gauge invariance might affect the situation. The idea I have been thinking about is roughly as follows. Suppose one has a gauge theory, and that field configurations are possible that are characterised by topological invariants. In any gauge theory, in order to do calculations, one must make a choice of gauge. Suppose the gauge choice had the effect of excluding the topologically characterised field configurations. Since the latter involve topological invariants of the theory, there must be some form of compensation for this gauge choice. In the usual discussion of non-

Abelian SU(n) gauge theories it is well known that some compensation has to be made for the usual covariant gauge choices, in order for the theory to remain unitary. Faddeev and Popov[7)] showed very clearly how this could be understood in terms of a suitable weighting that had to be applied to path configurations, to compensate for the asymmetry caused by a specific gauge choice. This led to the idea of "ghost" fermions.

I feel that there is a possibility that a suitable gauging of a theory admitting stable topological structures may enforce the addition of source particles to make it consistent. In this case, these extra source particles should correspond to physical states, since they have to supply the "missing" topological quantum numbers that were carried by those "topological" gauge field configurations which were excluded by the gauge choice.

At present I have only managed to make a little progress in the special case of U(n) gauge theories. There we have n^2 gauge fields $W^{(\alpha)\mu}$, $\alpha=1, 2, \ldots n^2$. There are n^2 redundant degrees of freedom, which have to be eliminated somehow by a gauge choice. They might be constrained as follows. Consider the matrix

$$W = \sigma_\mu \lambda_\alpha W^{(\alpha)\mu}$$

where

$$\sigma_\mu = \gamma_\mu(1-\gamma_5)$$

The σ_μ are taken to be 2x2, using two- component spinors; the λ_α are the usual generators of U(n) in the regular representation. W is a 2n x 2n matrix. If we demand that it have rank n this imposes exactly n^2 conditions, which seems to fix the gauge. If we follow the arguments of Faddeev and Popov[7)], this requires the introduction of n fermionic fields (to represent the weighting), coupled in a neutrino-like way, with chiral symmetry, to the gauge fields. These fermionic fields are in the fundamental representation of the U(n) gauge group. The idea seems attractive - the difficulty is to understand what the gauge constraint implies as to how calculations should be made.

The only simple case is unfortunately rather too trivial to be interesting or informative. If we apply this idea to U(1) electromagnetic theory in 1+1 dimensions, a theory which has some other interest, in that it shows confinement, the gauging condition simply means that we are left with ordinary electromagnetic waves propagating in one direction and with neutrinos having the opposite chirality - but of course there is no interaction between them.

REFERENCES

1. cf. Thompson, S.P., Life of Lord Kelvin, p.517 (London: Macmillan, 1910). The lecture referred to was the reading of his paper "On Vortex Atoms" to the Royal Society of Edinburgh on 18 February 1867. The reference for the published paper is Proc. Roy. Soc. Edinb. 6, 94 (1869)
2. Notes of Lectures on Molecular Dynamics and the Wave Theory of Light delivered at the Johns Hopkins University, Baltimore, by Sir William Thomson, stenographic report by A.S. Hathaway. (Baltimore: Johns Hopkins University, 1884). A revised version of these lectures appeared later as Lord Kelvin's Baltimore Lectures on Molecular Dynamics and the Wave Theory of Light. (London: Clay & Sons, 1904)
3. cf. lecture XX, pp. 270-1 of the stenographic report[2)]. The quotation used by Tony Skyrme was that given on pp. 835-6 of ref.[1)]. However, the revised version of lecture 20, appearing in the 1904 book[2)] of Lord Kelvin's lectures at Baltimore, does not include these remarks.
4. cf. paper No. 271, "The Tide Gauge, Tidal Harmonic Analyses and Tide Prediction", in Mathematical and Physical Papers of Lord Kelvin, vol.6. Ed. J. Larmor (Cambridge: C.U.P. 1911). The reference for the published paper is Proc. Inst. Civil Engrs. 65, 2 (1881)
5. Skryme, T.H.R., Proc. Roy. Soc. A260, 127 (1961)
6. Pauli, W., Meson Theory of Nuclear Forces. (New York: Interscience, 1946)
7. cf. Taylor, J.C., Gauge Theories of Weak Interactions, pp. 89-91 (Cambridge: C.U.P., 1976)

International Journal of Modern Physics A, Vol. 7, No. 1 (1992) 1–40
© World Scientific Publishing Company

GENESIS AND EVOLUTION OF THE SKYRME MODEL FROM 1954 TO THE PRESENT*

VALERY I. SANYUK[†]
Department of Physics, Syracuse University, Syracuse, NY 13244-1130, USA

Received 12 February 1991

Not widely known facts on the genesis of the Skyrme model are presented in a historical survey, based on Skyrme's earliest papers and on his own published remembrance. We consider the evolution of Skyrme's model description of nuclear matter from the "Mesonic Fluid" model up to its final version, known as the baryon model. We pay special tribute to some well-known ideas in contemporary particle physics which one can find in Skyrme's earlier papers, such as: Nuclear Democracy, the Solitonic Mechanism, the Nonlinear Realization of Chiral Symmetry, Topological Charges, Fermi–Bose Transmutations, etc. It is curious to note in the final version of the Skyrme model gleams of Kelvin's "Vortex Atoms" theory. In conclusion we make a brief analysis of the validity of Skyrme's conjectures in view of recent results and pinpoint some questions which still remain.

1. Introduction

As it is possible to find out from the only published biography of T.H.R. Skyrme,[1] after graduation from Trinity College, Cambridge, U.K. in 1943, the distinguished English physicist Tony Hilton Royle Skyrme spent some time in the U.S.A., where he took part in and made a notable contribution to the success of the Manhattan Project. He returned to Britain and from 1950 started his work in the Division of Theoretical Physics of the Atomic Center (A.E.R.E) in Harwell. It is rather difficult to determine when the earliest of Skyrme's papers were written. According to Dalitz[1]: "it was common knowledge" among colleagues at A.E.R.E that the Senior Principal Scientific Officer, Tony Skyrme, "had a desk drawer full of manuscripts awaiting completion and submission for publication." In 1952 B. Flowers returned to A.E.R.E from Birmingham, to became Head of the Theoretical Division, learned of this drawer and "made it one of his priorities to move as many of these papers as possible into scientific journals." Only due to the persistence manifested by B. Flowers may we approximate the initial date of the Skyrme model as 1954, when Skyrme's first paper[2] on the model description of nuclear matter was published. In this paper Skyrme proposed the

* The enlarged version of the introductory lecture of the course on "Skyrme Model in Hadron Physics" given in October 1990 at Syracuse University Physics Department under the Fulbright Scholar Program.
† On leave of absence from the Experimental Physics Department, People's Friendship University, Moscow, USSR.

"Mesonic Fluid" model in order to answer a question raised by experimental data of nuclear radii measurements. We will discuss this problem in more detail in the next section. Here we would just like to point out that already in this first of the series of his pioneering papers[2–10] he brought to the attention of nuclear physicists some hydrodynamical ideas for the description of non-point-like objects.

As far as it is known from his own writings,[11] Skyrme never believed in the validity of any point-like description of such particles as protons in the framework of a linear field theory with any conceivable renormalisation scheme, and because of this he was looking for a nonlinear field theory, which admits a description of the aforementioned particles as extended objects. In that very point Skyrme agreed with W. Heisenberg that a real particle theory must be strongly nonlinear. But at the same time he strongly disagreed with another opinion of Heisenberg's, who considered fermion fields to be the most fundamental. The main motive in Heisenberg's Unified Field Theory was that everything should be made from the so-called "profield"—the fundamental self-interacting fermion field. According to his talk,[11] Skyrme's point of view was that fermions do not have any real physical meaning, but rather are useful in some purely mathematical constructions. Despite his great intuition, which he demonstrated in a good number of physical problems, Skyrme "always felt uneasy about quantum-mechanical concepts that do not have clear classical analogs." This was particularly so in the case of fermions. Already in the early 1950s Skyrme held the firm opinion that bosonic fields could not be less fundamental than fermionic ones, that these two sorts of fields should be in some sense interchangeable and by no means considered on an equal footing. One of the main motives, which led him to the now well-known baryon model, was that:[11] "... it would be fun to see if I could get everything out of self-interacting boson field theory..."

Skyrme found unexpected support for the ideas mentioned above in comparatively ancient papers by Kelvin.[12,13] [a] He noticed that in the previous century, Kelvin was deeply concerned with similar problems arising in the description of atomic structure and "... was very reluctant to accept the idea of infinitely rigid pointlike atoms." As a more reasonable alternative Kelvin suggested the "Vortex Atom" model.[12] This peculiar "Vortex Atom" idea was inspired by Helmholtz's discovery of the so-called "Wirbelbewegung"—an absolutely unalterable quality in the motion of any portion of a perfect fluid.[b] Kelvin considered this "Wirbelbewegung" as a desirable fluid-mechanical model of the atom, as he also could not understand any physical phe-

[a] It is curious that, again according to Dalitz's paper,[1] T. Skyrme was acquainted with, as he put it, the "eminent Victorian, Sir William Thomson," not only from literature, but also from stories about Skyrme's great-grandfather on the maternal side, Edward Roberts. His great-grandfather "was in 1868 appointed Secretary to the Tidal Commitee of the British Association for the Advancement of Science, being made responsible later for the construction of the first Tidal Predicter, which had been designed by Lord Kelvin for this Commitee".[1] The first model of this machine, which was held in Roberts' house at 7 Blessington Road, made a strong impression on the young Tony and greatly influenced the development of his later ideas, as Skyrme himself recounted in Ref. 11.

[b] Here the term "perfect fluid" means a fluid perfectly destitute of viscosity. Later on, this concept was developed into the basic one in the ether theory.

nomenon without a simple mechanical model. He did not accept the usual way of explaining incomprehensible things by introducing conceptions that are even less understandable. On the basis of experiments with smoke rings, conducted by Professor Tait, and some analytical results on vortex description, Kelvin pulled out the hypothesis that *all bodies are composed of vortex atoms in a perfect homogeneous liquid.* He thought that this approach had some advantages in comparison with the traditional one at that time, *as vortex rings do not require any other property in the matter whose motion composes them than inertia and incompressible occupation of space.* According to his writings, Kelvin had far-reaching plans to develop a new kinetic theory of gases and to found a theory of elastic solids and liquids on the dynamics of vortex atoms. But as far we know those plans were not realized by Kelvin and his collaborators and were later on rejected by most physicists entirely along with the ether theory.

The paper "On Vortex Atoms" see Ref. 12, p. 1–12) demonstrates that Kelvin was a real pioneer in introducing topological concepts into physics: different sorts of atoms were to differ from each other in his theory in accordance with the number of intersections of vortex rings. Let us just quote this paper, dated 1867: "*It is to be remarked that two ring atoms linked together or one knotted in any manner with its ends meeting, constitute a system, which however it may be altered in shape, can never deviate from its own peculiarity of "multiple continuity," it being impossible for the matter in any line of vortex motion to go through the line of any other matter in such motion or any other part of its line. In fact, a closed line of vortex core is literally indivisible by any action resulting from vortex motion.*" In Fig. 1, which is taken from Kelvin's next paper "On Vortex Motion" (see Ref. 12, p. 13–66), one can find displayed knotted or knitted vortex atoms "*. . . the endless variety of which is infinitely more than sufficient to explain the varieties and allotropies of known simple bodies and their mutual affinities.*"

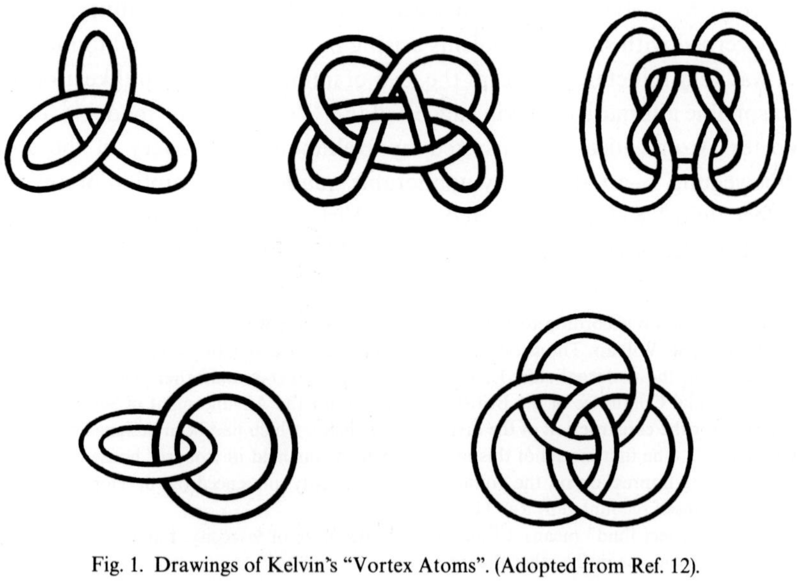

Fig. 1. Drawings of Kelvin's "Vortex Atoms". (Adopted from Ref. 12).

Now one can easily recognize that Kelvin was thinking about a concept similar to a topological invariant and, in particular, about the Hopf index, which is widely used in contemporary particle physics models. Kelvin complained in the same paper that he was not sufficiently acquainted with Riemann's "Lehrsätze aus der Analysis situs," which was known to him only through Helmholtz, but nevertheless he understood that such topological properties as "multiple continuity" or connectedness in present terminology may be important for a physical description of matter.

As a result of this historical survey one can come to see that it was quite natural for Skyrme to start with a hydrodynamical model of nuclear matter to describe extended particles (nucleons) as a sort of nuclear "Wirbelbewegung." It is probably a mistake to expect this model to provide us with answers to all possible questions in hadron physics, but the Skyrme model possesses so many interesting features, mostly originating in its hydrodynamical context, that any efforts taken to better understand its genesis are justified.

2. The "Mesonic Fluid" Model

This model was developed in order to solve the experimental data puzzle, which arose as a result of the 1953 fast electron scattering experiments, conducted in order to study the charge distribution within the nucleus. From those *electromagnetic type experiments*, as well as from μ-mesonic atoms spectroscopy data or from the isotopic shift data, the value for the radius of the charge distribution within the nucleus was suggested to be

$$\bar{R} = 1.2A^{1/3}\ \mathrm{fm}, \tag{2.1}$$

where, as usual, A is the atomic number. At the same time it was already known from specifically *nuclear type experiments* such as α decay or light nuclei reactions, from the penetrability factors, calculated for the scattering of neutrons, or from the evidence of the difference in binding energies of mirror nuclei, that one can get the value for the radius of the nucleus close to

$$R = 1.5A^{1/3}\ \mathrm{fm}. \tag{2.2}$$

These results signified that the electrical charge of the nucleus lies within the radius $\bar{R}$, but that the specifically nucleonic interactions extend out to the radius R. From the successful prediction of the magnetic moments on the basis of the shell model (B. Flowers, 1952), it was clear that the charge of the nucleus must be quite closely confined to the neighborhood of the protons. This means that protons should occupy the region of radius $\bar{R}$ and if we accept the charge-independence of nuclear interactions, so must the neutrons.

This picture conflicted with the description of the nucleonic interactions as short-range interactions, centered on the constituent nucleons, adopted at that time. It was difficult to understand how it could lead to the effective radius (2.2), which is also proportional to $A^{1/3}$. Skyrme[2] agreed with Yukawa's picture that nuleonic interactions are most likely to be due to the intermediary of the π-meson fields, so that π mesons should be regarded as constituents of the nucleus. To explain the nuclear type experiments π mesons should occupy the region with radius R, but if they are electrically charged, the electromagnetic experiments would give the same value R for the radius of charge distribution. Skyrme concluded that some model description is needed, in the framework of which the mesonic fields would be condensed into "Mesonic Fluid," so that at any point the densities of π^+ and π^- components would be equal to each other. As a result of interactions between nucleons and mesonic fluid, the nucleons would occupy only a smaller interior region with radius $\bar{R}$. The main positions of the "Mesonic Fluid" model proposed by Skyrme in order to meet the abovementioned requirements were as following:

1. The nucleus is considered to be a classical, electrically neutral incompressible "Mesonic Fluid," which occupies the region with radius R. In the standard hydrodynamical manner the state of this fluid in any point can be described by some scalar density and a direction (a vector in isospace).
2. The nucleons are immersed into the "Mesonic Fluid", which saturates them and they are free to move inside the region with the radius $\bar{R}$ with residual interactions that result from spin-charge fluctuations in the fluid. (As a very crude analog, it is possible to represent the nucleonic source as a sort of "oscillatory pump" free to move within some region of the mesonic "liquid" drop.)

In his next paper[3] Skyrme proposed a mathematical formulation of the ideas given in Ref. 2, starting from the usual expression for the Lagrangian density of nucleon fields interacting with a symmetrical pseudo-scalar meson field[14]:

$$\mathscr{L} = \tfrac{1}{2}[(\partial_\mu \underline{\phi})^2 - k^2 \underline{\phi}^2] + i\bar{\psi}(\gamma^\mu \partial_\mu + g\gamma_5 \underline{\tau} \cdot \underline{\phi})\psi \,. \tag{2.3}$$

Here the convention is accepted that underlined quantities as $\underline{\phi}$ are vectors in the isotopic spin space; k is a reciprocal length, $\underline{\phi}$ are mesonic fields; $\underline{\tau}$ are the Pauli matrices and spinor–isospinor fields ψ describe nucleons. Note that there is no "inertial" mass term of bare nucleons in the density (2.3). This is not an accident but a special assumption by Skyrme, namely that *all the nucleon's mass is of mesonic origin, i.e. it arises as a result of mesonic fluctuations.* This suggestion looks like a first mention of the idea of the "Solitonic Mechanism," expressed in a clearer form later on by L. Faddeev[15] as the following: "*The strong-interacting particles could be described in the framework of a nonlinear field theory as collective excitations in the system of weakly coupled fundamental fields. The corresponding Lagrangian should be expressed in terms of fundamental fields only and possesses the existence of particlelike (solitonic) solutions. Those solitons are to describe the evident spectroscopy of heavy particles, as well as interactions among them.*"

To realize the idea of the mesonic origin for the nucleon's mass, Skyrme proposed the following method: write down the mesonic field vector in polar coordinates $\underline{\phi} = \underline{\mathbf{n}}\phi$, where $\phi = +|\underline{\phi}|$, so that $\underline{\mathbf{n}}$ is a unit pseudo-vector and ϕ is a scalar; then perform the unitary transformation

$$\psi \to \exp\left(-i\frac{\pi}{4}\gamma_5 \underline{\tau}\cdot\underline{\mathbf{n}}\right)\psi = \frac{1}{\sqrt{2}}(1 - i\gamma_5 \underline{\tau}\cdot\underline{\mathbf{n}})\psi. \tag{2.4}$$

In Skyrme's terminology this is "a limiting form of the Foldy transformation, appropriate to zero inertial mass." In the transformed Lagrangian

$$\mathscr{L}' = \tfrac{1}{2}[(\partial_\mu \underline{\phi})^2 - k^2\underline{\phi}^2] + i\bar{\psi}[\gamma^\mu\partial_\mu + ig\phi - \tfrac{1}{2}(i\gamma_5 - \underline{\tau}\cdot\mathbf{n})\gamma^\mu\partial_\mu(\underline{\tau}\cdot\mathbf{n})]\psi, \tag{2.5}$$

the additional "mass" term arises in the form $\bar{\psi}g\phi\psi$ adjustable for real nucleons. Here g is the mesonic coupling constant. It is also possible to regard the transformation (2.4) as a chiral transformation. This is a remarkable fact, because this particular symmetry was chosen later by Skyrme in order to modify the initial version of his model so that it would be possible to take into account new motives emerging in particle physics after the discovery of parity violation in weak interactions. To conclude this section we note that the "Mesonic Fluid" model managed to explain the evident difference in experimental data, and its consequences agreed substantially with well established facts about the shell structure of nuclei, the collective motions in nuclei, etc. For those who are interested in more detailed information we refer to Skyrme's original papers.

3. The Chiral Modification

The real acknowledgment of the Chiral Symmetry ideas in particle physics can be dated to 1957, when it was discovered that the parity conservation law could be violated in weak interactions. The long-standing question, known as the "$\Theta - \tau$" puzzle, was at last resolved and "strange" Θ and τ mesons were united into the no less "strange" K meson. Those particles, we recall, were called "strange" because although they arose through the strong interactions, their decays were mediated through the weak channels. The Θ mesons decayed, as a rule, into two π mesons and τ mesons decayed into three π mesons. These hereditary strange features of K mesons demanded for their description an enlargement of the isotopic internal symmetry group SU(2) in such a way that the parity-violating transformations were to be presented in a new internal symmetry group of strong interactions. Some of those possibilities were already studied by A. Pais[16,17] and it was shown how to join three SU(2) generators of isotopic rotations $\mathbf{I}_k$, which do not change the parity of states, with three generators of chiral boosts $\mathbf{K}_j$, which do mix up the states with different parities. Thus the problem was to construct a representation of at least the six-parameter symmetry group, but it is well-known that there is no linear representation of such a group in the three-dimensional isotopic space. The possible solutions are two-fold: it is possible either to extend the isotopic space by introducing additional components or to look for a nonlinear group realization. The first way is analogous to the extension of the SO(3) rotation group to the homogenous Lorentz group: the superfluous coordinate x_0 could be introduced and in addition to the three generators of rotation about space axes $\mathbf{J}_k$, we also consider three Lorentz-boost generators in (x_0, x_i) planes. In the same manner one can supply the three-isovector $\underline{\phi}$ with the fourth component ϕ_0 and consider the resultant four-vector as an element of four-isospace. In that case generators $\mathbf{I}_k$ would mix components of $\underline{\phi}$ only and would not affect ϕ_0. At the same time the chiral-boost generators $\mathbf{K}_j$ would mix ϕ_0 with components of $\underline{\phi}$. Those generators would form the following algebra:

$$[\mathbf{I}_i, \mathbf{I}_j] = i\varepsilon_{ijk}\mathbf{I}_k; \qquad [\mathbf{I}_i, \mathbf{K}_j] = i\varepsilon_{ijk}\mathbf{K}_k; \qquad [\mathbf{K}_i, \mathbf{K}_j] = i\varepsilon_{ijk}\mathbf{I}_k, \tag{3.1}$$

which is locally isomorphic to the Lie algebra of the O(4) group. If we introduce the left and right generators

$$\mathbf{L}_i = \tfrac{1}{2}(\mathbf{I}_i - \mathbf{K}_i); \qquad \mathbf{R}_i = \tfrac{1}{2}(\mathbf{I}_i + \mathbf{K}_i), \tag{3.2}$$

the algebra (3.1) takes the form

$$[\mathbf{L}_i, \mathbf{L}_j] = i\varepsilon_{ijk}\mathbf{L}_k; \qquad [\mathbf{R}_i, \mathbf{R}_j] = i\varepsilon_{ijk}\mathbf{R}_k; \qquad [\mathbf{L}_i, \mathbf{R}_j] = 0. \tag{3.3}$$

The algebra has split into two independent subalgebras, each isomorphic to a *su*(2) algebra. That was the reason that for the corresponding chiral group the notation $SU(2)_L \otimes SU(2)_R$[c] was accepted.

Skyrme in 1958 realized the necessity for a chiral modification of the "Mesonic Fluid" model and performed it in Ref. 4, choosing from the aforementioned possibilities the second one—*the Nonlinear Realization of the Chiral Symmetry*. To our knowledge it was the first construction of a nonlinear σ model. In his approach, Skyrme leaned much upon already known results of Schwinger's,[18] namely, the possibility of constructing a desirable chiral generalization starting from the familiar *PS–PS* theory of pions and nucleons by replacing the nucleon mass by $g\phi_0$, where g is the pseudo-scalar coupling constant and ϕ_0 is another meson field of similar character to the three ϕ_i. In that case the interaction would be really symmetric between four fields. Skyrme added to the isotriplet of meson fields $\underline{\phi}$ an additional field ϕ_0, and, in order to prevent the unphysical enlargement of isospace degrees of freedom, he imposed the following constraint on the fields $\phi_\rho = (\phi_0, \underline{\phi})$, with $\rho = 0, 1, 2, 3$:

$$\phi_0{}^2 + \phi_a\phi^a = 1; \qquad a = 1, 2, 3. \tag{3.4}$$

But under restrictions (3.4) the natural generalization of the meson mass term in (2.3),

[c] It is curious to note that the chiral terminology was introduced in physics again by Lord Kelvin, who used it for the description of vortex orientations and for the description of light polarizations (see Ref. 13, lecture 20). Kelvin's definition of "chiral" and his geometrical theory of chirality can be found in Appendix H of Ref. 13.

namely $\frac{1}{2}k^2\sum_i \phi_i^2 \to \frac{1}{2}k^2\sum_\rho \phi_\rho^2$, is no longer a mass term, so to avoid this obstacle Skyrme considered two possibilities: to introduce into the Lagrangian a fourth-order term with respect to mesonic fields $\frac{1}{4}k^2\sum_\rho \phi_\rho^4$, or to admit that "*...the meson mass originates from the nucleon coupling.*" The latter proposal together with the previous hypothesis on the meson generation of the nucleon mass is perhaps one of the earliest suggestions of the "*Nuclear Democracy*" *Principle.*[d]

In his paper,[4] Skyrme chose the first possibility. Thus, after the chiral modification the Lagrangian may be written in the following form:

$$\mathscr{L}_{\mathrm{Ch}} = \frac{1}{2}\sum_\rho \left[(\partial_\mu \phi_\rho)^2 + \frac{1}{2}k^2\phi_\rho{}^4 \right] + \bar{\psi}[i\gamma^\mu\partial_\mu + g(\phi_0 + i\gamma_5 \underline{\tau}\cdot\underline{\phi})]\psi . \qquad (3.5)$$

The detailed analysis of the symmetry properties and features of the modified theory with the Lagrangian (3.5), undertaken by Skyrme in Refs. 4 and 5, in spite of some interesting and promising results, led him to the conclusion that in order to test the model further, some estimation of particle masses and of interaction coupling constants were required. A deep understanding of the model necessitated such practicalities. As a first step towards this, Skyrme considered a simplified version of his model "...of the same general character..., in which the configuration space–time and the isotopic spin space have both been reduced to two dimensions."[4] This simplified model will be the subject of our next section.

4. The Two-Dimensional Simplified (Sine-Gordon) Model

Going to a two-dimensional model was the most crucial step in the evolution of the baryon model. Here we pay special attention to this topic and also remind ourselves of some facts, included already in some modern textbooks on field theory, such as Ref. 20. If we consider the simpler proposed case, where the only two components ϕ_0 and ϕ_1 are nontrivial and both are functions on the $(1 + 1)$-dimensional space–time with coordinates x and t, then it is possible to fulfill the constraint relation (3.4) by setting

$$\phi_0 = \cos\alpha(x,t); \qquad \phi_1 = \sin\alpha(x,t). \qquad (4.1)$$

The direct substitution of (4.1) into (3.5) after some redefinitions of variables gives rise to the following form of the mesonic part of the Lagrangian density

$$\mathscr{L}_{\mathrm{S-G}} = \frac{1}{2}[(\partial_t\alpha)^2 - (\partial_x\alpha)^2 - k^2(1 - \cos\alpha)], \qquad (4.2)$$

with the corresponding Euler–Lagrange equation:

[d] According to G. F. Chew[19] the term "Nuclear Democracy" was suggested by M. Gell-Mann. In its full extent it expresses the conjecture that each strongly interacting particle is a composite (bound) state of those S-matrix channels with which is communicates. It means that each nuclear particle generates other particles, which in turn generate it. Thus no one nuclear particle is more fundamental than another.

$$\partial_x^2 \alpha - \partial_t^2 \alpha - k^2 \sin \alpha = 0 . \tag{4.3}$$

Known as the Sine–Gordon equation, this equation gained most of its popularity after the Inverse Scattering Transform Method discovery in the mid-1960s. Equation (4.3), by itself, was known in the previous century and investigated in great detail by geometers in connection with the problem of an enclosure of the Lobatschevskii plane into Euclidean space $\mathbf{R}^3$. Solutions of (4.3) correspond to different types of possible enclosures. It looks as though Skyrme did not know these results, so he tried to get possible solutions through his own methods[e] in Refs. 4, 7 and 8 and managed to find all types of nontrivial solutions of the Sine–Gordon equation. Those solutions are now known as 2π kinks, bions and breathers and were rediscovered by different methods in Ref. 21. In principle, Skyrme himself[4,7] and later with Perring[8] were the first investigators in the new branch of mathematical physics, now called *Field-Theoretic Integrable Models.* The complete integrability of the Sine–Gordon model was proved in 1974 by Russian physicists[22] and this work became a point of departure for this new branch. As for Skyrme, it seems that the reason he did not proceed in this direction was the discovery of a new type of conserved quantities in physics—"*Topological Charges*"—which strongly supported his long-held hope of developing a *Unified Field Theory of Mesons and Baryons.* To explain the idea of the Topological Charge it is best to use the geometrical language, suggested in 1959 by D. Finkelstein and C. Misner in Refs. 23 and 24.

4.1. *The Topological Classification of Solutions in Field-Theoretic Models*

The classical fields ϕ_0 and ϕ_1 in (4.1) may be considered, at any fixed moment of time, as a map

$$\phi(x) = \exp[i\alpha(x)] : \mathbf{R}^1 \to \mathbf{S}^1 , \tag{4.1.1}$$

so that $\mathbf{S}^1$ is *the field manifold* of the Sine–Gordon model. If we are interested only in field configurations with a finite energy

$$\mathscr{E}_{\mathrm{S-G}} = \int_{-\infty}^{\infty} dx \left[\frac{1}{2}(\partial_t \alpha)^2 + \frac{1}{2}(\partial_x \alpha)^2 + k^2(1 - \cos \alpha) \right], \tag{4.1.2}$$

then we have to impose the following boundary conditions:

$$\alpha(x) \to 0 (\mathrm{mod}\, 2\pi), \qquad \text{as} \qquad |x| \to \infty . \tag{4.1.3}$$

[e] According to his biography[1], Skyrme was consistently outstanding in Mathematics "...gaining the Hawtrey Prize in 1938, the Tomline Prize in 1939 and the Russell Prize in 1940..." during the years of study at Eton College. Later at Trinity College he gained "...1st Class Honors in Part II of the Mathematical Tripos in 1942, proceeding to Part III on Senior Scholarship in the following year." During his participation in the Manhattan Project he was known among colleagues as an expert in finding solutions of differential equations both analytically and numerically.

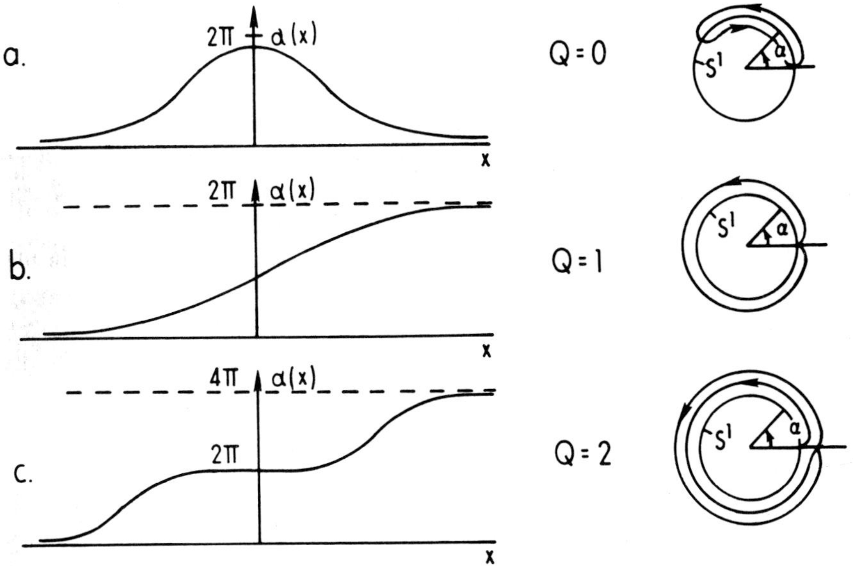

Fig. 2. Graphic topological classification of the Sine–Gordon solutions. Left-hand side profiles represent behavior of the angle variable $\alpha(x)$, as the right-hand side pictures display the images of axis $\mathbf{R}^1$ on the field manifold $\mathbf{S}^1$ under the maps (4.4). Situations: (a) corresponds to travelling waves or breathers with $\mathbf{Q} = 0$; (b) represents 2π kinks with $\mathbf{Q} = 1$; (c) represents 4π kinks or bions with $\mathbf{Q} = 2$. (Adopted from Ref. 46.)

The form of conditions (4.1.3) means that the anglelike field variable $\alpha(x)$ is a multi-valued function and to obviate it one should identify angles, which differ by $2\pi n$ in value, where $n \in \mathbf{Z}$. Under condition (4.1.3) the real axis $\mathbf{R}^1$ is compactified as the points at infinity ($x = \pm\infty$) are mapped into one and the same point—the "north pole" of the circle $\mathbf{S}^1$. Thus $\mathbf{R}^1 \cup \{\infty\} = \mathbf{S}^1$ and the maps (4.1.1) are effectively the maps: $\mathbf{S}^1 \to \mathbf{S}^1$.

If the angle variable $\alpha(x)$ takes the zero value at both boundaries $x = \pm\infty$ (one may say, at both vacua), then the corresponding solutions[f] are the travelling waves or the breathers and we can display the corresponding profile as shown in Fig. 2a. The image of axis $\mathbf{R}^1$ on the field manifold $\mathbf{S}^1$ under the map (4.1.1) would be a closed loop on $\mathbf{S}^1$ which does not cover the whole circle $\mathbf{S}^1$. Such a loop could be shrunk in a continuous way to a point on $\mathbf{S}^1$. These maps are called *topologically trivial.*

But we can just as well have a situation in which a time-independent $\alpha(x)$ interpolates from one vacuum at $x = -\infty$ ($\alpha(-\infty) = 0$) to another at $x = +\infty$ ($\alpha(+\infty) = 2\pi$). The corresponding class of solutions of (4.3) may be presented in the form

$$\alpha(x) = 4\tan^{-1}[\exp\{\pm k(x - x_0)\}]. \tag{4.1.4}$$

[f] We are not going to write down the explicit forms of solutions of the Sine–Gordon equation, except perhaps the 2π-kink solution, as it will just enlarge the paper and they are not supposed to be used in further discussion. One can find them in Ref. 20 (Chapter 2, Sec. 2.5).

In this case the image of $\mathbf{R}^1$ on $\mathbf{S}^1$ would be the loop, which covers the whole circle $\mathbf{S}^1$ and cannot be shrunk in any continuous way to a point on $\mathbf{S}^1$. These solutions are called 2π kinks and they are topologically nontrivial. There exists one more class of solutions, known as bion solutions, which satisfy the boundary conditions in the form: $\alpha(-\infty) = 0, \alpha(+\infty) = 4\pi$. The closed loop, still the image of axis $\mathbf{R}^1$, in that case covers the circle twice and the corresponding maps are topologically different from the previous two. The bions can never be continuously deformed into breathers or 2π kinks. All other possible cases just differ by the number of times the image of the axis $\mathbf{R}^1$ winds around the circle $\mathbf{S}^1$. The maps with the same number of closed loops around the field manifold $\mathbf{S}^1$ could be considered as belonging to the same class of solutions. Thus we have gained the main idea of *the Topological* (or, more precisely, *Homotopical*) *Classification* of maps. Two maps are called *homotopically equivalent* if they can be continuously deformed into one another. This means that they belong to the same class of solutions of an equation of motion, as, for example, Eq. (4.3), and satisfy the same type of boundary conditions. It is straightforward to see that this text is just another way of describing the evolution of any dynamical system, where maps are the states of the system and continuous deformations (homotopies) are just trajectories between different states.

4.2. *The Idea of Topological Charge*

From the general logic it follows that there should be a topological invariant of the maps (4.1.1) with values which differ for different classes of solutions, as those classes are topologically different. Such an invariant was long ago known in mathematics under different names: the winding number, the Browder degree of mapping, the Chern–Pontryagin number, etc. They are all applicable in the Sine–Gordon case, in spite of some difference in their definitions. Skyrme himself found in Ref. 4 the explicit expression for the topological invariant. He paid attention to the existence in the Sine–Gordon model of the conserved current J^μ, $\mu = 0, 1$; with components:

$$J^0 = \frac{1}{2\pi}\partial_x\alpha; \qquad J^1 = \frac{1}{2\pi}\partial_t\alpha, \tag{4.2.1}$$

and with the usual conservation law $\partial_\mu J^\mu = 0$, which holds independent of Eq. (4.3) because of the continuity of the functions $\alpha(x, t)$. The corresponding integral conserved quantity is:

$$\mathbf{Q} = \int_{-\infty}^{\infty} dx\, J^0 = \frac{1}{2\pi}\int_{-\infty}^{\infty} dx \frac{\partial\alpha}{\partial x} = \frac{1}{2\pi}[\alpha(+\infty) - \alpha(-\infty)], \tag{4.2.2}$$

and it is easy to note, that the integrand in the second term of equality (4.2.2) is just the Jacobian of the change of the Cartesian variable x into the angle variable α with values on the circle $\mathbf{S}^1$. This is just the explicit form of the winding number in the case of the Sine–Gordon model, which takes only integer values from $\mathbf{Z}$. Each integer value can be ascribed to each class of solutions of Eq. (4.3), as shown in Fig. 2 and is easily derived from the expression (4.2.2). The solutions (maps) with the same value of $\mathbf{Q}$ form the so-called *homotopy classes*, which are usually denoted by $[\mathbf{R}^1, \mathbf{S}^1]_i$. The space of all maps (4.1.1) with boundary conditions (4.1.3), denoted by $\mathbf{Map}^0(\mathbf{R}^1, \mathbf{S}^1)$, one says is split into homotopy classes, as

$$\mathbf{Map}^0(\mathbf{R}^1, \mathbf{S}^1) = \bigcup_i [\mathbf{R}^1, \mathbf{S}^1]_i . \tag{4.2.3}$$

The homotopy classes from (4.2.3) could be considered as elements of the first *homotopy group*, denoted by $\pi_1(\mathbf{S}^1)$ (or the *fundamental Poincaré group* of the circle $\mathbf{S}^1$) with an appropriate composition law imposed on them. Then one can say that the winding number $\mathbf{Q}$ from (4.2.2) establishes an isomorphism:

$$\mathbf{Q}: \pi^1(\mathbf{S}^1) \to \mathbf{Z} . \tag{4.2.4}$$

Pursuing his initial motivations, which we have already mentioned in the introduction, Skyrme also tried to get an answer to the question of whether it is possible in the framework of such a field theory as the Sine–Gordon model to treat the solutions, which could be massive in comparison with underlying fundamental fields,[21] as *fermion states*. This will be the topic of our next subsection.

4.3. *The Fermi–Bose Transmutations*

Here we will, following Skyrme, frequently use the term *particle* in the sense of *fermion*. As it follows from (4.1.4), there are two types of kinks, corresponding to the $\pm$ signs, which can be considered as representing a particle and an antiparticle respectively. When there are a number of x_0 values [let us remember, that x_0 can be interpreted as the localization point of the soliton (antisoliton)] which are sufficiently far apart from each other, then the net function, say $\bar{\alpha}(x)$, may be well approximated by

$$\bar{\alpha}(x) = \sum_i \alpha^+(x - x_0^i) + \sum_j \alpha^-(x - x_0^j) . \tag{4.3.1}$$

We denote here as $\alpha^\pm$ solutions (4.1.4), corresponding to $\pm$ sign. Each term of (4.3.1) contributes a kink or an antikink to the net function $\bar{\alpha}(x)$ and the net sum, which is

$$\bar{\alpha}(+\infty) - \bar{\alpha}(-\infty) = \sum_i n_i^+ + \sum_j n_j^- , \tag{4.3.2}$$

will remain constant throughout the motion, even though the kinks overlap.[g] The last equation is identical with the statement that the number of particles minus antiparticles is a conserved quantity, usually seen only in the fermionic type field theories. Thus it

[g] This consideration was partly borrowed from the Dalitz paper, Ref. 1.

is possible to say that Skyrme discovered in the purely bosonic Sine–Gordon model the existence of a conserved quantity, usually attributed to fermionic fields (one may be reminded of the electric charge, baryon and lepton numbers, etc.).

To proceed further, Skyrme, in Ref. 7, considered a possible quantization scheme for the Sine–Gordon model and was able to write down the explicit form for a creation operator of a fermion particle, which was neutrino-like (massless) and obeyed the anticommutation relation in the case where the energy-scale factor ε satisfied the condition $\varepsilon = \frac{\hbar c}{2}$. That gave him cause to suggest that the quantum solitons of the Sine–Gordon model may be equivalent to fermions interacting through a four-fermion interaction. This suggestion was rigorously proved only in 1975 by S. Coleman,[25] who showed that the quantum Sine–Gordon model is equivalent to the zero-charge sector of the massive Thirring model in (1 + 1) dimensions with the Lagrangian density:

$$\mathscr{L}_{\mathrm{MT}} = i\bar{\psi}\gamma_\mu \partial^\mu \psi - \mathrm{m_f}\bar{\psi}\psi - \tfrac{1}{2}g(\bar{\psi}\gamma^\mu\psi)(\bar{\psi}\gamma_\mu\psi). \tag{4.3.3}$$

In the framework of the perturbation theory, Coleman compared the so-called n-point functions of the composite operator $[\bar{\psi}\psi]$ in the massive Thirring model (4.3.3) and that of the term $[\cos\alpha]$ in (4.2) and found that these two sets are equal for all n.

It is obvious from Skyrme's subsequent papers—see Refs. 8, 10 and 11—that he was not satisfied with the result of Ref. 7 and undertook different and rather resourceful attempts to improve it. Thus in Ref. 10 he proposed the way of introduction collective coordinates, with the help of which he demonstrated that solitons of the Sine–Gordon equations propagate according to one-dimensional "neutrino-like" equations. Though the rigorous results in Fermi–Bose transmutations were gained in particle physics only recently, one can say that Skyrme's name ought to be listed among those of the first rank of physicists who launched this very promising branch of modern investigation. We shall return to this topic in Sec. 6 and will there discuss in brief its present state.

5. The Baryon Model—Topological Skyrmions

Encouraged by the nontrivial results obtained in the simplified (Sine–Gordon) model, especially with the discovery of the conserved topological integral of motion, Skyrme came to his final modification of the (3 + 1)-dimensional model in such a way that the latter should also possess a conserved quantity of topological character. As we remember, the starting point of his constantly modified sequence of models was just the nuclear matter model, but in the process of evolution he gradually came to look more at the idea of a single nucleon as a nuclear "Wirbelbewegung" in some ether-like fluid. In that case, an analogous topological conserved quantity, if any, could possess an interpretation as *the baryon number*. The nature of that conserved quantity, introduced into particle physics by E. Wigner et al. as a quite formal characteristic just in order to restrict the possible types of reactions between baryons, was and still remains the point of attraction for many investigators. The conservation law of the baryon number holds to the present day with mysterious accuracy (in spite of predictions of the Grand Unification Theory). It is clear that the nature and role of this number in hadron physics is quite different to that of the electric charge. The baryon number does not define in any sense the value of the coupling constant of strong interactions, contrary to the role of the electric charge in electrodynamics. That is why it seems unnatural to derive the baryon-number conservation law from the invariance of the Lagrangian under gauge transformations, as it is usually adopted in close analogy with the derivation of the electric charge conservation law. This suggestion looks too artificial and that was Skyrme's own rigorous conviction, shared by a number of thoughtful researchers.

5.1. *The Left Chiral Currents as* (3 + 1)*-dimensional "Angular" Variables*

As could be learned from the simplified (Sine–Gordon) lesson, to be able to construct a conserved quantity of the topological character in a more realistic physical model it is necessary to find a (3 + 1)-dimensional analog of the angular variable $\alpha(x)$. Skyrme managed to do this when he noticed that from a geometrical point of view the condition (3.4) is just the equation of the $\mathbf{S}^3$ manifold in the isotopic spin space. The analogous condition in the (1 + 1)-dimensional case is the equation ${\phi_0}^2 + {\phi_1}^2 = 1$, which is just the $\mathbf{S}^1$-manifold (the circle) equation. It is possible to represent the complex-valued field $\phi(x) = \exp\{i\alpha(x)\}$ from Eq. (4.1.1) in the form $\phi = \phi_0 + i\phi_1$. Because of this, one can write down the Jacobian of the change of variables [the integrand of the second term in Eq. (4.2.2)] in terms of the fields ϕ as follows

$$\frac{\partial\alpha}{\partial x} = -i\phi^* \frac{\partial\phi}{\partial x}, \tag{5.1.1}$$

where ϕ^* is the complex conjugate of ϕ, which coincides with ϕ^{-1}. To perform the analogous change to "angle" variables in the (3 + 1)-dimensional case it is suitable to use the corresponding generalization of complex numbers—the quaternionic representation of the group SU(2), which is topologically isomorphic to the sphere $\mathbf{S}^3$. Let us choose a unit quaternion $\mathbf{U}$, parametrized by mesonic fields $\phi_\rho = (\phi_0, \underline{\phi})$:

$$\mathbf{U}(\mathbf{x}, t) = \mathbf{I}\phi_0(\mathbf{x}, t) + i\underline{\tau}\underline{\phi}(\mathbf{x}, t), \tag{5.1.2}$$

where $\mathbf{I}$ is a unit 2×2 matrix; $\underline{\tau}$ are isospin Pauli matrices and $\mathbf{x} \in \mathbf{R}^3$. The condition (3.4) in terms of $\mathbf{U}$ can be re-expressed as $\mathbf{U}\mathbf{U}^{-1} = \mathbf{I}$. The field $\mathbf{U}(\mathbf{x}, t)$ is called *the chiral field* and at any fixed moment of time it is possible to think of this field as the map

$$\mathbf{U}(\mathbf{x}): \mathbf{R}^3 \to \mathbf{S}^3. \tag{5.1.3}$$

The other useful parametrization of the field $\mathbf{U}$ is:

$$\mathbf{U}(\mathbf{x}, t) = \exp\{i(\underline{\mathbf{n}} \cdot \underline{\tau})\Theta(\mathbf{x}, t)\}; \qquad \underline{\mathbf{n}} = \frac{\underline{\phi}}{|\underline{\phi}|}; \qquad |\underline{\phi}| = \pm\sin\Theta. \tag{5.1.4}$$

In what follows we will write down formulas for dynamical quantities using both of these parametrizations.

The next step is to find the form of the boundary conditions one should impose on the mesonic fields ϕ_ρ in order to provide a description of extended objects localized in space with finite dynamical characteristics. It is obvious that those conditions should be analogous to Eq. (4.1.3) in the Sine–Gordon model and it means that we are to identify the points at spatial infinity, so that as $|\mathbf{x}| \to \infty$ the fields ϕ_ρ on $\mathbf{S}^3$ should tend to a fixed point on $\mathbf{S}^3$:

$$(\phi_0, \underline{\phi})(\mathbf{x}) \to (1, 0), \quad \text{as} \quad |\mathbf{x}| \to \infty. \tag{5.1.5}$$

From definitions (5.1.2) and (5.1.4), it is clear that corresponding boundary conditions on $\mathbf{U}$ and on Θ look like the following:

$$\mathbf{U}(\mathbf{x}) \to \mathbf{I}; \qquad \Theta(\mathbf{x}) \to 0, \quad \text{as} \quad |\mathbf{x}| \to \infty. \tag{5.1.6}$$

One can easily come to the conclusion that because of conditions (5.1.5) or (5.1.6) the space $\mathbf{R}^3$ is effectively compactified, so that $\mathbf{R}^3 \cup \{\infty\} \simeq \mathbf{S}^3$. This conclusion would be an important one for the right definition of the topologically conserved quantity below.

Now we are in the position to write down the "angular" variable in the (3 + 1) model as the straightforward generalization of Eq. (5.1.1) in the case of quaternionic fields

$$\mathbf{B}_\mu = \mathbf{U}^{-1}\partial_\mu \mathbf{U} = i\tau_a \mathbf{B}^a_\mu; \qquad a = 1, 2, 3, \tag{5.1.7}$$

where, in terms of $\mathbf{U}$, ϕ_ρ and correspondingly in terms of $\mathbf{n}$ and Θ, the components of $\mathbf{B}^a_\mu$ are:

$$\begin{aligned}\mathbf{B}^a_\mu &= \frac{1}{2i}\mathrm{Tr}(\mathbf{U}^{-1}\tau^a\partial_\mu \mathbf{U}) \\ &= \phi_0 \partial_\mu \phi^a - \phi^a \partial_\mu \phi_0 + \varepsilon^{abc}\phi_b \partial_\mu \phi_c \\ &= n^a \partial_\mu \Theta + \sin\Theta \cos\Theta \partial_\mu n^a + \varepsilon^{abc} n_b \partial_\mu n_c .\end{aligned} \tag{5.8}$$

To be well defined the variable $\mathbf{B}^a_\mu$ should satisfy the compatibility conditions of the form:

$$\partial_\nu \mathbf{B}^a_\mu - \partial_\mu \mathbf{B}^a_\nu - 2\varepsilon_{abc}\mathbf{B}^b_\mu \mathbf{B}^c_\nu = 0, \tag{5.1.9}$$

which are just another expression for the natural conditions $\partial_\mu \partial_\nu \mathbf{U} = \partial_\nu \partial_\mu \mathbf{U}$. The condition (5.1.9), from the other side, means that the covariant curl of these "angular" variables $\mathbf{B}^a_\mu$ should be identical to zero. Later on, Skyrme's "angular" variables $\mathbf{B}_\mu$ were reintroduced by H. Sugawara and C. Sommerfield in their field theory of currents.[26] As a Lagrangian density they considered the simpliest chiral invariant, which in terms of $\mathbf{B}_\mu$ may be written in the form:

$$\mathscr{L} = -\tfrac{1}{4}\operatorname{tr}(\mathbf{B}_\mu \mathbf{B}^\mu),$$

and is a straightforward generalization of quadratic functionals in the linear field theory. Different aspects of those functionals have been studied before by F. Gürsey, J. Cronin, S. Weinberg and other authors in the Phenomenological Lagrangian approach to the strong interactions. The Lagrangian is manifestly invariant under the left transitions and $\mathbf{B}_\mu$ are conserved currents in the Sugawara–Sommerfield theory. Hence they were called *the left chiral currents* (with the notation $\mathbf{L}_\mu$, which has been admitted in all recent papers).[h] In what follows we shall also use the Sugawara–Sommerfield notation. From a geometrical point of view the chiral currents $\mathbf{L}_\mu$ are vector fields defined on the $\mathbf{S}^3$ manifold with values in the $su(2)$ algebra. Then the condition (5.1.9) in an equivalent form

$$\partial_\mu \mathbf{L}_\nu - \partial_\nu \mathbf{L}_\mu + [\mathbf{L}_\mu, \mathbf{L}_\nu] = 0, \tag{5.1.10}$$

where [,] is the commutator of the Lie algebra, could be considered as the Maurer–Cartan structural equations or the zero-curvature conditions. They are necessary and sufficient conditions for the reconstruction of the chiral field $\mathbf{U}$ (an element of the Lie group) out of the form of the left chiral current $\mathbf{L}_\mu$ (an element of the Lie algebra).[27]

5.2. *The topological charge and topological current*

To write down an explicit form for the topologically conserved quantity Skyrme, in Ref. 6, used the natural generalization of the expression for the Jacobian (4.2.2) in the following form $\partial\alpha/\partial x \to \det\{\mathbf{L}_i^a\}$ and as the coefficient in (4.2.2) the reverse "volume" of the $\mathbf{S}^1$ sphere. Then the following expression for the conserved quantity looks straightforward:

$$\begin{aligned}\mathbf{Q} &= -\frac{1}{2\pi^2}\int d^3x \det\{\mathbf{L}_i^a\} = -\frac{\varepsilon^{ijk}}{48\pi^2}\int d^3x \operatorname{Tr}(\mathbf{L}_i\cdot[\mathbf{L}_j, \mathbf{L}_k])\\ &= -\frac{\varepsilon^{ijk}\varepsilon_{\alpha\beta\gamma\delta}}{12\pi^2}\int d^3x\, \phi^\alpha \partial_i \phi^\beta \partial_j \phi^\gamma \partial_k \phi^\delta\\ &= -\frac{\varepsilon^{ijk}\varepsilon_{abc}}{4\pi^2}\int d^3x \sin^2\Theta\, \partial_i \Theta n^a \partial_j n^b \partial_k n^c,\end{aligned} \tag{5.2.1}$$

where $2\pi^2$ is the $\mathbf{S}^3$ "volume." There are many ways to check directly from one of expressions (5.2.1) that $\mathbf{Q}$ really is a conserved quantity, but the preferable way to do this is to write the corresponding conserved topological current:

[h] Let us hope that no confusion will occur with the notations for left chiral generators, denoted by the same letter in Sec. 3.

$$\mathbf{J}^\mu = -\frac{\varepsilon^{\mu\nu\lambda\rho}\varepsilon_{abc}}{12\pi^2}\mathbf{L}_\nu^a\mathbf{L}_\lambda^b\mathbf{L}_\rho^c$$

$$= -\frac{\varepsilon^{\mu\nu\lambda\rho}\varepsilon_{\alpha\beta\gamma\delta}}{12\pi^2}\phi^\alpha\partial_\nu\phi^\beta\partial_\lambda\phi^\gamma\partial_\rho\phi^\delta$$

$$= -\frac{\varepsilon_{abc}\varepsilon^{\mu\nu\lambda\rho}}{4\pi^2}\sin^2\Theta\,\partial_\nu\Theta n^a\partial_\lambda n^b\partial_\rho n^c. \tag{5.2.2}$$

The conservation law of this current $\partial_\mu \mathbf{J}^\mu = 0$ is just a consequence of the Jacobi identity for Lie algebra commutators. Another easy way to derive it is from the second expression of (5.2.2), keeping in mind that fields ϕ_ρ are not independent because of the constraint (3.4). The conservation of the topological charge **Q**, which has the current component $\mathbf{J}^0$ as the density, is then a straightforward result. It is clear that any derivation of this conservation law is independent of the dynamics of the model, which we are going to specify only now. In the same manner as in Sec. 4 for the (1 + 1)-dimensional model one may say that the topological charge **Q** establishes an isomorphism between the third homotopy group $\pi_3(\mathbf{S}^3)$ and the group of integers **Z**, so that the space of all maps (5.1.3) with the boundary conditions (5.1.5) or (5.1.6), denoted as $\mathbf{Map}^0(\mathbf{R}^3, \mathbf{S}^3)$, is split into homotopy classes with definite values of **Q**. So we again come up to the topological classification of fields but now in the (3 + 1)-dimensional model.

5.3. *The Skyrme model dynamics*

Skyrme suggested that the Lagrangian density for this modified model ought to be[6]:

$$\mathscr{L} = \frac{\varepsilon}{4\pi^2}\left\{\kappa^2\mathbf{L}_\mu^a\mathbf{L}_\mu^a - \frac{1}{2}[(\mathbf{L}_\mu^a\mathbf{L}_\mu^a)^2 - (\mathbf{L}_\mu^a\mathbf{L}_\nu^a)^2]\right\}$$

$$= -\frac{1}{4\lambda^2}\operatorname{Tr}\mathbf{L}_\mu{}^2 + \frac{\varepsilon^2}{16}\operatorname{Tr}[\mathbf{L}_\mu, \mathbf{L}_\nu]^2$$

$$= \frac{F_\pi^2}{16}\operatorname{Tr}(\partial_\mu\mathbf{U}\partial_\mu\mathbf{U}^+) + \frac{1}{32e^2}\operatorname{Tr}[\partial_\mu\mathbf{U}\mathbf{U}^+, \partial_\nu\mathbf{U}\mathbf{U}^+]^2. \tag{5.3.1}$$

Here, in addition to Skyrme's original Lagrangian density given in Ref. 6 [the first expression in (5.3.1)] we have also written the other forms one can find in recent papers, in order to provide an easy passage from one form to another, as well as to get the relations between values of different constants. The summation convention is the usual one, so that, for example,

$$\mathbf{L}_\mu^a\mathbf{L}_\mu^a = \mathbf{L}_0^a\mathbf{L}_0^a - \mathbf{L}_k^a\mathbf{L}_k^a; \qquad k = 1, 2, 3. \tag{5.3.2}$$

The latter term in all variants of expression (5.3.1) for the Lagrangian density is called in the recent literature *the Skyrme term* and there are many discussions of its nature. From our point of view[28] it is easy to understand this term if one remembers the roots of Skyrme's approach in Kelvin's work. If it is possible to consider the left chiral currents $\mathbf{L}_\mu$ as *generalized velocities of the Mesonic Fluid*, then the Skyrme term may be regarded as a squared *generalized vorticity* and in accordance with relations (5.1.10), an explicit expression for this quantity may be written in the form:

$$\mathbf{C}^a_{\mu\nu} \equiv \tfrac{1}{2}(\partial_\nu \mathbf{L}^a_\mu - \partial_\mu \mathbf{L}^a_\nu) = \varepsilon^{abc}\mathbf{L}^b_\mu \mathbf{L}^c_\nu . \qquad (5.3.3)$$

This hydrodynamical analogy could be extended as well to explain the existence of localized vortex-like structures in such kinds of models from a rather general topological point of view. As $\mathbf{L}_\mu$ is the vector field, defined on the $\mathbf{S}^3$ manifold, then according to the "hairy ball" theorem such a field has to contain an irregular point, where its direction is ill defined. At an intuitive level it is clear that the "hairy ball" could not be "combed" without a top in its "hairdo" and this is the main conclusion of the previously mentioned theorem, expressed in a colloquial manner. Those naive speculations could be supported in a more rigorous way due to H. Tze and Z. Ezawa,[29] who gave necessary and sufficient conditions for the existence of vortex-like structures on manifolds, but we shall not proceed in that direction in the frame of this paper. The intuitive level is enough for our purpose to demonstrate that in the final version of the Skyrme model the old ideas of Helmholtz and Kelvin are still vital and one can consider the Skyrmion as just a "nuclear Wirbelbewegung."

Now we may write down the Hamiltonian of the model:

$$\mathbf{H} = -\int d^3x \left\{ \frac{1}{4\lambda^2}\mathrm{Tr}(\mathbf{L}_0{}^2 + \mathbf{L}_i{}^2) + \frac{\varepsilon^2}{8}\mathrm{Tr}\left([\mathbf{L}_0, \mathbf{L}_i]^2 + \frac{1}{2}[\mathbf{L}_i, \mathbf{L}_k]^2 \right) \right\}, \qquad (5.3.4)$$

and following Skyrme[6] obtain the estimate of $\mathbf{H}$ from below through the topological charge $\mathbf{Q}$ from Eq. (5.2.1)

$$\begin{aligned}
\mathbf{H} &\geq -\int d^3x \left\{ \frac{1}{4\lambda^2}\mathrm{Tr}\,\mathbf{L}_i{}^2 + \frac{\varepsilon^2}{16}\mathrm{Tr}[\mathbf{L}_i, \mathbf{L}_k] \right\} \\
&= -\int d^3x \left\{ \frac{1}{4\lambda^2}\mathrm{Tr}\,\mathbf{L}_i{}^2 + \frac{\varepsilon^2}{32}\mathrm{Tr}(\varepsilon_{ijk}[\mathbf{L}_j, \mathbf{L}_k])^2 \right\} \\
&> \frac{\varepsilon}{4\sqrt{2}\lambda}\int d^3x\, |\mathrm{Tr}(\varepsilon_{ijk}\mathbf{L}_i[\mathbf{L}_j, \mathbf{L}_k])| \\
&\geq \frac{\varepsilon}{4\sqrt{2}\lambda} 48\pi^2 |\mathbf{Q}| ,
\end{aligned}$$

where we have used the standard triangle inequality for vectors $\bar{a}^2 + \bar{b}^2 \geq 2|(\bar{a}\cdot\bar{b})|$.

Thus the following estimate holds

$$\mathbf{H} > 6\sqrt{2}\pi^2 \frac{\varepsilon}{\lambda} |\mathbf{Q}|. \tag{5.3.5}$$

On behalf of this estimate Skyrme suggested that the Euler–Lagrange equation of the model

$$\partial_\mu \left(\mathbf{L}^\mu - \frac{\varepsilon^2 \lambda}{2} [\mathbf{L}_\nu, [\mathbf{L}^\mu, \mathbf{L}^\nu]] \right) = 0 \tag{5.3.6}$$

possesses stable solutions with finite dynamical characteristics. One may easily derive Eq. (5.3.6) from the standard variational principle for Lagrangian (5.3.1), taking into account the left chiral currents definition (5.1.7) together with an obvious relation $\delta \mathbf{U}^{-1} = -\mathbf{U}^{-1} \cdot \delta \mathbf{U} \cdot \mathbf{U}^{-1}$ for chiral fields $\mathbf{U}$, so that

$$\delta S = \delta \int d^4 x \mathscr{L} = \int d^4 x \operatorname{Tr} \left\{ \mathbf{U}^{-1} \delta \mathbf{U} \left(\frac{1}{2\lambda^2} \partial_\mu \mathbf{L}^\mu - \frac{\varepsilon^2}{4} \partial_\mu [\mathbf{L}_\nu, [\mathbf{L}^\mu, \mathbf{L}^\nu]] \right) \right\} = 0.$$

The equation (5.3.6) has the form of a local conservation law and to find the corresponding conserved current one has to remember that the Lagrangian (5.3.1) is invariant against the chiral $SU(2)_L \otimes SU(2)_R$ transformations and because of this we have at least two conserved Noether currents:

$$\mathbf{N}^k_{\mu, L} = i \operatorname{Tr} \left\{ \tau^k \left(-\frac{1}{2\lambda^2} \mathbf{L}_\mu + \frac{\varepsilon^2}{2} [\mathbf{L}_\mu, \mathbf{L}_\nu] \mathbf{L}^\nu \right) \right\}, \tag{5.3.7}$$

which corresponds to left generators in (3.2), and the analogous current for the right transformations

$$\mathbf{N}^k_{\mu, R} = i \operatorname{Tr} \left\{ \tau^k \left(-\frac{1}{2\lambda^2} \mathbf{R}_\mu + \frac{\varepsilon^2}{2} [\mathbf{R}_\mu, \mathbf{R}_\nu] \mathbf{R}^\nu \right) \right\}, \tag{5.3.8}$$

where $\mathbf{R}_\mu = \partial_\mu \mathbf{U} \cdot \mathbf{U}^{-1}$ are the right chiral currents by definition. In general in the framework of the Lagrangian formalism all Noether currents can be derived from the formula:

$$\mathbf{N}^k_\mu = \operatorname{Tr} \left(\frac{\partial \mathscr{L}}{\partial (\partial_\mu \mathbf{U})} \cdot \delta^k \mathbf{U} + \frac{\partial \mathscr{L}}{\partial (\partial_\mu \mathbf{U}^{-1})} \cdot \delta^k \mathbf{U}^{-1} \right), \tag{5.3.9}$$

where by $\delta^k \mathbf{U}$ we denote the variations of the chiral field under respective transformations. For example, the variations of $\mathbf{U}$ under isotopic rotations with generators $\mathbf{I}^k$ are as follows

$$\delta_V^k \mathbf{U} = i\left[\frac{\tau^k}{2}, \mathbf{U}\right], \tag{5.3.10}$$

and one can get the conserved isovector (or isospin) current by substitution of (5.3.10) into (5.3.9) in the form

$$\mathbf{V}_\mu^k = i\,\mathrm{Tr}\left\{\tau^k\left(-\frac{1}{2\lambda^2}\mathbf{L}_\mu + \frac{\varepsilon^2}{4}[\mathbf{L}^\nu, [\mathbf{L}_\mu, \mathbf{L}_\nu]] + (\mathbf{L} \to \mathbf{R})\right)\right\}, \tag{5.3.11}$$

so that the expressions in parentheses of Eq. (5.3.6) and of Eq. (5.3.11) are identical. To summarize this consideration, one can say that the Euler–Lagrange equations of the Skyrme model express the conservation of the isotopic spin current.

5.4. *The Skyrme conjectures*

Without going into a more detailed treatment of the final version of the Skyrme model, let us outline to what extent Skyrme managed to study the proposed theory by himself. The other no less important aim of this section is to list the main suggestions and hypotheses of Skyrme, which were expressed in Refs. 6, 9 and 10 and became the subject of great interest among particle physicists during the previous decade. The reason for the revival of interest in Skyrme's ideas is well-known. The first substantial step in that direction has been done by G. t'Hooft.[30] Looking for an expansion parameter in QCD he found the so called $1/N_c$ expansion, where N_c is the number of colors. The next discovery was that QCD is drastically simplified in the framework of this expansion in the limit $N_c \to \infty$. If one assumes confinement, then QCD might possess an effective description of the low-energy events (the Hadron Physics region) in terms of meson fields and glueballs, with the meson coupling constant of order $1/N_c$. Then E. Witten in Ref. 31 showed that for large N_c, baryon masses are of order N_c, while the baryon size is of order one and baryon–baryon and baryon–meson cross-sections are of the same order one. He also suggested that the effective meson theory should be of the σ-model type with a spontaneously broken internal symmetry, which admits soliton solutions. It happened that the Skyrme model, which substantially predates the formulation of the main principles of QCD, is one of the simplest variants of such a theory, compatible with most of the aforementioned requirements.

After this step aside, let us return to Skyrme's proposals.

1. First of all Skyrme suggested that the mesonic fields could take their values on the $\mathbf{S}^3$ manifold. As a result of this assumption he discovered the conserved quantity—the topological charge or the winding number (5.2.1) and after some debates he expressed the hypothesis that this quantity might be interpreted as the baryon number.
2. Skyrme's next important suggestion was that it is possible to search for solutions

of the field equations (5.3.6) in the form:

$$\phi_0 = \cos\Theta(r); \qquad \phi_i = \frac{x_i}{r}\cdot\sin\Theta(r); \qquad i = 1, 2, 3, \tag{5.4.1}$$

where r is the radial variable. Nowadays this type of solution is known as *the Skyrme's or the "hedgehog" ansatz.*[i]

3. On the behalf of the availability of the already demonstrated estimate (5.3.5) of the energy from below, Skyrme presupposed that solutions (5.4.1) could describe *a stable extended particle with the unit topological charge and all finite dynamical characteristics.*

To proceed further, Skyrme (with the help of a vacation student at A.E.R.E., A.J. Leggatt) undertook computer calculations in search of numerical solutions of the equation (5.3.6) on the "hedgehog" ansatz (5.4.1) and found the behavior of the profile function $\Theta(r)$, reported in Ref. 9. In the same article Skyrme checked if it is possible to obtain a description of bound states in the framework of his model. From the one side, he proved that the topological charge $\mathbf{Q}$ is the additive conserved number, when the two-particle state was considered, described by the field $\mathbf{U}_{12}(\mathbf{x})$ which Skyrme suggested to take as a product

$$\mathbf{U}_{12}(\mathbf{x}) = \mathbf{U}(\mathbf{x} - \mathbf{x}_1)\cdot\mathbf{U}(\mathbf{x} - \mathbf{x}_2). \tag{5.4.2}$$

Here $\mathbf{U}(\mathbf{x} - \mathbf{x}_i)$ are chiral fields, describing single particles with locations $\mathbf{x}_1$ and $\mathbf{x}_2$. But from the numerical analysis it was found that the total energy of the state (5.4.2) with $\mathbf{Q} = 2$ was about three times larger than that of the energy of one-particle configurations: $E^{(2)} \simeq 3E^{(1)}$. Those results Skyrme interpreted as showing the non-existence of stable particles with $\mathbf{Q} = \mathbf{2}$ or, from an alternative point, that there exists a short-range repulsion between particles equal in order of magnitude to a particle mass.

4. As a possible description of the interaction picture among two particles Skyrme suggested the so-called "*product ansatz*" (5.4.2), which he considered as a good approximation for when the particles are far apart from each other.
5. The last, but not the least important suggestion was that solutions of the form (5.4.1) with the unit topological charge *should be quantized as fermions*, so that it would be possible to identify the states with equal isotopic I and total J spins with the nucleon doublet in the case $I = J = \frac{1}{2}$, and with the Δ resonance in the case $I = J = \frac{3}{2}$.

[i] Skyrme in Ref. 6 and later on in his talk, Ref. 11, mentioned, that this form of solutions was inspired by the great influence of Pauli's semiclassical picture of spin and isospin locked together, but as one can find in the obituary on the occasion of Professor T.H.R. Skyrme's death in 1987, one of his characteristic features was to find a person to whom it was possible to ascribe his own findings.

It is interesting to note that Skyrme was one of the first to adopt Finkelstein's ideas of a topological treatment of fermion states in nonlinear field theories. We have already mentioned in Sec. 4.3. that he was not satisfied with his own results in that field and may be that was the reason that as the theme of a thesis for his graduate student J. Williams, he suggested a rigorous proof of the existence of Finkelstein's spinorial structures in the $(3 + 1)$-dimensional model.[32] Later on, the particle states with all foregoing features were called *Skyrmions*[j] and nowadays this term has acquired a nominal meaning and is widely used for description of solitonic states with a nontrivial topological index, which arise in purely bosonic theories but obey the Fermi–Dirac statistics. To conclude this section let us take one more quotation from Skyrme's talk, Ref. 11: "*I still entertain the hope that some type of nonlinear theory will yield an explanation of elementary particles that can be visualized in a semiclassical way, and that quarks or leptons introduced as sources in most theories will be seen to be mathematical constructs helpful in the understanding, rather than fundamental constituents, just as the idea of vortices in a fluid is an indispensible way of talking about certain types of fluid motion.*"

6. The Validity of Skyrme's Conjectures and Some Unsolved Problems

The novelty of Skyrme's approach in the Particle Physics of the 1960s was so impressive that all his proposals were largely ignored for almost two decades. Only a very few amateurs "were splashing in the calm basin" of his theory which seemed to be rather far away from the mainstream of progress in particle physics. The situation changed drastically in the early 1980s after the understanding that the Skyrme model could be considered as a possible low-energy limit of QCD. It became the time of the professionals who "crowded the basin, showed the amateurs some good swimming styles" and then, after a few years of enormous activity, were carried away by the main stream of progress.[k] To be straightforward we have to note that some professionals, mostly nuclear physicists, are still keeping an eye on the flounderings of the amateurs in this not yet formerly quiet basin with a willingness to stretch out a helping hand on the occasion of a good catch.

In chronological order, the first comprehensive papers after Skyrme's were written by N. Pak and H. Tze[33] and J. Gipson and H. Tze,[34] where the Skyrme soliton was studied at the semiclassical level in the Sugawara–Sommerfield current algebra approach and possible implications of Skyrme's ideas for weak interactions were suggested. But the real explosion of interest in the Skyrmion features was aroused after the series of papers by A. P. Balachandran et al. (Refs. 35 and 36) and E. Witten et al. (Refs. 37 and 38). For this part of the Skyrmion history one may consult good reviews

[j] Thus after E. Fermi and S. Bose, T. Skyrme became the third person in Particle Physics' history whose last name, by adding the suffix "on", was used for a designation of the special type of particles, which are fermions composed from bosons.

[k] This descriptive picture of relations among amateurs and professionals in any branch of science is due to F. Dyson in his well-known talk called "The Lost Opportunities."

(Refs. 38–42), lecture notes (Refs. 43–45), and proceedings of conferences and workshops (Refs. 47 and 48). Before we enter into a more detailed examination of the validity of Skyrme's suggestions, conjectures and hypotheses, we can declare at once that most of them gained strong support from QCD fundamental principles. Here we are not going to review all of the recent results in support of Skyrme's conjectures, as it would lead to an enormous enlargement of the paper. We choose only those results which either support or refute Skyrme's main ideas and suggestions listed in Sec. 5.4. We will consider the further development and modification of the Skyrme model in the final section.

6.1. *Chiral fields on* SU(N) *and Baryon number from QCD fundamentals*

According to the order in which Skyrme's conjectures were listed in previous sections, let us first consider the suggestion that the mesonic fields could be valued on the $\mathbf{S}^3$ manifold. This suggestion could be supported from the QCD point of view in the following sense: Assume that it is possible to ignore the "bare" or "current" quark masses in comparison with the pion masses or the QCD scale parameter $\Lambda \simeq 300$ MeV. Following t'Hooft's ideology,[30] consider the generalized QCD fundamental Lagrangian in the $\mathrm{U}(N_f)_L \otimes U(N_f)_R$-invariant form:

$$\mathscr{L}_{\mathrm{QCD}} = -\frac{N_c}{4g^2}\mathbf{G}^a_{\mu\nu}\cdot\mathbf{G}^a_{\mu\nu} + i\bar{q}\gamma^\mu D_\mu q, \tag{6.1.1}$$

where the quark field q^a_i is in the fundamental representation of both the $\mathrm{SU}(N_c)$ color and the $\mathrm{SU}(N_f)$ flavor groups ($a = 1,2,..N_c$; $i = 1,2,..N_f$). The field strength $\mathbf{G}^a_{\mu\nu}$ of the gauged gluon fields is taken in the adjoint representation of $\mathrm{SU}(N_c)$; here $D_\mu = \partial_\mu + gA^a_\mu\lambda^a$ is the usual covariant derivative; λ^a being the $\mathrm{SU}(N)$ analogs of the Gell-Mann matrices; g is the color coupling constant. The invariance group of (6.1.1) may be presented in a more precise form as

$$\begin{aligned} U(N_f)_L \otimes U(N_f)_R &= \mathrm{SU}(N_f)_L \otimes \mathrm{SU}(N_f)_R \otimes \mathrm{U}(1)_L \otimes \mathrm{U}(1)_R \\ &= \mathrm{SU}(N_f)_L \otimes \mathrm{SU}(N_f)_R \otimes \mathrm{U}(1)_A \otimes \mathrm{U}(1)_V, \end{aligned} \tag{6.1.2}$$

where $\mathrm{U}(1)_V$ and $\mathrm{U}(1)_A$ are the vector and axial subgroups with generators which could be constructed from the generators of left- and right-handed transformations like (3.2) as the half-sum and half-difference, correspondingly. At the quantum level the $\mathrm{U}(1)_A$ symmetry is explicitly broken by the Adler–Bell–Jackiw anomaly, leaving us with the symmetry group $\mathrm{SU}(N_f)_L \otimes \mathrm{SU}(N_f)_R \otimes \mathrm{U}(1)_V$, with transformations of the type

$$q \to \exp\left\{\frac{i}{2}\gamma_5\pi^k\lambda^k\right\}q, \tag{6.1.3}$$

which interchange states with different parities; here π^k are chiral phases. For the left- and right-handed quark fields:

$$q = q_L + q_R; \qquad q_{L(R)} = (1 \pm \gamma_5)q, \tag{6.1.4}$$

the transformations (6.1.3) can be written in the form

$$q \to \mathbf{U}^{1/2} \cdot q_L + \mathbf{U}^{-1/2} \cdot q_R; \qquad \mathbf{U} = \exp\{i\pi^k \lambda^k\}. \tag{6.1.5}$$

The spontaneous breaking of the global $SU(N_f)_L \otimes SU(N_f)_R$ symmetry means that the relative chiral phases of the left- and right-handed quarks became locally fixed so that phase functions $\pi^k(x)$ might be related with the multiplet of pseudo-Goldstone bosons (with the octet of pseudo-scalar mesons $[\pi^0, \pi^\pm, \eta, K^0, K^\pm, \bar{K}^0]$ in the case of three flavors). As was already discussed in Sec. 5.1, at spatial infinity $|x| \to \infty$ the pion fields $\pi^k(x) \to 0$ or $\mathbf{U}(x) \to \mathbf{I}$, where $\mathbf{I}$ is an $N \times N$ unit matrix. Under the transformations from the chiral group $G = SU(N_f)_L \otimes SU(N_f)_R$ the field $\mathbf{U}(x)$ is transformed as $\mathbf{U} \to V \cdot \mathbf{U} \cdot W^{-1}$, where $V \in SU(N_f)_R$ and $W \in SU(N_f)_L$. But the vacuum $\mathbf{U} = \mathbf{I}$ is invariant under this type of transformations iff $V = W$, or in other words under transformations from the group

$$G_V = \text{diag}[SU(N_f)_L \otimes SU(N_f)_R] \simeq SU(N_f)_I, \tag{6.1.6}$$

where $SU(N_f)_I$ is the analog of the isospin group. The field $\mathbf{U}$ defines the orbit of the group G passing through the unit element $\mathbf{I}$ and thus takes the values on the homogenous coset space $G/G_V \simeq SU(N)$. In other words a one-to-one correspondence between the weak (phase) vacuum excitations, which parametrize the field $\mathbf{U}(x)$, and elements of the $SU(N_f)$ group manifold emerges. That is why it is reasonable to describe the low-energy QCD dynamics in terms of chiral fields $\mathbf{U}(x)$ and at the same time it means that only pionic degrees of freedom are essential in that region of QCD. Thus in the case of three-flavored QCD the field manifold might be taken as a SU(3) group manifold, and in the case of two-flavored theory the SU(2) manifold, which is homeomorphic to the sphere $\mathbf{S}^3$. This completes the available justification of the first of Skyrme's suggestions.

The following remarks are in order. All these considerations are carried through on the so-called physical level of accuracy. First of all we still do not have a good understanding or even a reasonable scenario for a number of phenomena involved in the preceding discussion, such as the spontaneous breaking of symmetry or the bosonization in a (3 + 1)-dimensional theory. Thus one can say that we support Skyrme's suggestions by producing new ones. The other remark is that in fact only masses of u and d quarks are really ignorable with respect to Λ, the mass of the s quark is already of the same order of magnitude, and the masses of heavy quarks $c, b, t, \ldots$ are much bigger than Λ, and the chiral limit could be taken as a good approximation only in the two-flavored case.

The demonstration that the topological charge (5.2.1) possesses the interpretation as the baryon number is due to Balachandran et al.[35] and is based on some results of Goldstone and Wilczek.[49] They considered the dynamics of the second quantized quark field in the presence of the externally prescribed classical chiral field $\mathbf{U}(x)$, i.e. the interacting system of quarks and solitons. Due to the presence of this external field, the Dirac sea of quarks is "polarized" and carries a baryonic current exactly equal to the topological current (5.2.2). The accessible exposition of this result can be found in Ref.[44]

6.2. *Existence and stability of hedgehog skyrmions*

Concerning the proposed form of the hedgehog ansatz (5.4.1) it is possible to say that in this case Skyrme demonstrated his really great mathematical intuition, as the rigorous proof of that result requires not only time and paper, but also the use of a huge mathematical apparatus of functional analysis and direct methods of the calculus of variations. In Ref. 50 it was shown that in the class of fields with $|\mathbf{Q}| = 1$, the Skyrme ansatz (5.4.1) exactly realizes an absolute minimum of the energy. In order to seek the absolute minimum of the functional (5.3.4) *the method of extending the phase space* has been used. This method consists of regarding the field functions and their derivatives as independent and then proving that the minimum found in this way is indeed a critical point of the original functional. The validity of this device is based on the fact that the minimum in the extended space will also be a true minimum since the imposition of constraints (between the functions and their derivatives, which reflect the fact of their dependence) can only raise the value of the energy functional. The existence of regular solutions for Eq. (5.3.6) on the Skyrme ansatz (5.4.1) has also been proved in Ref. 50 and it was shown that the profile function $\Theta(r) \in C^{\infty}[0, \infty]$, or in other words the solution $\Theta(r)$, is a real analytic function.[1] The modernized exposition of this proof with more detail is given in Ref. 46. This question has also been considered in Ref. 52 by a method based on P. Lions's concentration-compactness principle, but the existence of the energy minimum has been proved under some rather restrictive assumptions.

The stability in the Lyapunov sense of Skyrmions in accordance with Skyrme's suggestion is in fact guaranteed by the estimate (5.3.5), which means that in any homotopy class the energy has the lower bound. If this lower bound is attainable on a solution of field equations, then such a solution is stable in the Lyapunov sense. The type of estimates (5.3.5) for nonlinear functionals was used in mathematical papers[53] for the proof of harmonic mappings stability. There is also a general statement due to Rybakov,[54] that in field theories with translationally invariant functionals there could not be any absolutely stable stationary solitons in the Lyapunov sense and the only possibility is the conditional stability or the so-called $\mathbf{Q}$ stability.[55] This statement

[1] This preprint[50] is still called "unpublished" in references, as in 1982 it was rejected from *Nuclear Physics B* on the basis of the following referee's conclusion, which merits to be quoted: "While the Skyrme model is a valid exercise for application of the method due to Isham[51], *it is not a model of physical interest* and also does not possess an interesting mathematical structure.... Therefore, this referee doubts the suitability of the paper, although it is apparently mathematically correct, for publication in *Nuclear Physics B*." We hope that now this referee has changed his or her opinion after the 1983 boom.

generalizes the "no-go" Hobart–Derrick theorem,[56] which establishes the negativeness of the second variation of the energy functional in the neighborhood of static solitons with respect to scale transformations. In Ref. 54 it was shown that the second variation of a translationally invariant functional does not have a definite sign in the neighborhood of stationary solitons. This, a more general than static class of solutions, includes as a special case periodic-in-time solutions. It means that stable solitons could arise in these kind of theories only under some restrictions on the form of initial perturbations. In the case of topological solitons these restrictions are a sort of natural condition. The availability of an attainable energy estimate from below by a positive definite function of a topological charge $\mathbf{Q}$ is sufficient for stability of solitons in topologically nontrivial models. This kind of stability gained the special name *The Topological Stability*. When such an estimate is exact it was shown by Bogomol'nyi[57] that it is possible to reduce the Euler–Lagrange equation for a given functional to a first-order (self-duality) equation. But this kind of trick does not work for Skyrme's functional, as the estimate (5.3.5) is not exact and the corresponding self-duality equations do not have any solution compatible with the integrability conditions (5.1.10) and satisfying simultaneously the boundary conditions (5.1.6). Thus to prove the stability of Skyrmions one has to show that the lower-energy bound (5.3.5) is attainable on solutions from the first homotopy class. This can be done by a direct method of the calculus of variations, as was demonstrated in Ref. 50.

6.3. *Bound states and multi-baryon configurations*

The problems of bound states and interactions among Skyrmions are closely related with the problem of existence of stable soliton structures in higher homotopy classes with $\mathbf{Q} \geq 2$. Those questions are of great importance as long as one considers possible applications of the Skyrme model in nuclear physics and for this reason they have been reexamined at once after the renewal of interest in 1983. At first the existence of stable bound states, called *dibarions*, was disclosed in the SU(3) generalized Skyrme model by A. P. Balachandran et al.[58] Then E. Braaten and L. Carson[59] found numerically the bound states in a modified product-ansatz approximation found in Ref. 60, computing the static energy as a function of the relative isospin orientation and the half-separation of two Skyrmions. Their results differed from Skyrme's own, cited in the previous section, as the static energy of a two-Skyrmion state was found to be $E^{(2)} - 2E^{(1)} = -24$ MeV, when the half-separation of Skyrmions $(\mathbf{x}_1 - \mathbf{x}_2)/2 = 1.1$ fm. In accordance with suggestions in Ref. 60, E. Braaten and L. Carson demonstrated that when the isospin axes of one Skyrmion are rotated through π about any axis perpendicular to the line of separation, this would be the most favored isospin orientation corresponding to the minimum of interaction energy. Yet again, contrary to Skyrme's original conclusions, two Skyrmions under that orientation exert an attractive force on each other and thus could bind together to form stable, localized multisoliton configurations. On behalf of obtained results it was proposed in Ref. 59 and confirmed by further analysis in Ref. 61, that the $\mathbf{Q} = 2$ soliton could be identified with the deuteron, as it has the correct quantum numbers and gives reasonable values for such observables as mean charge radius, magnetic and quadrupole moments. The values obtained for $\mathbf{Q} = 2$ solitons agreed with experiments to within 30%. Later on the crude variational ansatz used in Ref. 59 was improved in a series of papers 62 and it was verified that the disclosed solution is indeed a $\mathbf{Q} = 2$ soliton with the lowest energy. The other important feature for our discussion underlined in Ref. 61 is that those $\mathbf{Q} = 2$ solitons are approximately toroidal in shape. We shall exploit this feature later on, but now let us explain the Braaten–Carson toroidal Skyrmions from the point of the Principle of symmetric criticality, following Ref. 63 and Chapter 7 in Ref. 46.

The principle of symmetric criticality, known among physicists as the Coleman–Palais theorem 64, in brief asserts that *critical symmetric points are symmetric critical points* or in an extended version: *extremals of symmetric functionals (if any) are realizable in the class of invariant functions.* In more detail this statement could be formulated as follows: Let the functional under study $\mathbf{H}[\phi]$ be invariant against transformations from a compact group G.[m] Let the group G include as a subgroup the group of spatial rotations $SO(3)_S$ or its subgroup $SO(2)_S$, as well as an internal symmetry group, such as the isotopic rotation group $SO(3)_I$ or its subgroup $SO(2)_I$. Then the principle makes it possible to narrow our search of extremals of $\mathbf{H}[\phi]$ to the class of invariant fields $\phi_0(\mathbf{x})$, defned by the relation

$$\phi_0(\mathbf{x}) = T_g \phi_0(g^{-1} \cdot \mathbf{x}); \qquad g \in G, \tag{6.3.1}$$

where T_g is an operator in a representation of group G. Two symmetry groups are of special interest in our discussion:

$$\begin{aligned} G_1 &= \mathrm{diag}[SO(3)_I \otimes SO(3)_S], \\ G_2 &= \mathrm{diag}[SO(2)_I \otimes SO(2)_S]. \end{aligned} \tag{6.3.2}$$

In Refs. 65 and 66 it has been proven that groups G_1 and G_2 are the only maximal compact groups which possess the invariant fields with nontrivial topological charges in realistic (3 + 1)-dimensional models. The fields invariant against transformations from G_1 are spherically-symmetric fields of the hedgehog type (5.4.1). It can be easily verified just by writing down the exact form of Eq. (6.3.1) in a concrete representation of G_1 and getting (5.4.1) as the only possible topologically nontrivial solution of those equations (the detailed calculations may be found in Refs. 44 and 46). The analogous calculations for the group G_2 give already two possible types of solutions: the one is still the hedgehog configuration (5.4.1) and another one is the axisymmetric ansatz:

$$\Theta = \Theta(r, \alpha); \qquad \beta = \beta(r, \alpha); \qquad \gamma = m\phi \tag{6.3.3}$$

[m] We choose one of the cases where the principle is valid. As it was demonstrated by R. Palais in Ref. 64 the principle is not valid in its whole generality, but the special cases where it is valid cover almost all situations of physical interest: the compact groups, unitary representations of noncompact groups, semi-simple groups and Riemannian manifolds.

where (r, α, ϕ) are spherical coordinates in $\mathbf{R}^3$, $m \in \mathbf{Z}$, and Θ, β, γ are variables corresponding to the parametrization (5.1.4) with

$$n^1 + in^2 = \sin\beta(\mathbf{x})\exp\{i\gamma(\mathbf{x})\}; \qquad n^3 = \cos\beta(\mathbf{x}). \tag{6.3.4}$$

It was demonstrated in Ref. 63 by direct minimization of the energy functional over extended variables that the situation in higher homotopy classes $\mathbf{Q} \geq 2$ essentially differs from that in the first homotopy class as the minimal-energy configurations are no longer spherically symmetric and the hedgehog ansatz might not be used for their description. In the paper refered to an attempt was undertaken to provide the analysis of an absolute minimum configuration similar to that in Ref. 50. The results are the following. In each higher homotopy class there are two aforementioned types of invariant fields, which are isotopic[n] in the first homotopy class. Thus we have a more complicated situation, as we do not have a unique class of extremals and it is necessary to compare the energy values for both configurations (5.4.1) and (6.3.3). This comparison could not any longer be carried out by analytical methods, as the minimization of functional (5.3.4) on the axisymmetric configuration (6.3.3) leads to a two-dimensional vector variational problem for the functions Θ and β. To our knowledge, there is no effective mathematical tool to deal with this kind of problem. That is why, in Ref. 63, the energy values for different symmetric configurations were estimated with the help of trial functions. As a result, the conclusions of Refs. 59, 61 and 62 were confirmed, i.e. that the minimum of functional (5.3.4) in the second homotopy class is realized by axisymmetric configurations. Because of the already mentioned mathematical difficulties the problem of existence of axisymmetric solutions and the related problem of stability of toroidal Skyrmions still remain unsolved.

Encouraged by the success in the description of the deuteron as the $\mathbf{Q} = 2$ soliton of the Skyrme model, Braaten and Carson proposed an approach, which might be considered as an alternative to conventional nuclear physics.[67] In their approach nuclei arise as quantum states of solitons in higher homotopy classes with the topological charge $\mathbf{Q}$, which now might be interpreted as the atomic number. It is too early to discuss the validity of this proposal as only first initial steps have been made in that direction. But some remarks of general character are in order. First of all it is possible to recognize that once again Skyrme's Mesonic Fluid idea developed itself at a new level of study of nuclei structure. From that point of view, the appearance of toroidally shaped solitons would not seem to be an exception, but rather a support to Skyrme's initial image of the nucleon as a vortex in the Mesonic Fluid. There are some, still nonrigorous considerations, that the Skyrmion itself might have a vortex origin[68] so that the hedgehog configuration can be regarded as a "frozen" vortex. From that position it might be possible to suggest a naive explanation of complex structures obtained by further computer calculations in Ref. 69. For the reader's convenience we

[n] Two fields (or mappings) ϕ_1 and ϕ_2: $\mathbf{R}^2 \to \mathrm{SU}(2)$ are called isotopic if there exist homeomorphisms η_1: $\mathbf{R}^3 \to \mathbf{R}^3$ and η_2: $\mathrm{SU}(2) \to \mathrm{SU}(2)$ such that $\phi_2 = \eta_2 \cdot \phi_1 \cdot \eta_1^{-1}$.

display in Fig. 3 those structures presenting surfaces of constant baryon number density in higher homotopy classes.

It is obvious, that for $\mathbf{Q} = 3, 4, 5, 6$ corresponding structures are not axisymmetric. A natural question is whether these computer results are in contradiction to the main statement of Ref. 63, that minimal-energy field configurations in higher homotopy classes with $\mathbf{Q} \geq 2$ should be realized on axially symmetric fields. To answer this question one should take into account the possibility that those structures might arise by the cause, chosen in Ref. 69, of the numerical formulation of the model on a cubic lattice, where continuum quantities $\mathbf{H}$ and $\mathbf{Q}$ were replaced by their discretized analogs. The whole space was divided into some sectors and the minimization of $\mathbf{H}$ was carried out in each sector separately. From Fig. 3 it is clear that the minimal configuration in each sector proves to be axially symmetric. The final minimal configurations were restored in each homotopy class by a continuation procedure. As it was reported by authors themselves, in that approach $\mathbf{Q}_{\text{discrete}}$ is not a topological invariant and so is not strictly conserved. The other objection is a not entirely clear correspondence between surfaces presented in Fig. 3 and minimal-energy configurations in the same homotopy classes. The simplest way to establish this correspondence is to use a discretized analog of the estimate (5.3.5). A one-to-one correspondence would hold in the case when such an estimate would be exact on used numerical solutions. As it follows from numerical data given in Ref. 69 (see Table 1) for $\mathbf{Q}_{\text{discrete}}$ and $M(\varepsilon)_{\text{discrete}}$, where M and ε are masses and static energies of solitons, this, however, is not the case. It is not difficult to check out that for the discretized analogs of topological charge in the N-th homotopy class $\mathbf{Q}^N_{\text{discrete}}$, the inequality

$$\mathbf{Q}^N_{\text{discrete}} \geq N \cdot \mathbf{Q}^1_{\text{discrete}} \tag{6.3.5}$$

holds with good accuracy. At the same time, for discretized analogs of static energies or masses of solitons, we have the inequality in the opposite direction

$$M(\varepsilon)^N_{\text{discrete}} \leq N \cdot M(\varepsilon)^1_{\text{discrete}} \,. \tag{6.3.6}$$

The only conclusion we can make about those inequalities is that this is an important problem that merits a special study, both numerically and analytically.

But anyway, we may prefer to believe that those structures reflect realities in higher homotopy classes, so it might be possible to understand configurations on Fig. 3 as *equilibrium states in interactions among vortices or as clusters of vortices.* For the abovementioned reasons it is difficult to expect a rigorous anaytical confirmation of this observation in the near future. That is why we are able to support our suggestion only by an analogy. Stable, localized structures of nonlinear fields are of great interest in various branches of physics, ranging from an intrinsic structure of elementary particles to the formation of such astrophysical objects as galaxies and clusters of galaxies. The computer study of possible spatially localized structures in the framework of a condensed matter dissipative model of a rather universal character, undertaken

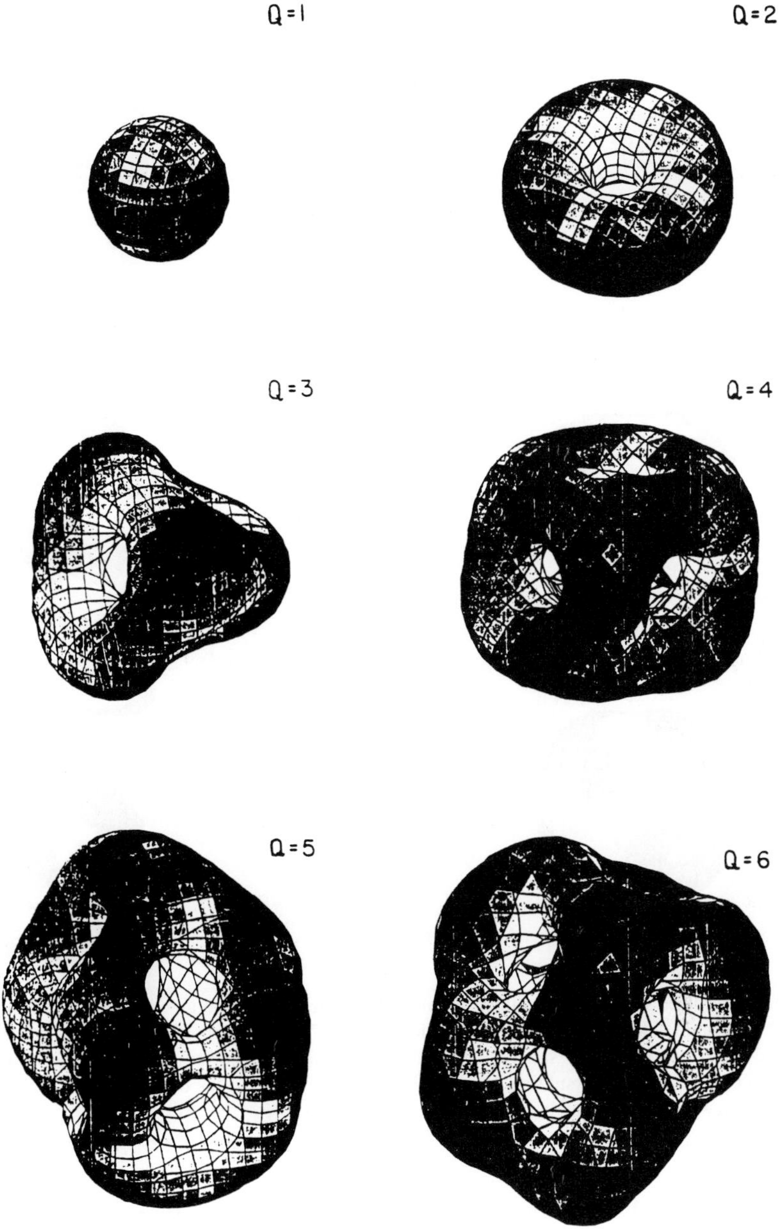

Fig. 3. Surfaces of constant baryon number density for solutions of baryon number $\mathbf{Q} = 1, 2, \ldots, 6$, obtained by computer simulations in Ref. 69. (Adopted from Ref. 69.)

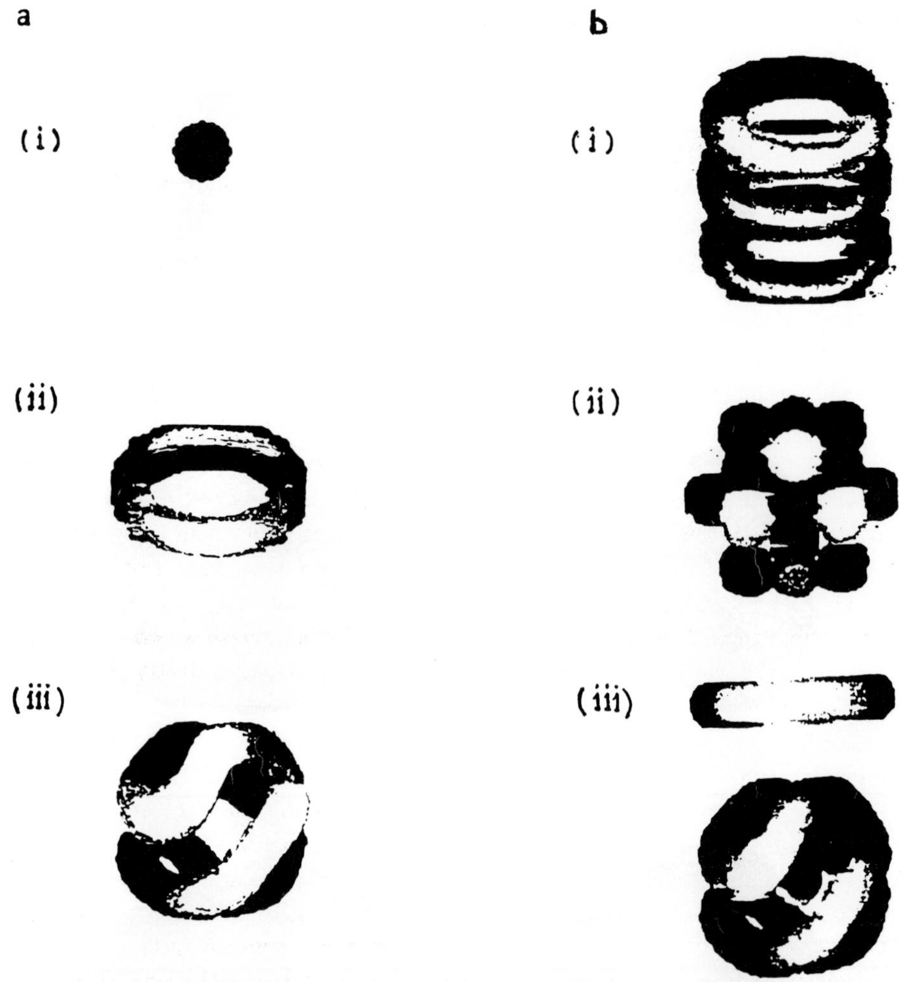

Fig. 4. Stable localized states of three-dimensional nonlinear field model, obtained by computer simulations in Ref. 70. Left-hand side pictures represent "building blocks" or "elementary particles": (i) a "ball", (ii) a "torus" and (iii) a "baseball" or a "spherical lattice." Right-hand side pictures represent stable bound states or clusters, constructed from "building blocks": (i) three-tori; (ii) a cluster of spheres; (iii) a torus and a spherical lattice with a sphere inside. (Adopted from Ref. 70.)

in,[70] demonstrated that there is a limited number of stable structures: a "ball," a "torus" and a "baseball" or a "spherical lattice," presented in Fig. 4a. (The last one may be considered as a "rolled-up torus.")

From those "building blocks" it is possible to construct a variety of observable structures as bound states or clusters, displayed in Fig. 4b. It is not difficult to note a similarity between Fig. 3 and Fig. 4a, and it would not be a surprise if the origin of stable, localized structures in nonlinear field models, applied in different branches of physics, and the formation of clusters from those building blocks were governed by a general rule of topological character. Probably it might be a proper formalization of Skyrme's scenario. It is even more amazing that structures in Figs. 3 and 4 are in some sense similar to the drawings in Fig. 1, representing different configurations of Kelvin's "Vortex Atoms". A possible way to understand the last similarity is to regard all those distinct physical situations as universal manifestations of Helmholtz's "Wirbelbewegung." Thus one can say that *old ideas never die*, especially when they are based on ancient ones.

6.4. *Spin and statistics of skyrmions*

The last of Skyrme's suggestions, listed in the previous section, by no means appears to be the most attractive. Recall, that Skyrme's central idea was to discover a possible way to regard a bose-field-theory-originated soliton as a fermionic state. We have already mentioned that this idea led to a new branch of study in nonlinear physics, called the Fermi–Bose transmutations. Localized structures with transmutations of spins and statistics have been studied in a variety of nonlinear models, applied in gravitation ("geons"),[71] in condensed matter physics for model description of such novel effects as high temperature superconductivity ("holons"),[72] in the fractional quantum Hall effect ("anyons"),[73] etc. We refer the reader to a self-contained review on related questions in Ref. 74. Thus this conjecture of Skyrme's appeared to be not only valid, but also productive.

To demonstrate in what sense Skyrmions are fermions it is better to use the topological formalization of the distinction between boson and fermion states, provided by D. Finkelstein and J. Rubinstein[24,75] (see also Refs 76, 77 for more modern and accessible expositions). As a keystone in their topological approach, they used the quantum-mechanical fact that a fermion wave-function changes its sign or picks up a phase (-1) when adiabatically rotated by 2π radians. Under double 2π rotation the fermionic wave function must attain the initial value. In the configuration space $\mathbf{Map}^0(\mathbf{S}^3;\mathbf{\Phi}) \equiv \mathbf{M}$ of maps ϕ: $\mathbf{S}^3 \to \Phi$, where ϕ are classical fields at a fixed moment of time and Φ is the field manifold of a classical theory, analogs of 2π rotations are closed paths or loops, starting and ending at the same field ϕ. The homotopical classification of classical fields (Secs. 4.1 and 5.2) and the topological charge $\mathbf{Q}$ are, by their definitions, independent of time. Therefore, they could be applied to classify quantum states.[77] The set $\pi_0(\mathbf{M})$ of connected components of $\mathbf{M}$ can be identified with $\pi_3(\Phi)$. Due to the isomorphism $\pi_k(\mathbf{M}) = \pi_{k+3}(\Phi)$, the elements of the fundamental group of the configuration space $\mathbf{M}$ [recall that $\pi_1(\mathbf{M})$ is the group of loops or closed paths on $\mathbf{M}$] might be associated to elements of the fourth homotopy group of Φ: $\pi_1(\mathbf{M}) = \pi_4(\Phi)$. In the case of the Skyrme model, the field manifold $\Phi = SU(2)$, so that $\pi_4(\Phi) = \pi_4(SU(2)) = \mathbf{Z}_2$, where $\mathbf{Z}_2 = \{1, -1\}$ is the Abelian group of integers modulo 2. It means that there are only two distinct types of loops in the Skyrme model configuration space: those that are trivial (or, in other words, contractable in a continuous way to a point on $\mathbf{M}$) and nontrivial (noncontractible) loops. A representation of a nontrivial path in the Hilbert space of quantum states should be an operator which changes the phase of wave functions by (-1). As for trivial paths, the corresponding operators should not effect phases of quantum states. Thus we can construct two distinct quantization schemes for Skyrmions: fermionic and bosonic. To choose one of them means to impose a constraint on the Hilbert space of physically allowed states, which would select only desirable states. It is clear from the aforegiven, that Skyrmions in the original SU(2) model *might be quantized as fermions*, if by one or another method we can exclude the possibility of trivial path representations in the Hilbert space. A possible form of this constraint was given by Finkelstein and Rubinstein,[24,75] and later on by Williams,[32] as necessary and sufficient conditions for the existence of the so-called double-valued functionals or "spinorial structures" on field manifolds in the framework of nonlinear field theory. It is clear that one can conceive those conditions as a generalization of the Pauli exclusion principle in the form applicable to quantized nonlinear theory.[61]

Some remarks are in order. First of all, the Finkelstein–Rubinstein scheme looks like a rather abstract consideration of general character which is model-independent. Remember that we did not use any other attributes of the Skyrme model, except the choice of the field manifold and the dimensionality of physical space–time. One may say that we appealed to the simpliest variant of the scheme, which in full extent is not only based on the formalization of 2π rotations, which is useful only for distinction between half-odd-integral and integral spin states. In Ref. 75 the procedure for exchange of particles has also been formalized and the normal connection between spin and statistics of the SU(2) Skyrmion has been established. But even in its full extent, the original Finkelstein–Rubinstein scheme is limited in applications, as, for example, it fails to give an answer in the case of the three and more flavors with $\Phi = SU(N)$ as a field manifold, $N \geq 3$. All closed paths are trivial in this case, because of isomorphism $\pi_4(SU(N)) = 0$. This fact was notified by E. Witten, who developed in Ref. 37 a dynamical realization of Finkelsten–Rubinstein ideas. Witten demonstrated that the statistics of states, corresponding to solitons in a chiral effective theory based on a $SU(N_c)$ gauge group, depends on the presence of the Wess–Zumino term. If the Wess–Zumino term is nontrivial, then the quantum states with the unit topological charge $\mathbf{Q} = 1$ (Skyrmions) *must* (*not only can*) *be quantized as bosons for N_c even and as fermions for N_c odd.* In that very sense the Wess–Zumino term breaks the ambiguity in the choice of a quantization scheme, pointing to one of them in accordance with the number of colors in the underlying gauge theory. As we restricted ourselves mostly to validity of Skyrme's suggestions, we will not reproduce Witten's results, which one can find in any review on Skyrmion physics.[37–45] We would like only to specify some points. In the case of three or more flavors the Wess–Zumino action might be written in the form $N_c\Gamma$, where $\Gamma = \pi\mathbf{Q}$ for the sequence of states, corresponding to a soliton, which is adiabatically rotated through 2π. Thus the amplitude for the rotated Skyrmion acquires an additional phase $\exp\{iN_c\Gamma\} = (-1)^{N_c Q}$ in comparison with the amplitude of an unrotated Skyrmion. It is not difficult to note that this is still another implementation of the Pauli exclusion principle. In its turn this approach could not be applied in the SU(2) case, as the Wess–Zumino action vanishes for two flavors.

A unified approach, applicable in both the aforementioned situations, was developed in a comprehensive paper.[78] It was demonstrated that the $\mathbf{Z}_2$ ambiguity in the SU(2) model can be regarded as a discrete Wess–Zumino term. That means, in general, that the Skyrme model with any number of flavors might be considered as the theory containing the Wess–Zumino term (in its ordinary or extended version). For a construction of wave functions in the presence of the Wess–Zumino term, a modernized Dirac's idea, used in his treatment of monopoles, was implemented.[79] The wave functions or state vectors of a quantum system, which are determined only up to a phase, might be regarded as not functions on the configuration space $\mathbf{M}$, but rather as functions on a U(1) bundle $\hat{\mathbf{M}}$ over $\mathbf{M}$. The bundle $\hat{\mathbf{M}}$ is obtained by associating a circle $\mathbf{S}^1$ to each point of $\mathbf{M}$. The bundle $\hat{\mathbf{M}}$ is trivial when circles or "fibers" $\mathbf{S}^1$ are glued together in the way that $\hat{\mathbf{M}} = \mathbf{M} \otimes \mathbf{S}^1$ and nontrivial or twisted in any other case. The latter possibility gives rise to a nontrivial Wess–Zumino term. The further development of this idea in Ref. 78 demonstrated that symmetries of quantum systems are not in general the symmetries of underlying classical action. The group $\hat{G}$, which acts on $\hat{\mathbf{M}}$, was constructed as a central extension of classical symmetry group G, and it was shown that in general $\hat{G}$ may not even contain G as a subgroup. The symmetries of the quantum system are transformations from either the group $\hat{G}$ itself, or its subgroup. It is because of this change in the group that the states might have half-odd-integral spin and change sign under 2π rotation, in spite of the fact that the symmetry group G of the underlying classical theory does not admit representations of half-odd-integral spin. The normal form of spin-statistics correlation for Skyrmions (for any number of flavors) has been established in Ref. 80. As a consequence, it was shown that this correlation will exist whenever the underlying theory incorporates the possibility of pair creation and annihilation. This is the present case in the problem of spin and statistics of Skyrmions.

As far as it is clear from the results listed here, the ambiguity of quantization procedure in the case of the SU(2) Skyrme model, might be improved only in a rather tricky way in the now-popular spirit of the "Cheshire cat." The conventional Wess–Zumino term, which uniquely picks out one of two possible quantization schemes, does not exist but, as was done in Ref. 78, it is possible to convince oneself that the same role is appropriate to its discrete analog or the "smile", which only remains in the two-flavor case. To do that it is necessary to consider the SU(2) model as a restriction of more general SU(N) theory.

As the main subject throughout this section was the validity of suggestions, it seems to be apologetic to make one more of them. Why not think that this tenacious ambiguity indicates a sort of *Fermi–Bose Duality*, which might be one of the features of a quantum description of extended objects? It might be that the proper distinction between fermions and bosons could be well established only in the quantum physics of point-like particles. From textbooks it is known that choosing the wrong quantization in description of point particles leads one to difficulties with the consistency of the theory. At present it is not evident whether or not we could suffer from inconsistencies in the quantum theory of extended objects if one suggests that both quantizations are valid. It may happen that a proper quantum theory of nonlinear fields will substantially differ from the semiclassical collective coordinates approach. This is an open question. Another open question is whether it is possible to construct a fermionic Lagrangian, which at the low energies would provide an equivalent description of Skyrmions (the "spinorization" problem). To get an answer we need to extend the Sine–Gordon–Thirring duality, established by Coleman and discussed in Sec. 4.3, on the (3 + 1)-dimensional model. This question has been studied in Ref. 81 and the main conclusion of those papers is that there is no Lagrangian with fermion fields alone which reproduces that of the Skyrme model. The Lagrangian, which was found in the second paper of Ref. 81 to be equivalent to Skyrme's at low energies, contains both fermionic and bosonic fields and, besides, fermions of that Lagrangian are not solitions. To sum up, one can say that there is no doubt that Skyrme's conjecture is valid, but there are still many problems, which merit better understanding.

The present state in a quantum description of Skyrmions corresponds, if it is possible to say, to a quantum mechanical level, when we do not think about the nature of particle attributes, such as spin. But the Skyrme model claims to describe the baryon structure. Then it is possible to expect that it should provide us with a deeper understanding of spin nature. It would be interesting to see if the spin, which was introduced as an intrinsic angular momentum of particle, could be related with a representation of the Skyrmion as a vortex in "Mesonic Fluid."

7. Concluding Remarks

Throughout all previous sections we were mostly concerned with, so to say, the "ideological" or conceptual consistency of the original Skyrme model. The present-day adopted conclusion is that this model provides us with a reasonable approach in the description of hadron physics, which does not contradict the QCD first principles. Now it is high time to turn to a practical use of Skyrme's ideas. Already at the early stage of recent studies it became clear that the model itself can be considered only as a rather crude approximation to realistic meson physics. The more realistic chiral Lagrangian must be dependent not only on pseudoscalar mesons, but also on all low-lying meson multiplets. The search for this Lagrangian was undertaken in both evident directions: to derive it from the QCD-generating functional and to determine its parameters in a more phenomenological approach from experiment.

The first approach, which in its full extent is equivalent to the problem of the low-energy QCD solvability, has been launched in Refs. 82 and 83. The obtained results are not so far-reaching and reflect the lack in a quantitative understanding of low-energy nonperturbative phenomena in QCD, the problem with planar diagrams of all orders of summation, etc. The only terms of the mesonic Lagrangian one can derive from QCD with full guarantee in this approach are the "kinetic" (quadratic in $\mathbf{L}_\mu$) term and the Wess–Zumino term. To derive higher-order terms further assumptions are needed, as, for example, a bosonization scenario and so on.[84] It looks like even if we do not succeed in deriving a realistic meson Lagrangian, nontheless the important advantage of this approach will be powerful new mathematical methods in calculating fermionic determinants, in dealing with anomalies, and it may also lend to a better understanding of the problem in low-energy QCD physics.

The other "trial and error" approach, launched by J. Schechter, et al., in the series of papers in Ref. 85 is based on a careful modeling of the underlying low-energy QCD phenomenon and gives more reliable answers. The resulting realistic pseudoscalar-vector chiral Lagrangian,[86] known also as the "Syracuse model," provides a reasonable description of the nucleon static properties in sufficient agreement with experimental data. The comprehensive review of achievements in the field, with a detailed exposition of the approach itself, can be found in Ref. 43. It seems possible to reinforce the predictive power of the Syracuse model by a more attentive examination of the topological structure of solitonic solutions. The results to compare with that of experiments were obtained in the standard manner by numerical analysis of field equations on Skyrme's hedgehog configurations. As it is clear from the analysis in Refs. 50, 63 and 65, there is no guarantee that the numerical results thus obtained really correspond to energetically preferable configurations, whose parameters we are supposed to measure in experiments. No doubt, to carry out an analogous analysis for the Syracuse model is a much more complicated task, as the functional depends on the larger number of fields and has a more complicated structure than the Skyrme functional. Nevertheless, to guarantee the competence of numerical calculations one has to be sure that one is looking for solutions which really exist, are stable and can provide a description of the nucleon.

Let us also mention two promising modifications of the Skyrme model suggested recently. The first one is related to the inherent possibility of the model to describe baryonic matter even when the latter is compressed to high densities. Recall that at its early stage of evolution[2] the "Mesonic Fluid" model was a model of nuclear matter. The other interesting fact is that the estimate for high density of baryonic matter, obtained by Skyrme himself in Ref. 9, is in rather good agreement with the recent estimate[87] in the quark-gluon plasma theory. The further development of the Skyrme model approach in application to condensed matter resulted in the S^3 reformulation of the model, pioneered by N. Manton.[88] The main idea was to consider compact manifold of the three-sphere of radius L as the domain of chiral fields U, instead of traditional $\mathbf{R}^3$ flat space. One of the advantages of this $S^3(L)$ modification is that the model is far simpler and allows for mathematically rigorous results.[89] It is well known that we still do not have an exact solution of the original Skyrme model, in spite of numerous efforts, as, for example, in Ref. 90. This is one of the reasons which limit further analytical study of Skyrmion physics. The $S^3(L)$ model possesses the so-called uniform exact solution, which corresponds to a constant map and thus describes a homogeneous distribution of baryonic matter on the S^3 sphere, e.g. a condensed phase. Unfortunately, this fact does not help to find an exact solution in the original model, as in the large L region, which corresponds to the ordinary Skyrmions, the uniform solution has no finite limit. This modification is very interesting from the point of a link between the soliton approach in hadron physics and localized structures in condensed matter physics. It is quite possible that methods of studying soliton structures in σ models would be effective in a description of localized condensed matter structures.

The second attractive modification, called the Composite Skyrme model, has been proposed by H. Cheung and F. Gürsey,[91] who considered the possibility of formulating the model in terms of composite chiral fields $\mathbf{U}^n$, which are composed from n copies of ordinary chiral fields **U** with the insertion of the constant SU(2) matrix between each pair of **U**s. The composite chiral field $\mathbf{U}^n$ transforms linearly under $SU(2) \otimes SU(2)$ transformations, but the more interesting fact is that this modification, without any tricks, gives for $n = 3$ an almost experimentally observed value for the pion-decay constant F_π to be 185 MeV. The fact that $n = 3$ was singled out led the authors to the hypothetic suggestion that this Composite Skyrme model describes effective quarks in a nucleon.

In Witten's already quoted report (the second paper of Ref. 31) the theory of the Callan–Rubakov effect[92] was taken as the only example to that date where the Skyrme model of baryon gives the most realistic approach to physical phenomena. Now it is possible to believe that the model demonstrated its consistency for the second time in solving the so-called "proton spin"puzzle.[93]

All the abovegiven examples and modifications testify that Skyrme's approach is not only original and interesting, but also shows a universal character as applicable to description of phenomena in the wide range of nonlinear physics from quarks to condensed matter localized structures. It is difficult to conclude now which of the listed suggestions is true, but anyway it is possible to expect in future the third wave of interest in the Skyrme model features. Let us express the hope that till that time we will acquire a better understanding of the nature of Helmholtz–Kelvin–Skyrme's "Wirbelbewegung" and that it will be possible to explain it in a much shorter paper.

Acknowledgments

It is a pleasure to thank Yu. P. Rybakov and V. G. Makhankov for stimulating discussions of the greater part of this text. I am indebted to J. Schechter, R. Sorkin and T. Allen for the careful reading of the manuscript and many helpful comments, to E. Braaten for the valuable discussion and permission to use in this paper computer pictures from Ref. 69, to J. McCracken and A. Subbaraman for numerous improvements in my English. I am also grateful to the Fulbright Program for the grant, which made possible these lectures, and to all members of the High Energy Group in Syracuse University for their kind attention and warm hospitality.

References

1. R. N. Dalitz, *Int. J. Mod. Phys.* **A3** (1988) 2719.
2. T. H. R. Skyrme, *Proc. Roy. Soc. London* **A226** (1954) 521.
3. T. H. R. Skyrme, *Proc. Roy. Soc. London* **A230** (1955) 277.
4. T. H. R. Skyrme, *Proc. Roy. Soc. London* **A247** (1958) 260.
5. T. H. R. Skyrme, *Proc. Roy. Soc. London* **A252** (1959) 236.
6. T. H. R. Skyrme, *Proc. Roy. Soc. London* **A260** (1961) 127.
7. T. H. R. Skyrme, *Proc. Roy. Soc. London* **A262** (1961) 237.
8. T. H. R. Skyrme, J. K. Perring, *Nucl. Phys.* **31** (1962) 550.
9. T. H. R. Skyrme, *Nucl. Phys.* **31** (1962) 556.

10. T. H. R. Skyrme, *J. Math. Phys.* **12** (1971) 1755.
11. T. H. R. Skyrme, *Int. J. Mod. Phys.* **A3** (1988) 2745.
12. Kelvin, William Thomson, 1st Baron, *Mathematical and Physical Papers*, vol. IV *Hydrodynamics and General Dynamics* (Cambridge University Press, Cambridge, 1910), pp. 1–69.
13. Kelvin, William Thomson, 1st Baron *Baltimore Lectures on Molecular Dynamics and the Wave Theory of Light* (Clay & Son, Baltimore, 1904), p. 703.
14. W. Pauli, *Meson Theory of Nuclear Forces* (Interscience, New York, 1946), p. 80.
15. L. D. Faddeev, *ZhETP Pis'ma* **21** (1975) 141 (in Russian).
16. A. Pais, *Proc. Nat. Acad. Sci.* (*USA*) **40** (1954) 484.
17. A. Pais, *Physica* **19** (1953) 869.
18. J. Schwinger, *Ann. Phys.* **2** (1957) 407.
19. G. F. Chew, *S-Matrix Theory of Strong Interactions* (W. A. Benjamin Inc., New York, Amsterdam, 1961), pp. 182; M. Jacob and G. F. Chew, *Strong-Interaction Physics* (W. A. Benjamin Inc., New York, Amsterdam, 1964), p. 153.
20. R. Rajaraman, *Solitons and Instantons* (North-Holland, Amsterdam, 1982), p. 409.
21. R. Hirota, *J. Phys. Soc. Jap.* **33** (1972) 1459; P. J. Caudrey, J. D. Gibbon, J. C. Eilbeck and R. K. Bullough, *Phys. Rev. Lett.* **30** (1973) 237; M. J. Ablowitz, D. J. Kaup, A. C. Newell and H. Segur, *Phys. Rev. Lett.* **30** (1973) 1262.
22. V. E. Zakharov, L. A. Takhtadzhian and L. D. Faddeev, *DAN of USSR* **219** (1974) 1334 (in Russian).
23. D. Finkelstein and C. Misner, *Ann. Phys.* **6** (1959) 230.
24. D. Finkelstein, *J. Math. Phys.* **7** (1966) 1218.
25. S. Coleman, *Phys. Rev.* **D11** (1975) 2088.
26. H. Sugawara, *Phys. Rev.* **170** (1968) 1659; C. Sommerfield, *Phys. Rev.* **176** (1968) 2019.
27. A. A. Slavnov and L. D. Faddeev, *Teor. Mat. Fiz.* (*Sov. Phys.*) **8** (1971) 397 (in Russian).
28. V. I. Sanyuk, in *Topological Phases in Quantum Theory*, Proc. of the Int. Seminar, Dubna, 1988, eds. S. I. Vinitsky and B. Markovski (World Scientific, Singapore, 1989), p. 316.
29. H. C. Tze and Z. F. Ezawa, *Phys. Rev.* **D14** (1976) 1006.
30. G. t'Hooft, *Nucl. Phys.* **B72** (1974) 461; **B75** (1974) 461.
31. E. Witten, *Nucl. Phys.* **B160** (1979) 57; see also Ref. 47, p. 306.
32. J. Williams, *J. Math. Phys.* **11** (1970) 2611.
33. P. K. Pak and H. Ch. Tze, *Ann. Phys.* **117** (1979) 164.
34. J. M. Gipson and H. Ch. Tze, *Nucl. Phys.* **B183** (1981) 524.
35. A. P. Balachandran, V. P. Nair, S. G. Rajeev and A. Stern, *Phys. Rev. Lett.* **49** (1982) 1124; *Phys. Rev.* **D27** (1983) 1153.
36. A. P. Balachandran, V. P. Nair and C. G. Trahern, *Phys. Rev.* **D27** (1983) 1369.
37. E. Witten, *Nucl. Phys.* **B223** (1983) 422, 433.
38. G. S. Adkins, C. R. Nappi and E. Witten, *Nucl. Phys.* **B228** (1983) 552.
39. I. Zahed and G. E. Brown, *Phys. Reports* **142** (1986) 1.
40. G. Holzwarth and B. Schwesinger, *Rep. Progr. Phys.* **49** (1986) 825.
41. G. E. Brown and M. Rho, *Comm. Nucl. Part. Phys.* **XV** (1986) 245.
42. U.-G. Meissner and I. Zahed, *Adv. Nucl. Phys.* **17** (1986) 143.
43. U.-G. Meissner, *Phys. Reports* **161** (1988) 213.
44. A. P. Balachandran, in *High Energy Physics* 1985, Proc. of the Yale Theoretical Advanced Study Institute, eds. M. J. Bowick and F. Gürsey (World Scientific, Singapore, 1986), Vol. I, pp. 1–82.
45. G. S. Adkins, in "*Chiral Solitons*," ed. K. F. Liu (World Scientific, Singapore, 1987), pp. 99–170.
46. V. G. Makhankov, Yu. P. Rybakov and V. I. Sanyuk, *The Skyrme Model and Solitons in the Hadron Physics*, Lectures for Young Scientists Series, No. 55 (JINR Publ., Dubna, 1989), p. 172 (in Russian).

47. A. Chodos, E. Hadjimichael and Ch. Tze (eds.), *Solitons in Nuclear and Elementary Particle Physics*, Proc. of the Lewes Workshop 1984. (World Scientific, Singapore, 1984), p. 314.
48. M. Jezabek and M. Praszalowicz (eds.), *Skyrmions and Anomalies*, Proc. of the Krakow Workshop 1987. (World Scientific, Singapore, 1987), p. 531.
49. J. Goldstone and F. Wilczek, *Phys. Rev. Lett.* **47** (1981) 986.
50. Yu. P. Rybakov and V. I. Sanyuk, *Topological Skyrmions*, The Niels Bohr Inst. Preprint NBI-HE-81-49 (1981), pp. 36.
51. C. J. Isham, *J. Phys.* **A10** (1977) 1397.
52. M. J. Esteban, *Commun. Math. Phys.* **105** (1986) 571.
53. J. Eells and J. H. Sampson, *Amer. J. Math.* **86** (1964) 109; J. Eells, L. Lamaire, *Bull. Lond. Math. Soc.* **10** (1978) 1.
54. Yu. P. Rybakov and S. Chakrabarti, *Int. J. Theor. Phys.* **23** (1984) 325.
55. V. G. Makhankov, *Phys. Reports* **35** (1978) 1; A. Kumar, V. P. Nisichenko and Yu. P. Rybakov, *Int. J. Theor. Phys.* **18** (1979) 425.
56. R. H. Hobart, *Proc. Phys. Soc. London* **82** (1963) 201; G. H. Derrick, *J. Math. Phys.* **5** (1964) 1254.
57. E. B. Bogomol'nyi, *Sov. J. Nucl. Phys.* **24** (1976) 861.
58. A. P. Balachandran, A. Barducci, F. Lizzi, V. G. J. Rogers and A. Stern, *Phys. Rev. Lett.* **52** (1984) 887.
59. E. Braaten and L. Carson, *Phys. Rev. Lett.* **56** (1986) 1897.
60. A. Jackson, A. D. Jackson and V. Pasquier, *Nucl. Phys.* **A432** (1985) 567.
61. E. Braaten and L. Carson, *Phys. Rev.* **D38** (1988) 3525; **D39** (1989) 838.
62. J. J. M. Verbaarschot, T. S. Walhout, J. Wambach and H. W. Wyld, *Nucl. Phys.* **A468** (1987) 520; V. B. Kopeliovich and B. E. Shtern, *JETP Pis'ma* **45** (1987) 165 (in Russian); J. J. M. Verbaarschot, *Phys. Lett.* **B195** (1987) 135; A. J. Schramm, Y. Dothan and L. C. Biedenharn, *Phys. Lett.* **B205** (1988) 151.
63. I. R. Kozhevnikov, Yu. P. Rybakov and M. B. Fomin, *Teor. Mat. Fiz. (Sov. Phys.)* **75** (1988) 353.
64. R. S. Palais, *Commun. Math. Phys.* **69** (1979) 19.
65. A. Kundu, Yu. P. Rybakov and V. I. Sanyuk, *Indian J. Pure & Appl. Phys.* **17** (1979) 673.
66. A. Kundu and Yu. P. Rybakov, *J. Phys.* **A15** (1982) 269.
67. E. Braaten and L. Carson, in *Workshop on Nonlinear Chromodynamics*, eds. S. Brodsky and E. Moniz (World Scientific, Singapore, 1986), pp. 454; in *Relativistic Dynamics and Quark-Nuclear Physics*, eds. M. B. Johnson and A. Picklesimer (Wiley, New York, 1986), pp. 854.
68. V. I. Sanyuk, in *Problems of High Energy Physics and Field Theory*, Proc. of the XII Workshop, Protvino, 1989, ed. V. A. Petrov (Nauka, Moscow, 1990), p. 254; in *Solitons and Applications*, Proc. of IV Int. Workshop, Dubna, 1989, eds. V. G. Makhankov, V. K. Fedyanin and O. K. Pashaev (World Scientific, Singapore, 1990), p. 374.
69. E. Braaten, S. Townsend and L. Carson, *Phys. Lett.* **B235** (1990) 147.
70. K. A. Gorshkov, A. S. Lomov and M. I. Rabinovich, *Phys. Lett.* **A137** (1989) 250.
71. R. D. Sorkin, *Phys. Rev. Lett.* **51** (1983) 87; **54** (1985) 86; *Phys. Rev.* **D27** (1983) 1787.
72. P. W. Anderson, *Science* **235** (1987) 1196; P. B. Wiegman, *Phys. Rev. Lett.* **60** (1988) 821; I. E. Dzyaloshinskii, A. M. Polyakov and P. B. Wiegman, *Phys. Lett.* **A127** (1988) 112.
73. F. Wilczek, *Phys. Rev. Lett.* **49** (1982) 957.
74. R. Mackenzie and F. Wilczek, *Int. J. Mod. Phys.* **A3** (1988) 2827.
75. D. Finkelstein and J. Rubinstein, *J. Math. Phys.* **9** (1968) 1762.
76. D. Ravenel and A. Zee, *Commun. Math. Phys.* **98** (1985) 239.
77. A. S. Schwarz, *Mod. Phys. Lett.* **A4** (1989) 403.
78. A. P. Balachandran, H. Gomm and R. D. Sorkin, *Nucl. Phys.* **B281** (1987) 573.
79. P. A. M. Dirac, *Proc. Roy. Soc. London* **A133** (1933) 60.

80. R. D. Sorkin, *Commun. Math. Phys.* **115** (1988) 421.
81. S. G. Rajeev, *Phys. Rev.* **D29** (1983) 2924; G. Bhattacharya and S. G. Rajeev, see Ref. 47, p. 299.
82. A. A. Andrianov, *Phys. Lett.* **B157** (1985) 425; A. A. Andrianov and Yu. V. Novozhilov, *Phys. Lett.* **B153** (1985) 422.
83. N. I. Karchev and A. A. Slavnov, *Teor. Mat. Fiz.* (*Sov. Phys.*) **65** (1985) 192 (in Russian); R. Ball, see Ref. 48, p. 54.
84. A. A. Andrianov, V. A. Andrianov, V. Yu. Novozhilov and Yu. V. Novozhilov, *Lett. Math. Phys.* **11** (1986) 217; *Teor. Mat. Fiz.* (*Sov. Phys.*) **70** (1987) 63; *Phys. Lett.* **B186** 401.
85. Ö. Kaymakcalan, S. Rajeev and J. Schechter *Phys. Rev.* **D30** (1984) 594; Ö. Kaymakcalan and J. Schechter, *ibid.* **D31** (1985) 594; H. Gomm, Ö. Kaymakcalan and J. Schechter, *ibid.* **D30** (1984) 2345; J. Schechter, *ibid.* **D34** (1986) 868; H. Gomm, P. Jain, R. Johnson and J. Schechter, *ibid.* **D33** (1985) 3476; P. Jain, R. Johnson and J. Schechter, *ibid.* **D34** (1986) 2230.
86. P. Jain, R. Johnson, Ulf-G. Meissner, N. W. Park and J. Schechter, *Phys. Rev.* **D37** (1988) 3252; Ulf-G. Meissner, N. Kaiser, H. Weigel and J. Schechter *ibid.* **D39** (1989) 1956.
87. C. J. Petchick, *Nucl. Phys.* **A434** (1985) 587c.
88. N. S. Manton and P. J. Ruback, *Phys. Lett.* **B181** (1986) 137; N. S. Manton, *Commun. Math. Phys.* **111** (1987) 469.
89. N. S. Manton, *Phys. Rev. Lett.* **60** (1988) 1960; A. D. Jackson, N. S. Manton and A. Wirzba, *Nucl. Phys.* **A495** (1989) 499; A. Wirzba and H. Bang, *Nucl. Phys.* **A515** (1990) 571.
90. V. I. Sanyuk, in *Solitons and Applications*, Proc. of IV Int. Workshop, Dubna, 1989. eds. V. G. Makhankov, V. K. Fedyanin and O. K. Pashaev (World Scientific, Singapore, 1990), p. 425.
91. N. Y. Cheung and F. Gürsey, *Mod. Phys. Lett.* **A5** (1990) 1685; S. Nam and R. L. Workman, *Phys. Rev.* **D41** (1990) 2323.
92. V. Rubakov, *Nucl. Phys.* **B203** (1982) 311; *ZhETP Pis'ma* **33** (1981) 658 (in Russian); C. G. Callan, Jr., *Phys. Rev.* **D25** (1982) 2141; **D26** (1982) 2058; C. G. Callan, Jr. and E. Witten, *Nucl. Phys.* **B239** (1984) 161.
93. S. Brodsky, J. Ellis and M. Karliner, *Phys. Lett.* **B206** (1988) 309; N. M. Park, J. Schechter and H. Weigel, *Phys. Rev.* **D41** (1990) 2836.

A non-linear theory of strong interactions

By T. H. R. Skyrme

Atomic Energy Research Establishment, Harwell

(*Communicated by B. F. J. Schonland, F.R.S.—Received 5 May* 1958)

A non-linear theory of mesons, nucleons and hyperons is proposed. The three independent fields of the usual symmetrical pseudo-scalar pion field are replaced by the three directions of a four-component field vector of constant length, conceived in an Euclidean four-dimensional isotopic spin space. This length provides the universal scaling factor, all other constants being dimensionless; the mass of the meson field is generated by a ϕ^4 term; this destroys the continuous rotation group in the iso-space, leaving a 'cubic' symmetry group. Classification of states by this group introduces quantum numbers corresponding to isotopic spin and to 'strangeness'; one consequence is that, at least in elementary interactions, charge is only conserved modulo 4. Furthermore, particle states have not a well-defined parity, but parity is effectively conserved for meson-nucleon interactions. A simplified model, using only two dimensions of space and iso-space, is considered further; the non-linear meson field has solutions with particle character, and an indication is given of the way in which the particle field variables might be introduced as collective co-ordinates describing the dynamics of these particular solutions of the meson field equations, suggesting a unified theory based on the meson field alone.

1. Introduction

In an earlier paper (Skyrme 1955) the author attempted a discussion of the usual symmetric pseudo-scalar theory of pions coupled with nucleons on the basis that there was no explicit mass term for the nucleon field, that the total field strength saturated to a constant value in the neighbourhood of a nucleon source, and that the directions of the meson field and of the source (in isotopic spin space) were strongly coupled. Developments in the field of strange particles, K-mesons and hyperons, have led to the view that there exist a set of 'strong interactions' between the π- and K-mesons,* and the baryons, i.e. nucleons and hyperons; and that these interactions conserve both isotopic spin and the quantum number variously called strangeness, attribute or hypercharge (Gell-Mann 1957); and furthermore it appears that the underlying symmetry of the isotopic space may be of four dimensional rather than three, as has been suggested among others by Polkinghorne & Salam (1955) and by Schwinger (1957). In another direction the discovery of parity violation in weak interactions has led to discussion of the γ_5 or mass-reversal transformations, normally possible only for a field with zero mass (i.e. the constant in the primitive equations of motion). These ideas have suggested a four-dimensional analogue of the theory mentioned first above.

This theory is obtained from the familiar symmetric *PS-PS* theory of pions and nucleons by replacing the nucleon mass by $g\phi_4$, where g is the pseudo-scalar coupling constant and ϕ_4 is another meson field of similar character to the other three ϕ_i; as has been noted by Schwinger (1957), such an interaction is really symmetric between all four fields. It is not, however, supposed that these four meson fields

* In this paper μ-mesons are not considered, nor are electrons, neutrinos and the electromagnetic field.

are independent, but that they must necessarily form a vector of constant length (that is, the sum of their squares is a universal constant); in other words the possible meson fields at any point of space-time span the surface of a hypersphere in four dimensions rather than an Euclidean space of three dimensions. This relation does not affect the symmetry of their interaction.

This assumption raises immediately another problem, that of the meson field mass, because this relation between the field strengths destroys the significance of the usual mass term, $\frac{1}{2}\kappa^2\Sigma\phi^2$. It might be that the meson mass were entirely generable through interaction with the nucleon field; an alternative possibility is considered here, that the basic Lagrangian contains a term involving the sum of the fourth powers of the meson fields. As has been pointed out by Schwinger (1957) such an interaction does not require the introduction of any dimensional constant, so that the theory postulated here contains only one dimensional constant (the total magnitude of the meson field).

This ϕ^4 interaction destroys the invariance under the continuous groups of rotations in four dimensions, leaving only the symmetry associated with rotations of $\frac{1}{2}\pi$ about any axis, that of the 'proper cubic' group in four dimensions. Isotopic spin and strangeness may be defined by the quantum numbers of the representations of the finite group, just as was done for the continuous group by Polkinghorne & Salam (1955); they can only take on small values and are only conserved modulo 2 or 4. In particular charge, as it appears naturally to be defined (though electromagnetic field interactions are not considered), is only conserved modulo 4. The elementary reaction violating charge conservation is of the form

$$\pi^+ + \pi^+ \rightleftarrows \pi^- + \pi^-$$

or as it might practically be realized $P + \pi^+ \rightarrow N + \pi^- + \pi^-$. Experiment does not strongly preclude such elementary phenomena, although charge is certainly conserved macroscopically to a high degree. It has not been possible in this paper to estimate the magnitude of either the primary or secondary effects of these reactions, so that further discussion is of little value.

An analysis of the symmetry properties of the fields indicates that the fields ϕ should be identified with the two K- and $\overline{\text{K}}$-doublets rather than with the pion field, which should be described by a three-dimensional representation of the group. Two such three-dimensional representations naturally appear, one an isotopic triplet with strangeness zero and the other a 'strangeness triplet' of zero isotopic spin; these can be represented, for example, by the two three-dimensional local rotation operators of the fields which arise from the four-dimensional continuous rotation group. The former of these may be identified with pions; no manifestation of the latter has been reported.

The Fermion field ψ has the usual four-component character as far as co-ordinate space is concerned, and has two isotopic spin components (upon which the τ matrices operate); a further doubling of the number of components is needed to prevent the necessity of extending the symmetry group to its double, which appears undesirable in the light of the interpretation offered (essentially because half-integral charges do not occur in nature). These do not have the symmetry necessary to be simple particle creation operators; it is, however, possible to introduce a unitary operator Z, depending upon the direction of the meson field, such that $Z\psi$ has suitable symmetry properties, describing four states forming isotopic spin and strangeness doublets. One of these doublets is identified with nucleons; the other might be associated either with anti-nucleons or with the Ξ^0, Ξ^- doublet. These two doublets must be degenerate within the framework of the present theory, so that the Ξ^0, Ξ^- interpretation is unattractive on that account; a consistent interpretation in terms of anti-nucleons can only be given by imposing a constraint on the Fermion fields in terms of a supplementary condition on the state-function. The latter possibility, though most unnatural from the usual point of view (independent fields) from which we started, might not be so unreasonable from the interpretation of the ψ which is suggested at the end of this paper.

The Λ and Σ hyperons may be associated with operators of the form $\phi Z\psi$; loosely speaking, the description is equivalent to saying that the hyperons are to be regarded as tightly bound states of a nucleon and a $\overline{\text{K}}$-meson; similarly the Ξ^0, Ξ^- doublet can be regarded as compound states involving two $\overline{\text{K}}$-mesons.

The form of the interaction of nucleons with the meson field may be got by making a point transformation to the new variable $Z\psi$. The part of the meson field with which they interact has, inevitably, the symmetry of the pion field. In a certain approximation the dominant interaction term is of the symmetrical pseudo-vector type, while the ϕ^2 terms that usually appear from transformation of the symmetrical *PS-PS* theory are here absent. Under inversion the fields $Z\psi$ and the pions are transformed out of recognition, in fact into quite different fields, but it can be seen that the scattering matrix for the nucleon-pion system will nevertheless conserve parity in the conventional sense.

Further test of this suggested theory requires some estimation of particle masses and of interaction coupling constants, and this quantitative analysis requires a deeper understanding of the form of the solutions than has been achieved. As a step towards this a rather simpler theory of the same general character has been considered, in which the configuration space-time and the isotopic spin space have both been reduced to two dimensions, thereby eliminating both ordinary spin and isotopic spin. The meson field is described by the value of an angle variable θ at each point, and the spinor fields have only two space-time components instead of four. The symmetry group is just the cyclic group of order 4, corresponding to increases of θ by multiples of $\frac{1}{2}\pi$.

The non-linear differential equation of the meson field, considered as a classical equation, has solutions which approximate to travelling waves (interpretable as 'pions'); in addition, under the assumption of a certain rather natural boundary condition, the equation has a static solution in which the field vanishes at infinity. In this solution the angle variable θ has values at $x = \pm\infty$ which differ by $\frac{1}{2}\pi$; consequently it has a particle character of indestructibility, the difference of the values of θ at $\pm\infty$ being a constant of the motion, having the natural interpretation of number of particles less anti-particles. This continuous 'particle' solution may be regarded as built up from a discontinuous step-function (to which it approximates) coupled to the meson field by a point-derivative interaction, the discontinuous step-function has an infinite energy which is cancelled by the infinite self-energy associated with the particle-field coupling.

This situation has an obvious analogy with the coupling of Fermion fields to the meson field in quantum field theory, and it is natural to ask whether the Fermion field operators ψ can be identified with the creation or destruction of such step-function states, in the model. The first difficulty in quantizing the theory is the nature of the ground, vacuum state: is there a unique such state, necessarily invariant under the cyclic group, or is there a degeneracy, and if so what is its meaning? A satisfactory answer to this has not been found. Nevertheless, it is of interest to see whether particle field variables can be introduced as collective co-ordinates for the description of the dynamics of the singularities. This can be done by following the usual analogies of second quantization procedure; the result is to add onto the original Hamiltonian a non-linear Fermion part, of structure similar to that which has been considered by Heisenberg (1957). This is supplemented by subsidiary conditions linking the particle and meson fields. A partial interpretation of these conditions may be given in the following way; the 'enlarged' state-function, depending upon both meson and particle field variables, has components with varying numbers of particles: one component has no particles but has singularities in its meson field, another has a singularity-free meson field and a number of particles, and so on. The subsidiary conditions relate together these different components, while the Hamiltonian describes the development of any one component if the generation of singularities in the meson field is disallowed; the component with a smooth meson field would then correspond to the familiar state-function.

In this form, however, the subsidiary conditions also contain the interaction between particles and field; by their use it can be seen that the new Hamiltonian is also formally identical with the model Hamiltonian postulated originally. But the significance of these results depends upon finding a satisfactory solution to the unanswered questions concerning the vacuum state and the subsidiary conditions.

If the interpretation suggested here can consistently be carried through, a unified description of particles and fields in the model will have been obtained in terms of the non-linear meson field equation alone. Since the real (i.e. classical) particle solutions have finite energy and finite extension, it is possible that the theory may really be free of singularities, although these arise when the particle field variables are introduced as extra co-ordinates. The considerable problem will then remain as to the possibility of generalizing the results to the real four-dimensional problem. The obvious difficulty is that there is no simple way of generalizing the concept of the particle number, because 'infinity' is not divided by the source. If an analogous concept exists it will presumably require the use of a multiple-valued solution which has some sort of branch point at the source.

The following sections contain the mathematical procedures underlying the picture sketched in this introduction.

2. The Lagrangian

The point of departure for this discussion is the usual theory of nucleons interacting with a symmetric pseudo-scalar pion field through a pseudo-scalar coupling, which is described by the following expression for the Lagrangian density,

$$\mathscr{L}_0 = -\tfrac{1}{2}\Sigma(\mathrm{d}\phi_i/\mathrm{d}x_\mu)^2 - \tfrac{1}{2}\kappa_0^2\Sigma\phi_i^2$$
$$-\psi^\dagger\beta\left(\gamma_\mu\frac{\partial}{\partial x_\mu} + M_0 + ig\gamma_5\tau_i\phi_i\right)\psi. \quad (1)$$

The meson field index i runs over the values 1, 2, 3, and as throughout this paper units are used in which $\hbar = c = 1$; $\psi^\dagger$ is the Hermitian conjugate of ψ, and other symbols have their usual meanings.

The first proposed change is that the 'bare mass' M_0 be replaced by $g\phi_4$, where the new meson field is generally similar to the ϕ_i; reasons for this are: (i) this introduces a four-dimensional symmetry into the isotopic spin space, hereafter called iso-space, which is rather indicated by strange-particle phenomena; (ii) the possibility of a mass-reversal type of transformation.

The second proposal is that these four fields* ϕ_ρ should not be independent but be constrained by a relation

$$\Sigma\phi_\rho^2(x) = Q^2, \quad \text{at all points } x, \quad (2)$$

where Q is a fundamental constant of the dimensions of a mass or reciprocal length. Reasons are (i) this is a possible way of introducing a dimensional scale into the system now that M_0 has been eliminated; (ii) in a naïve interpretation this choice removed the ϕ^2 interaction term that would otherwise appear on transforming the pseudo-scalar interaction to a pseudo-vector form.

The third proposal concerns the meson mass term. The natural generalization $-\frac{1}{2}\kappa_0^2\Sigma\phi_\rho^2$ has no significance if the fields are constrained by the condition (2). One possibility would be to argue that the meson mass originates from the nucleon coupling. The alternative chosen here is to add a fourth-order term

$$+\tfrac{1}{4}\gamma^2\sum_\rho\phi_\rho^4. \quad (3)$$

Reasons for this choice are:

(i) It is the only simple possibility that does not introduce a new dimensional constant, and is suggested by conventional renormalization procedures.

(ii) It reduces the symmetry in iso-space to a finite group, thereby restricting the quantum numbers of elementary particles to small values.

(iii) It introduces an interesting non-linearity into the meson field with consequences suggested in the last section of this paper.

(iv) In a naïve interpretation, the new interaction term is approximately equivalent to the former meson mass term, together with an attractive meson-meson interaction.

* In §§ 2 to 5, indices running from 1 to 4 will generally be denoted by greek letters, and those from 1 to 3 by latin.

The obvious disadvantage is the abolition of arbitrary rotations about the 3-axis in iso-space, with the consequence that, at least microscopically, charge is only conserved modulo 4.

The theory proposed here proceeds therefore from the following expression for the Lagrangian density

$$\mathscr{L} = -\tfrac{1}{2}\Sigma(\mathrm{d}\phi_\rho/\mathrm{d}x_\mu)^2 + \tfrac{1}{4}\gamma^2\Sigma\phi_\rho^4$$

$$-\psi^\dagger\beta\left(\gamma_\mu\frac{\partial}{\partial x_\mu} + g(\phi_4 + \mathrm{i}\gamma_5\tau_i\phi_i)\right)\psi \qquad (4)$$

with the constraint given by equation (2). The naïve interpretation mentioned above consists in supposing that in some approximation the fields ϕ_i are all small compared with Q, and that ϕ_4 is nearly equal to Q. Then the new Lagrangian $\mathscr{L}$ is roughly equivalent to $\mathscr{L}_0$ with

$$M_0 = gQ \quad \text{and} \quad \kappa_0 = \gamma Q. \qquad (5)$$

This Lagrangian (4), and a simplified model version of it, will be discussed in §§ 3 to 6 on the usual basis that the ϕ and ψ fields represent completely independent degrees of freedom. Consideration of the non-linear meson field, in the simplified model, suggests that the spinor fields may be interpretable as collective co-ordinates describing the motions of singularities of solutions of the meson field equations alone. This point of view will not be followed with the realistic meson field described above on account of its complexity, and the aim of the following sections is to show that the hypothetical structure of the meson fields is, in some respects, a reasonable one.

3. Symmetry

The constraint, equation (2), is invariant under any rotation in the Euclidean four-dimensional iso-space. The infinitesimal rotation operators of the meson field will be denoted by $j_{\mu\nu}(x)$

$$j_{\mu\nu} = -\mathrm{i}\phi_\mu\frac{\partial}{\partial\phi_\nu} - \phi_\nu\frac{\partial}{\partial\phi_\mu} \quad (\mu, \nu = 1, \dots 4, \text{ any } x). \qquad (6)$$

The interaction term in $\mathscr{L}$, equation (4), remains invariant if a suitable transformation is also made on the spinor field; this is defined by adding to j_{12} the operator $\frac{1}{2}(\psi^\dagger\tau_3\psi)$, and to j_{34} the operator $-\frac{1}{2}(\psi^\dagger\gamma_5\tau_3\psi)$, and to the other j according to cyclic permutation of the suffixes 1, 2, 3. The first term of $\mathscr{L}$ is obviously also invariant under rotation, but the ϕ^4 term is invariant only under rotations through $\frac{1}{2}\pi$, about any axis.

These rotations, of the form $\exp(\frac{1}{2}\mathrm{i}\pi j_{\mu\nu})$, generate a group of order 192, the analogue of the proper cubic group in three dimensions,* of which every element r satisfies $r^4 = 1$. If to $j_{\mu\nu}$ were simply added the contributions mentioned above for the spinor field, it would be necessary to consider the double group. However, with the interpretation of charge, which will be given in the next section, it is desirable

* This group is the proper subgroup of the group of order 384 whose character table is given by Littlewood (1950, p. 278).

to retain $r^4 = 1$, so that half-integral charges do not appear; this can be achieved by doubling the number of components of ψ in the following way.

In addition to the usual four spinor indices, and to the isotopic spin label (the eigenvalue of τ_3), the 16 components of the new ψ will also be labelled by the eigenvalue ± 1 of ρ_3, the third component of another set of 2×2 spin matrices ρ_i. Since $\mathscr{L}$ does not involve this degree of freedom explicitly it is possible to associate an arbitrary rotation in the ρ-space with the other necessary rotations of the fields, and this can be done in such a way that the inclusion of the ψ-fields do not enlarge the symmetry group of the meson fields. The complete infinitesimal operators can be defined suitably in the following way,

$$\left.\begin{aligned} \mathscr{J}_{12} &= j_{12} + \tfrac{1}{2}\psi^\dagger(\tau_3 + \rho_3)\,\psi, \\ \mathscr{J}_{34} &= j_{34} - \tfrac{1}{2}\psi^\dagger(\gamma_5\tau_3 + \rho_3)\,\psi \end{aligned}\right\} \tag{7}$$

and their cyclic permutations.

It will appear that the most natural interpretation of ρ_3 is that of particle-anti-particle label, and the invariant

$$\int \psi^+ \rho_3 \psi \,\mathrm{d}^3x = N \tag{8}$$

is the Baryon number, particles less anti-particles.

The symmetry group in iso-space is that generated by the $\frac{1}{2}\pi$ rotation operators $R_{\mu\nu}$, defined by

$$\left.\begin{aligned} R_{\mu\nu} &= \exp\left(\tfrac{1}{2}\mathrm{i}\pi J_{\mu\nu}\right), \\ J_{\mu\nu} &= \int \mathscr{J}_{\mu\nu}\,\mathrm{d}^3x, \end{aligned}\right\} \tag{9}$$

and the explicit transformations of the field variables are illustrated by

$$\left.\begin{aligned} R_{12}\phi_1 R_{12}^{-1} &= -\phi_2, \quad R_{12}\phi_2 R_{12}^{-1} = \phi_1, \\ R_{12}\psi R_{12}^{-1} &= \exp\left(-\tfrac{1}{4}\mathrm{i}\pi(\tau_3 + \rho_3)\right)\psi, \\ R_{34}\psi R_{34}^{-1} &= \exp\left(+\tfrac{1}{4}\mathrm{i}\pi(\gamma_5\tau_3 + \rho_3)\right)\psi. \end{aligned}\right\} \tag{10}$$

As mentioned previously $R^4_{\mu\nu} = 1$ for all μ, ν. There is one element of the group, apart from the identity, which commutes with all others; this is the reflexion operator M, which transforms ϕ_ρ into $-\phi_\rho$ and ψ into $\gamma_5\psi$, and

$$M = R_{12}^2 R_{34}^2 = R_{23}^2 R_{14}^2 = R_{31}^2 R_{24}^2, \tag{11}$$

with eigenvalues ± 1.

The theory has symmetry under the charge conjugation operation C, which is defined by

$$C\phi_2 C^{-1} = -\phi_2; \quad C\phi_\rho C^{-1} = \phi_\rho, \ \rho \neq 2; \quad C\psi C^{-1} = \psi^\dagger, \tag{12}$$

where the matrices τ are represented in the conventional way, and where a Majorana representation is adopted for the γ_μ-matrices.

There is also symmetry under inversion; the operation P is defined by

$$\left.\begin{aligned} Px_i P^{-1} &= -x_i; \quad P\phi_i P^{-1} = -\phi_i, \\ P\phi_4 P^{-1} &= \phi_4; \quad\ \ P\psi P^{-1} = \beta\psi. \end{aligned}\right\} \tag{13}$$

Neither C nor P commute with the rotations $R_{\mu\nu}$, nor do the separate transformations of R by them have any simple expression. However, under the combined transformation CP the operators $R_{\mu\nu}$ transform in the same way as under the transformation by R_{31}^2 or R_{24}^2. This is evident for the meson fields, because for them the effect of CP is identical with that of R_{13}^2, but this simple identification is not true for the Fermion fields, because $\psi^\dagger$ and $\rho_2\tau_2\psi$ are not the same thing. Nor can they simply be equated as an additional constraint, because their difference does not commute with the Hamiltonian. It is, however, possible to apply a constraint involving also the meson fields (cf. equation (22) below) to remove the redundant description of anti-particles.

4. Interpretation

There is assumed to be a well-defined physical vacuum state $|0\rangle$, generated out of the bare vacuum, for which all the $R_{\mu\nu}$ have the eigenvalue unity,

$$R_{\mu\nu}|0\rangle = |0\rangle. \tag{14}$$

The multiplets formed by states describing particles with the same mass will form representations of the symmetry group, and the individual members may be characterized by the eigenvalues of a commuting set of $R_{\mu\nu}$. These latter may be taken to be R_{12} and R_{34}, and in analogy with the definitions suggested by Polkinghorne & Salam (1955) the isotopic spin T_3 and its strange counterpart S_3 are defined by

$$\left.\begin{aligned} R_{12}R_{34} &= \exp(\mathrm{i}\pi T_3), \\ R_{12}R_{43} &= \exp(\mathrm{i}\pi S_3). \end{aligned}\right\} \tag{15}$$

The charge $Q = T_3 + S_3$, so that we may consistently define

$$R_{12} = \exp(\tfrac{1}{2}\mathrm{i}\pi Q). \tag{16}$$

The properties of the operators mean that Q must be integral, and that T_3 and S_3 may (simultaneously) be either integral or half integral. Further, T_3 and S_3 are only defined modulo 2, and the charge Q modulo 4. The reflexion operator M has the value $+1$ for integral T_3, iso-Bosons, and the value -1 for iso-Fermions.

These multiplets may conveniently be discussed by considering the operators which might generate them out of the vacuum, for these operators must form an equivalent representation of the symmetry group.

The first point to note is that the meson field variables ϕ_ρ, cannot on the above interpretation, describe the creation or destruction of pions, for the ϕ_ρ form a representation with four components having $T_3 = \pm\frac{1}{2}$ and $S_3 = \pm\frac{1}{2}$; the obvious association is with the K-, $\overline{\text{K}}$-doublets, which becomes in detail,

$$\left.\begin{aligned} &\phi_1 - \mathrm{i}\phi_2: \quad \text{create } \mathrm{K}^-, \text{ absorb } \mathrm{K}^+, \\ &\phi_4 - \mathrm{i}\phi_3: \quad \text{create } \overline{\mathrm{K}}^0, \text{ absorb } \mathrm{K}^0, \\ &\phi_4 + \mathrm{i}\phi_3: \quad \text{create } \mathrm{K}^0, \text{ absorb } \overline{\mathrm{K}}^0, \\ &\phi_1 + \mathrm{i}\phi_2: \quad \text{create } \mathrm{K}^+, \text{ absorb } \mathrm{K}^-. \end{aligned}\right\} \tag{17}$$

The four fields ϕ_ρ are not independent, but they are *linearly* independent so that it is sensible to speak of four such distinct states.

The field operators associated with the pion field must form a triplet with $T_3 = \pm 1, 0$; $S_3 = 0$. The simplest possibility is afforded by the triplet of rotation operators,

$$\dot{j}_{14}+\dot{j}_{23}, \quad \dot{j}_{24}+\dot{j}_{31}, \quad \dot{j}_{34}+\dot{j}_{12}. \tag{18}$$

An equivalent representation in terms of the fields only is exhibited by allowing these operators to operate upon the invariant $\Sigma\phi_\rho^4$ that must occur in the vacuum state. This gives

$$\phi_1\phi_4(\phi_4^2-\phi_1^2)+\phi_2\phi_3(\phi_3^2-\phi_2^2) \quad \text{and permutations.} \tag{19}$$

In the approximation in which ϕ_4 is nearly equal to Q, these are simply proportional to the original fields ϕ_i.

Degenerate with these states would be a triplet with $T_3 = 0$ and $S_3 = \pm 1, 0$. These would comprise another neutral meson with zero strangeness, and two others that were strange particles; no evidence of these has been found.

The effect of the operation CP is to permute these states in the expected way, interchanging K and $\overline{\text{K}}$ and changing the signs of the 1 and 3 components of the pion field. Neither operation by itself, C nor P, has such a simple effect; K^0 appears, as might have been expected, as a mixture of parity states, while inversion is accompanied by an interchange of the pion triplet with their degenerate companions. It must be remembered, however, that parity is only a relative concept, and this rather startling result about the pion field is meaningless until some discussion of the interactions has been made, and it will be seen in the following section that parity is nevertheless conserved in practice in meson-nucleon interactions.

Considering now the Fermion particles, it is evident from the form of equation (10) that the field operator ψ cannot by itself create eigenstates of T_3 and S_3. Nucleons, being iso-Fermions, must belong to the eigenvalue -1 of M, and bearing this in mind a suitable form for the nucleon operator is $Z\psi$, where Z is the unitary operator of the meson field,

$$Z = \tfrac{1}{2}(1-\gamma_5)+\tfrac{1}{2}(1+\gamma_5)(\phi_4+\mathrm{i}\gamma_5\tau_i\phi_i)/Q = \exp\left(\tfrac{1}{2}\mathrm{i}(1+\gamma_5)(\tau_i\phi_i/|\boldsymbol{\phi}|)\theta\right), \tag{20}$$

where θ is the angle defined by $\phi_4 = Q\cos\theta$, $|\boldsymbol{\phi}| = Q\sin\theta$. The operator has the properties

$$Z^+Z = 1, \quad Z^+\beta Z = (\phi_4-\mathrm{i}\gamma_5\tau_i\phi_i)/Q\,.\,\beta. \tag{21}$$

The four states $Z\psi$ form a representation of the symmetry group with $T_3 = \frac{1}{2}\tau_3$ and $S_3 = \frac{1}{2}\rho_3$. On the assumption that the states with $\rho_3 = +1$ correspond to the nucleon doublet, two possibilities appear for the interpretation of the other doublet. It might in the first place correspond to the Ξ^0, Ξ^- doublet; but in that case an explanation would be needed for the large mass difference between the two doublets outside the theory suggested here, in contrast to the attractive suggestion made by Gell-Mann (1957) and by Schwinger (1957) that this can be ascribed to K-meson interactions. The second possibility is that this doublet provide an alternative description of the anti-nucleon pair; but, as was noted above in the previous section, it is not possible simply to identify the components of the second doublet with the conjugates of those of the former; the identification must involve the meson field, and the simplest possibility is a subsidiary condition on the state function, of the form

$$(\psi^{\dagger}-\rho_2\tau_2\psi I)\,\Phi = 0, \tag{22}$$

where I is the operator that changes the signs of the ϕ_i. This condition commutes with the Hamiltonian, and with the Baryon number N, equation (8).

From the point of view with which we started, that the ϕ and the ψ are two entirely independent sets of fields, such a constraint appears very unnatural; but from the point of view indicated in the final section of this paper it may be quite reasonable. We shall assume that the interpretation in terms of anti-nucleons is the correct one, in which case the quantum number $\int\psi^{+}\rho_3\psi\,\mathrm{d}^3x$ may be interpreted as the number of particles less anti-particles.

As with the pion field, inversion alone leads to a quite different set of particles; in this case four particles with $T = 0$ and $S_3 = \pm 1, 0, 0$.

In the present theory the hyperons are most naturally regarded as compound states, of nucleons and $\overline{\mathrm{K}}$-mesons. The corresponding operators are of the form $\phi Z\psi$. In particular those for the Σ^0 and Λ^0 hyperons will have the form

$$2^{-\frac{1}{2}}(\phi_1+\mathrm{i}\phi_2)/Q(Z\psi)_p \pm 2^{-\frac{1}{2}}(\phi_3-\mathrm{i}\phi_4)/Q(Z\psi)_n. \tag{23}$$

5. Interactions

The Lagrangian $\mathscr{L}$, equation (4) does not exhibit in a direct way the interactions between particles and field because ψ has no simple interpretation, according to the scheme adopted above. The interactions can be exhibited in a more significant form by transforming to $Z\psi$ as a new field variable; accordingly ψ is replaced by $Z^{+}\psi$ in the Lagrangian and the transformed Fermion part has the form

$$-\psi^{+}\beta\left[\gamma_{\mu}\frac{\partial}{\partial x_{\mu}}+gQ+\gamma_{\mu}\gamma_5 V_{\mu}\right]\psi, \tag{24}$$

where the interaction

$$V_{\mu} = Z\mathrm{d}Z^{+}/\mathrm{d}x_{\mu} = (1+\gamma_5/2Q^2)(\phi_4+\mathrm{i}\gamma_5\boldsymbol{\tau}.\boldsymbol{\phi})\,\mathrm{d}/\mathrm{d}x_{\mu}(\phi_4-\mathrm{i}\gamma_5\boldsymbol{\tau}.\boldsymbol{\phi}) \tag{25}$$

$$= (\mathrm{i}/2Q^2)(1+\gamma_5)\sum_{\text{cyc.}}\tau_1\left\{\phi_1\frac{\partial\phi_4}{\partial x_{\mu}}-\phi_4\frac{\partial\phi_1}{\partial x_{\mu}}+\phi_2\frac{\partial\phi_3}{\partial x_{\mu}}-\phi_3\frac{\partial\phi_1}{\partial x_{\mu}}\right\}. \tag{26}$$

This quantity has mixed vector and pseudo-vector character, but the 'wrong parity' vector component will *not* give rise to parity violating elements in the scattering matrix for the following reason. These 'wrong' components may be eliminated by taking out the parity-violating part of the operator Z, i.e. the factor $\exp(\frac{1}{2}\mathrm{i}(\boldsymbol{\tau}.\boldsymbol{\phi}/|\,\phi\,|)\,\theta)$, and since this factor commutes with the γ-matrices it will not affect the scattering matrix; it may also be verified that in perturbation theory, the contributions of the parity violating part vanish for transitions between real (as opposed to virtual) particle states.

The fact that the theory proposed here has this type of parity symmetry although it is not manifest in the formalism, because the classification has been made by the rotation operators with which inversion does not commute, is a rather objectionable feature but seems to be inherent in this approach. This apparent difficulty appears also, for example, in the work of Gell-Mann & Feynmann (1958); since the interaction only affects the part of ψ for which $\gamma_5 = 1$, the equation of motion for ψ can easily be transformed into one of second order for $\chi = \frac{1}{2}(1+\gamma_5)\psi$ as in their work, giving

$$[\gamma_\mu\gamma_\nu(\mathrm{d}/\mathrm{d}x_\mu)(\mathrm{d}/\mathrm{d}x_\nu + V_\nu) - g^2Q^2]\chi = 0,$$

where $\gamma_\mu\gamma_\nu$ can then be replaced by 2×2 spin matrices.

The interaction has the isotopic symmetry of $\tau_i\pi_i$, where π_i has the symmetry of operators, such as (18) or (19), characteristic of the pion field. In the approximation where $\phi_4 \sim Q$, the interaction is of the familiar pseudo-vector type, if the non-effective vector interaction is removed; the coupling constant is $(\frac{1}{2}Q)$, but this cannot be identified with any empirical value without discussion of the renormalizations needed in this theory. It suggests, however, that Q may be of the order of the pion mass, or that Q^{-1} is a length of the order of 10^{-13} cm.

Discussion of the interactions of the hyperons requires the introduction of new fields for the description of the compound states. This can be done with the help of the usual devices of subsidiary conditions, but will not be pursued in this paper.

6. A simplified model

The theory suggested in the earlier part of this paper cannot be developed in greater or more quantitative detail without a clear understanding of the nature of the states described by the Lagrangian. As a step towards this a simplified model of the theory has been considered in which the complications of spin and isotopic spin have been removed; this has been done by restricting both the configuration space-time and the iso-space to two dimensions instead of four. The spatial coordinates will be x and t, and the two directions in iso-space will be labelled 3 and 4 to keep as close an analogy as possible with the realistic problem; the spinor field has two spatial components, acted upon by the 2×2 anti-commuting matrices, α, β and $\mathrm{i}\alpha\beta$, and another doubling labelled by the value ± 1 of ρ.

The Lagrangian density has the form

$$\mathscr{L}_m = \tfrac{1}{2}\Sigma(\mathrm{d}\phi_\lambda/\mathrm{d}x_\mu)^2 + \tfrac{1}{4}\gamma^2\Sigma\phi_\lambda^4 - \psi^\dagger[-\mathrm{i}\,\mathrm{d}/\mathrm{d}t - \mathrm{i}\alpha\,\mathrm{d}/\mathrm{d}x + g\beta(\phi_4 + \mathrm{i}\alpha\phi_3)]\psi, \quad (27)$$

where λ is summed over the values 3 and 4, with the constraint

$$\phi_3^2 + \phi_4^2 = Q^2. \quad (28)$$

The simplicity of two dimensions in iso-space consists in the possibility of a simple representation of the constrained fields by an angle variable θ,

$$\phi_3 = Q\sin\theta, \quad \phi_4 = Q\cos\theta. \quad (29)$$

The explicit introduction of (29) into (27) gives

$$\mathscr{L}_m = \tfrac{1}{2}Q^2[(\mathrm{d}\theta/\mathrm{d}t)^2 - (\mathrm{d}\theta/\mathrm{d}x)^2 - (\tfrac{1}{8}\kappa^2)(1-\cos 4\theta)] + \psi^\dagger(\mathrm{i}\,\mathrm{d}/\mathrm{d}t + \mathrm{i}\alpha\,\mathrm{d}/\mathrm{d}x - gQ\beta\,\mathrm{e}^{\mathrm{i}\alpha\theta})\psi, \quad (30)$$

where $\kappa = \gamma Q$, a reciprocal length.

The Hamiltonian is formed in the usual way, by introducing a canonical co-ordinate conjugate to θ, which will be denoted by η. It is also convenient to introduce a two-dimensional element of volume ω, and to write for the total Lagrangian

$$L_m = \int \mathscr{L}_m \omega \, dx \tag{31}$$

so as to preserve correct dimensionality. Then the Hamiltonian has the form

$$H_m = \int \{(2\omega Q^2)^{-1} \eta^2 + (\tfrac{1}{2}\omega Q^2)(d\theta/dx)^2 + (\omega Q^2 \kappa^2/16)(1-\cos 4\theta)\} + \int \psi^\dagger(-i\alpha \, d/dx + gQ\beta \, e^{i\alpha\theta}) \psi \, dx. \tag{32}$$

α and β are taken as real and imaginary matrices, respectively; α also corresponds to γ_5 in the 4-D theory.

This system has the following symmetries:

Isotopic rotation: of the rotation operators in the four-dimensional case only R_{34} survives. It will be denoted by R, with the properties

$$\left.\begin{aligned} R\theta R^{-1} &= \theta - \tfrac{1}{2}\pi, \\ R\psi R^{-1} &= \exp(\tfrac{1}{4}i\pi(\alpha+\rho))\,\psi. \end{aligned}\right\} \tag{33}$$

Charge conjugation: this operation C is defined by

$$C\theta C^{-1} = -\theta; \quad C\psi C^{-1} = \psi^\dagger. \tag{34}$$

Inversion: this operation P is defined by

$$P \times P^{-1} = -x; \quad P\theta P^{-1} = -\theta; \quad P\psi P^{-1} = \beta\psi. \tag{35}$$

Under the combined operation CP, θ is unaltered while the effect on R is equivalent to a change of sign of ρ.

There is no charge, $R^4 = 1$, and an isotopic spin or equivalent strangeness may be defined by

$$R = \exp(i\pi T) = \exp(-i\pi S). \tag{36}$$

This theory can be given an interpretation closely similar to that suggested in four dimensions, and for the sake of analogy the same particle names will be applied to corresponding states. K-mesons are characterized by the operators $\exp(\pm i\theta)$, with $T = \pm\frac{1}{2}$. In analogy with (18) and (19), pions might be described by the rotation operator η, or by $\sin 4\theta$; a closer analogy with conventional linear theory would be provided by $\tan 2\theta$, which also has the right symmetry property, and ranges from $-\infty$ to $+\infty$.

The analogue of Z, equation (20), is here defined by

$$Z = \tfrac{1}{2}(\rho+\alpha) + \tfrac{1}{2}(\rho-\alpha)\exp(i\alpha\theta) = \rho \exp(\tfrac{1}{2}i(\alpha-\rho)\theta), \tag{37}$$

so that for the state characterized by $Z\psi$, S has the value $\frac{1}{2}\rho$. In this model Z now depends upon ρ, unlike equation (20); moreover, in this model it is really unnecessary to introduce ρ at all, except for the sake of symmetry and analogy.

The analogue of the Λ, Σ hyperons is described by $\exp(i\rho\theta) Z\psi$.

18-2

If the states with $\rho = -1$ are to be interpreted as anti-particles, it is necessary to impose a subsidiary condition analogous to equation (22), namely

$$(\psi^{\dagger}_{\rho=1} - I\psi_{\rho=-1})\,\Phi = 0, \tag{38}$$

where I is the operator changing θ into $-\theta$.

The transformation to the variable $Z\psi$ leads to the following expression for the Fermion part of the transformed Lagrangian:

$$\psi^{\dagger}[\mathrm{i}\,\mathrm{d}/\mathrm{d}t + \mathrm{i}\alpha\,\mathrm{d}/\mathrm{d}x - gQ\beta + \tfrac{1}{2}(1-\alpha\rho)\,(\mathrm{d}\theta/\mathrm{d}x + \alpha\,\mathrm{d}\theta/\mathrm{d}t)]\,\psi. \tag{39}$$

The ρ-dependent part of the interaction violates parity, but as before the parity violating terms can be removed by the transformation $\exp(-\tfrac{1}{2}\mathrm{i}\theta\rho)$. The interaction is predominantly therefore a straightforward pseudo-vector coupling to the variable θ.

This exposition has been made in complete analogy with the treatment of the realistic fields in four dimensions, assuming that the Fermion fields are quite independent of the θ-field, apart from the explicit interaction term in the Hamiltonian. A closer investigation of the structure of the non-linear meson fields indicates that they also describe particle-like entities, in a way not envisaged above, with many of the properties of the particles explicitly described by ψ. This is examined in the following sections, in a tentative way, and leads to the suggestion that all the physical features described by the Lagrangian (30), when regarded from the point of view of perturbation theory, may really be described by its mesonic part alone when solutions are also included that do not arise naturally in perturbation theory.

7. The classical meson field

The classical field equation is

$$\mathrm{d}^2\theta/\mathrm{d}t^2 - \mathrm{d}^2\theta/\mathrm{d}x^2 + \tfrac{1}{4}\kappa^2 \sin 4\theta = 0. \tag{40}$$

One class of solutions will be of the form

$$\theta = \tfrac{1}{2}n\pi + A\exp(\mathrm{i}kx - \mathrm{i}\omega t), \tag{41}$$

where $\omega^2 = k^2 + \kappa^2$; provided A is a small amplitude (compared with unity) the non-linear term will introduce corrections, but these will be negligible for small A. This type of wave for which $\sin 4\theta$, or $\tan 2\theta$, is everywhere small corresponds to the pions of the quantized theory.

Another formal solution of wave character has the form

$$\theta = \tfrac{1}{2}n\pi + \tfrac{1}{2}m\pi f(x - t - x_0), \tag{42}$$

where $f(z)$ is the step-function,

$$f(z) = 0, \quad (z<0); \qquad f(z) = 1 \quad (z>0). \tag{43}$$

It is questionable whether these singular solutions, moving with the velocity of light, are significant.

The possibility of static solutions is of especial interest. This raises the question of the boundary conditions that should be imposed on the differential equation; a natural form that preserves the essential symmetry of the theory is

$$\sin 2\theta = 0, \quad \text{for} \quad x = \pm\infty \tag{44}$$

and this will be assumed hereafter.

With this condition there is just one possible type of static, everywhere continuous, solution, given by

$$\tan(\theta - \tfrac{1}{2}n\pi) = \exp(\pm\kappa(x/x_0)). \tag{45}$$

This has the character of a smoothed-out step-function, and the limiting values of θ at $\pm\infty$ differ by $\frac{1}{2}\pi$. The energy of the field is $\frac{1}{2}\omega Q^2\kappa$.

The character of this solution suggests that it should be identified with 'a particle at x_0', for:

(i) It represents a localized disturbance in the field, capable of moving with arbitrary uniform velocity.

(ii) It exists in complementary particle and anti-particle forms, corresponding to the $\pm$ sign on the right side of equation (45).

(iii) There is a conservation law for the number of particles less anti-particles; this number N is a constant of the motion given by

$$\theta(\infty) - \theta(-\infty) = \tfrac{1}{2}N\pi. \tag{46}$$

(iv) The process of pair creation is seen to be possible as the result of a mesonic fluctuation of amplitude $\frac{1}{2}\pi$, which can then broaden out to describe two particles travelling in opposite directions.

This particle solution is a unified description of source and the field bound to it. An interesting correspondence with the usual ideas of local field theory may be obtained by separating the solution into a singular step-function, describing the source, and a residual field coupled to the source. This suggests the substitution

$$\theta = (\tfrac{1}{2}\pi) f(x - x_0) + \theta' \tag{47}$$

and if the resulting equation for θ', which should be small on *both* sides of the singularity, is linearized we obtain

$$\mathrm{d}^2\theta'/\mathrm{d}x^2 - \kappa^2\theta' = -(\tfrac{1}{2}\pi)\,\delta'(x - x_0), \tag{48}$$

the source being the derivative of a δ-function. The solution of these equations gives

$$\theta \sim (\tfrac{1}{2}\pi) f(x - x_0) + (\tfrac{1}{4}\pi)\,\mathrm{sgn}\,(x - x_0)\exp(-\kappa\,|\,x - x_0\,|), \tag{49}$$

which is indeed a fair approximation to the true solution, given by equation (45). The infinite energy associated with the step-function is almost entirely cancelled by the self-energy of coupling to the meson field.

The motion of the singularity can be discussed in a similar way, by the familiar device of introducing extra co-ordinates x_0 and p_0 to describe the position and momentum of the singularity. For the sake of analogy with the quantized theory this will be expressed in terms of transformations of the Hamiltonian, to which will be adjoined originally the condition

$$p_0 = 0. \tag{50}$$

The first transformation made is the same as (47),

$$\left.\begin{aligned} \theta &\to \theta + (\tfrac{1}{2}\pi) f(x - x_0), \\ p_0 &\to p_0 + (\tfrac{1}{2}\pi)\, \eta(x_0). \end{aligned}\right\} \tag{51}$$

The second transformation is designed approximately to remove p_0 explicitly from the transformed condition; the main effect is

$$\eta \to \eta - (2/\pi)\, p_0 \delta(x - x_0)/\delta(0). \tag{52}$$

(The rather singular form of this expression should be interpreted by replacing the step-function f by some smooth approximation to it.) In the transformed Hamiltonian there appears a kinetic energy term for the particle which may be written as

$$p_0^2/2M_0 \quad \text{with} \quad M_0 = (\tfrac{1}{4}\pi^2)\, \omega Q^2 \delta(0), \tag{53}$$

where M_0 is the (infinite) inertial mass associated with the singular source. Then the interaction energy has the form

$$(\tfrac{1}{2}\pi^2 \omega Q^2)\, [\mathrm{d}\theta/\mathrm{d}x + (p_0/M_0)\, \mathrm{d}\theta/\mathrm{d}t]_{x=x_0} \tag{54}$$

which strongly resembles that occurring in the expression (39).

It would, of course, be perfectly possible to discuss the interactions of the finite particle, described by equation (45), with the meson field without any reference to these singular sources, but the main object of this investigation is to see whether the conventional description of particles with local interactions is contained within the non-linear meson theory.

8. The quantized field theory

A number of new problems arise when the non-linear meson field is quantized. The first concerns the magnitude of ω (cf. equation (31)); the comparison of the expressions (54) and (39) suggests that ωQ^2 ought to have the value $1/\pi$, and this necessity will also appear in the course of the following analysis. This problem is of course peculiar to the two-dimensional model problem that we are considering, and its significance is obscure; *a priori* this particular value is quite reasonable, because we might expect ω to be related to a cut-off momentum K, such that by the uncertainty principle $\pi K^2 \omega = 4\pi^2$, so that the necessary value of ω correspond with $K = 2\pi Q$; on the other hand this leads to a value of the 'mass' of the particle described by (45) equal to $(\frac{1}{2}\pi)\,\kappa$, much smaller than the apparent mass κ of the meson field; however, these masses will certainly be changed by self-mass corrections in the quantum theory.

The next problem lies in the nature of the vacuum state, especially whether a unique non-degenerate ground state exists. The correct answer to this has not been found, and the following point of view will be adopted as a working hypothesis. If the Hamiltonian (of the meson field) is expressed in terms of the variable $\tan 2\theta$, or a one-valued function of this variable, it will have a non-linear but rational algebraic form; any state-function that is a function only of this variable will have the 'pion' symmetry, so this transformed Hamiltonian will be regarded as that of the uncoupled pion field. Nucleons and strange particles cannot be described by such state-functions, and additional co-ordinates will be needed to describe these corresponding with the fact that the original fields ϕ_3, ϕ_4 are multiple-valued functions of the variable $\tan 2\theta$. Since this multiplicity of value can occur independently at any point of space, additional fields will in fact be needed. The objective then is to define such additional fields, and add on to the Hamiltonian such new terms involving them that the original Hamiltonian can legitimately be expressed in terms of the $\tan 2\theta$ variable, with the multiple-valuedness taken care of by the new fields.

The classical field does not appear to contain any solutions corresponding with the K-meson or hyperon states. These will not be considered further here, and the remainder of this paper will be devoted to the problem of introducing field variables to describe the nucleonic particles, for which there is a clear classical analogy.

The classical analysis suggested that it would be natural to identify the 'bare' particles of field theory with step-function singularities in the variable θ. When the number of particles is regarded as fixed the quantum theory can be discussed in exactly the same way as the classical. Redundant co-ordinates x_n, p_n are introduced to describe the positions of the particles, and conditions $p_n = 0$ are adjoined to the Hamiltonian as subsidiary conditions on the state-function

$$p_n \Phi = 0. \tag{55}$$

Canonical transformations of the system are then made by unitary matrices S, such that any function F transforms according to

$$F \to S^{-1} F S,$$

i.e. the *new* variables F' are expressed in terms of the *old* by

$$F' = S F S^{-1}. \tag{56}$$

The first transformation to be made corresponds to (51), with

$$S = \exp\left(-\tfrac{1}{2}\mathrm{i}\pi \sum_n \int \eta(x) f(x - x_n)\, \mathrm{d}x\right). \tag{57}$$

(It will generally be assumed that $f(z)$ is not actually the discontinuous step-function, but some close continuous approximation to it.) The subsidiary conditions become

$$\left[p_n + (\tfrac{1}{2}\pi) \int \eta(x) f'(x - x_n)\, \mathrm{d}x\right] \Phi = 0. \tag{58}$$

The second transformation will be taken to be given by

$$S = \exp\left(-\tfrac{1}{2}\mathrm{i} \sum_n \int [(p_n/2M_0) f'(x - x_n) + \text{conj.}]\, \theta(x)\, \mathrm{d}x\right), \tag{59}$$

where
$$M_0 = (\tfrac{1}{4}\pi) \int (f'(x))^2\, \mathrm{d}x, \tag{60}$$

which is the same as equation (53) with $\omega Q^2 = 1/\pi$. This transformation is somewhat untidy because the p_n and x_n in the exponent do not commute.

In the limit in which $f(x)$ approaches a step-function, these transformations will not alter the $\cos 4\theta$ term in the Hamiltonian. The interaction terms that arise, in the first approximation in which the momenta p_n are small, are exactly the same as in the classical case, and the non-relativistic analogue of the interaction term in equation (39).

A field-theoretic description of the singularities is of course essential to describe the pair-creation processes that can occur. Appropriate fields can be introduced by following the usual methods adopted in second quantization. The subsidiary condition (55) says that $\mathrm{d}\Phi/\mathrm{d}x_n = 0$. This suggests the introduction of particle fields ψ_λ, with the number of components yet unspecified, satisfying the subsidiary conditions,

$$(\mathrm{d}\psi_\lambda/\mathrm{d}x)\,\Phi = 0, \quad \text{for each } \lambda \text{ and all } x. \tag{61}$$

The fields will have conjugates $\psi_\lambda^\dagger$ such that $\psi_\lambda^\dagger\,\psi_\lambda$ has integral eigenvalues (when the ψ are normalized appropriately). The generalization of the first transformation (57) will be taken to be

$$S_1 = \exp\{-\tfrac{1}{2}\mathrm{i}\pi \sum_\lambda \iint \eta(x)\,[\psi_\lambda^\dagger(y)\,\rho_\lambda\,\psi_\lambda(y)]\,f(x-y)\,\mathrm{d}x\,\mathrm{d}y\}, \tag{62}$$

where the coefficient ρ_λ might take on the values ± 1 corresponding to steps 'up' or 'down' in θ, i.e. to particles or anti-particles.

For the second transformation we notice that (p/M_0) occurring in (59) represents the velocity of the particle. If now we assume that the fields will satisfy an equation of Dirac's type, we can represent the velocity by $\psi^\dagger\alpha\psi$, and the transformation will assume a comparatively simple form in which the complications in (59) due to the non-commutativity of p_n and x_n do not arise. Accordingly we take

$$S_2 = \exp\{-\tfrac{1}{2}\mathrm{i} \sum_\lambda \iint \theta(x)\,[\psi_\lambda^\dagger(y)\,\rho_\lambda\,\alpha_\lambda\,\psi_\lambda(y)]\,f'(x-y)\,\mathrm{d}x\,\mathrm{d}y\}, \tag{63}$$

where the coefficients α_λ may again take on the values ± 1.

The new subsidiary condition has the form

$$\mathrm{d}/\mathrm{d}x(U_\lambda^+(x)\,\psi_\lambda(x))\,\Phi = 0, \tag{64}$$

where the unitary operator $U_\lambda(x)$ is defined by

$$U_\lambda(x) = \exp[\tfrac{1}{2}\mathrm{i}\rho_\lambda\alpha_\lambda \textstyle\int \theta(z) f'(z-x)\,\mathrm{d}z]\exp[\tfrac{1}{2}\mathrm{i}\pi\rho_\lambda \int(\eta(z) - \tfrac{1}{2}\int \psi^\dagger\rho\alpha\psi f'))f(z-x)\,\mathrm{d}z]. \tag{65}$$

The transformed Hamiltonian can be written in a simple form by adding on to it a multiple of the subsidiary condition, namely

$$\int \psi^\dagger(x)\,U_\lambda(x)(-\mathrm{i}\alpha_\lambda\,\mathrm{d}/\mathrm{d}x)\,(U_\lambda^+(x)\,\psi_\lambda(x))\,\mathrm{d}x.$$

The final result for the equivalent transformed Hamiltonian is

$$\begin{aligned} H' = H &+ \int \psi_\lambda^+(x)\,(-\mathrm{i}\alpha_\lambda\,\mathrm{d}/\mathrm{d}x)\,\psi_\lambda(x)\,\mathrm{d}x \\ &+ (\tfrac{1}{8}\pi)\int[(\psi_\lambda^+(x)\,\rho_\lambda\,\psi_\lambda(x))^2 - (\psi_\lambda^+(x)\,\rho_\lambda\,\alpha_\lambda\,\psi_\lambda(x))^2]\,\mathrm{d}x, \end{aligned} \tag{66}$$

where in the last term $f'(z)$ has been replaced by its limiting δ-function value.

The problem of solving this Hamiltonian (66) with the subsidiary conditions (64) raises many difficulties, and no more will be attempted beyond this point than to sketch some lines of thought and possible development, and it must be remembered that this is a model problem, the extent of whose analogy to the real one is quite uncertain.

It seems reasonable, with a backward glance at the original model Hamiltonian equation (32), to allow four components for the fields ψ; corresponding with the values ± 1 of α and ρ, and these products will now be written in the obvious matrix form. The subsidiary condition (64) suggests that $\psi_\lambda(x)$ is equivalent to a constant multiple of $U_\lambda(x)$ acting upon the state-function; however, the $U_\lambda(x)$ do not have the proper anti-commutation relations, but something rather more complicated.* Supposing that some such interpretation can consistently be made the subsidiary condition then says that the part of the state-function which has an extra particle at some point must have a mate in which there is an extra step or kink in the variable θ. All these different parts of the state-function will describe the same physical system in different ways, a particle being described either explicitly in terms of the ψ field or by a singularity in the meson field. As far as this is the content of the subsidiary conditions, they would appear to be ignorable provided only non-singular states of the meson field are allowed, that is in equation (66) the original Hamiltonian H would be regarded as a function of some variable such as $\tan 2\theta$.

This, however, is not the whole content of the subsidiary conditions, which also provide relations between the different components ψ_λ, and must also express the coupling between particle and meson fields. The components of ψ will be labelled by the eigenvalues of α and ρ, i.e. $\psi_{1,1}$ has $\alpha = \rho = +1$. The subsidiary conditions almost imply and might in part be replaced by the simpler conditions.†

$$\left.\begin{aligned}(\psi_{1,1} - \mathrm{i}\,\mathrm{e}^{\mathrm{i}\theta}\,\psi_{-1,1})\,\Phi &= 0,\\ (\psi_{1,-1} + \mathrm{i}\,\mathrm{e}^{-\mathrm{i}\theta}\,\psi_{-1,-1})\,\Phi &= 0.\end{aligned}\right\} \tag{67}$$

As far as (38) is concerned, we notice that a subsidiary condition of the form

$$(\psi^+_{1,1} - I\psi_{1,-1})\,\Phi = 0. \tag{68}$$

would be preserved in the same form for the transformed system because U is unaltered by a combined change of sign of ρ and of θ. Some such condition is not unnatural because $U_{\alpha,-1}$ is *in fact* the complex conjugate of $U_{\alpha,1}X$. However, the conditions (67) and (68) are not mutually compatible. If we could make this identification, the effect would be to replace the products $\psi^+\psi$ that occur in the Hamiltonian by the symmetrized form $\frac{1}{2}(\psi^+\psi - \psi\psi^+)$, and the anti-commuting property of the ψ, which we have assumed *ad hoc*, would be essential to prevent the vanishing of the kinetic energy term.

Finally, it is possible to see some equivalence between the non-linear Fermion term in the Hamiltonian (66) and the meson-nucleon coupling term in equation (32). Suppose that the terms with $\rho = -1$ can be suppressed by symmetrizing the products, and write for brevity $\psi_{\pm 1,1}$ simply as $\psi_\pm$. Then the integrand of the last term in (66) is equal to

$$\tfrac{1}{8}\pi(\psi^+_+\psi_+ - \psi_+\psi^+_+)\,(\psi^+_-\psi_- - \psi_-\psi^+_-). \tag{69}$$

* The operators $U^2(x)$ 'almost' anti-commute, and one combination of them has exactly the right properties, but the significance of this is not understood.

† The phase relation has been chosen arbitrarily.

We use the facts that $\psi^2(x) = 0$, and $\psi^+(x)\,\psi(x) + \psi(x)\,\psi^+(x) = \delta(0) = (4/\pi)\,M_0$, according to the definition (60). The second factor, acting upon the state-function, is then equal to, assuming the condition (67)

$$-2i(\psi_-^+\,e^{-i\theta}\,\psi_+) - (4M_0/\pi). \tag{70}$$

The first factor is written as $(2\psi_+^+\psi_+ - 4M_0/\pi)$ and a second application of the condition (67) eventually shows that (69) acting on the state-function is equal to

$$iM_0(\psi_-^+\,e^{-i\theta}\,\psi_+ - \psi_+^+\,e^{i\theta}\,\psi_-) + (2M_0^2/\pi).$$

Apart from the odd constant this is just

$$M_0(\psi^+\beta\,e^{-i\alpha\theta}\psi), \tag{71}$$

of the same form as the interaction term in (32), apart from an accidental change of sign in θ. This interesting result is not, however, significant until it is explained why the subsidiary conditions may entirely be neglected after this final transformation.

It is interesting to note that the Fermion terms in equation (66) are of a form similar to that hypothesized by Heisenberg (1957). In this model it so happens that no dimensional constant appears in the expression; but *if* a four-dimensional analogue of our analysis exists and leads to a similar result a constant proportional to Q^{-2} would have to appear before the fourth-order Fermion terms. In our discussion the Boson field has been fundamental, and the Fermion field is brought in to describe some of the non-linear effects. This point of view is completely opposite to that of Heisenberg, who regards the spinor field as fundamental. It is possible that the analysis given here may admit of this alternative interpretation, and that the original Hamiltonian H is that which must be added to the Fermion part to describe the Bosons, whose field is expressed implicitly in terms of the Fermion field by the same supplementary conditions (64).

References

Gell-Mann, M. 1957 *Proceedings of the 7th Rochester Conference.*
Gell-Mann, M. & Feynmann, R. P. 1958 *Phys. Rev.* **109**, 193.
Heisenberg, W. 1957 *Rev. Mod. Phys.* **29**, 269.
Littlewood, D. E. 1950 *The theory of group characters.* Oxford University Press.
Polkinghorne, J. C. & Salam, A. 1955 *Nuovo Cim.* **2**, 685.
Schwinger, J. 1957 *Annal. Phys.* **2**, 407.
Skyrme, T. H. R. 1955 *Proc. Roy. Soc.* A, **230**, 277.

A unified model of K- and π-mesons

By T. H. R. Skyrme*

University of Pennsylvania

(*Communicated by B. F. J. Schonland, F.R.S.—Received* 12 *February* 1959)

On the foundation of an antecedent non-linear meson field theory it is suggested that the π-meson fields may be described in terms of collective motions of the K-meson fields. A particular model of the K-nucleon interaction is considered whose collective π-modes have symmetrical PV coupling with the nucleon system; parity is conserved to a great extent for the π-nucleon system in the absence of strange particles. The direct K-nucleon interactions do not conserve parity; their sign and symmetry are qualitatively acceptable. The masses and coupling constants of the meson fields are determinate in terms of one universal coupling constant and a cut-off. The structure of this model suggests a natural way for the introduction of the 'spurion', describing weak interactions that violate strangeness.

1. Introduction

In a previous paper (Skyrme (1958), referred to here as NLT) the author has considered the qualitative implications of a particular type of non-linear meson field. In that model the three pion field amplitudes were replaced by four fields interacting symmetrically with the nuclear field and constrained to have constant mean-square amplitude.

The symmetry properties of that system indicated that the four fields should be identified with those of the K- and $\bar{K}$-mesons, while certain *ratios* of fields might describe π-mesons. However, there is difficulty in interpreting the field amplitudes as those for emission or absorption of K-mesons on account of the rigid constraint, and this suggests the adoption of the following more general viewpoint.

The K- and $\bar{K}$-fields are described in terms of real fields ϕ_ρ; unspecified non-linearities are assumed to introduce an effective cut-off such that the mean-square sum $\Sigma\phi_\rho^2$ has a finite vacuum expectation value Q^2 with relatively small fluctuation; the presence of a single particle (in a stationary state) only alters this value by an infinitesimal amount.

The relation of particles to vacuum is imagined as analogous to that of fluctuations in a fluid endowed with density and direction at each point of space. K-mesons correspond to waves involving density fluctuations, while π-mesons are likened to waves of directional oscillation; it is possible, further, that the Fermion sources of the fields may also be describable as singularities with determinate laws of motion and interaction but this point of view, adumbrated in NLT, is not further pursued here.

The particular interaction discussed in this paper is a development from that used in NLT, modified to accord with the present more general philosophy and to remove the excessive symmetry of that theory; the new Lagrangian has a different but high degree of symmetry and involves only one coupling constant, a length F of the order

* On leave of absence from Atomic Energy Research Establishment, Harwell.

of magnitude of $2f$. The masses of the K- and π-fields both arise from interaction with the nucleon field; a very simple estimate yields the ratio of 3 or 4 to 1. The character of the π-N and K-N interactions are in general accord with experiment.

2. The interaction

The K- and $\bar{K}$-fields may alternatively be described in terms of four real fields ϕ_ρ, according to the following conventions:

$$\left.\begin{aligned} K &= \begin{pmatrix} K^+ \\ K^0 \end{pmatrix} = \frac{1}{2}\begin{pmatrix} \phi_2 + \mathrm{i}\phi_1 \\ \phi_4 - \mathrm{i}\phi_3 \end{pmatrix}, \\ \bar{K} &= \begin{pmatrix} -\bar{K}_0 \\ K^- \end{pmatrix} = -\mathrm{i}\tau_2 K^\dagger. \end{aligned}\right\} \tag{1}$$

The square sum
$$K^\dagger . K = \tfrac{1}{2}\Sigma\phi_\rho^2 \tag{2}$$
is postulated to have the vacuum expectation value $\frac{1}{2}Q^2$.

In NLT we assumed a symmetrical interaction with a baryon field

$$g\bar{\psi}(\phi_4 + \mathrm{i}\gamma_5\tau_i\phi_i)\psi \tag{3}$$

and made a transformation of the Foldy type on ψ, to replace the pseudoscalar coupling by a derivative one and to facilitate the identification of the Fermion fields with nucleons; this led to the interaction

$$\frac{1}{2Q^2}[\bar{\psi}\gamma_\mu(1+\gamma_5)\tau_i\psi]\left[\frac{\partial K^\dagger}{\partial x_\mu}\tau_i K - K^\dagger\tau_i\frac{\partial K}{\partial x_\mu}\right] \tag{4}$$

when expressed in terms of K, according to (1).

This interaction, however, has undesirable symmetry under the interchange $K \leftrightarrow \bar{K}$; the degeneracy can be removed by adding to (4) a corresponding term in the isobaric scalars:

$$\frac{1}{2Q^2}[\bar{\psi}\gamma_\mu(1+\gamma_5)\psi]\left[\frac{\partial K^\dagger}{\partial x_\mu}.K - K^\dagger.\frac{\partial K}{\partial x_\mu}\right]. \tag{5}$$

The sum of (4) and (5) can be written in the symmetrical form

$$F^2\left\{(\chi^\dagger.K)\beta\gamma_\mu\left(\frac{\partial K^\dagger}{\partial x_\mu}.\chi\right) - \left(\chi^\dagger.\frac{\partial K}{\partial x_\mu}\right)\beta\gamma_\mu(K^\dagger.\chi)\right\}, \tag{6}$$

where we have introduced the abbreviation

$$\chi = \tfrac{1}{2}(1+\gamma_5)\psi, \tag{7}$$

which may be regarded as a 2-spinor (in space) instead of a 4-spinor, as discussed by Feynman & Gell-Mann (1958). We have also replaced $1/Q$ by a constant F of similar dimensions, length or reciprocal mass, since contact has now been lost with the original PS form of interaction, (3).

The form of (6) suggests that we write the Lagrangian of the K-fields and of the interactions in the combined form

$$L_K = -\left(\frac{\partial K^\dagger}{\partial x_\mu} - F^2\overline{[\chi^\dagger\beta\gamma_\mu(K^\dagger.\chi)]}\right)\left(\frac{\partial K}{\partial x_\mu} + F^2\overline{[(\chi^\dagger.K)\beta\gamma_\mu\chi]}\right), \tag{8}$$

where the bar implies charge-symmetrization of the Fermion fields. The phenomenological mass of the K-field has been replaced by a self-mass arising from the Fermion interactions; to estimate its magnitude we introduce a cut-off or quantization of space such that the anticommutator at one point

$$\psi_\alpha^\dagger(x)\,\psi_\beta(x)+\psi_\beta(x)\,\psi_\alpha^\dagger(x)=\frac{\delta_{\alpha\beta}}{\tau}=\delta_{\alpha\beta}\sum_{\vec{k}}1, \tag{9}$$

where τ is a small volume related to the allowed volume of k-space by the uncertainty principle; for a square cut-off at Λ

$$\tau(\tfrac{4}{3}\pi\Lambda^3)=8\pi^3. \tag{10}$$

The non-derivative part of L_K then has the expectation value

$$-2\frac{F^4}{\tau^2}(K^\dagger.K)-\frac{F^4}{\tau^2}[(K^\dagger.K)\,(\chi^\dagger.\chi)-2(\chi^\dagger.K)\,(K^\dagger.\chi)] \tag{11}$$

if not more than one Fermion is present. If no Fermions are present the expectation value of the second part vanishes leaving for the K-field an effective mass m, where*

$$m^2=2F^4/\tau^2. \tag{12}$$

There is no mass term in the χ-field arising from the K-vacuum.

The magnitude of $Q^2=2(K^\dagger K)$ can be estimated with the same cut-off; the result is

$$Q^2=2\Sigma\frac{1}{\omega}=\frac{2}{\overline{\omega}_\tau}. \tag{13}$$

The condition $F^2Q^2=1$ must be maintained to ensure parity conservation for pion interactions (see §3); this imposes a condition on the cut-off

$$\Sigma\frac{1}{\omega}=\Sigma\frac{1}{m\sqrt{2}}, \tag{14}$$

giving numerically

$$\Lambda\sim 1{\cdot}4m=0{\cdot}7M,$$
$$F\sim 1{\cdot}5f. \tag{15}$$

In this paper the nucleon field will be treated in the usual phenomenological way, with the total Lagrangian density

$$L=L_K-\overline{\psi}(\gamma_\mu\partial/\partial x_\mu+M)\,\psi. \tag{16}$$

No attempt is here made to include a description of hyperons; on occasion they will be treated as compounds of nucleons and $\overline{K}$. The Lagrangian is invariant under CP; it conserves baryon number, isobaric spin and hypercharge. The currents of the latter have the form

$$J_\mu(\rho)=\overline{\psi}\gamma_\mu\rho\psi+\left(\frac{\partial K^\dagger}{\partial x_\mu}\rho K-K^\dagger\rho\frac{\partial K}{\partial x_\mu}\right)$$
$$-F^2[(\overline{\psi}\rho K)\,\gamma_\mu(1+\gamma_5)\,(K^\dagger\psi)+(\overline{\psi}K)\,\gamma_\mu(1+\gamma_5)\,(K^\dagger\rho\psi)], \tag{17}$$

where $\rho=\frac{1}{2}\boldsymbol{\tau},\ \frac{1}{2}$ for isobaric spin and hypercharge, respectively. The electromagnetic current vector, apart from terms involving the potentials A_μ, is $\mathrm{i}eJ_\mu(\frac{1}{2}+\frac{1}{2}\tau_3)$.

* See also the footnote after equation (34), p. 241.

3. The pion field

To describe the waves of directional fluctuation assumed to correspond with π-mesons we must introduce angular co-ordinates to describe the direction of K. These will be introduced as redundant collective co-ordinates by the formal substitution

$$K \to UK = e^{\frac{1}{2}i\tau_3\phi}\, e^{\frac{1}{2}i\tau_2\theta}\, e^{\frac{1}{2}i\tau_1\psi}\, K. \tag{18}$$

θ, ϕ, ψ and K are now all functions of position; we assume that, when a cut-off is introduced, the two sets of variables describe approximately independent modes of vibration, corresponding to π- and K-mesons, but the high-frequency components must be closely linked together. The formal problem of introducing such co-ordinates has been discussed by many authors, e.g. Villars (1957), but there are no simple rules of procedure. The analysis in this paper is consequently based on the tentative assumption that a separation of this kind is practicable. The coupling terms that appear between the collective and particle modes are not significant until the two sets of variables are clearly separated; they will be disregarded here.

We easily find

$$\begin{aligned}\frac{\partial}{\partial x_\mu} UK - U\frac{\partial K}{\partial x_\mu} &= \tfrac{1}{2}iU\bigg[\tau_1\frac{\partial\psi}{\partial x_\mu} + (\tau_2\cos\psi + \tau_3\sin\psi)\frac{\partial\theta}{\partial x_\mu} \\ &\quad + (\tau_1\sin\theta - \tau_2\cos\theta\,\sin\psi + \tau_3\cos\theta\,\cos\psi)\frac{\partial\phi}{\partial x_\mu}\bigg]K \\ &= \tfrac{1}{2}i\bigg[(\tau_1\cos\theta\,\cos\phi - \tau_2\cos\theta\,\sin\phi + \tau_3\sin\theta)\frac{\partial\psi}{\partial x_\mu} \\ &\quad + (\tau_1\sin\phi + \tau_2\cos\phi)\frac{\partial\theta}{\partial x_\mu} + \tau_3\frac{\partial\phi}{\partial x_\mu}\bigg]UK, \end{aligned} \tag{19}$$

so that the kinetic energy

$$\begin{aligned}\frac{\partial}{\partial x_\mu}(K^\dagger U^\dagger)\frac{\partial}{\partial x_\mu}(UK) &= \frac{\partial K^\dagger}{\partial x_\mu}\frac{\partial K}{\partial x_\mu} + \text{coupling terms} \\ &\quad + \tfrac{1}{4}(K^\dagger . K)\left[\left(\frac{\partial\theta}{\partial x_\mu}\right)^2 + \left(\frac{\partial\phi}{\partial x_\mu}\right)^2 + \left(\frac{\partial\psi}{\partial x_\mu}\right)^2 + 2\sin\theta\frac{\partial\phi}{\partial x_\mu}\frac{\partial\psi}{\partial x_\mu}\right]. \end{aligned} \tag{20}$$

The new terms that appear in the hypercharge current are all 'coupling' terms and proportional to $K^\dagger\boldsymbol{\tau}K$, which has vanishing expectation value. For the isobaric spin current, however,

$$\begin{aligned}&\left(\frac{\partial}{\partial x_\mu}K^\dagger U^\dagger\right)\boldsymbol{\tau}UK - K^\dagger U^\dagger\boldsymbol{\tau}\left(\frac{\partial}{\partial x_\mu}UK\right) \\ &\quad = \frac{\partial K^\dagger}{\partial x_\mu}\boldsymbol{\tau}K - K^\dagger\boldsymbol{\tau}\frac{\partial K}{\partial x_\mu} + \text{coupling terms} - i(K^\dagger . K)\,\mathbf{V}_\mu, \end{aligned} \tag{21}$$

where the current of the pion field

$$\mathbf{V}_\mu = \left\{\begin{array}{l}\dfrac{\partial\psi}{\partial x_\mu}\cos\theta\,\cos\phi + \sin\phi\,\dfrac{\partial\theta}{\partial x_\mu}, \\[2ex] \dfrac{\partial\theta}{\partial x_\mu}\cos\phi - \dfrac{\partial\psi}{\partial x_\mu}\cos\theta\,\sin\phi, \\[2ex] \dfrac{\partial\phi}{\partial x_\mu} + \sin\theta\,\dfrac{\partial\psi}{\partial x_\mu}.\end{array}\right\} \tag{22}$$

Thus if we ignore the K-modes and all coupling to them the effective Lagrangian for the nucleon and pion fields is

$$-\tfrac{1}{8}Q^2\left[\left(\frac{\partial\theta}{\partial x_\mu}\right)^2+\left(\frac{\partial\phi}{\partial x_\mu}\right)^2+\left(\frac{\partial\psi}{\partial x_\mu}\right)^2+2\sin\theta\,\frac{\partial\phi}{\partial x_\mu}\frac{\partial\psi}{\partial x_\mu}\right]$$
$$-\bar{\psi}\left[\gamma_\mu\frac{\partial}{\partial x_\mu}+M-\tfrac{1}{4}F^2Q^2\mathrm{i}\gamma_\mu(H\gamma_5)\boldsymbol{\tau}.\mathbf{V}_\mu\right]\psi. \tag{23}$$

The nucleon-pion interaction current may be written directly in terms of the rotation operator U,

$$\mathrm{i}\boldsymbol{\tau}.\mathbf{V}_\mu=\frac{\partial U}{\partial x_\mu}U^\dagger-U\frac{\partial}{\partial x_\mu}U^\dagger=-2U\frac{\partial U^\dagger}{\partial x_\mu}, \tag{24}$$

as might be seen by taking expectation values directly in L_K, (8).

Under *inversion* the pion fields will be postulated so to change that

$$U\to U^\dagger=U^{-1};$$

then the parity-violating terms in (23) can be removed by the additional transformation

$$\psi\to U^{\frac{1}{2}}\psi \tag{25}$$

provided that $Q^2F^2=1$. Indeed* they would also be removed if $(1+\gamma_5)$ were replaced by $(1+A\gamma_5)$ with arbitrary A; however, A must be of the order of unity to explain $N-\pi$ interactions, so simplicity (coupled with the historical derivation) suggests the choice $A=1$. The fact that parity is not violated in low order in the meson fields is partly a consequence of the fact that the theory is invariant under CP and is charge independent; it is interesting, however, that this invariance is exact within the framework of no coupling to K-modes.

The transformations will be given in an explicit power series form using a more symmetrical set of variables than the Eulerian angles θ,ϕ,ψ. We define $\boldsymbol{\alpha}=(\alpha,\beta,\gamma)$, pseudoscalars, such that

$$U=\exp\left[\mathrm{i}\boldsymbol{\tau}.\boldsymbol{\alpha}(1+\lambda\alpha^2+\ldots)\right]. \tag{26}$$

Then

$$\left.\begin{aligned}\tfrac{1}{2}\psi&=\alpha-\beta\gamma+2\beta^2\alpha+\tfrac{2}{3}\alpha^3+(\lambda-\tfrac{2}{3})\,\alpha(\alpha^2+\beta^2+\gamma^2)+\ldots,\\ \tfrac{1}{2}\theta&=\beta+\gamma\alpha+\tfrac{2}{3}\beta^3+(\lambda-\tfrac{2}{3})\,\beta(\alpha^2+\beta^2+\gamma^2)+\ldots,\\ \tfrac{1}{2}\phi&=\gamma-\alpha\beta+2\beta^2\gamma+\tfrac{2}{3}\gamma^3+(\lambda-\tfrac{2}{3})\,\gamma(\alpha^2+\beta^2+\gamma^2)+\ldots,\end{aligned}\right\} \tag{27}$$

giving

$$\tfrac{1}{2}\mathbf{V}_\mu=\frac{\partial\boldsymbol{\alpha}}{\partial x_\mu}-\left(\boldsymbol{\alpha}\times\frac{\partial\boldsymbol{\alpha}}{\partial x_\mu}\right)+(\lambda+\tfrac{1}{3})\,\boldsymbol{\alpha}\frac{\partial}{\partial x_\mu}(\alpha^2)+(\lambda-\tfrac{2}{3})\,\alpha^2\frac{\partial}{\partial x_\mu}\boldsymbol{\alpha}+\ldots. \tag{28}$$

The parity-violating terms are removed by the transformation

$$\psi\to\exp\left[\tfrac{1}{2}\mathrm{i}\boldsymbol{\tau}.\boldsymbol{\alpha}(1+\lambda\alpha^2+\ldots)\right]\psi. \tag{29}$$

The choice of λ is in principle arbitrary and makes no difference after all renormalization effects are included. It seems natural, however, to choose λ so that the

* The possibility of this was pointed out to the author by Professor K. Brueckner.

renormalizations of lowest order vanish; this condition gives $\lambda = \frac{1}{15}$ and then the Lagrangian takes on the following form in the variables $\boldsymbol{\alpha}$,

$$-\frac{1}{2F^2}\left[\left(\frac{\partial\boldsymbol{\alpha}}{\partial x_\mu}\right)^2+\frac{3}{5}\left(\boldsymbol{\alpha}.\frac{\partial\boldsymbol{\alpha}}{\partial x_\mu}\right)^2-\tfrac{1}{5}(\alpha^2)\left(\frac{\partial\boldsymbol{\alpha}}{\partial x_\mu}\right)^2+\ldots\right]$$
$$-\bar{\psi}\left[\gamma_\mu\frac{\partial}{\partial x_\mu}+M-\tfrac{1}{2}i\gamma_\mu\gamma_5\boldsymbol{\tau}.\frac{\partial\boldsymbol{\alpha}}{\partial x_\mu}+\tfrac{1}{4}i\gamma_\mu\boldsymbol{\tau}.\boldsymbol{\alpha}\times\frac{\partial\boldsymbol{\alpha}}{\partial x_\mu}\right.$$
$$\left.-\tfrac{3}{20}i\gamma_\mu\gamma_5(\boldsymbol{\tau}.\boldsymbol{\alpha})\left(\boldsymbol{\alpha}.\frac{\partial\boldsymbol{\alpha}}{\partial x_\mu}\right)+\tfrac{1}{20}i\gamma_\mu\gamma_5(\alpha^2)\left(\boldsymbol{\tau}.\frac{\partial\boldsymbol{\alpha}}{\partial x_\mu}\right)\right]\psi. \tag{30}$$

No explicit pion mass-terms appear in this expression. Non-linearity in the K- or π-fields alone probably cannot create a mass different from zero; relativistic perturbation calculations of the self-mass give contributions proportional to the assumed mass. The reason for this appears to be that the isobaric current of the K-field cannot create pairs out of the vacuum. On the other hand, the current of the nuclear field *can* create pairs and so create a self-mass for the pion fields.

The interaction current responsible in lowest order is

$$-\tfrac{1}{4}\bar{\psi}i\gamma_\mu\boldsymbol{\tau}\psi.\left(\boldsymbol{\alpha}\times\frac{\partial\boldsymbol{\alpha}}{\partial x_\mu}\right) \tag{31}$$

(no cross-terms between the linear and cubic interactions on account of the choice of λ). We estimate the effect of this term in second order by considering the terms which will appear in the Hamiltonian due to the derivative coupling in (31), the significant terms are

$$\frac{F^2}{32}(\psi^\dagger\boldsymbol{\tau}\times\boldsymbol{\alpha}\psi)^2. \tag{32}$$

The expectation value of this gives an overestimate of the second-order effects of the coupling (31) but will serve as a qualitative guide. This expectation value is estimated by the use of (9) and is equal to

$$F^2\alpha^2/8\tau^2 = \mu^2\alpha^2/2F^2, \tag{33}$$

if μ is the effective pion mass. Consequently

$$\mu^2 = F^4/(4\tau^2) = m^2/8, \tag{34}$$

with the use of (12). This simple argument shows that the pion mass should be smaller than the K mass by a factor* of the order of 3; the experimental ratio is 3·6.

If now we identify the fields $\boldsymbol{\alpha}$ with the usual pion-fields $\boldsymbol{\pi}$ the relation ought to be

$$\boldsymbol{\alpha} = (2f/\mu)\,\boldsymbol{\pi}, \tag{35}$$

where f is the derivative coupling constant. It is known that the linear and quadratic interaction terms in (30) give a fair account of low-energy pion phenomena with $f \sim 1$. At the same time the free field part of the Lagrangian indicates $\boldsymbol{\alpha} = F\boldsymbol{\pi}$, and the combination of these conditions implies

$$F \sim 2f/\mu \sim 2{\cdot}8f. \tag{36}$$

* A similar estimate of m from the Hamiltonian (39) rather than the Lagrangian replaces the factor 2 by 3 in equation (12); this would give $\mu/m = 1/\sqrt{12}$.

This value is rather larger than that given by (15), but there are many possible renormalization effects which have not been considered here.

The fourth-order terms in the free-pion Lagrangian indicate a direct π-π interaction; in the static limit the matrix element (volume integral of the potential) has the values

$$\left.\begin{aligned} T &= 0:\ -\tfrac{1}{2}F^2, \\ T &= 2:\ +\tfrac{1}{10}F^2. \end{aligned}\right\} \tag{37}$$

4. K-NUCLEON INTERACTIONS

The interaction terms in L_K, equation (8), describe a direct $NNKK$ interaction. For clarity we transform to the Hamiltonian, introducing the conjugate variables

$$\left.\begin{aligned} L &= \frac{\partial K^\dagger}{\partial t} - \mathrm{i}F^2\chi^\dagger(K^\dagger.\chi), \\ L^\dagger &= \frac{\partial K}{\partial t} + \mathrm{i}F^2(\chi^\dagger.K)\chi. \end{aligned}\right\} \tag{38}$$

The Hamiltonian density is

$$H = LL^\dagger + (F^2/\mathrm{i})\left[(\chi^\dagger.K)(L.\chi) - (\chi^\dagger.L^\dagger)(K^\dagger.\chi)\right]$$
$$+ [\vec{\nabla}K^\dagger - \mathrm{i}F^2\chi^\dagger\vec{\sigma}(K^\dagger.\chi)][\vec{\nabla}K + \mathrm{i}F^2(\chi^\dagger.K)\vec{\sigma}\chi] + \psi^\dagger(\vec{\alpha}.\vec{p} + \beta M)\psi. \tag{39}$$

We examine the quality of the interactions by considering the *static limit* in which only the second term contributes with $\chi = \frac{1}{2}\psi$; parity violation is a relativistic effect in this term. The diagonal matrix elements for the various states $N + K$ are given in the following table:

hyper-charge	total isospin	state	matrix element
0	0	$(pK^- + n\bar{K}^0)/\sqrt{2}$	$-F^2$
0	1	nK^-, etc.	0
1	0	$(pK^0 + nK^+)$ 2	$-\frac{1}{2}F^2$
1	1	pK^+, etc.	$+\frac{1}{2}F^2$

The interaction is most strongly attractive in the $S = T = 0$ state; this agrees qualitatively with the fact that the Λ^0 hyperon is the lowest in energy. The interaction in the pK^+ system is repulsive, in agreement with experiment; for an order of magnitude estimate consider the scattering length in Born approximation:

$$a = \frac{m_{\text{red.}}}{2\pi}\int V\,\mathrm{d}\tau = \frac{F^2}{4\pi}m_{\text{red.}}. \tag{40}$$

The experimental value $a \sim 0{\cdot}34f$ indicates $F \geqslant 1{\cdot}7f$, in qualitative agreement with previous estimates. The pK^- scattering is dominated by absorption effects so that no simple quantitative comparison is possible.

The spin-gradient interaction terms in (39) describe processes in which a nucleon creates $K, \bar{K}$ pair. If, tentatively, we regard a Λ or Σ hyperon as a compound of N and $\bar{K}$ this will lead also to terms in which $N \to K + \Lambda, \Sigma$ by capture of the $\bar{K}$ to form the hyperon. The effective coupling constant g_K for this process will be

$$g_K/2M = \tfrac{1}{2}F^2A/\sqrt{(2m)}, \tag{41}$$

where A is the amplitude for the dissociation $\Lambda \leftrightarrow N + \bar{K}$ at the origin.

We can obtain a speculative estimate for A by supposing that the Σ^0, Λ^0 mass difference is entirely due to the static interaction considered above; then

$$75\,\text{MeV} = \Delta E = A^2F^2. \tag{42}$$

Now also, in the present model, $g_\pi/2M = \frac{1}{2}F$; so that combining these relations we obtain

$$g_\kappa^2/g_\pi^2 = A^2F^2/2m = \Delta E/2m \sim 0{\cdot}08, \tag{43}$$

which is the order of magnitude suggested by several experimental facts (Gell-Mann 1957).

Other K-nucleon interactions will arise through the intermediary of pion modes, and this raises the question of the K-π interaction. If we could carry out a clean separation of the K and π variables we might expect to obtain a Hamiltonian in which the derivative cross-terms had been completely eliminated (cf. Villars 1957); there are also terms which apparently introduce parity violation. In any event, however, there will remain coupling through the dependence of the 'inertia parameter' $K^\dagger.K$, that scales the pion fields, upon the K-fields. This term describes a *repulsive* $KK\pi\pi$ interaction with matrix element $(\mu/4m)F^2$; there is no parity violation from such terms.

Interaction sufficiently strong to form bound hyperonic states of a nucleon and a $\bar{K}$ is not apparently described by the present model; we conjecture that such effects might appear when the phenomenological mass M of the nucleon field is replaced by self-mass.

5. Weak interactions

Weak interactions in the baryon-meson system involving a change of strangeness may be fitted into the present model in a rather simple way, after the idea of a 'spurion' introduced by Wentzel (1956).

Weak interactions of this kind may be introduced into our system by postulating the substitution

$$\phi_4 \to \phi_4 + \epsilon/F \tag{44}$$

in L_K, where ϵ is a small dimensionless constant. This obviously leads to the rules $\Delta S = \Delta T = \frac{1}{2}$ for the interactions proportional to ϵ.

Pionic decay of the K-mesons is described directly by the terms that arise in this way from the mass-term of the K-field, namely

$$\frac{\epsilon}{F} m_\kappa^2 \phi_4 = \frac{\epsilon}{F\sqrt{2}} m_\kappa^2 (K^0 + \bar{K}^0). \tag{45}$$

The physical interpretation must be made after the explicit introduction of the pion modes by the transformation (18). We use the form (26) for U and replace $\boldsymbol{\alpha}$ by $F\boldsymbol{\pi}$. The interaction terms up to the third order in the pion field are

$$\begin{aligned} &-\tfrac{1}{2}\epsilon F m^2[\pi^0 + 2\pi^+\pi^-]\,K^0, \\ &-\mathrm{i}\epsilon m^2[1 - \tfrac{1}{10}F^2(\pi^0 + 2\pi^+\pi^-)]\,[\pi^0K_2^0 + \pi^+K^- + \pi^-K^+], \end{aligned} \tag{46}$$

where, as usual, K_1^0, K_2^0 stand for the combinations $(K^0 \pm \bar{K}^0)/\sqrt{2}$.

The first term describes the two-pion decay of the K_1^0. The matrix element for decay into charged pions with energies ω equals

$$-\tfrac{1}{2}\epsilon F m^2\{1/\omega\sqrt{(2m)}\} \tag{47}$$

and so the decay rate for the process $K^0 \rightarrow \pi^+ + \pi^-$ should be

$$\frac{1}{\tau} = \frac{p\omega}{2\pi}|M|^2 = \frac{\epsilon^2 F^2 \mu^2}{16\pi}\frac{m^3}{\mu^3}\frac{cp}{\omega}\frac{\mu c}{\hbar^2}. \tag{48}$$

The data $cp/\omega = 0{\cdot}82$, $1/\tau = 0{\cdot}9 \times 10^{10}\,\mathrm{s}^{-1}$, require

$$\epsilon F \mu = 2{\cdot}4 \times 10^{-7}. \tag{49}$$

The three-pion decay may be discussed in a similar way. The matrix element for $K \rightarrow \pi^+ + \pi^+ + \pi^-$ is

$$\tfrac{1}{5}i\epsilon F^3 m^2\left[\frac{1}{\sqrt{(2m)}\,\{\sqrt{(2\mu)}\}^3}\right] \tag{50}$$

in non-relativistic approximation for the pions. In the same approximation the density of final states

$$2\iint \frac{\mathrm{d}^3k_2\,\mathrm{d}^3k_3}{(2\pi)^6}\,\delta(E_1 + E_2 + E_3 - \Delta E) = \frac{\mu}{(2\pi\sqrt{3})^3}(2\mu\Delta E)^2. \tag{51}$$

This gives for the process the decay rate

$$\frac{1}{\tau} = \left(\frac{1}{400\pi^2 3\sqrt{3}}\right)(\epsilon F^2\mu^2)^2\left(\frac{m}{\mu}\right)^3\left(\frac{\Delta E}{\mu}\right)^2\left(\frac{\mu c^2}{\hbar}\right). \tag{52}$$

The data $\Delta\epsilon = 84\,\mathrm{MeV}$, $1/\tau = 4{\cdot}6 \times 10^6\,\mathrm{s}^{-1}$ require

$$\epsilon F^2 \mu^2 = 1{\cdot}6 \times 10^{-7}. \tag{53}$$

Agreement between (49) and (53) requires a rather small value $F \sim 1f$. However, this simple calculation takes no account of any renormalization effects or final state interactions which might modify the rates appreciably.

The $\Lambda \rightarrow N + \pi$ decay can be discussed in a similar way; the result is in qualitative agreement with those for K-decay, but a quantitative comparison requires a knowledge of the dissociation amplitude $\Lambda \rightarrow N + \bar{K}$ for the particular momentum involved in the decay, for which only an order-of-magnitude estimate can as yet be made in this model.

6. Conclusions

The qualitative discussions in this paper show that the Lagrangian density (16), with the postulate $\langle K^\dagger . K\rangle = \frac{1}{2}F^{-2}$ and a value of F around $2f$, may provide a satisfactory unified model of the interacting N-, K- and π-fields. Quantitative verification depends on more detailed calculations of the various interaction effects, upon a more complete understanding of the Lagrangian with its postulated condition, and most importantly upon the inclusion of hyperons within the same framework. In particular the following points need to be investigated:

(i) Renormalizations in the pion-nucleon system described by (25); at present the worst discrepancy in the model is the large value (36) for F indicated by the pion coupling constant.

(ii) Alternatively, the discrepancy might arise from error in the coefficient $K^\dagger . K$, in its appearance as an inertia parameter scaling the pion fields. This necessitates a discussion of how the different K- and π-modes of the meson field are to be separated and defined.

(iii) While electromagnetic interaction can be introduced in the usual gauge-invariant way into the expression (16) some complications arise with the introduction of the collective co-ordinates which have not been clarified; a simultaneous change of gauge is probably required.

(iv) The extent of parity violation. This appears explicitly in the direct, but relatively weak, $NNKK$ interactions and will also appear to some extent in the $KK\pi$ couplings.

(v) The possibility that there may be some N-K interactions strong enough to bind together N and $\bar{K}$ into hyperons.

It is apparent from the formalism of this model that the pion fields, the collective modes of isobaric rotation, are closely related to the concept of a gauging field (Yang & Mills 1954) for locally variable rotations in isobaric spin space. When we treat these *angle* variables as if they were linear we are neglecting effects which can arise from twists through 2π and ought to be described by the use of a many-valued state function, whose several branches correspond to the multiple values of an angle at a point. We conjecture that the Fermion fields form an equivalent way of describing this multiple-valuedness, and correspond with locally variable *discontinuous* rotations introduced to keep the angle variables within one period; the motions of these singularities are determined by those of the fields. This description, begun tentatively in NLT, should remove the arbitrary postulate $\langle K^\dagger . K\rangle = \frac{1}{2}F^{-2}$ and clarify the status of hyperons in the model.

The author is grateful for the hospitality of the Physics Department of the University of Pennsylvania, where this work was completed.

References

Feynman, R. P. & Gell-Mann, M. 1958 *Phys. Rev.* **109**, 193.
Gell-Mann, M. 1957 *Phys. Rev.* **106**, 1296.
Skyrme, T. H. R. 1958 *Proc. Roy. Soc.* A, **247**, 260.
Villars, F. 1957 *Nucl. Phys.* **3**, 240.
Wentzel, G. 1956 *Proc. 6th Annual Rochester Conference.*
Yang, C. N. & Mills, R. L. 1954 *Phys. Rev.* **96**, 191.

A non-linear field theory

By T. H. R. Skyrme
Atomic Energy Research Establishment, Harwell

(*Communicated by Sir Basil Schonland, F.R.S.—Received* 5 *September* 1960)

A unified field theory of mesons and their particle sources is proposed and considered in its classical aspects. The theory has static solutions of a singular nature, but finite energy, characterized by spin directions; the number of such entities is a rigorously conserved constant of motion; they interact with an external meson field through a derivative-type coupling with the spins, akin to the formalism of strong-coupling meson theory. There is a conserved current identifiable with isobaric spin, and another that may be related to hypercharge. The postulates include one constant of the dimensions of length, and another that is conjectured necessarily to have the value $\hbar c$, or perhaps $\frac{1}{2}\hbar c$, in the quantized theory.

1. Introduction

This paper considers, in some classical aspects, a non-linear field theory (tentatively describing the strongly interacting elementary particles) similar to those that have been suggested in two previous communications (Skyrme 1958 and 1959, referred to here as I and II, respectively). The essential feature of the theory is the representation of the fundamental field quantities in terms of angular variables rather than linear ones. The periodicity of these variables means that they are not uniquely determined (in a classical sense) by the physical state of the system. In regions of weak field the different determinations of angle generate a set of equivalent descriptions, forming separate sheets of a multiple-valued system; but when the fields become strong these sheets may cross one with the other, forming singularities.

A simplified model was considered in I, characterized by a single dimension of space and one angular variable. The periodicity introduced a new constant of motion, identified with the number of sources ('baryon number'). The sources might for convenience of description, be idealized as point singularities, whose dynamical behaviour was similar to that of particle coupled to the gradient of the residual field.

The extension of this model to a realistic three-dimensional world is described here. It appears only to be possible when there are also three angular variables, a condition that is fortunately satisfied by the pion fields of nature. An analogous constant of the motion can then be defined, which measures the number of times that space (three dimensions) is mapped by the fields onto the elementary volume of angular space.

A singularity must be characterized, at least, by some direction which determines the way in which the sheets cross. A possibility for the field configuration is suggested by the forms used in the strong-coupling theory of the symmetrical pseudoscalar pion fields (Pauli 1946), which is the nearest classical description of the physical situation that we are investigating. The source is then characterized by an orthogonal matrix describing the angles between spin and isospin. Such a singularity contributes one unit to the baryon number, a constant of motion.

The dynamics must be more complicated than in the one-dimensional model; the Lagrangian must contain terms of the fourth degree, at least, in the fields. A simple possibility is considered, not involving field derivatives beyond the first. For this system there is a static solution of the above type. The field energy is finite, and then the singularity is coupled to the residual field by an interaction of symmetrical derivative form. An estimate is made of the moment of inertia for rotation of the spin-isospin system.

The free meson waves in this theory are massless; it would not be possible to introduce a mass term without destroying the basic symmetry of the theory, which must be preserved for the description of the pion fields at least. It is conjectured that mass arises as a quantum-mechanical effect through the creation of virtual pairs. The representation of the particle as an idealized point singularity coupled to the meson field is investigated, and leads to a description having affinities with Dirac's equation.

2. The angular fields

There are postulated four real fields ϕ_ρ, constrained to satisfy

$$\sum_{\rho=1}^{4} \phi_\rho^2 = 1. \qquad (1)$$

They may be considered either as components of a vector in a four-dimensional iso-space, or as a spinor $K = (\phi_2+\mathrm{i}\phi_1, \phi_4-\mathrm{i}\phi_3)$. The angles may conveniently be defined, as in II, by specifying the rotation that generates K from a standard spinor $(0,1)$; namely

$$\left.\begin{aligned} K &= U(0,1) = \exp(\tfrac{1}{2}\mathrm{i}\tau_1\phi)\exp(\tfrac{1}{2}\mathrm{i}\tau_3\theta)\exp(\tfrac{1}{2}\mathrm{i}\tau_2\psi)\,(0,1), \\ \text{or} \quad \phi_4+i\boldsymbol{\tau}.\boldsymbol{\phi} &= U(\phi,\psi,\theta). \end{aligned}\right\} \qquad (2)$$

The rotation operators in the 4-space form two independent three-dimensional rotation groups, one of which is identified with isobaric spin; the other may be related to the hypercharge or 'strangeness' quantum number, as was discussed in I. The infinitesimal operators of isobaric spin I^α have, in the four-dimensional ϕ-space, the representation $\frac{1}{2}\mathrm{i}T^\alpha$; the matrices T^α are antisymmetric and have components

$$(T^\alpha)_{\rho\sigma} = \delta_{\rho 4}\delta_{\sigma\alpha} - \delta_{\sigma 4}\delta_{\rho\alpha} - \epsilon_{\alpha\rho\sigma} \qquad (3)$$

and satisfy

$$T^1T^2 = T^3, \quad \text{etc.} \qquad (4)$$

The operators $\mathrm{i}T^\alpha$ are equivalent to Pauli spin matrices τ^α acting on K. Similarly the other angular momentum $\tilde{I}\alpha$ is represented by matrices $\frac{1}{2}\mathrm{i}\tilde{T}^\alpha$, where

$$(\tilde{T}^\alpha)_{\rho\sigma} = -\delta_{\rho 4}\delta_{\sigma\alpha} + \delta_{\sigma 4}\delta_{\rho\alpha} - \epsilon_{\alpha\rho\sigma}. \qquad (5)$$

A useful relation that will be used frequently is

$$\Sigma_\alpha(AT^\alpha B)(CT^\alpha D) = (AC)(BD) - (AD)(BC) + \det(A,B,C,D), \qquad (6)$$

where A, B, C, D, are any four 4-vectors; the summation over matrix indices has been suppressed. This convention, and the ordinary summation convention, will be employed throughout this paper.

The gradients of U may conveniently be expressed in terms of three-dimensional representations of the isobaric spin; these are defined by

$$\partial U/\partial x_\mu = \mathrm{i}\tau^\alpha B^\alpha_\mu U, \tag{7}$$

so that

$$B^\alpha_\mu = \phi T^\alpha \partial\phi/\partial x_\mu.$$

Alternatively the B may be expressed in terms of an orthogonal matrix $g_{\alpha\beta} = g^{\beta\alpha}$, where

$$B^\alpha_\mu = -\tfrac{1}{4}\epsilon_{\alpha\beta\gamma} g^{\beta\delta}\partial/\partial x_\mu g_{\gamma\delta}$$

$$g_{\alpha\beta} = \phi(I^\beta \tilde{I}^\alpha)\phi = 2\phi_\alpha\phi_\beta - 2\epsilon_{\alpha\beta\gamma}\phi_\gamma\phi_4 + \delta_{\alpha\beta}(2\phi_4^2 - 1). \tag{8}$$

g is the rotation operator for vectors analogous to U for spinors.

The B^α_μ describe the way in which the orientation of the isospin axes varies from point to point of space. If α be regarded as an additional co-ordinate label, then as suggested by equation (8), B may be regarded as a Christoffell symbol in an affine geometry, with

$$\Gamma^\alpha_{\beta\mu} = \Gamma^\alpha_{\mu\beta} = -\epsilon_{\alpha\beta\gamma} B^\gamma_\mu. \tag{9}$$

The covariant curl of B vanishes identically,

$$\partial B^\alpha_\mu/\partial x_\nu - \partial B^\alpha_\nu/\partial x_\mu - 2\epsilon_{\alpha\beta\gamma} B^\beta_\mu B^\gamma_\nu = 0 \tag{10}$$

expressing the integrability conditions for equations (7). For given B that satisfy (10), U is completely determined by (7) apart from an arbitrary post-multiplying factor, equivalent to a constant rotation $\tilde{I}$.

The (ordinary) curl of B will be written as

$$\left.\begin{aligned} C^\alpha_{\mu\nu} &= \tfrac{1}{2}(\partial B^\alpha_\mu/\partial x_\nu - \partial B^\alpha_\nu/\partial x_\mu) = (\partial\phi/\partial x_\nu)\, T^\alpha(\partial\phi/\partial x_\mu) \\ &= \epsilon_{\alpha\beta\gamma} B^\beta_\mu B^\gamma_\nu, \quad \text{by (10).} \end{aligned}\right\} \tag{11}$$

It is convenient also to separate the space and time parts of this tensor.

$$\left.\begin{aligned} C^\alpha_i &= \tfrac{1}{2}\epsilon_{ijk} C^\alpha_{jk}, \\ D^\alpha_i &= \mathrm{i}C^\alpha_{4i} = \epsilon_{\alpha\beta\gamma} B^\beta_i B^\gamma_0. \end{aligned}\right\} \tag{12}$$

3. The particle number

In the model considered in I the new constant of motion identified with particle number was proportional to the total change in angle,

$$\theta(\infty) - \theta(-\infty) = \int (\partial\theta/\partial x)\, \mathrm{d}x.$$

The natural generalization to three dimensions involves the space integral of a Jacobian expression such as $\partial(\phi, \psi, \theta)/\partial(x, y, z)$.

A symmetrical definition independent of the particular angular representation is obtained by defining the current,

$$N_\lambda = (\mathrm{i}/12\pi^2)\,\epsilon_{\alpha\beta\gamma\delta}\epsilon_{\lambda\mu\nu\rho}\phi_\alpha(\partial\phi_\beta/\partial x_\mu)(\partial\phi_\gamma/\partial x_\nu)(\partial\phi_\delta/\partial x_\rho). \tag{13}$$

The conservation law is satisfied identically,

$$\partial N_\lambda/\delta x_\lambda = 0 \tag{14}$$

because the Jacobian of the four fields ϕ with respect to the four space-time co-ordinates must vanish (they are functionally dependent).

In particular the fourth component $N_4 = \mathrm{i}N$, where

$$N = -(1/2\pi^2)\det(\phi, \partial\phi/\partial x, \partial\phi/\partial y, \partial\phi/\partial z),$$
$$= -(1/2\pi^2)\det(B_i^\alpha). \tag{15}$$

Also, from (11) and (12), $$B_i^\alpha C_j^\alpha = -2\pi^2 N\delta_{ij}. \tag{16}$$

In terms of the angular representation (2),

$$N = (1/16\pi^2)\,\partial(\phi, \psi, \sin\theta)/\partial(x, y, z). \tag{17}$$

The factor $2\pi^2$ in the definitions, equation (15), is included for the convenience of subsequent interpretation; it arises naturally as the volume of angular space (surface of a unit sphere in four dimensions).

4. The particle nature

In the one-dimensional model the particle could be pictured as a small region of space wherein the angle θ increased rapidly from 0 to 2π. If in three dimensions one of the angles suffers a similar increase on a path through the particle, then it must decrease on a path approaching from the opposite direction to ensure that the different sheets describing the system are equivalent. However, in three dimensions, unlike one, it is possible to reach a starting point on the far side of the particle without traversing it; and it is natural to demand that on a path distant from any such singularities there should be no crossing of sheets (i.e. that there should be no 'cuts' emanating from any of the singularities; such a situation was however considered by Dirac (1948), in his discussion of magnetic poles). The particle must then be characterized by some intrinsic direction in relation to which the path direction is significant; this characteristic appears to be a feature of the particles that might so be represented, the baryons with spin $\frac{1}{2}$.

The semi-classical strong-coupling theory of the symmetrical pseudoscalar meson fields (Pauli 1946) invokes a description of the source and of the field in its vicinity by an orthogonal matrix e_i^α which measures the relative orientations of the spin and isospin co-ordinate frames. The interpretation, suggested in I, of the B_i^α as the gradients of the pion fields, and the nature of the formula (15) conspire to indicate the following form typical of a particle singularity.

$$\phi_\alpha = (e_i^\alpha x_i/r)\sin\tfrac{1}{2}\omega; \quad \phi_4 = \cos\tfrac{1}{2}\omega. \tag{18}$$

Here ω is some function of the radial distance r from the particle centre, tending to zero at large distances, and to 2π at the origin.

The behaviour of the angular variables can be obtained by the comparison of (18) and (2). On a straight path through, or passing close to, the centre the angle ϕ changes by 4π if

$$(e_i^1 x_i/r)^2 > (e_i^2 x_i/r)^2;$$

for the opposite inequality it is the angle ψ which changes by 4π.

With these fields ϕ given by (18)

$$\left.\begin{aligned} B_i^\alpha &= (e_j^\alpha/2r)\,[(\delta_{ij} - x_i x_j/r^2)\sin\omega + (x_i x_j/r^2)(r\omega') - \epsilon_{ijk}(x_k/r)(1-\cos\omega)],\\ C_i^\alpha &= (e_j^\alpha/4r^2)\,[(\delta_{ij} - x_i x_j/r^2)(r\omega')(\sin\omega) + (x_i x_j/r^2)\,2(1-\cos\omega)\\ &\qquad - \epsilon_{ijk}(x_k/r)(r\omega')(1-\cos\omega)]. \end{aligned}\right\} \tag{19}$$

From the orthogonality relation $e_i^\alpha e_j^\alpha = \delta_{ij}$ it follows that

$$B_i^\alpha C_j^\alpha = (1/4r^2)(1-\cos\omega)(\mathrm{d}\omega/\mathrm{d}r)\,\delta_{ij},$$

and so, by comparison with (16),

$$N = -(1/8\pi^2 r^2)(1-\cos\omega)(\mathrm{d}\omega/\mathrm{d}r), \tag{20}$$

and the volume integral

$$\int N\mathrm{d}^3x = (1/2\pi)\int_{r=\infty}^{r=0}(1-\cos\omega)\,\mathrm{d}\omega \tag{21}$$

is equal to unity provided only that ω satisfies the boundary conditions stated (equation (18)).

5. The dynamics

It was suggested in I and II that the Lagrangian density of the fields should contain, in the present notation, the term

$$-\tfrac{1}{2}Q^2(\partial\phi_\rho/\partial x_\mu)^2 = -\tfrac{1}{2}Q^2(B_\mu^\alpha)^2. \tag{22}$$

The corresponding static energy is $\tfrac{1}{2}Q^2\int(B_i^\alpha)^2\,\mathrm{d}^3x$. For the fields (18) this will have a value of the order of magnitude of aQ^2, where a measures the extension of the source; at the same time the strength of coupling between the particle and an external field will be of the order of a^2Q^2. In the limit of a point-source, $a\to 0$, both of these will vanish, and it is unlikely that the system described by (22) could really contain any singular configurations of this type.

A satisfactory Lagrangian should give an energy divergent as $1/a$ (at least) and a coupling constant tending to a finite value. The dimensions indicate terms of the fourth degree in the field derivatives. The simplest expression of this type that

(i) does not contain derivatives beyond the first order, and

(ii) allows the weak free fields in which B_μ^α is a gradient, is obtained by regarding (22) as the mass term for a vector meson field and adding thereto a term in $(C_{\mu\nu}^\alpha)^2$.

This selected expression for the Lagrangian density will be written

$$\mathscr{L} = -(\epsilon/2\pi^2)\,[\tfrac{1}{4}(C_{\mu\nu}^\alpha)^2 + \tfrac{1}{2}\kappa^2(B_\mu^\alpha)^2], \tag{23}$$

where κ is a reciprocal length (of the order of $10^{13}\,\mathrm{cm}^{-1}$), and ϵ is a constant of the order of unity in the natural units in which $\hbar = c = 1$. The constant Q scaling the meson fields to the conventional normalization is equal to $\kappa(\epsilon/2\pi^2)^{\frac{1}{2}}$. It is conjectured that in the quantized theory the value of ϵ will be determinate, possibly unity, for reasons similar to those which made $\omega Q^2 = 1/\pi$ in the model of I.

From the relations (7) and (11) we obtain the alternative forms

$$\begin{aligned}(C_{\mu\nu}^\alpha)^2 &= (B_\mu^\alpha)^2(B_\nu^\alpha)^2 - (B_\mu^\alpha B_\nu^\alpha)^2\\ &= (\partial\phi_\rho/\partial x_\mu)^2(\partial\phi_\sigma/\partial x_\nu)^2 - (\partial\phi_\rho/\partial x_\mu\,.\,\partial\phi_\rho/\partial x_\nu)^2.\end{aligned} \tag{24}$$

The variables ϕ_ρ may be regarded as independent provided that a term $\lambda(x)(\phi_\rho^2-1)$ be added to $\mathscr{L}$.

9-2

The energy-momentum tensor-density is

$$\mathscr{T}_{\mu\nu} = \Pi_{\rho\mu}(\partial\phi_\rho/\partial x_\nu) - \delta_{\mu\nu}\mathscr{L} \tag{25}$$

where
$$\begin{aligned}\Pi_{\rho\mu} &= \delta\mathscr{L}/\delta(\partial\phi_\rho/\partial x_\mu) \\ &= -(\epsilon/2\pi^2)\,[(\partial\phi_\rho/\partial x_\mu)\,[(B^\alpha_\nu)^2+\kappa^2]-(\partial\phi_\rho/\partial x_\nu)\,(B^\alpha_\mu B^\alpha_\nu)].\end{aligned} \tag{26}$$

The energy density is

$$\mathscr{E} = (\epsilon/4\pi^2)\,[(C^\alpha_i)^2+(D^\alpha_i)^2+\kappa^2\{(B^\alpha_i)^2+(B^\alpha_0)^2\}]. \tag{27}$$

The field equations of motion, and the Hamiltonian formulation, may be carried through in the usual way; some results are mentioned in the next section.

It is easily seen that there is a static solution of the form (18) for some suitable shape $\omega(r)$. The differential equation for ω is the Eulerian equation that arises from minimizing the energy, and it is therefore sufficient to consider the latter problem. Substitution from (19) gives

$$E = (\epsilon/4\pi)\int_0^\infty \left\{\frac{(1-\cos\omega)}{r^2}(1-\cos\omega+4\kappa^2r^2)+(\mathrm{d}\omega/\mathrm{d}r)^2(1-\cos\omega+\kappa^2r^2)\right\}\mathrm{d}r. \tag{28}$$

At large distances ω must fall off as $1/r^2$, as would be expected for a massless meson field with the derivative coupling implied by the form (18).

It is easy to obtain a lower bound for E, by using

$$[(1-\cos\omega)^2/r^2+\kappa^2r^2(\mathrm{d}\omega/\mathrm{d}r)^2] > 2\kappa(1-\cos\omega)\,|\mathrm{d}\omega/\mathrm{d}r|$$

So
$$\begin{aligned}E &> (\epsilon/4\pi)\int_0^\infty (1-\cos\omega)\,[4\kappa^2+2\kappa\,|\mathrm{d}\omega/\mathrm{d}r|+(\mathrm{d}\omega/\mathrm{d}r)^2]\,dr \\ &= (\epsilon/4\pi)\int_0^{2\pi}(1-\cos\omega)\left(2\kappa+|\omega'|+\frac{4\kappa^2}{|\omega'|}\right)\mathrm{d}\omega.\end{aligned}$$

The seond factor in the integrand is greater than 6κ, and this gives for E the lower bound $3\epsilon\kappa$.

An upper bound may be obtained by a trial form, for example a linear variation of ω with r; if

$$\left.\begin{aligned}\omega &= 0 && (r > a),\\ \omega &= 2\pi(1-r/a) && (r < a),\end{aligned}\right\} \tag{29}$$

then the integrals can easily be evaluated, giving

$$E_a = \epsilon\kappa(3{\cdot}83/\kappa a+1{\cdot}36\kappa a) \tag{30}$$

with a minimum value of $4{\cdot}5\epsilon\kappa$ when $\kappa a = 1{\cdot}7$.

6. Constants of motion

The Lagrangian is invariant against any rotation in ordinary space or in isospace. With the former is associated the angular momentum tensor, whose spatial components have the density

$$\begin{aligned}\mathscr{J}_i &= -\epsilon_{ijk}x_j\mathscr{T}_{0k} \\ &= B^\alpha_0\, I_{\alpha\beta}(\epsilon_{ijk}x_j)\,B^\beta_k,\end{aligned} \tag{31}$$

where the inertia coefficient

$$I_{\alpha\beta} = (\epsilon/2\pi^2)\,[(B^2+\kappa^2)\,\delta_{\alpha\beta} - (B_i^\alpha B_i^\beta)]. \tag{32}$$

The combination

$$B_0^\alpha\, I_{\alpha\beta} = \phi T^\beta \Pi, \tag{33}$$

where $\Pi = \Pi_0$ is the canonical conjugate of ϕ.

Under isobaric spin rotations the B transform like vectors in a three-dimensional isospace. The isobaric spin operator density is just

$$\mathscr{I}^\alpha = -\tfrac{1}{2}\phi T^\alpha \Pi = \tfrac{1}{2} I_{\alpha\beta} B_0^\beta. \tag{34}$$

The current $\mathscr{I}_\mu^\alpha$, of which this operator is the time component, satisfies the conservation equation

$$\partial \mathscr{I}_\mu^\alpha / \partial x_\mu = 0. \tag{35}$$

Since the fields have in fact only three degrees of freedom at each point, these equations must be equivalent with the full equations of motion of the field, and this can easily be verified.

Similar currents and operators $\tilde{\mathscr{I}}$ may be defined for the second group of rotations in the iso-4-space. Under a *constant* (in space and time) rotation of this group not only the Lagrangian but all the fields B_μ^α are unaltered. All the constants of motion are therefore unaltered except the direction of this second spin.

If it be tentatively assumed that all physical phenomena are describable in terms of the B_μ^α alone, then these constant $\hat{\mathscr{I}}$-rotations might be regarded as non-significant gauge transformations; then only the total $\tilde{\mathscr{I}}$-spin would remain as a characteristic constant of the physical system. Electromagnetic interactions which destroy the full symmetry will leave only I^3 and $\tilde{I}^3$ as constants, forcing a particular choice of this 'gauge'.

The current N_λ, equation (13), is conserved independently of the dynamics of the system. It may, however, also be obtained from a modified Lagrangian as the current associated with changes of scale of ϕ. If the Lagrangian be written as a homogeneous function of degree zero in the ϕ, by suitable divisions by powers of ϕ^2 then it will be invariant against changes of by a constant (real) factor, and there will be a corresponding conserved current $(\phi\Pi_\lambda)$.

For (23) this will vanish identically on account of (1). It is permissible, however, to add to $\mathscr{L}$ a multiple of the invariant $C_i^\alpha D_i^\alpha$, since this also vanishes identically on account of the constraint; this term gives a contribution proportional to N_λ. In particular if we take

$$\mathscr{L}' = -(\epsilon/4\pi^2)\,[(C_i^\alpha + \mathrm{i}D_i^\alpha)^2 + \kappa^2(B_\mu^\alpha)^2], \tag{36}$$

involving only the combination $C+\mathrm{i}D$, analogous to $B+\mathrm{i}E$ in electromagnetic theory (cf. Feynmann & Gell-Mann 1958), then

$$\phi\delta\mathscr{L}'/\delta(\partial\phi/\partial x_\lambda) = -3\mathrm{i}\epsilon N_\lambda. \tag{37}$$

7. Particle rotation

It is interesting to consider possible rotations of the particle state. There is no simple exact solution of the equations of motion, but for sufficiently small angular velocities an adiabatic approximation may be permissible. Then

$$\phi = \mathrm{e}^{\Omega t}\phi_0, \tag{38}$$

9-3

where ϕ_0 is the static solution, and Ω is a constant antisymmetric rotation operator. At time $t = 0$, the B^α have their static values, and

$$B_0^\alpha = (\phi T^\alpha \partial\phi/\partial t)_0 = \phi_0 T^\alpha \Omega \phi_0. \tag{39}$$

The angular momentum, isobaric spin and hypercharge spin are given by the formulae in the preceding section. We find that

$$\left.\begin{aligned} J_i &= -e_i^\alpha (I^\alpha + \bar{I}^\alpha), \\ I^\alpha + \bar{I}^\alpha &= \tfrac{1}{2} A \epsilon_{\alpha\beta\gamma} \Omega_{\beta\gamma}, \\ I^\alpha - \bar{I}^\alpha &= I_{\alpha\beta} \Omega_{\beta 4}. \end{aligned}\right\} \tag{40}$$

The inertia coefficient A has the value

$$A = (\epsilon/6\pi) \int_0^\infty (1-\cos\omega)\,[2(1-\cos\omega) + (r\omega')^2 + 4\kappa^2 r^2]\,\mathrm{d}r, \tag{41}$$

while $I_{\alpha\beta}$ is divergent. This last arises because a rotation coupling the 4-direction introduces a field at large distances, reflecting the inadequacy of the adiabatic approximation there where the effects of retardation must be important. We are mainly interested in rotations that leave $\bar{I}$ unaltered; these can be described formally by allowing $\Omega_{\beta 4}$ to be infinitesimal, adding a negligible contribution to the energy.

Then the rotation state would have equal values of the total angular momentum and isobaric spin. The additional energy is given by the kinetic energy terms

$$\mathscr{E}_{\text{kin.}} = (\epsilon/4\pi^2)\,[(D_i^\alpha)^2 + \kappa^2 (B_0^\alpha)^2]. \tag{42}$$

This gives

$$E_{\text{rot.}} = \tfrac{1}{4} A (-\Omega^2)_{\alpha\alpha} = J^2/2A. \tag{43}$$

For the approximate field given by (29) and (30), we find

$$A \sim 1{\cdot}35\epsilon/\kappa. \tag{44}$$

Thus the ratio of rotational energy to particle mass ($M \sim 4{\cdot}5\epsilon\kappa$) is

$$1/2AM \sim 0{\cdot}08/\epsilon^2; \tag{45}$$

this is comparable with the experimental value of 0·11, for the position of the 33-resonance in pion nucleon scattering, when $\epsilon = 1$.

8. The particle as source

It is doubtful whether these considerations of the complete solutions can be carried much further. Following the programme envisaged in I, the next step is to separate the system into a 'non-singular' meson field (i.e. confined to one sheet of the angular representation) and into 'bare' point sources coupled with it. These should correspond, roughly speaking, with the terms of $\mathscr{L}$ quadratic and quartic in the fields. It is true that the quadratic terms alone appear to describe a massless meson field, but it would seem possible, as suggested in II, that the observed meson mass may arise as a quantum effect from the creation of virtual particle antiparticle pairs. The classical analysis indicates that the particles have non-zero masses scaled by the constant κ.

The precise definition of the source is arbitrary, variations being compensated by changes in the coupling to the residual field. The formulae (19) point clearly however to one especially simple choice; this arises when the dependence of ω on r is such that the B_i^α and C_i^α are proportional, and each therefore proportional to an orthogonal matrix, on account of their reciprocity. The condition is

$$(r\omega')^2 = 2(1-\cos\omega) \quad \text{or} \quad \tan\tfrac{1}{4}\omega = (a/r); \tag{46}$$

the point source will correspond with the limit $a \to 0$.

In addition to position the source must also be characterized by its spin properties; these may be defined in terms of contributions to the isospin currents $\mathscr{I}_\mu^\alpha$. For the static point source

$$\mathscr{I}_i^\alpha = -\epsilon N E_i^\alpha = \epsilon N e_j^\alpha S_{ij}; \tag{47}$$

the orthogonal matrix

$$\begin{aligned} S_{ij} &= \delta_{ij}\cos\tfrac{1}{2}\omega - (x_i x_j/r^2)(1+\cos\tfrac{1}{2}\omega) - \epsilon_{ijk}(x_k/r)\sin\tfrac{1}{2}\omega \\ &= [\delta_{ij}(r^2-a^2) - 2x_i x_j - 2\epsilon_{ijk} a x_k]/(r^2+a^2) \\ &= -\tfrac{1}{2}\,\text{trace}\left\{\frac{\boldsymbol{\sigma}.\mathbf{r} - ia}{(r^2+a^2)^{\frac{1}{2}}}\sigma_i \frac{\boldsymbol{\sigma}.\mathbf{r}+ia}{(r^2+a^2)^{\frac{1}{2}}}\sigma_j\right\}, \end{aligned} \tag{48}$$

and the volume integral gives

$$I_i^\alpha = \epsilon e_j^\alpha \int N S_{ij}\,\mathrm{d}^3x = -\tfrac{1}{3}\epsilon\, e_i^\alpha. \tag{49}$$

This quantity describes the coupling of the source to a residual field ΔB_i^α, for

$$\delta\mathscr{L}/\delta B_i^\alpha = \phi T^\alpha \delta\mathscr{L}/\delta(\partial\phi/\partial x_i) = -2\mathscr{I}_i^\alpha, \tag{50}$$

and in the limit of a point source the coupling to higher powers of the residual field will vanish.

The surviving coupling term is therefore

$$-2I_i^\alpha \Delta B_i^\alpha = (2\epsilon/3)\, e_i^\alpha \Delta B_i^\alpha. \tag{51}$$

The coefficient $2I_i^\alpha/Q$, for the conventionally normalized field, is to be compared with the coupling $(g/2M)\,\sigma_i\tau_\alpha$ of symmetrical pseudoscalar theory. For the approximate solution found in § 5, $gQ/M \sim (g/18\epsilon^{\frac{1}{2}})$; this is of the order of unity for the empirical value of g and for $\epsilon = 1$; it was identically equal to unity from the point of view adopted in I, towards the starting point of which the present analysis is directed. The relation between e_i^α and the operator $\sigma_i\tau_\alpha$ is rather ambiguous; precise numerical comparisons cannot be made between the proper quantized theory and a classical approximation.

The introduction of the source may conveniently, and in anticipation, be described in quantum-mechanical terms as a unitary transformation of the state of the system. The position of the source, x_0 (here taken at the origin for convenience), and its spin matrix e_i^α will be taken as auxiliary variables. In the original field description these do not enter, formally equivalent to saying that the canonically conjugate variables vanish. The transformation* which removes or creates a singularity described by the fields (18) is

$$S_1 = \exp\left[-i\theta\int (e_i^\alpha x_i/r)\,\omega\mathscr{I}^\alpha\,\mathrm{d}^3x\right], \tag{52}$$

with $\theta = \pm 1$.

* Defined by S, such that new variables F' are equal to SFS^{-1}, or $F = S^{-1}F'S$.

The original fields are expressed in terms of the new fields ϕ' by

$$\phi = \exp\left[-\tfrac{1}{2}\theta\omega(e_i^\alpha x_i/r)\, T^\alpha\right]\phi'. \tag{53}$$

The isobaric spin density is rotated so that

$$\mathscr{I}^\alpha = \mathscr{I}'^\alpha \cos\omega - \theta\epsilon_{\alpha\beta\gamma}(e_i^\beta x_i/r)\,\mathscr{I}'^\gamma \sin\omega + (e_i^\alpha x_i/r)\,(e_j^\beta x_j/r)\,\mathscr{I}'^\beta(1-\cos\omega). \tag{54}$$

The f_i^α, the originally vanishing canonical conjugates of the e_i^α, are transformed so that

$$0 = f_i^\alpha = f_i'^\alpha + \theta\int[\mathscr{I}'^\alpha \sin\omega - \theta\epsilon_{\alpha\beta\gamma}(e^\beta x_j/r)\,\mathscr{I}'^\gamma(1-\cos\omega)$$

$$+ (e_j^\alpha x_j/r)\,(e_k^\beta x_k/r)\,\mathscr{I}'^\beta(\omega - \sin\omega)]\,(x_i/r)\,\mathrm{d}^3x. \tag{55}$$

Lorentz invariance suggests that we should add to the exponent of S_1 equation (52) a term involving $e_4^\alpha x_4$. We can avoid the introduction of such time-dependent transformations by subtracting from $(e_\mu^\alpha x_\mu)\,\mathscr{I}_4^\alpha$ the term $(x_\mu\,\mathscr{I}_\mu^\alpha)\,e_4^\alpha$. The additional terms form a second transformation

$$S_2 = \exp\left[+i\int(t^\alpha x_i/r)\,\mathscr{I}_i^\alpha\omega\,\mathrm{d}^3x\right], \tag{56}$$

where $i\theta t^\alpha$ has been written instead of e_4^α.

We treat this transformation as subsequent to the first, and neglect the time derivatives in $\mathscr{I}_i^\alpha$ (of the second order). The fields are unaltered by this transformation, but the isobaric spin becomes $\mathscr{I}''^\alpha$, where

$$\mathscr{I}'^\alpha = \mathscr{I}''^\alpha - \epsilon_{\alpha\beta\gamma}(x_i/r)\,\omega t^\beta \mathscr{I}_i^\gamma. \tag{57}$$

From (46) and (47)

$$(x_i/r)\,\mathscr{I}_i^\gamma = -\epsilon\theta N(e_i^\alpha x_i/r) \tag{58}$$

so that the source contributes to the isobaric spin the terms

$$\mathscr{I}^\alpha_{\text{source}} = \epsilon\theta\omega N[\epsilon_{\alpha\beta\gamma}t^\beta(e_i^\gamma x_i/r)\cos\omega - \theta(t^\alpha - t^\beta(e_i^\beta x_i/r)\,(e_j^\alpha x_j/r))\sin\omega]. \tag{59}$$

The volume integral is

$$I^\alpha_{\text{source}} = -(\tfrac{2}{3})\,\epsilon t^\alpha\int\omega\sin\omega N\,\mathrm{d}^3x$$

$$= \tfrac{1}{2}\epsilon t^\alpha. \tag{60}$$

Similarly the source terms in the transformation of f_i^α give in (55),

$$f_i'^\alpha = \tfrac{1}{4}\epsilon\theta\epsilon_{\alpha\beta\gamma}t^\beta e_i^\gamma. \tag{61}$$

so that

$$\theta\epsilon_{\alpha\beta\gamma}e_i^\beta f_i'^\gamma = \tfrac{1}{2}\epsilon t^\alpha. \tag{62}$$

The equality of (60) and (62) exhibits the isobaric spin explicitly as angular momentum of the rotating spin matrix, a result familiar in ordinary strong-coupling theory.

The ordinary angular momentum can be handled in a similar way. From (31) and (34),

$$J_i = \int\mathscr{I}^\alpha 2\epsilon_{ijk}x_j B_k^\alpha\,\mathrm{d}^3x; \tag{63}$$

from (17),

$$2\epsilon_{ijk}x_j B_k^\alpha = -e_j^\alpha[(\delta_{ij} - x_i x_j/r^2)\,(1-\cos\omega) + \theta\epsilon_{ijk}(x_k/r)\sin\omega]. \tag{64}$$

Combined with (59) this gives

$$J_i = -\tfrac{1}{2}\epsilon e_i^\alpha t^\alpha = \theta\epsilon_{ijk} e_j^\alpha f_k'^\alpha, \tag{65}$$

the expected strong-coupling results.

The linear momentum may be evaluated similarly,

$$\begin{aligned} P_i &= -\int \mathscr{I}^\alpha B_i^\alpha \, \mathrm{d}^3x \\ &= -\theta J_i(4\epsilon/3) \int \omega(1-\cos\,\omega)^2 \, \mathrm{d}\omega/2\pi r. \end{aligned} \tag{66}$$

These transformations are very similar to those for the model discussed in I. There the transformation function involved integrals of $\partial\theta/\partial x$ and $\partial\theta/\partial t$, forming the analogue of an isobaric spin vector, with a step-function. Here the integrals are weighted with $(x_i/r)\,\omega$, which has a similar effect on the field ϕ to a step function, but does not exhibit explicitly the crossing of the sheets near a particle.

Just as $-\partial\theta/\partial t$ and $\partial\theta/\partial x$ also form the components of the current N_λ, so also here it is possible to write the transformation as integrals of N_λ. Equation (58) exhibits $\mathscr{I}_i^\alpha$ in terms of N, and $e_i^\alpha \mathscr{I}^\alpha$ can similarly be expressed in terms of N_i. Thus the integrand of the exponent of $S_1 S_2$,

$$\omega(t^\alpha x_i/r)\,\mathscr{I}_i^\alpha - \theta\omega(e_i^\alpha x_i/r)\,\mathscr{I}^\alpha \tag{67}$$

can also be written as $$-\epsilon\omega(e_i^\alpha t^\alpha x_i/r)\,N - \epsilon\theta\omega(x_i N_i/r). \tag{68}$$

This form is less useful, however, as it does not exhibit explicitly the auxiliary variables e_i^α in the second term.

The formulae strongly suggest that there will be a natural passage, in the quantized theory, to a Dirac's equation for the description of the source singularities. There is an obvious correspondence between θ and γ_5, and between the spin matrix e_μ^α and $\mathrm{i}\gamma_\mu\gamma_5\tau_\alpha$. The integrand (67) may be regarded as the 'real part' of $(\gamma_i x_i/r)\,\tau_\alpha(\gamma_\nu \mathscr{I}_\nu^\alpha)$.

The result (66), analogous to $\alpha = -\gamma_5\sigma$, shows that the simple transformation introduced here really describes a completely polarized, or helicity, particle state; a particle at rest must be described by a mixture of right- and left-handed spin-matrices.

9. Summary

The field theory suggested here, defined by (1), (6), (10) and (22), has many of the attributes of a unified theory of mesons and their sources, so far as these can be described classically. There are particle-like states of the system, whose number is rigorously conserved; there are three fields of massless spinless particles, which may be identified with pions if it be assumed that the meson mass is a quantum effect. There is a conserved isobaric spin current, and another which might be related to the conservation of hypercharge for the strongly interacting particles. The particle states may be regarded as built from a 'bare' point source, a field singularity, coupled to the meson field through a derivative coupling with the source spins.

The classical Lagrangian contains two constants, the dimensionless (in units of $\hbar c$) ϵ and the constant κ which determines the mass or length scale. The value of

ϵ will almost certainly be fixed in the quantized theory by considerations of single-valuedness and Lorentz invariance. Some of the results in § 8, e.g. (58), suggest that ϵ will have the value unity. This is supported by the very tentative estimates made (in §§ 7 and 8) of the moment of inertia and of the coupling constant; there are other indications however that $\epsilon = \frac{1}{2}$ may be the correct condition.

The next step in the development of this theory will be the quantum-mechanical expression of the operators that create or remove singularities of the field. The analysis given here suggests that this will probably lead (as of course is wanted) to Dirac's equation for their description. The derivation of the anticommunication laws for the source field appears, however, essentially more difficult than for the one-dimensional model, where they arise naturally from the structure of the S_i.

References

Dirac, P. A. M. 1948 *Phys. Rev.* **74**, 817.
Feynmann, R. P. & Gell-Mann, M. 1958 *Phys. Rev.* **109**, 193.
Pauli, W. 1946 *Meson theory of nuclear forces.* New York: Interscience Publishers, Inc.
Skyrme, T. H. R. 1958 *Proc. Roy. Soc.* A, **247**, 260.
Skyrme, T. H. R. 1959 *Proc. Roy. Soc.* A, **252**, 236.

Particle states of a quantized meson field

By T. H. R. Skyrme
Atomic Energy Research Establishment, Harwell

(*Communicated by Sir William Penney, F.R.S.—Received* 27 *January* 1961)

A simple non-linear field theory is considered as the model for a recently proposed classical field theory of mesons and their particle sources. Quantization may be made according to canonical procedures; the problem is to show the existence of quantum states corresponding with the particle-like solutions of the classical field equations. A plausible way to do this is suggested.

1. Introduction

In previous communications (Skyrme 1958, 1959, and 1961, referred to here as I, II and III, respectively) the author has considered a particular class of non-linear field theories, that may be relevant to the description of the strongly interacting elementary particles, mesons and baryons. As a contribution towards understanding the quantized form of the field theory proposed in III, a much simpler model is considered here.

This model is essentially the same as that considered in the latter part of I, with some alteration of detail. The aim here is to formulate the problems as precisely as possible and to suggest the direction in which solutions may be found. Difficulties are associated with various limiting operations and these have not been resolved.

The classical model field theory is simple and well defined with solutions of mesonic and particle types. There is a conserved particle current associated with the multiple-valuedness of an angular representation of the field. A Hamiltonian can be found and the theory quantized formally in the usual way. Meson-like quantum states can be constructed in a perturbation series but there is no simple correspondence with the classical theory for particle-like states. If such states exist, and *a priori* this seems reasonable, they will be anomalous solutions with some resemblance to 'superconductor' solutions that have been considered by Nambu (1960) and Goldstone (1961).

The transformation that would be expected to give such solutions leads naturally to particle operators almost of neutrino character, interacting indefinitely weakly with the rest of the field. These operators can be introduced as auxiliary variables describing the multiple valuedness of the angular field representation. A further transformation introduces coupling to the meson field, leading to a Hamiltonian very similar to that for particles in the classical theory. From this Hamiltonian particle-like solutions, with non-zero mass, can apparently be constructed in a self-consistent perturbation series; these provide the closest correspondence with the classical solution.

The field is defined in § 2, and the classical aspect reviewed in § 3. Some analogies are discussed in § 4 and these lead to a tentative analysis of the quantized field theory outlined in §§ 5 and 6.

2. Field definitions

The model involves one angular variable α in one dimension x of space (apart from time), analogous to the three angular variables and three space-dimensions of the theory formulated in III. In terms of two real field variables

$$\phi_3 = \sin\alpha \quad \text{and} \quad \phi_4 = \cos\alpha, \tag{1}$$

the postulated Lagrangian density is

$$\mathscr{L} = -\frac{\epsilon}{2\pi}\left[\frac{1}{2}\left(\frac{\partial\phi_\rho}{\partial x_\mu}\right)^2 + \kappa^2(1-\phi_4)\right]. \tag{2}$$

$$= -\frac{\epsilon}{4\pi}\left(\frac{\partial\alpha}{\partial x_\mu}\right)^2 - \frac{\epsilon\kappa^2}{2\pi}(1-\cos\alpha). \tag{3}$$

Here the indices ρ and μ are summed over the values 3 and 4, for the sake of analogy, with $x_3 = x$ and $x_4 = ict$; subsequently c is put equal to unity. κ is a constant reciprocal length, and ϵ an energy scaling constant of dimensions $\hbar c$ if $\int\mathscr{L}\,\mathrm{d}x$ is an energy. The two terms correspond to the two terms in equation (23) of III, but the difference in numbers of dimensions and of degree in derivatives means that the second term here destroys the rotational symmetry of the former. This is an imperfection of the model which seems inevitable: it is not now suggested, as in I, that the rotational symmetry of the full theory should be limited, at least for the strongest interactions.

The boundary condition will be, as in III, that

$$\phi_4 = 1 \quad \text{at infinity} \quad (x = \pm\infty). \tag{4}$$

The field equation derived from (3) is

$$\Box\alpha - \kappa^2\sin\alpha = 0, \tag{5}$$

where here

$$\Box = \frac{\partial^2}{\partial x^2} - \frac{\partial^2}{\partial t^2}.$$

The energy-momentum tensor of the field is defined in the usual way, with the usual conservation laws. In addition there is a conserved particle current j_μ, with

$$j_3 = -\frac{\mathrm{i}}{2\pi}\frac{\partial\alpha}{\partial x_4}, \quad j_4 = \frac{\mathrm{i}}{2\pi}\frac{\partial\alpha}{\partial x_3} = \mathrm{i}j_0, \tag{6}$$

obviously satisfying

$$\partial j_\mu/\partial x_\mu = 0. \tag{7}$$

The particle number

$$N = \int j_0\,\mathrm{d}x = (1/2\pi)[\alpha(\infty) - \alpha(-\infty)]. \tag{8}$$

The canonically conjugate field variable is

$$\beta = \frac{\partial\mathscr{L}}{\partial\dot\alpha} = \frac{\epsilon}{2\pi}\int\frac{\partial\alpha}{\partial t}, \tag{9}$$

and the Hamiltonian has the density

$$\mathscr{H} = \frac{\pi}{\epsilon}\beta^2 + \frac{\epsilon}{4\pi}\left(\frac{\partial\alpha}{\partial x}\right)^2 + \frac{\epsilon\kappa^2}{2\pi}(1-\cos\alpha). \tag{10}$$

3. Classical solutions

The field equation (5) has wave-like solutions in which α is a function of $(kx-\omega t)$ with $\omega^2-k^2 > 0$; these correspond to the elementary solutions of the linearized equation

$$\Box\alpha-\kappa^2\alpha = 0. \tag{11}$$

As was noted in I, equation (5) also has particle-like solutions in which α is a function of $(x-vt)/(1-v^2)^{\frac{1}{2}}$. In particular the static solution

$$\tan\tfrac{1}{4}\alpha = \exp\left[\pm\kappa(x-x_0)\right] \tag{12}$$

describe a particle $(+)$ or anti-particle $(-)$ at the point $x = x_0$. For these the particle number N, equation (8), has the values ± 1.

The field equation is sufficiently simple that it is practicable to examine numerically the interaction of wave-packets of these types, and computations will be reported elsewhere. The aim of this paper, however, is the formal elucidation of a similar solution in the quantized theory.

The characteristic of a particle in this theory is that α changes by 2π in a small distance, and it is natural to introduce such steps explicitly into the field equations. Suppose that $\omega(x)$ is some function varying smoothly from 0 to 1 around the point $x = 0$, tending in a limit to the step-function $\theta(x)$ equal to 1 or 0 according as $x >$ or < 0.

To consider the motion of a particle at the point $x = X(t)$ the transformation

$$\alpha \to \alpha+2\pi\omega(x-X) \tag{13}$$

is made, with particle velocity

$$v = \mathrm{d}X/\mathrm{d}t. \tag{14}$$

To the derivative terms in the Lagrangian $L_0 = \int \mathscr{L}\,\mathrm{d}x$ must then be added

$$L' = -\epsilon\int\left(\frac{\partial\alpha}{\partial x}+v\frac{\partial\alpha}{\partial t}\right)\omega'(x-X)\,\mathrm{d}x-\epsilon\pi(1-v^2)\int\omega'^2(x-X)\,\mathrm{d}x. \tag{15}$$

The $\cos\alpha$ term will also be modified but the change will become infinitesimal as $\omega(x)$ tends to the step-function $\theta(x)$, and will not be written explicitly. L' gives a source term for the α-field arising from the derivative interaction. The variational equation for X is, however, satisfied identically when substitution is made from the field equation, as is to be expected because additional equations cannot be found without imposing some constraint on the variation of X.

A natural condition to impose in the classical theory is

$$\alpha(X) = \pi. \tag{16}$$

Time derivatives of this equation, $\partial x/\partial t+v(\partial\alpha/\partial x) = 0$ at $x = X$, etc., then give reasonable equations of motion for the particle co-ordinate X. The inertia of the source is proportional to $(\partial\alpha/\partial x)_X$; when this vanishes the particle will be on the verge of annihilation or creation with a partner antiparticle at the same point.

The terms (15) are not Lorentz covariant because of the non-relativistic form-factor $\omega(x)$, but the result of any calculation will be invariant as is the basic theory;

the form factor is merely introduced for calculational convenience. In the limit $\omega \to \theta$ and $\omega' \to \delta$, the first term is covariant and the second may formally so be made by division with $(1-v^2)^{\frac{1}{2}}$ to allow for the contraction of the moving source. The terms may then be written

$$L' = -(1-v^2)^{\frac{1}{2}}\left\{M_0 + \frac{\epsilon}{(1-v^2)^{\frac{1}{2}}}\left(\frac{\partial\alpha}{\partial x} + v\frac{\partial\alpha}{\partial t}\right)_X\right\}, \tag{17}$$

where M_0 is an invariant bare mass constant, formally infinite; the divergence is cancelled by that of the self-mass which arises from the coupling to the meson field as was seen in I, § 7.

With this addition to the Lagrangian the conjugate variable becomes

$$\beta = \frac{\epsilon}{2\pi}\frac{\partial\alpha}{\partial t} - \epsilon v \delta(x - X) \tag{18}$$

and formally the Hamiltonian is

$$\begin{aligned} H &= vP + \int \beta \frac{\partial\alpha}{\partial t}\,\mathrm{d}x - L_0 - L' \\ &= H_0 + vP + M_0(1-v^2)^{\frac{1}{2}} + \left[\epsilon\frac{\partial\alpha}{\partial X} + 2\pi v\beta\right]_{x=X} + \frac{M_0 v^2}{(1-v^2)^{\frac{1}{2}}}, \end{aligned} \tag{19}$$

the last term arising from the square of the additional term in (18) associated with the derivative interaction. This indeed is the correct expression for the total field energy if the momentum

$$P = M_0 v/(1-v^2)^{\frac{1}{2}} \tag{20}$$

but this relation could not be obtained in the usual way because X is a redundant variable of the system.

4. The quantized field

The Hamiltonian (10) with the canonical commutation relations

$$[\alpha(x), \beta(x')] = \mathrm{i}\hbar\,\delta(x - x') \tag{21}$$

may be solved by a formal perturbation series starting with the linearized form describing a meson field of mass $\hbar\kappa/c$. The only (logarithmic) divergences arise from the meson-meson interactions in the last term which could be handled by the usual renormalization methods; these are not particularly significant, however, as the structure of this term in the model is arbitrary. The important interesting question is whether states exist analogous to the classical particle solutions with $N = \pm 1$. By symmetry the vacuum should have $N = 0$, above which there should exist some lowest state with $N = \pm 1$, interpretable as a particle. The plan is to look for a transformation of the Hamiltonian which will exhibit some particle state as a start for a perturbation analysis; the following analogies suggest a possible approach.

In two dimensions the analogue of Dirac's equation can be written in the usual form

$$\left(\gamma_\mu \frac{\partial}{\partial x_\mu} + \frac{Mc}{\hbar}\right)\psi = 0, \tag{22}$$

where ψ is a two-component field, the index μ is summed only over 3 and 4, and $\gamma_3, \gamma_4, \gamma_5$ form a triad of anticommuting Pauli matrices. The field Hamiltonian is

$$\int \psi^\dagger \left\{\gamma_5 \left(-i\hbar c \frac{\partial}{\partial x}\right) + \gamma_4 Mc^2\right\} \psi \, dx. \tag{23}$$

γ_5 is both the velocity and the analogue of its namesake in 4 dimensions (where $\boldsymbol{\alpha} = -\gamma_5 \boldsymbol{\sigma}$).

The exponential type of interaction discussed in I and II arises when M is replaced by $M \exp(-i\gamma_5 \alpha)$; after the transformation $\psi \rightarrow \exp(\frac{1}{2}i\gamma_5 \alpha)\,\psi$ the particle Lagrangian density is

$$-\bar{\psi}\left(\gamma_\mu \hbar c \frac{\partial}{\partial x_\mu} + Mc^2 + \tfrac{1}{2} i\hbar c \gamma_\mu \gamma_5 \frac{\partial \alpha}{\partial x_\mu}\right)\psi \tag{24}$$

with an interaction term similar to (15).

The transformation (13) introduced to describe a classical particle suggests the general separation of the field

$$\alpha(x) = \hat{\alpha}(x) + 2\pi N(x), \tag{25}$$

where $\hat{\alpha}(x)$ lies in the standard zone, $-\pi < \hat{\alpha} < \pi$, and $N(x)$ is an integer at any point x. Then the meson Lagrangian

$$L_0(\alpha) = L_0(\hat{\alpha}) - \epsilon \frac{\partial \hat{\alpha}}{\partial x_\mu} \frac{\partial N}{\partial x_\mu} - \epsilon\pi \left(\frac{\partial N}{\partial x_\mu}\right)^2. \tag{26}$$

Comparison of (24) and (26) indicates a correspondence, that is

$$\left.\begin{aligned} \frac{\partial N}{\partial x_\mu} &\leftrightarrow \frac{\hbar c}{2\epsilon} \bar{\psi} i \gamma_\mu \gamma_5 \psi, \\ \frac{\partial N}{\partial x} &\leftrightarrow \frac{\hbar c}{2\epsilon} \psi^\dagger \psi, \\ \frac{\partial N}{\partial t} &\leftrightarrow \frac{\hbar c}{2\epsilon} \psi^\dagger \gamma_5 \psi. \end{aligned}\right\} \tag{27}$$

Now $\partial N / \partial x$ is the particle number density of the meson field; the correspondence (27) would be natural if the constant $\epsilon = \frac{1}{2}\hbar c$. This leads to the conjecture that the quantized field has particle solutions if and only if this condition is satisfied.

The interaction terms arise from the changes

$$\left.\begin{aligned} \alpha &\rightarrow \alpha + 2\pi \int^x \psi^\dagger \psi \, dx', \\ \beta &\rightarrow \beta - \tfrac{1}{2} \psi^\dagger \gamma_5 \psi, \end{aligned}\right\} \tag{28}$$

which would follow from unitary transformation by

$$S = \exp\left\{\tfrac{1}{2} i \int \alpha \psi^\dagger \gamma_5 \psi \, dx + 2\pi i \int \beta \left(\int^x \psi^\dagger \psi \, dx'\right) dx\right\}. \tag{29}$$

On the ψ-field this transformation would introduce the factor

$$\exp\left\{-\tfrac{1}{2} i \gamma_5 \alpha(x) - 2\pi i \int_x^\infty \beta(x') \, dx'\right\}. \tag{30}$$

5. Particle operators

Consider therefore in connexion with the quantum Hamiltonian (10) the operators

$$U_\gamma = \exp\left\{ i\epsilon\gamma \int \alpha(x')\,\omega'(x'-x)\,dx' + 2\pi i \int \beta(x')\,\omega(x'-x)\,dx' \right\}, \tag{31}$$

where γ has the eigenvalues ± 1 of γ_5 and $\omega(x)$ is an approximation to the step-function as in § 3; both $\hbar$ and c will now be put equal to unity.

They satisfy the equation

$$\frac{i\,\partial U}{\partial t} = [U, H] = -i\gamma\frac{\partial U}{\partial x} + \frac{\epsilon\kappa^2}{2\pi}\int [\cos\alpha(x') - \cos\{\alpha(x') + 2\pi\omega(x'-x)\}]\,dx'. \tag{32}$$

As ω tends to the step-function the last term becomes proportional to

$$\int [1 - \cos 2\pi\omega(t)]\,dt,$$

tending to zero. In this limit then the operators satisfy a neutrino-like equation.

The commutation relations of the U are easily evaluated

$$U_\gamma(x)\,U_{\gamma'}(x') = U_{\gamma'}(x')\,U_\gamma(x)\exp(2\pi i\epsilon G), \tag{33}$$

with

$$G = \int [\gamma'\omega(t+x'-x) - \gamma\omega(t+x-x')]\,(d\omega/dt)\,dt. \tag{34}$$

For $\gamma = -\gamma'$, $G = \pm 1$, independent of the shape of ω; for $\gamma = \gamma'$, $G \to \pm 1$ as ω tends to the step-function, except for $x = x'$ when $G = 0$. Therefore in the limit the U 'almost always' anticommute, if $\epsilon = \frac{1}{2}\hbar c$, providing another motive for this choice.

Again with $\epsilon = \frac{1}{2}$, U is a double-valued function only of the basic fields ϕ_ρ, but for a general choice it would be indefinitely many valued. From such a point of view the particle states of the system arise when the state-function is allowed to be a double-valued function of the field, at all points of space independently. Whatever may be the basic reason for this determination of the energy scale, it will now be assumed fixed thus.

In the limit, if the last term in (31) is negligible, the operators U may be used to construct states with additional 'neutrinos'; for if ϕ is any eigenstate of H with energy E and particle number N, the state

$$\left[\int U_\gamma^\dagger(x)\,e^{ikx}\,dx\right]\phi$$

has energy $E + \gamma k$ and particle number $N+1$. Whether these limiting states really 'exist' is in fact rather academic because in this context the 'neutrinos' are completely non-interacting and so effectively unobservable.

These operations seem to be the obvious ones needed to describe the possible multiple-valuedness of the angular variables; the next problem is to find how to use them to build up massive particle states. The operators can be introduced as additional co-ordinates for the system; if $\psi_\pm$, labelled by the eigenvalues of γ_5, are the anticommuting field operators of an assumed neutrino-like field with Hamiltonian

$$H_\nu = \int \psi^\dagger\left(-i\gamma_5\frac{\partial}{\partial x}\right)\psi\,dx, \tag{35}$$

then the commuting products $(U^\dagger\psi)$ of corresponding operators satisfy

$$[H+H_\nu, U^\dagger\psi] = \pm\mathrm{i}(\partial/\partial x)(U^\dagger\psi) \tag{36}$$

so that $H^* = H+H_\nu$ commutes with the operators $\int U^\dagger\psi\,\mathrm{d}x$. Similarly

$$N^* = N+\int\psi^\dagger\psi$$

commutes with $U^\dagger\psi$.

If ψ is any eigenstate of H^* then by operation with a function F of the $\int U^\dagger\psi\,\mathrm{d}x$ other states can be generated with different numbers of neutrinos, replacing steps in the meson field. By projection on to the neutrino vacuum a state is formed which will then be an eigenstate of H. For if χ is the neutrino vacuum and

$$\phi = \chi^\dagger F\psi, \tag{37}$$

then

$$H\phi = \chi^\dagger(H+H_\nu)F\psi = \chi^\dagger FH^*\psi \tag{38}$$

and the particle number is likewise the eigenvalue of N^*.

Since H^* is separated the transformation is in some way trivial but it provides a convenient framework for the representation of anomalous solutions. The form of F does not appear to be important, so long as the projected state ϕ is not null.

6. Quantum particles

The aim is to find an eigenstate of H that describes a particle. Such a state could be generated from various states ψ with Hamiltonian H^* by the projection (37). It should be possible to find one such state that is a functional only of the reduced variables $\hat{\alpha}(x)$, of equation (25), with $N = 0$, and of the 'neutrino' field ψ.

This will be sought by transformation of H^* into a form similar to that which arises naturally in the classical description of a particle, as equation (19). The formal possibility of a (divergent) bare mass for the sources may be introduced by the canonical transformation of the particle field:

$$\left.\begin{aligned}\psi_+ &\to \exp\left[\mathrm{i}\pi\int_{-\infty}^{x}\psi_-^\dagger\psi_-\,\mathrm{d}x\right]\psi_+,\\ \psi_- &\to \exp\left[\mathrm{i}\pi\int_{x}^{\infty}\psi_+^\dagger\psi_+\,\mathrm{d}x\right]\psi_-,\end{aligned}\right\} \tag{39}$$

where $\psi_\pm$ are the components of ψ with $\psi_5 = \pm 1$. This sends

$$H_\nu \to H_\nu + \tfrac{1}{2}\pi\int\psi^\dagger(1+\gamma_5)\psi\psi^\dagger(1-\gamma_5)\psi\,\mathrm{d}x. \tag{40}$$

The additional contact interaction term equivalent with

$$\tfrac{1}{4}\pi\int\psi\gamma_\mu(1+\gamma_5)\psi\bar{\psi}\gamma_\mu(1-\gamma_5)\psi\,\mathrm{d}x \tag{41}$$

is still consistent with zero mass, but embraces also the possibility of a divergence.

Next the coupling to the meson field is introduced by the second canonical transformation

$$\left.\begin{aligned}\psi &\to \exp(\tfrac{1}{2}\mathrm{i}\gamma_5\hat{\alpha})\psi,\\ \beta &\to \beta+\tfrac{1}{2}\psi^\dagger\gamma_5\psi.\end{aligned}\right\} \tag{42}$$

16-2

This gives the new Hamiltonian, a transform of H^*,

$$H_1 = H_0(\alpha) + \int \psi^\dagger \gamma_5 \left(-\mathrm{i}\frac{\partial}{\partial x}\right) \psi \,\mathrm{d}x + \tfrac{1}{2}\pi \int \psi^\dagger(1+\gamma_5)\,\psi\psi^\dagger(1-\gamma_5)\,\psi \,\mathrm{d}x$$
$$+ \int \psi^\dagger \left(\frac{1}{2}\frac{\partial \alpha}{\partial x} + 2\pi\beta\gamma_5\right) \psi \,\mathrm{d}x + \tfrac{1}{2}\pi \int (\psi^\dagger \gamma_5 \psi)^2 \,\mathrm{d}x. \tag{43}$$

There is a term-by-term correspondence with the classical form (19); γ_5 replaces v and the third and fifth terms can be combined to give

$$\tfrac{1}{2}\pi \int (\psi^\dagger \psi)^2 \,\mathrm{d}x, \tag{44}$$

corresponding to the bare mass term $M_0(1-v^2)^{-\frac{1}{2}}$.

The Hamiltonian H_1 might be derived in the conventional way from an extended Lagrangian density

$$\mathscr{L}_1 = -\frac{1}{8\pi}\left(\frac{\partial \hat{\alpha}}{\partial x_\mu} + 2\pi\bar{\psi}\mathrm{i}\gamma_\mu\gamma_5\psi\right)^2 - \frac{\kappa^2}{4\pi}(1-\cos\alpha) - \bar{\psi}\gamma_\mu \frac{\partial}{\partial x_\mu}\psi. \tag{45}$$

This Lagrangian defines a theory that is at least renormalizable in the conventional manner since the worst possible types of divergency are logarithmic (the derivative interaction is compensated by the fewer number of space dimensions). It has 'γ_5-invariance' because there is no explicit mass-like term, and indeed it still describes massless non-interacting neutrino-like particles as the normal type of particle solution. But it is now possible in this form to consider also whether self-consistent solutions are possible with non-zero mass.

Suppose that the field ψ can describe particles with mass M, and that μ is the renormalized meson mass. Then, in second order, the mass correction terms will equate $\int \bar{\psi}\Delta M \psi$ with the one-particle part of

$$\tfrac{1}{2}\pi \int (\bar{\psi}\mathrm{i}\gamma_\mu\gamma_5\psi)^2 \,\mathrm{d}x - \tfrac{1}{8}\mathrm{i}\left[\int \left(\bar{\psi}\mathrm{i}\,\psi_\mu\gamma_5\psi\frac{\partial \alpha}{\partial x_\mu}\right)\mathrm{d}x\right]^2. \tag{46}$$

If p is the momentum of the external particle line, q that of the internal particle and $k = p - q$ that of the meson, the usual methods give

$$\Delta M = \pi \int \frac{\mathrm{d}^2 q}{(2\pi)^2} \mathrm{i}\gamma_\mu\gamma_5[\epsilon + \mathrm{i}(M + \mathrm{i}\gamma q)]^{-1}\mathrm{i}\gamma_\mu\gamma_5$$
$$-\tfrac{1}{4}\mathrm{i} \int \frac{\mathrm{d}^2 k}{(2\pi)^2}[4\pi/\epsilon + \mathrm{i}(k^2+\mu^2)]\,\mathrm{i}\gamma k\gamma_5[\epsilon + \mathrm{i}(M+\mathrm{i}\gamma q)]^{-1}\mathrm{i}\gamma k\gamma_5. \tag{47}$$

Since
$$\mathrm{i}\gamma k(M + \mathrm{i}\gamma q)\,\mathrm{i}\gamma k = -k^2(M - \mathrm{i}\gamma p) - \mathrm{i}\gamma k(q^2 - p^2), \tag{48}$$
the second term is equal, for a real state with $M + \mathrm{i}\gamma p = 0$, to

$$2\pi\mathrm{i}M \int \frac{\mathrm{d}^2 k}{(2\pi)^2} k^2[k^2+\mu^2-\mathrm{i}\epsilon]^{-2}[q^2+M^2-\mathrm{i}\epsilon]^{-1}. \tag{49}$$

The divergent part of this integral is cancelled exactly by the first term of (46), just as in the classical theory the interaction cancels the divergence of the bare mass (§ 3). The remainder gives

$$\Delta M/M = -2\pi\mathrm{i}\mu^2 \int \frac{\mathrm{d}^2 k}{(2\pi)^2}[k^2+\mu^2-\mathrm{i}\epsilon]^{-1}[q^2+M^2-\mathrm{i}\epsilon]^{-1}$$
$$= \left(\frac{4M^2}{\mu^2-1}\right)^{-\frac{1}{2}} \tan^{-1}\left(\frac{4M^2}{\mu^2-1}\right)^{\frac{1}{2}}. \tag{50}$$

Thus to 'second order' the consistency condition $\Delta M = M$ is satisfied either by $M = 0$, the neutrino-like solutions, or by $M = \frac{1}{2}\mu$; for comparison the classical theory, with $\epsilon = \frac{1}{2}\hbar c$, has $M = (2/\pi)(\hbar\kappa/c)$. In the same order the meson mass correction is generally finite though it diverges for the particular value $M = \frac{1}{2}\mu$ because then the meson is almost unstable against disintegration into a particle pair. There are divergent corrections that arise from the cos α term but these are peculiar to the model, and probably not of fundamental significance. Apart from the latter it is plausible that higher-order renormalizations may be finite and lead to a sensible description of a massive particle state.

7. Comments

I have considered the system described by the Hamiltonian (10). Classically there are meson-like and particle-like solutions; the fundamental problem is to study the nature of the quantized solutions, and in particular to show the existence of particle-like states. I have outlined a method for doing this which needs critical investigation on a number of points:

(1) Whether the condition $\epsilon = \frac{1}{2}\hbar c$ is essential for the existence of particle-like states; and if it is, the significance. It may be related either to the nature, geometrical interpretation, of the fields ϕ, or to some deeper interpretation of quantum theory.

(2) The nature of the limiting operations in which the form-factor $\omega(x)$ tends to a step-function, and the existence of the 'neutrino' operators U in this limit.

(3) The introduction of additional neutrino co-ordinates ψ with the projection operation F, and its compatibility with the above limiting operation.

(4) The significance of the transformation (40) which introduces the divergent self-mass associated with a point-singularity.

The final stage, the effective Lagrangian (46) and Hamiltonian (44), has an attractive resemblance to the classical theory; I have not, however, found a satisfactory proof that its solutions will project into ones of the original mesonic Lagrangian (3) and Hamiltonian (10).

An alternative discussion of the quantization that seems particularly appropriate to the problem of relating quantum and classical solutions might involve the use of Feymann path integrals for the field variables separated as in equation (25); the condition $\epsilon = \frac{1}{2}\hbar c$ would then enter in a simple and central way. I have, however, found considerable difficulty in formulating the limiting operations that are needed.

References

Goldstone, J. 1961 *Nuovo Cimento*, **19**, 154.
Nambu, Y. 1960 *Chicago Rep.* no. 60/21.
Skyrme, T. H. R. 1958 *Proc. Roy. Soc.* A, **247**, 260.
Skyrme, T. H. R. 1959 *Proc. Roy. Soc.* A, **252**, 236.
Skyrme, T. H. R. 1961 *Proc. Roy. Soc.* A, **260**, 127.

8.C
Nuclear Physics **31** (1962) 550—555; © *North-Holland Publishing Co., Amsterdam*
Not to be reproduced by photoprint or microfilm without written permission from the publisher

A MODEL UNIFIED FIELD EQUATION

J. K. PERRING and T. H. R. SKYRME †
Atomic Energy Research Establishment, Harwell, England

Received 29 September 1961

Abstract: The classical solutions of a unified field theory in a two-dimensional space-time are considered. This system, a model of interacting mesons and baryons, illustrates how the particle can be built from a wave-packet of mesons and how reciprocally the meson appears as a tightly bound combination of particle and antiparticle.

1. Introduction

One author has proposed the consideration of non-linear meson fields with periodic properties for the unified description of mesons and their particle sources [1, 2]). In this and the following paper [6]) some properties of these systems are examined further.

A simple model of this type was defined in sect. 2 of ref. [2]) by the field equation

$$\partial^2 \alpha/\partial x^2 - (1/c^2)\partial^2 \alpha/\partial t^2 - \kappa^2 \sin \alpha = 0, \tag{1}$$

and the boundary condition

$$\cos \alpha = 1 \quad \text{at} \quad x = \pm\infty, \tag{2}$$

to be satisfied by the "angular" field variable $\alpha(x, t)$.

Units are chosen such that $c = \kappa = 1$, and such that the energy density is

$$\mathscr{E}(x) = (1/8\pi)[(\partial\alpha/\partial x)^2 + (\partial\alpha/\partial t)^2] + (1/4\pi)(1-\cos\alpha). \tag{3}$$

The particle number is

$$N = (1/2\pi)[\alpha(+\infty) - \alpha - (-\infty)], \tag{4}$$

the integral of the particle number density $(1/2\pi)(\partial\alpha/\partial x)$, and necessarily an integer on account of the boundary condition (2).

2. Simple Solutions

Simple solutions may be found in which α is a function only of $s = (x-vt)/(1-v^2)^{\frac{1}{2}}$ and the field equation reduces to the pendulum-like equation

$$\mathrm{d}^2\alpha/\mathrm{d}s^2 = \sin \alpha. \tag{5}$$

† Now at Dept. of Mathematics, University of Malaya, Pantai Valley, Kuala Lumpur, Malaya.

If the phase-velocity $v > 1$ the solution is a travelling wave

$$\sin \tfrac{1}{2}\, \alpha = \pm k \text{ sn } (i(s-s_0), k), \tag{6}$$

reducing to a plane wave as the amplitude $k \to 0$. This can be made to fit the boundary condition (2) in a large box; the condition is evidently equivalent to $\alpha = 0$ in this case, and all these solutions have particle number $N = 0$. They are interpreted as meson waves.

For a real physical velocity $v < 1$, there are the solutions

$$\sin \tfrac{1}{2}\, \alpha = \pm \text{sech } (s-s_0), \tag{7}$$

which have $N = \pm 1$, and total energy $E = (2/\pi)(1-v^2)^{-\frac{1}{2}}$. They are interpreted as the fields associated with a particle (or antiparticle) of mass $(2/\pi)$ centred at $s = s_0$ and moving with velocity v.

These are the only simple solutions with finite N. There are also solutions, analogous to those for a pendulum making complete revolutions, with infinite N describing a lattice of uniformly spaced particles.

Certain other analytical solutions have been found and will be described below; their existence depends, however, upon the specially simple nature of sin α in the field equation. If sin α is replaced by some other periodic function there are always solutions like (6) and (7), but usually numerical analysis will be needed beyond that point.

The structure of the wave-equation (1) is sufficiently simple to allow direct numerical integration on a computer, and programmes were written to follow the interaction of wave-packets, such as (6) or (7), that are initially far apart. The method of integration is outlined in the appendix.

3. Particle-Particle Interactions

The problem of two colliding particles is equivalent, by symmetry, to that of a single particle moving in the half-space $x > 0$, with the boundary conditions $\alpha(0) = 0$ and $\alpha(\infty) = 2\pi$. To our initial surprise the numerical integration showed that the scattering off the boundary at $x = 0$ was purely elastic, almost like hard-sphere scattering. An analytical solution was then found with these properties; this has

$$\text{tg } \tfrac{1}{4}\alpha = v \sinh(x/(1-v^2)^{\frac{1}{2}})/(\cosh(vt/(1-v^2)^{\frac{1}{2}}) \tag{8}$$

and describes two particles centred approximately at $x = \pm vt$ and colliding elastically at $x = 0$ at time $t = 0$. The positions of the particles are defined more precisely by the condition that there $\cos \alpha = -1$; this then gives the 'scattering length' in the collision equal to $2(1-v^2)^{\frac{1}{2}} \log (1/v)$, in units of $1/\kappa$.

A similar solution exists describing a particle-antiparticle collision, with

$$\text{tg } \tfrac{1}{4}\alpha = (1/v) \sinh (vt/(1-v^2)^{\frac{1}{2}})/\cosh(x/(1-v^2)^{\frac{1}{2}}). \tag{9}$$

The scattering length is the same, describing here an attractive process in which particle and antiparticle are accelerated through one another.

4. The Potential

Although we have found an explicit solution of the two-particle scattering problem in this model, it is interesting also, for the better understanding of the nature of the interactions, to construct a potential.

The definition of potential energy is somewhat arbitrary, but that most natural in this problem is the energy of the static field associated with particles held in fixed positions. This is found from the static solution of the field equation (1) which satisfies in addition to (2) the conditions

$$\begin{gathered}\cos\alpha = -1 \quad \text{at particle positions,} \\ (\partial\alpha/\partial x) > 0 \text{ for particle,} \quad (\partial\alpha/\partial x) < 0 \quad \text{for antiparticle.}\end{gathered} \tag{10}$$

In general (i.e. except for a free particle) the field gradient will be discontinuous across the particle position; this corresponds with the force that has to be applied at a point to keep the particle fixed.

Between two particles separated by a distance r the field will be given by an elliptic function whose parameter k is determined by

$$\kappa r = 2k\,K(k). \tag{11}$$

The energy of the fields between the particles is then

$$E = (2/\pi)k^{-1}\,[E(k) - \tfrac{1}{2}(1-k^2)\,K(k)], \tag{12}$$

and for two particles the fields outside would be the same as for free particles. For large r, one has $k \to 1$ and E, given by eq. (12), becomes equal to $(2/\pi)$. For finite r the difference from this value measures the potential, with the approximate formulae

$$\begin{aligned} V &\sim (8\kappa/\pi)\,e^{-\kappa r}, \quad && \text{as } r \to \infty, \\ V &\approx (\pi/2r) && \text{as } r \to 0. \end{aligned} \tag{13}$$

It is interesting to see the origins of these two limits. At large distances the interaction may be treated as a perturbation and will satisfy the linearised wave-equation

$$(\nabla^2 - \kappa^2)\,\alpha = \text{const} \sum \delta'(x - x_n), \tag{14}$$

giving the exponential dependence in (13); the coefficient $(8/\pi)$ expresses the effective coupling constant appropriate to the detailed source structure. The sign (repulsion) is that expected for a derivative interaction between similar particles.

At small distances there is also repulsion, but its origin is distinct, It comes from the nature of the boundary conditions: α has to change by 2π over the small distance r, implying a large field gradient and a large addition to the gradient term in the energy density (3).

A similar analysis can be made of the potential between a particle and an antiparticle; at large distances there is simply a change in sign as would be expected from the asymptotic field. At small distances the boundary conditions 'match' so that α assumes a nearly constant value between the sources and $E \to 0$; there is then a potential well of maximum depth equal to one particle mass.

5. The Meson as a Particle Pair

The most natural aspect of the field equation (1) is that of a self-interacting meson field, and the particles appear as particular types of non-dispersive wave-packets that can be constructed on account of the periodic nature of the boundary conditions. The analytical solution (9) leads to the converse picture of a meson as a tightly bound state of the particle-antiparticle system [3]).

For real $v < 1$, this solution (9) describes the scattering of a pair with asymptotic velocities $\pm v$, and the total energy is $(4/\pi)(1-v^2)^{-\frac{1}{2}}$. Evidently it continues to satisfy the field equation when v is continued through zero to imaginary values, giving a state of total energy less than two particle masses, i.e. a bound state. The sinh function is replaced by a sine, so that the field falls off exponentially at large distances at all times; it describes a localised but oscillating meson wave-packet with $N = 0$. In the limit as $v \to i\infty$, one has

$$\alpha = \varepsilon \exp(i\kappa t),$$

an infinitesimal plane meson wave of zero momentum. By a Lorentz transformation states with arbitrary momentum appear similarly as the limits of particle-pair systems with this centre-of-mass momentum.

6. Meson-Particle Interactions

The existence of analytical expressions for meson wave-packets of finite amplitude makes it easier to handle numerically the problem of the interaction of mesons and particles in this model. A number (7) of integrations of this type have been made. In these the particle was initially at rest at the origin, and the meson packet approached it from a distance with chosen values of velocity, amplitude and phase. The results were again unexpectedly simple in character. In all cases the final state consisted of a displaced particle at rest and a transmitted meson wave, and appeared to be independent of the choice of phase for the meson wave.

We have no explanation of these results except that they appear again to be connected with accidental features of the equation with $\sin \frac{1}{2}\alpha$. In the case of a small amplitude meson wave we can use perturbation theory, writing

$$\alpha(x) = \alpha_p(x) + A(x, t),$$

where $\alpha_p(x)$ is the solution for a particle at rest at the origin (eq. (7)):

$$\sin \tfrac{1}{2}\alpha_p = \operatorname{sech} x.$$

Then the linearised equation for A is

$$(\Box - \kappa^2 \cos \alpha_p) A = (\Box - \kappa^2 + 2\kappa^2 \operatorname{sech}^2 \kappa x) A = 0, \tag{15}$$

with periodic solutions

$$A = e^{i\omega t} e^{\pm ikx} (1 \pm (i/k) \tanh x), \tag{16}$$

where $\omega^2 = \kappa^2 + k^2$. This describes meson scattering in which there is no reflected wave and the transmitted wave has a phase change δ, $\operatorname{tg} \delta = (1/k)$. The absence of reflection arises from the particular form of the scattering potential and is thus an accidental feature arising from the shape of $\sin \alpha$.

Appendix

NUMERICAL INTEGRATION PROCEDURE

Two methods of integration were tried. First, a simple leapfrogging scheme [4]) was programmed for the Harwell Mercury Computer, writing

$$\begin{aligned} \alpha(x, t+\tau) &= \alpha(x, t) + \tau\dot{\alpha}(x, t+\tfrac{1}{2}\tau) \\ \dot{\alpha}(x, t+\tfrac{1}{2}\tau) &= \dot{\alpha}(x, t-\tfrac{1}{2}\tau) + \tau(\partial^2 \alpha/\partial x^2 - \sin \alpha)_{x,t}, \end{aligned}$$

with $\partial^2 \alpha/\partial x^2$ given as

$$h^{-2}[\alpha(x+h) - 2\alpha(x) + \alpha(x-h)].$$

In solving such equations stability considerations are always important; the method was found to be wildly unstable for $\tau = h$, but a reduction of τ to $0.95\, h$ was sufficient to give stability.

The first results, on particle-particle scattering, showed that any inelasticity was extremely small, and it became clear that greater accuracy and a larger lattice were required than was possible with the use of the Mercury core store. A new programme was therefore written for the IBM 7090, at Aldermaston, which used the slightly more complex method of integration along the characteristics [5]). The method is adequately described in this reference; the characteristics are known but the presence of the source term makes iteration for α at each step of the integration necessary. However, error analysis shows that this method is, on the average, four times as fast as the preceding one.

We are indebted to Mr. A. R. Curtis for valuable discussions.

References

1) T. H. R. Skyrme, Proc. Roy. Soc. **260** (1961) 127
2) T. H. R. Skyrme, Proc. Roy. Soc. **262** (1961) 237
3) E. Fermi and C. N. Yang, Phys. Rev. **76** (1949) 1739
4) R. D. Richtmeyer, Difference methods for initial value problems (Interscience, New York, 1957) p. 166
5) National Physical Laboratory, Modern computing methods (1957) p. 66
6) T. H. R. Skyrme, Nuclear Physics **31** (1962) 556

Kinks and the Dirac Equation

T. H. R. SKYRME
Department of Mathematical Physics, University of Birmingham, Birmingham, England

(Received 2 November 1970)

In a model quantum theory of interacting mesons, the motion of certain conserved particlelike structures is discussed. It is shown how collective coordinates may be introduced to describe them, leading, in lowest approximation, to a Dirac equation.

1. INTRODUCTION

It has often been suggested that the particle sources of fields might really be very localized bound states of the same fields. There are at least two important reasons for seeking theoretical models of this type; first they should reduce the number of independent fundamental variables, and second they might lead to a theory of interactions that is free from the difficulties associated with point interactions. In such a model the localized particlelike states may behave like point sources for interaction with weak external fields, but exhibit a structure in strong fields.

This paper is mainly concerned with a model theory of a self-interacting meson-like field. Considered classically, the theory has particlelike solutions; in particular there are solutions that describe static localized concentrations of field strength. The problem is to find the analogs of these in the quantized theory; the analysis suggests that these may appear as states with fermion symmetry characters, and a systematic method of solution is developed starting from an approximation in which the particlelike states are described by independent variables satisfying a Dirac equation.

Although the starting point is a relativistic theory, the discussion is presented in a form that is not manifestly covariant, following the canonical quantization procedures. At this stage a nonrelativistic description exhibits more clearly the physical ideas inherent in the model.

The problem is that of describing certain collective motions of a quantized system. That is analyzed in the usual way by introducing auxiliary redundant coordinates to describe the collective modes and imposing constraints through supplementary conditions that link them to the original variables.

2. THE ONE-DIMENSIONAL MODEL (A)

A simple one-dimensional model was considered earlier, and it will be useful to refer to it as Model A. The field variable is an angle $\alpha(x, t)$ and all physically observable quantities depend only on $\alpha \bmod (2\pi)$; the Lagrangian density is taken to be

$$-\frac{1}{8\pi}\left[\left(\frac{\partial\alpha}{\partial x}\right)^2 - \left(\frac{\partial\alpha}{\partial t}\right)^2\right] - \frac{K^2}{4\pi}(1 - \cos\alpha),$$

where K is a mass constant [units with $\hbar = c = 1$ will be used throughout].

The vacuum condition is taken as $\cos\alpha = 1$, and all solutions are assumed to satisfy the boundary condition that

$$\cos\alpha \to 1 \quad \text{as} \quad x \to \infty.$$

There is then a conserved current giving rise to the quantum number

$$N = \frac{1}{2\pi}\int_{-\infty}^{\infty}\left(\frac{\partial\alpha}{\partial x}\right) dx = \frac{1}{2\pi}[\alpha(+\infty) - \alpha(-\infty)].$$

This is interpreted as a particle number. There is a localized static solution of the classical field equation for which $\alpha(-\infty) = 0$ and $\alpha(+\infty) = 2\pi$.

This field theory may be quantized in the canonical manner, with the conjugate variable

$$\beta(x, t) = \frac{1}{4\pi}\frac{\partial\alpha}{\partial t}.$$

So long as the fields are weak, $1 - \cos\alpha \sim \frac{1}{2}\alpha^2$, and the theory describes "mesons" with mass K, in states with $N = 0$. A state with $N = 1$ may be created by operating upon an $N = 0$ state, e.g., the vacuum, with the (singular) operator,

$$K = \exp\left[2\pi i\int_{x_0}^{\infty}\beta\, dx\right],$$

that sends $\alpha(x) \to \alpha(x) + 2\pi\theta(x - x_0)$, a step at $x = x_0$; this is interpreted as an ideal particle singularity at x_0, but is not by itself a stationary state. It was shown, however, that the operators FK, where $F = \exp(\pm\frac{1}{2}i\alpha(x_0))$, satisfy the equations

$$\left[\pm\frac{\partial}{\partial x_0} - \frac{\partial}{\partial t}\right]FK = 0,$$

so that FK may be identified with a particle creation or annihilation operator at the point x_0; furthermore, these operators associated with different x_0 anticommute. This "particle" is a massless neutrinolike object that can be introduced as a starting point for the description of massive particles.

3. THE THREE-DIMENSIONAL MODEL (B)

A three-dimensional model was suggested that has rather similar classical properties§[1]; this will be called model B and is the main subject of this paper. The field variables are four, $\phi_\rho(\mathbf{x}, t)$, $\rho = 0, 1, 2, 3$, constrained to satisfy $\sum \phi_\rho^2 = 1$ everywhere; it is convenient to describe them alternatively by a quaternion

$$U(\mathbf{x}, t) = \phi_0 + i\tau_\alpha \phi_\alpha .$$

(Indices such as α are summed from 1 to 3; τ_α denote a standard set of Pauli matrices, which will often but *not* always be associated with an "isobaric spin.")

The fields are described by the gradients $B_{\alpha\mu}$ defined by

$$\partial_\mu U = i\tau_\alpha B_{\alpha\mu} U$$

and so satisfy conditions

$$\partial_v B_{\alpha\mu} - \partial_\mu B_{\alpha v} = 2\epsilon_{\alpha\beta\gamma} B_{\alpha\mu} B_{\gamma v},$$

the $\epsilon_{\alpha\beta\gamma}$ being the structure constants associated with the quaternion algebra. The Lagrangian density of the model B is taken as

$$-(\epsilon/8\pi^2)[(B_{\alpha\mu}B_{\alpha\mu})^2 - (B_{\alpha\mu}B_{\alpha v})(B_{\beta\mu}B_{\beta v}) + 2K^2(B_{\alpha\mu}B_{\alpha\mu})],$$

where K is a mass constant and ϵ a dimensionless parameter, whose value should probably be $\frac{1}{2}(\hbar c)$.

The vacuum condition is taken to be $U = 1$, and there is the boundary condition

$$U(\mathbf{x}, t) \to 1 \quad \text{as} \quad \mathbf{x} \to \infty.$$

There is a conserved current that leads to the quantum number

$$N = \frac{1}{2\pi^2} \int \det (B_{\alpha i})\, d^3x.$$

There are localized solutions of the classical field equations for which $N = 1$ and which have properties that are classical analogs of particles with spin and isobaric spin $\frac{1}{2}$. Weak field solutions describe a triplet of massless mesons. Meson mass could be introduced by an additional term in the Lagrangian destroying its high symmetry, but this is not particularly relevant to the problem of the existence of particlelike structures, whose finite mass is governed classically by the constant K.

The field theory may be quantized in the canonical manner. The conjugate variables are (apart from a numerical factor) the local rotation operators (fourth components of spin currents)

$$I_\alpha(\mathbf{x}) = \frac{\epsilon}{4\pi^2}[(B_{\gamma i}B_{\gamma i} + K^2)\delta_{\alpha\beta} - (B_{\alpha i}B_{\beta i})]B_{\beta 0}$$
$$= G_{\alpha\beta}B_{\beta 0}$$

(i denotes the space indices, summed from 1 to 3, 0 is the real time index), with the fundamental commutation relations

$$[I_\alpha(\mathbf{x}'), U(\mathbf{x})] = -\tfrac{1}{2}\tau_\alpha U(\mathbf{x})\delta(\mathbf{x}' - \mathbf{x}),$$

from which

$$[I_\alpha(\mathbf{x}'), B_{\beta i}(\mathbf{x})] = i\epsilon_{\alpha\beta\gamma}B_{\gamma i}(\mathbf{x})\delta(\mathbf{x}' - \mathbf{x}) + \tfrac{1}{2}i\delta_{\alpha\beta}\partial_i\delta(\mathbf{x} - \mathbf{x}').$$

In terms of these variables I and B, the physical quantities are:

Hamiltonian:

$$H(I, B) = \int I_\alpha(\mathbf{x})G^{-1}_{\alpha\beta}I_\beta(\mathbf{x})\, d^3x + \frac{\epsilon}{8\pi^2}\int \{[(B_{\alpha i}B_{\alpha i})^2 - (B_{\alpha i}B_{\alpha j})(B_{\beta i}B_{\beta j})] + 2K^2(B_{\alpha i}B_{\alpha i})\}\, d^3x,$$

where $G_{\alpha\beta}(B)$ is the inertia tensor relating I_α with $B_{\beta 0}$, defined above;

momentum:

$$P_i = 2\int I_\alpha B_{\alpha i}\, d^3x;$$

angular momentum:

$$J_i = 2\epsilon_{ijk}\int I_\alpha x_j B_{\alpha k}\, d^3x.$$

There are two internal symmetries leading to the conserved quantities

$$H_\alpha = \int I_\alpha\, d^3x$$

and

$$\tilde{H}_\alpha = \int \tilde{I}_\alpha\, d^3x;$$

here $\tilde{I}_\alpha$ rotates U "on the right":

$$[\tilde{I}_\alpha(\mathbf{x}'), U(\mathbf{x})] = U(\tfrac{1}{2}\tau_\alpha)\delta(\mathbf{x}' - \mathbf{x}).$$

The $\tilde{I}_\alpha$ are an orthogonal transformation of the I_α, defined by

$$\tau_\alpha \tilde{I}_\alpha = U^+ \tau_\beta I_\beta U,$$

and commute with the I_α.

4. TOPOLOGY

A particular field distribution, such as $U(\mathbf{x})$, is a mapping from coordinate space onto the space of field values. In model A the latter is a circle; in model B it is a three-dimensional sphere S^3. In both cases "infinity" is mapped onto a fixed point; it is then possible to regard the coordinate space as a spherical rather than Euclidean space, so that in both models the mapping is from one sphere to another of the same dimension, with a fixed point of one mapping into a fixed point of the other. The quantum number N is then just the degree of the mapping, which is the only topological characteristic of the mapping of one sphere onto another.

The topological significance of N means that the mapping space (the space of all maps with a fixed pair of corresponding points) is composed of subspaces labeled by N, so that maps in one subspace are all deformable into one another but not into those of another subspace. The classical evolution of the system may be pictured as a deformation of the map, thus showing alternatively how N must be a classical constant of the motion.

A quantum state may be regarded as a functional of $U(\mathbf{x})$, i.e., a function defined on the mapping space. As N commutes with H, a state initially defined on just one subspace will remain on that subspace.

A map with $N = 1$ will have, by Brouwer's theorem, at least one point where $U = -1$ and $\det B_{\alpha i} > 0$; this point may be identified with the "center" of the particle. More generally there may be $m > 1$ points at which $U = -1$ and $\det B > 0$; then there will also be $(m - 1)$ points at which $U = -1$ and $\det B < 0$; this situation would be identified with a configuration containing m particles and $(m - 1)$ antiparticles.

The difficulty in trying to associate these states with fermions having spinor symmetries is that the latter could not be (proper single-valued) functions defined on the mapping space. In model A indeed, the operators associated with particles involved $F = \exp(\frac{1}{2}i\alpha)$, which is a double-valued function on the circle of values $\exp(i\alpha)$.

To define such functions properly, we must go to the covering space of the mapping space, which in the case of model A means using the angle α instead of the basic field variables $\cos\alpha$, $\sin\alpha$. It does not seem possible to give such a simple explicit representation in the three-dimensional case. However, it is known that the covering space for maps from S^3 to S^3 is double sheeted; in particular, a 2π rotation of a map of degree one is nontrivial, corresponding to a path from a point on one sheet to its mate on the other. This gives the possibility of defining spinor quantities on the covering space.

A map with $N = 1$ may be deformed so that the region where U is significantly different from 1 is very small; the field distribution is then generally called a "kink,"[2] e.g., a steep step of 2π in model A. Maps with $N = 2$ may be constructed by putting two kinks at different points, and so on. These could be added in either order giving the same maps, but the operations correspond to opposite points in the covering space, as it is known[2] that a path interchanging two kinks is nontrivial; it is therefore equivalent to a path corresponding with a 2π rotation, so that antisymmetry of the particle operators is naturally associated with change of sign on 2π rotation. Williams[3] showed functions of this type could be formed for the model B; these constructs, however, were not evidently associated with particle structures, and the object of this paper is to show how more realistic structures may be defined.

5. PARTICLE COORDINATES

If quantum particlelike states exist they will be characterized by quantum numbers such as momentum, spin, etc; we introduce some corresponding variables.

Evidently one such variable should be a position, denoted by $\mathbf{x}_0$ with a corresponding momentum $\mathbf{p}_0 = -i\nabla_0$. In model A this is all; there are just particles and antiparticles, steps propagating to the right and to the left, respectively; the operators FK depend only on the parameter x_0.

In model B there must be some internal spin variables. To find the appropriate choice, consider the static classical field solutions. The only ones that have been found are characterized by the central position $\mathbf{x}_0$ and by a proper orthogonal matrix $e_{\alpha i}$ that links the space and isospin directions. The classical theory, moreover, has many similarities to the semiclassical theory of the symmetrical pseudoscalar meson field[4] in which the spin operators $\sigma_i\tau_\alpha$ (of the usual meaning) become replaced by $e_{\alpha i}$. This suggests that we should introduce such a matrix as an internal coordinate. Conjugate to $e_{\alpha i}$ are the rotation operators t_α, s_i that rotate the corresponding coordinate labels and are related by the identity

$$s_i + e_{\alpha i}t_\alpha = 0,$$

so that $s_is_i = t_\alpha t_\alpha$, and s_i, t_α have eigenstates with the same total spin values (in terms of Eulerian angles these are just the solid harmonics).

A particle operator depending on $\mathbf{x}_0$ and $e_{\alpha i}$ would then describe not just one fermion or doublet of fermions but a family of isobars; the operator would be a generating function from which the different states could be projected. This seems both reasonable and quite desirable.

6. COLLECTIVE COORDINATES

We seek to introduce the particle variables as collective coordinates, identifying them with appropriate functions of the original field variables.

In a general notation the standard procedure is this. We want to find solutions of the Schrödinger equation

$$H\psi(q) = E\psi(q),$$

where q denotes the set of independent physical variables. We introduce auxiliary variables q_a and supplementary conditions A_k, acting only in the q_a space, with

$$[A_k, A_l] = 0, \quad [H, A_k] = 0,$$

so that the system

$$H\psi(q, q_a) = E\psi(q), \quad A_k\psi(q, q_a) = 0$$

is trivially equivalent to the original one. We then make a suitable canonical transformation in the complete space; denote this by S so that

$$H' = SHS^{-1}, \quad A' = SAS^{-1}, \quad \psi' = S\psi,$$

with new dynamical equations $H'\psi' = 0$, $A'\psi' = 0$.

Suppose now $H^* = H' + \sum A_k'\lambda_k$, where the λ_k are some multipliers, arbitrary functions of the variables, and that ψ^* is an eigenstate of H^*,

$$H^*\psi^* = E\psi^*;$$

then we can form an eigenstate of H' by projecting onto the subspace allowed by the supplementary conditions

$$\psi' = \delta(A')\psi^*,$$

for, since $[H', A'] = 0$,

$$H'\psi' = \delta(A')H'\psi^* = \delta(A')H^*\psi^* = E\psi'.$$

In the original description the projected state is

$$\psi = S^\dagger\delta(A')\psi^* = \delta(A)S^\dagger\psi^*.$$

This analysis is quite general; it is useful if we can find multipliers λ_k such that H^* has the form $H + H_a + H_{\text{int}}$, where H_a is a suitable Hamiltonian describing motion of the collective coordinates and H_{int} can be treated as a perturbation. There will be different states ψ^* that project into the same state; this redundancy arises because the same original state may be described either in the original variables or with the help of the auxiliary ones; this is not generally a serious problem when the collective states are clearly physically distinguishable.

In our application we can describe particle states either by kinks in the meson fields or by the auxiliary coordinates, and there will not be any difficulty provided that in the H^* description we avoid any kinks in the meson fields, as indeed we should like to do since they cannot be generated by perturbations.

The structure of the transformations has the following typical form. Suppose we want to identify q_a with q and that p_a, p are the corresponding conjugate variables. We start with

$$H(q, p)\psi = E\psi, \quad \delta(q_a)\psi = 0.$$

The first step is to make the supplementary condition identify q_a with q; this is achieved by $S_1 = \exp(-ip_a q)$, giving

$$H(q, p + p_a) \quad \text{and} \quad \delta(q_a - q)\psi = 0.$$

The second step is to replace q by $q + q_a$; $S_2 = \exp(iq_a p)$ gives

$$H(q + q_a, p_a) \quad \text{with} \quad \delta(-q)\psi = 0.$$

The complete transformation is

$$S = S_2 S_1 = \exp(iq_a p)\exp(-ip_a q).$$

7. APPLICATION TO MODEL A

The original variables are $\alpha(x)$, $\beta(x)$, with Hamiltonian

$$H = \int_{-\infty}^{\infty}\left[2\pi\beta^2 + \frac{1}{8\pi}\left(\frac{\partial\alpha}{\partial x}\right)^2 + \frac{K^2}{4\pi}(1 - \cos\alpha)\right]dx.$$

Introduce the auxiliary variables x_0, p_0 with the original supplementary condition

$$A\psi = p_0\psi = 0.$$

An appropriate transformation is

$$S = \exp\left[i\int_{x_0}^{\infty} 2\pi\beta\, dx\right]\exp[\tfrac{1}{2}i\alpha(x_0)],$$

which gives

$$H(\beta, \alpha) \to H(\beta - \tfrac{1}{2}\delta(x - x_0), \alpha + 2\pi\theta(x - x_0)) = H'$$

and

$$p_0 \to p_0 + 2\pi\beta(x_0) - \tfrac{1}{2}\left(\frac{\partial\alpha}{\partial x}\right)_{x_0} - \pi\delta(0) = p_0'.$$

Then

$$H' = H + \tfrac{1}{2}\alpha'(x_0) - 2\pi\beta(x_0) + \pi\delta(0),$$

and

$$H_+^* = H' + p_0' = H + p_0$$

exhibits a complete separation of the modes.

A similar transformation may be made with the factor $e^{\frac{1}{2}i\alpha}$ replaced by $e^{-\frac{1}{2}i\alpha}$; then

$$H_-^* = H' - p_0' = H - p_0.$$

These transformations introduce variables describing massless right- and left-going particles. The coordinate description of the particles can easily be generalized to a field description, and then, as described before, they can be used as a basis for a self-consistent description of a massive particle.

This analysis is somewhat unsatisfactory because of the singular nature of the operators S. Alternatively the following method may be followed which provides a better analog of that appropriate in three dimensions. The sharp step described by $\theta(x - x_0)$ is replaced by some suitable smooth approximation to it, $\theta_1(x - x_0)$, which has a small finite width

$$S = \exp\left(i2\pi \int \beta(x)\theta_1(x - x_0)\right) \times \exp\left(i\tfrac{1}{2}\int \alpha(x)\theta_1'(x - x_0)\right).$$

The only difference in the analysis is that now there is a contribution from the "potential energy" term in H, so that

$$H_+^* = H' + p_0' = H + p_0 + \frac{K^2}{4\pi}\int [\cos\alpha - \cos(\alpha + 2\pi\theta_1)]\, dx.$$

If we treat the last term in perturbation, the first approximation to it is a constant

$$\frac{K^2}{4\pi}\int (1 - \cos 2\pi\theta_1)\, dx = m$$

proportional to the width of the step. However, the form $(p_0 + m)$ is not a relativistic particle Hamiltonian, and so this does not give a suitable starting point. To describe a massive particle, we need a two-component description with a Hamiltonian of the form $(\alpha p_0 + \beta m)$, where α and β are anticommuting matrices.

The two-component description can be introduced quite naturally by considering together the two transformations that lead to H_+^* and H_-^*. Introduce an additional auxiliary variable which labels these, so that

$$S(x_0, \rho) = \exp\left[i2\pi\int \beta\theta_1\right] \exp\left[\rho i\tfrac{1}{2}\int \alpha\theta_1'\right],$$

where ρ can take on the values ± 1. This gives

$$H' + p_0'\rho = H + p_0\rho + m$$

in the first approximation. This is not a suitable form for H^*; however, we are free to add to it multiples of the supplementary conditions, and there must be a condition associated with the new variable ρ; we take this to have the form

$$(\xi + 1)\psi = 0$$

in the original description, where ξ anticommutes with ρ. Then we can add to H' the term $-(\xi' + 1)m$ giving

$$H^* = H + p_0\rho - m\xi'.$$

Here

$$\xi' = S\xi S^\dagger = \exp\left(-\rho i \int (\alpha + 2\pi\theta_1)\theta_1'\, dx\right) = \xi \exp\left(-\rho i\int \alpha\theta_1'\right)$$

since $\int \theta_1\theta_1'\, dx = \frac{1}{2}$. So in lowest order $H^* = H + p_0\rho + m\xi e^{-\rho i\int\alpha\theta'_1}$, which has a relativistic form [ξ, ρ are the analogs of $\gamma_4 \cdot \gamma_5$ in the usual Dirac equation notation].

The first approximation to an eigenstate of H^* has the form

$$\psi^* = \psi_0(\alpha)u_k e^{ikx_0},$$

where $(k\rho + m\xi)u_k = Eu_k$ and $\psi_0(\alpha)$ is the vacuum state of the meson field. In the original description the projected state is

$$\psi(\alpha) = \delta(p_0)\delta(\xi + 1)S^\dagger(x_0, \rho)\psi^* = \int [v^+ S^\dagger(x_0, \rho)u_k e^{ikx_0}]\, dx_0\psi_0,$$

where v satisfies $(\xi + 1)v = 0$. This shows how $S^\dagger$ is a generating function for particle creation.

To develop a proper self-consistent theory, the analysis must be generalized to a field description of the particles, introducing an auxiliary two-component field $\psi(x_0)$. The transformation is similar but leads to additional two-particle interaction terms which in a one-particle state give a mass contribution proportional to a^{-1}, just as in the classical theory the kinetic energy terms give a similar contribution to be balanced against the K^2a contribution of the potential energy.

8. THE THREE-DIMENSIONAL PROBLEM

In a three-dimensional problem, such as posed by model B, we seek a similar type of transformation with two factors, $S = S_2S_1$, where S_2, analogous to $\exp(i\int 2\pi\beta\theta)$, introduces a kink and S_1 identifies the auxiliary momentum and spin variables with field quantities.

It is easy to construct a suitable S_2. Since the field values U form a group (here SU_2), a transformation $U \to U' = U_0U$, where U_0 is any one-kink field configuration, suffices. It is natural and convenient to choose a U_0 that has a symmetry similar to that of the classical field solutions; a specially simple choice is

$$U_0(\mathbf{x}; \mathbf{x}_0, e_i) = \frac{a + i\tau_\alpha e_{\alpha i}(\mathbf{x} - \mathbf{x}_0)_i}{-a + i\tau_\alpha e_{\alpha i}(\mathbf{x} - \mathbf{x}_0)_i},$$

depending on the auxiliary variables $\mathbf{x}_0$ and $e_{\alpha i}$; a is a length measuring the size of the kink. The transformation is described by the operator

$$S_2 = \exp\left(i\int I_\alpha(\mathbf{x})\frac{e_{\alpha i}(\mathbf{x} - \mathbf{x}_0)_i}{r}\,\omega(r)\, d^3x\right),$$

where $r = |\mathbf{x} - \mathbf{x}_0|$ and $\omega(r)$ is a suitable angle function.

S_1, analogous to $\exp(\frac{1}{2}i\alpha)$ of model A, should identify the auxiliary momentum vector $\mathbf{p}_0$ with a suitable field expression, and the $e_{\alpha i}$ with a matrix characterizing the field orientation. It must also introduce the spinor character. For U_0, the fields $B_{\alpha i}$ near $\mathbf{x} = \mathbf{x}_0$ are proportional to $e_{\alpha i}$, suggesting that $e_{\alpha i}$ should be identified with the "direction" of $B_{\alpha i}$ near $\mathbf{x} = \mathbf{x}_0$. In general, however, $B_{\alpha i}$ will not be proportional to an orthogonal matrix, but it can be used to define one uniquely by noting that any matrix with positive determinant can be written uniquely as the product of a real symmetric positive definite matrix with a proper orthogonal matrix, that is,

$$B_{\alpha i} = c_{\alpha\beta} f_{\beta i},$$

with $c_{\alpha\beta}$ symmetric and $f_{\beta i}$ orthogonal; $f_{\alpha i}$ is also the orthogonal matrix such that $B_{\alpha i} f_{\alpha i}$ is maximal. We can then try to identify the auxiliary variables $e_{\alpha i}$ with the local values of the $f_{\alpha i}$.

It is convenient to describe the orthogonal matrices by corresponding quaternions $\hat{E}$, $\hat{F}$, which are uniquely defined, apart from sign, by

$$\tau_\alpha e_{\alpha i} = \hat{E}\tau_i \hat{E}^\dagger, \quad \tau_\alpha f_{\alpha i} = \hat{F}\tau_i \hat{F}^\dagger;$$

(that is, we are describing O_3 by its double-valued SU_2 representation). Then $\hat{F}$ is defined, apart from sign, in terms of $B_{\alpha i}$ by the eigenvalue equation

$$\tau_\alpha B_{\alpha i} \hat{F} \tau_i = \lambda \hat{F},$$

with the maximum value of λ. Alternatively we may write the 2×2 matrix $\hat{F}$ as a four-component vector ϕ, and the corresponding equation is

$$\sigma_i \tau_\alpha B_{\alpha i} \phi = -\lambda \phi,$$

where σ and τ are two sets of Pauli matrices acting on the four-component ϕ, which is therefore seen to be a spinor in both the spaces labeled by α and i.

The quantity ϕ satisfies the reality condition $\phi^* = \sigma_2 \tau_2 \phi$, in the usual conventions, which is consistent with the eigenvalue equation. It is this that enables us to define a spinor quantity uniquely apart from sign in terms of the fields B. In contrast, a spinor quantity cannot be defined by a single vector; for a vector $\mathbf{a}$, the equation

$$(\boldsymbol{\tau} \cdot \mathbf{a})\chi = a\chi$$

might seem to define χ, but its phase cannot be assigned in any way independently of the coordinates.

$\hat{F}$ is a double-valued function of the field, but it is single valued on the covering space; the sign of $\hat{F}$ must be fixed arbitrarily for one particular map taken as reference point, and is then uniquely defined for a map obtained by a given path from the reference one, *provided* that we avoid points for which $\det B = 0$.

We can then write $\hat{F} = e^{i\frac{1}{2}\tau_\alpha \theta_\alpha}$, where $\boldsymbol{\theta}$ is a function of the map in the covering space. The transformation $S = \exp(it_\alpha \theta_\alpha)$, where t_α are the rotation operators for the $e_{\alpha i}$, then gives

$$S\hat{E}S^{-1} = e^{-i\frac{1}{2}\tau_\alpha \theta_\alpha}\hat{E} = \hat{F}^\dagger \hat{E}.$$

So a supplementary condition $\hat{E} = 1$ is transformed into $\hat{E} = \hat{F}$ as wanted. At the same time, as will be seen below, S identifies the momentum $\mathbf{p}_0$ with an appropriate field quantity.

9. APPLICATION TO MODEL B

We apply these ideas to the model B defined in Sec. 3. The transformation considered is $S_+ = S_2 S_1$ where S_2 was defined above to introduce a kink so that $U(\mathbf{x}) \to U_0(\mathbf{x}; \mathbf{x}_0, e_{\alpha i}) U(\mathbf{x})$.

S_2 induces the following consequential changes on the field variables:

$$B_{\alpha i} \to B^0_{\alpha i} + W_{\alpha\beta} B_{\beta i} = W_{\alpha\beta}(-\tilde{B}^0_{\beta i} + B_{\beta i}),$$
$$I_\alpha \to W_{\alpha\beta} I_\beta, \quad \tilde{I}_\alpha \to \tilde{I}_\alpha.$$

Here $B^0_{\alpha i}$ is the field due to the source U_0

$$B^0_{\alpha i} = b(r) e_{\alpha j} R_{ji}(\mathbf{x} - \mathbf{x}_0), \quad r = |\mathbf{x} - \mathbf{x}_0|,$$

$b(r) = 2a(a^2 + r^2)^{-1}$, and R_{ji} is the orthogonal matrix

$$R_{ji}(\mathbf{x}) = [\delta_{ji}(a^2 - r^2) + 2x_i x_j + 2a\epsilon_{jik} x_k](a^2 + r^2)^{-1},$$

with

$$\tau_j R_{ji} = \hat{R}\tau_i \hat{R}^\dagger, \quad \hat{R} = \frac{a + i\tau_\alpha x_\alpha}{(a^2 + r^2)^{\frac{1}{2}}} = [-U_0(\mathbf{x}; 0, 1)]^{\frac{1}{2}}.$$

$W_{\alpha\beta}(\mathbf{x} - \mathbf{x}_0)$ is an orthogonal matrix describing the rotation of the fields induced by S_2:

$$W_{\alpha\beta} = e_{\alpha j}(R^2)_{ji} e_{\beta i}.$$

It is convenient to introduce also $\tilde{B}^0_{\alpha i}$ so that

$$B^0_{\alpha i} = -W_{\alpha\beta}\tilde{B}^0_{\beta i}, \quad \tilde{B}^0_{\alpha i} = -W_{\beta\alpha} B^0_{\beta i} = -b(r) e_{\alpha j} R_{ij};$$

the significance of $\tilde{B}$ will be discussed further below.

We define S_1 as in the last section, using for the $B_{\alpha i}$ a suitable average of the field values near the sources. For the particular shape of U_0 used in S_2, it appears best to define an average by

$$V_{\alpha i}(B; \mathbf{x}_0) = \frac{1}{2\pi^2} \int B_{\alpha j}(\mathbf{x}) R_{ij}(\mathbf{x} - \mathbf{x}_0) b^2(r)\, d^3x;$$

then, in particular, when $B = B^0$,

$$V_{\alpha i} = e_{\alpha i}.$$

So $S_1 = \exp(it_\alpha \theta_\alpha)$, where $\hat{F} = e^{i\frac{1}{2}\tau_\alpha \theta_\alpha}$ satisfies

$$\tau_\alpha V_{\alpha i} \hat{F} \tau_i = \lambda \hat{F}, \quad \text{with maximum } \lambda.$$

The transformation S_1 makes $\hat{E} \rightarrow \hat{F}^\dagger \hat{E}$, as shown above, and rotates the vector t_α; it leaves unchanged the vector s_i and the fields $B_{\alpha i}$. The field rotation operators $I_\alpha(\mathbf{x})$ transform to

$$I_\alpha(\mathbf{x}) + A_{\alpha\beta}(\mathbf{x}, B; \mathbf{x}_0)t_\alpha,$$

where

$$A_{\alpha\beta}t_\beta = \int_0^1 e^{\lambda i t\cdot\theta_\gamma/\gamma} e^{-\lambda i t\cdot\theta}\, d\lambda[i\theta_\gamma(B, \mathbf{x}_0), I_\alpha(\mathbf{x})];$$

the commutator at the end is a function only of the B and $\mathbf{x}$, $\mathbf{x}_0$.

In particular $\int A_{\alpha\beta}(\mathbf{x}, B, \mathbf{x}_0)\, d^3x = \delta_{\alpha\beta}$. This follows from a simple argument. It is clear from the definition of $\hat{F}$ that it must behave as a spinor for rotation of the $B_{\alpha i}$ induced by the total spin operator

$$H_\alpha = \int I_\alpha(x)\, d^3x,$$

that is,

$$[H_\alpha, \hat{F}] = -\tfrac{1}{2}\tau_\alpha \hat{F} \quad \text{or} \quad \hat{F}H_\alpha F^\dagger = H_\alpha + \tfrac{1}{2}\tau_\alpha.$$

Then

$$S_1 H_\alpha S_1^\dagger = H_\alpha + t_\alpha$$

since, if we expand the left side in a series of commutators, all terms after the second involve only the mutual commutators of the t_α, which are similar to those of the $\frac{1}{2}\tau_\alpha$.

The total effect of the transformation S_2S_1 on the field variables is then

$$B'_{\alpha i} = SB_{\alpha i}S^{-1} = B^0_{\alpha i} + W_{\alpha\beta}B_{\beta i} = W_{\alpha\beta}(-\tilde{B}^0_{\beta i} + B_{\beta i}),$$

$$I'_\alpha = SI_\alpha S^{-1} = W_{\alpha\beta}I_\beta + A_{\alpha\beta}(\mathbf{x}, B^0 + WB, \mathbf{x}_0)\left[t_\beta + \int(\delta_{\beta\gamma} - W_{\beta\gamma})I_\gamma\right].$$

The transformation of t_β by S_2 exhibited in the last factor is derived by noting that the total spin $t_\beta + H_\beta = t_\beta + \int I_\beta\, d^3x$ commutes with S_2.

For the auxiliary variables the transformation of $\mathbf{p}_0$ is most easily found by noting that S is translationally invariant and must therefore commute with the total linear momentum $(\mathbf{p}_0)_i + 2\int I_\alpha B_{\alpha i}\, d^3x$.

$\hat{E}$ is transformed into $\hat{F}^\dagger(B^0 + WB, \mathbf{x}_0)\hat{E}$; but $\hat{F}(B^0, \mathbf{x}_0) = \hat{E}$ so that the transformed supplementary condition $\hat{E}' = I$ is identically satisfied in its leading term (compare the end of Sec. 6).

The consequences of this transformation combined with the supplementary conditions

$$\mathbf{p}_0\psi = 0, \quad (\hat{E} - I)\psi = 0$$

are considered in the following sections.

10. MOMENTA AND SPINS

As mentioned above, $p'_{0i} + 2\int I'_\alpha B'_{\alpha i}\, d^3x = p_{0i} + 2\int I_\alpha B_{\alpha i}\, d^3x$, so that the condition $\mathbf{p}'_0 = 0$ gives

$$p_{0i} = 2\int [I'_\alpha B'_{\alpha i} - I_\alpha B_{\alpha i}]\, d^3x.$$

The term independent of the new fields is then

$$2\int A_{\alpha\beta}(\mathbf{x}, B^0, \mathbf{x}_0)B^0_{\alpha i}\, d^3x t_\beta.$$

To evaluate $A_{\alpha\beta}$, we calculate the commutator of $I_\alpha(\mathbf{x})$ with $\hat{F}(\mathbf{x}_0)$ from the eigenvalue equations, using the selected definition of the average $V_{\alpha i}$, and express the result in the form

$$[I_\alpha(\mathbf{x}), \hat{F}(\mathbf{x}_0)] = -A_{\alpha\beta}\tfrac{1}{2}\tau_\beta\hat{F}.$$

Putting $B = B^0$ in the formula, we obtain the simple result

$$A^0_{\alpha\beta} = A_{\alpha\beta}(\mathbf{x}, B^0, \mathbf{x}_0) = (1/2\pi^2)b^3(r)\delta_{\alpha\beta}, \quad r = |\mathbf{x} - \mathbf{x}_0|.$$

So the leading term for $\mathbf{p}_{0i}$ is

$$\frac{1}{\pi^2}\int e_{\alpha j}R_{ji}b^4(r)\, d^3x t_\alpha = \frac{1}{a}e_{\alpha i}t_\alpha = -\frac{1}{a}s_i.$$

The total angular momentum of the system is, originally,

$$J_i = 2\epsilon_{ijk}\int I_\alpha x_j B_{\alpha k}\, d^3x.$$

To see how this transforms, it is convenient to use the supplementary conditions to begin instead with

$$J_i + L_i + s_i + t_i,$$

where $\mathbf{L} = \mathbf{x}_0 \times \mathbf{p}_0$ is the auxiliary orbital angular momentum. The additional terms are zero because of the conditions $\mathbf{p}_0 = 0$ and $e_{\alpha i} = \delta_{\alpha i}$. By a similar argument to that used for H_α, we see that

$$[J_i + L_i, \hat{F}] = \hat{F}\tfrac{1}{2}\tau_i$$

and so, as before, $S_1(J_i + L_i + t_i)S_1^\dagger = J_i + L_i$. Also S_1 commutes with s_i, and so does S_2 with $J_i + L_i + s_i$ since it is rotationally symmetric; thus the transformed angular momentum is

$$S(J_i + L_i + s_i + t_i)S^\dagger = J_i + L_i + s_i.$$

The internal symmetry operators are $H'_\alpha = H_\alpha + t_\alpha$, $\tilde{H}'_\alpha = H_\alpha$.

11. HAMILTONIAN

The transformed Hamiltonian is

$$H'(I, B) = H(I', B').$$

Consider the terms independent of the new fields.

These are

$$\int A^0_{\alpha\beta}t_\beta(G_0^{-1})_{\alpha\gamma}A^0_{\gamma\delta}t_\delta\, d^3x + \frac{\epsilon}{8\pi i}\int [(B^0_{\alpha i}B^0_{\alpha i})^2 - (B^0_{\alpha i}B^0_{\alpha j})(B^0_{\beta i}B^0_{\beta j}) + 2K^2(B^0_{\alpha i}B^0_{\alpha i})]\, d^3x,$$

where

$$(G_0)_{\alpha\beta} = (\epsilon/4\pi^2)[(B^0_{\gamma i}B^0_{\gamma i} + K^2)\delta_{\alpha\beta} - (B^0_{\alpha i}B^0_{\beta i})].$$

Now $B^0_{\alpha i}B^0_{\beta i} = b^2(r)\delta_{\alpha\beta}$, and so

$$G^0_{\alpha\beta} = (\epsilon/4\pi^2)(2b^2(r) + K^2)\delta_{\alpha\beta},$$

while, as above,

$$A^0_{\alpha\beta} = (1/2\pi^2)b^3(r)\delta_{\alpha\beta}.$$

So the leading terms in H' are

$$\left(\frac{1}{2\pi^2}\int\frac{b^6(r)}{b^2(r)+\frac{1}{2}K^2}\, d^3x\right)t_\beta t_\beta + \frac{\epsilon}{4\pi^2}\int(3b^4 + K^2b^2)\, d^3x.$$

For comparison with the usual field theories, it may be convenient to introduce a point particle as auxiliary (although this may not give the best scheme of approximation), i.e., to regard a as very small. Then we may neglect K in the integrals to give

$$(1/\epsilon a)t_\alpha t_\alpha + 3\epsilon/2a.$$

Suppose also that we are only concerned with spin-$\frac{1}{2}$ states; then we should want H^* to contain a term $\pm p_{0i}(2s_i)$. If we took $H^* = H' - p'_{0i}2s_i$, the leading terms in H^* would be

$$H + (1/\epsilon a)t_\alpha t_\alpha - (2/a)s_is_i + 3\epsilon/2a.$$

If $\epsilon = \frac{1}{2}$, the first two terms cancel, leaving a mass term from the potential energy, whose divergence should be largely cancelled by the self-energy corrections.

However, the terms $(-p'_{0i}2s_i + m)$ are not a relativistic particle Hamiltonian; to achieve that, we must introduce an extra variable like ρ for model A, associated with the parity symmetry.

12. PARITY

Consider the symmetry properties of the system under inversion. The original meson field theory is obviously invariant for the transformations

$$R\colon U(x)\to U(-\mathbf{x}) \quad\text{and}\quad C\colon U(\mathbf{x})\to U^\dagger(\mathbf{x}).$$

C is a charge conjugation operation and R appears to correspond to "CP." In connection with C it is convenient to introduce the charge conjugate fields $\tilde{B}$ defined by $\partial_\mu U^\dagger = i\tau_\alpha\tilde{B}_{\alpha\mu}U^\dagger$ so that $\tilde{B}_{\alpha i} = -g_{\alpha\beta}B_{\beta i}$, where $U\tau_\alpha U^\dagger = g_{\alpha\beta}\tau_\beta$; corresponding to them are the conjugates $\tilde{I}_\alpha = -g_{\alpha\beta}I_\beta$ (see also Sec. 3). Then C gives $B\to\tilde{B}$, $I\to\tilde{I}$, in particular the particle density $\det B\to|\tilde{B}| = -\det B$, since g is proper orthogonal, and so, in the suggested interpretation, will change particles into antiparticles.

The space inversion defined by R makes $B_{\alpha i}(\mathbf{x})\to -B_{\alpha i}(-\mathbf{x})$ and therefore also changes the sign of $\det B$, suggesting that it should be regarded as a "CP" operation.

The combined operation RC appears to be the proper parity operation:

$$RC\colon U(\mathbf{x})\to U^\dagger(-\mathbf{x}),\quad B_{\alpha i}(\mathbf{x})\to -\tilde{B}_{\alpha i}(-\mathbf{x}),$$
$$I_\alpha(\mathbf{x})\to\tilde{I}_\alpha(-\mathbf{x}).$$

Consider the effect of RC on the transformation S, supposing that R also transforms $\mathbf{x}_0$ to $-\mathbf{x}_0$. In S_2 the effect is to change I_α into $\tilde{I}_\alpha$ and to change the sign of the factor $(\mathbf{x}-\mathbf{x}_0)$; it is the operator which makes $U^\dagger(\mathbf{x})\to U_0^\dagger(\mathbf{x},\mathbf{x}_0,e)U^\dagger$, i.e. $U\to UU_0$. It is obviously as reasonable to introduce a kink by multiplying U on the right as on the left, and we would expect the two operations to enter symmetrically into a theory.

On S_1 the effect is to transform it to $\tilde{S}_1$, defined similarly in terms of a local field average, which becomes

$$\tilde{V}_{\alpha i} = \frac{1}{2\pi^2}\int[-\tilde{B}_{\alpha j}(\mathbf{x})]R_{ji}(\mathbf{x}-\mathbf{x}_0)b^2(r)\, d^3x.$$

The new transformation $\tilde{S}$ then gives $\hat{B}'_{\alpha i} = \tilde{B}^0_{\alpha i} + W_{\beta\alpha}\tilde{B}_{\beta i}$, and $\tilde{V}_{\alpha i}$ evaluated for $\tilde{B} = \tilde{B}^0$ is again equal to $e_{\alpha i}$.

In the transformed Hamiltonian, the leading additional terms are the same as with S. The leading term in the momentum, however, changes sign, as it should for a parity transformation, so that with this S we should naturally consider $H^*_- = H' + p'_{0i}(2s_i)$, of which the leading terms are $H + p_{0i}(2s_i) + m$.

13. DIRAC HAMILTONIAN

We can treat these two equally good types of transformation in a symmetrical manner by introducing extra variables ρ, ξ just as for model A; these are anticommuting matrices with eigenvalues ± 1. It is natural to use a representation in which ρ is diagonal,

$$\rho = \begin{pmatrix}1 & 0\\ 0 & -1\end{pmatrix},\quad \xi = \begin{pmatrix}0 & 1\\ 1 & 0\end{pmatrix}.$$

In this representation the transformations are combined as $\binom{S\,0}{0\,\tilde{S}}$, which we now simply write as S. We introduce the supplementary condition $(\xi - 1)\psi = 0$.

ξ transforms into

$$\xi' = S\xi S^\dagger = \begin{pmatrix} 0 & S\tilde{S}^\dagger \\ \tilde{S}S^\dagger & 0 \end{pmatrix}.$$

$\tilde{S}S^\dagger$ differs from unity because $U_0 U \neq UU_0$, but the difference is small when the transformed fields can be regarded as small: This can easily be verified by detailed study of the operator.

We therefore consider $H^* = H' - p'_{0i}(2s_i\rho) + (\xi' - 1)m$ so that, for a suitable choice of m,

$$H^* = H - p_{0i}(2s_i\rho) + m\xi + \text{interaction terms}.$$

For states with spin $\frac{1}{2}$, the expression $[-p_{0i}(2s_i\rho) + m\xi]$ has exactly the structure of the Dirac Hamiltonian $(\alpha \cdot \mathbf{p} + \beta m)$.

We have thus shown that a Dirac Hamiltonian can be introduced in a quite natural way to describe the particlelike modes of the meson field.

14. COMMENTS

To investigate whether this analysis does give a starting point from which valid solutions describing particles can be constructed, the transformations must first be written with auxiliary fields for the particles rather than coordinates; then the interaction terms can be investigated by standard field theoretic methods, and it is possible that, for some suitable choice of kink U_0, perturbation theory might lead to a solution free from divergences.

The model should also contain isobaric states of higher spin; the choice made for H^* is evidently inappropriate for these as it is certainly not then relativistic.

[1] T. H. Skyrme, Proc. Roy. Soc. (London) **A260**, 127 (1961).
[2] D. Finkelstein, J. Math. Phys. **7**, 1218 (1966).
[3] J. G. Williams, J. Math. Phys. **11**, 2611 (1970).
[4] W. Pauli, *Meson Theory of Nuclear Forces* (Interscience, New York, 1946).

IV. THE REDISCOVERY OF SKYRMIONS

For nearly two decades following Skyrme's work, not much was done on skyrmions, although Finkelstein and Rubinstein (D. Finkelstein and J. Rubinstein, *J. Math. Phys.* **9** (1968) 1762) showed that the skyrmion could be quantized as either a fermion or boson. (In one of the articles which follow, Witten shows that for an odd number of colors, it *must* be quantized as a fermion.) As sketched in Sanyuk's article, N. Pak and H. Tze took up the study of the skyrmion in 1979, at the classical level. However, the "real explosion of interest in the Skyrmion features was aroused after the series of papers by A. P. Balachandran *et al.* and E. Witten *et al.*" References are given by Sanyuk. Although Balachandran's work came first, and strongly motivated the upsurge, I have included here three papers of Witten, and the paper by Adkins, Nappi and Witten, also one by Adkins and Nappi, since they form a nice pedagogical unit. It should be remarked that Fadeev kept the idea of solitons alive in the years between 1960 and 1980, as can be seen from the Sanyuk and Witten papers, and doubtless Witten's interest was stimulated by Fadeev's lectures at Harvard. This section ends with a general review, as of 1986, of the Skyrme model by Ismail Zahed, who wrote most of the review, and myself. I hope that the reader will not be overly disturbed by the editor exercising his prerogative to include one of his own papers.

Witten's "Baryons in the $1/N$ expansion", on page 235 is based on 't Hooft's work, Witten's Ref. 1. 't Hooft showed that there was a systematic expansion of caricature versions of QCD in power of $1/N_c$, where N_c is the number of colors. Of course, it would be nice if, already for $N_c = 3$, the number of colors in the real world, this expansion parameter were small enough so that the expansion gave quantitatively useful results, but this is by no means necessary, since the behavior of the theory for large N_c tells us a lot, anyway. Bosonization of fermion operators is employed in many branches of physics. Thinking of the quark and antiquark put together to form a meson, then this bosonization becomes exact as $N_c \to \infty$. This is the familiar situation for nuclear physicists that as the degeneracy Ω goes to infinity, the RPA approximation — which replaces bilinear forms of fermion operators by bosons — becomes exact.

In the large N_c limit, then, QCD goes over to a meson theory, "a weakly coupled field theory of mesons. It is a theory of effective local fields with effective local interactions, of order $1/N_c$", from E. Witten, "Baryons in the $1/N_c$ expansion", p. 113.

The nucleon mass is of order N_c, and becomes heavy as N_c becomes large. This follows simply from the fact that N_c quarks, which can be thought of as constituent quarks for our purpose, make up the nucleon. How can a heavy nucleon be made up out of a weakly coupled theory of (massless) mesons? Most reasonably, it can be made as a soliton. As a soliton, its mass can be written as proportional to $1/(1/N_c)$, analogous to the Polyakov–'t Hooft monopole, the mass of which is of order $1/\alpha$, where α is the strong coupling constant.

Having come to the conclusion that in large N_c, QCD can be bosonized, resulting in a weakly coupled meson theory, and that the nucleon emerges from this theory as a soliton, it was natural for Witten to look at the Skyrme theory.

In "Global aspects of current algebra", Witten discusses the need for adding the Wess–Zumino term in the $SU(3) \times SU(3)$ invariant Lagrangian. By an extension to five dimensions and in analogy with the Dirac monopole, the Wess–Zumino term is quantized in terms of the number of colors N_c. The Wess–Zumino term is a product of the Skyrme baryon density times two-dimensional disc, the latter comprised of the time and the additional dimension introduced in the extension. The quantization shows that for odd N_c, the baryon must be quantized as a fermion, for even N_c as a boson. The Wess–Zumino term is gauged, so as to incorporate electromagnetic interaction into the anomaly.

Although formally there is no Wess–Zumino term in $SU(2) \times SU(2)$, the Skyrme baryon number density remains.

In "Current algebra, baryons and quark confinement", Witten returns to the Skyrme theory, elucidating in what sense baryons are solitons. Skyrme's Jacobian for the mapping of the three pion fields onto the three spatial coordinates becomes the winding number in $\pi_3(SU(3))$. The excitation spectrum of the nucleon, for N_c odd, is found to be

$$I = J = \frac{1}{2}, \frac{3}{2}, \frac{5}{2}, \frac{7}{2}, \cdots$$

although, with three flavors, it cuts off with $I = J = \frac{3}{2}$, the isobar. Quantization of the soliton is similar to that of an isotropic rigid rotor, bringing us back to the Bohr–Mottelson quantization in nuclear physics.

In a note added in proof, Witten discusses how the soliton quantum numbers can be calculated if there are three flavors, thinking of adding the strange quark to the up and down ones. This extension to three flavors has been carried out recently in many different ways, most of them interesting; but the discussion of them goes beyond the extent of this book.

In "Static properties of nucleons in the Skyrme model", by Adkins, Nappi, and Witten, one can say that much of the Skyrme program was carried out to completion. Had Tony had any encouragement in his earlier efforts, he might have carried through most of this. The Skyrme hedgehog equation is solved in terms of a profile function $F(r)$. Quantization by slow rotation picks out the nucleon and isobar. Currents, charge radii, and magnetic moments are calculated. Because of the slow rotation, an isoscalar component in the magnetic moments arises, essentially the magnetic moment of a slowly rotating charged rigid body. The Goldberger–Treiman relation is shown to hold (it is a tautology for a chiral invariant theory), and the large distance behavior of the radial profile function is shown to determine the g_A in this relation.

I have included a paper in this series, "Stabilization of chiral solitons via vector mesons" by Adkins and Nappi, chiefly because it explains a lot of the physics of soliton stabilization in terms of a picture similar to the one Lorentz invented for his extended electron. In this paper, the ω-meson is used to stabilize the skyrmion, as we describe below.

Skyrme had to introduce a fourth-order term in his equation in order to stabilize the soliton. It was understood later that this fourth-order term originated from the ρ-meson. Although one might say that the ρ-meson exchange between the different pieces of the finite size soliton gives the repulsion which stabilizes the skyrmion, the physical picture is somewhat lost because the Skyrme fourth-order term results when the zero range approximation is made in this exchange. This is made clear in the Adkins and Nappi paper, where the repulsion results from ω-meson exchange, and the zero range approximation is not made.

There is some similarity with the turn of the century Lorentz theory of the electron. Lorentz introduced a finite size electron, also rods to hold the various pieces of the electron together. He was never able to make a covariant description of his electron however. In the case of the skyrmion, the meson field provides attraction, not repulsion as the electric field gives in the case of the electron, and ω-meson exchange between different pieces of the soliton provides the repulsion which keeps the soliton from collapsing.

The final paper, "The Skyrme model", by Zahed and Brown gives a review of the status and applications of the Skyrme model as of 1986.

Nuclear Physics B160 (1979) 57–115
© North-Holland Publishing Company

BARYONS IN THE $1/N$ EXPANSION

Edward WITTEN *
Lyman Laboratory of Physics, Harvard University, Cambridge, Massachusetts 02138

Received 3 April 1979

In this paper the existing results concerning mesons and glue states in the large-N limit of QCD are reviewed, and it is shown how to fit baryons into this picture.

1. Introduction

By now there are strong reasons to believe that the SU(3) gauge theory of quarks and gluons, quantum chromodynamics, is the theory of hadronic physics. Yet many of the essential properties that the theory is presumed to have, including confinement, dynamical mass generation, and chiral symmetry breaking, are only poorly understood. And apart from the low-lying bound states of heavy quarks, which we believe can be described by a non-relativistic Schrödinger equation, we are unable to derive from the basic theory even the grossest features of the particle spectrum, or of traditional strong interaction phenomenology – particle decay rates and scattering amplitudes.

Because of the complexity of phenomena which this theory describes – it includes, for instance, all of nuclear physics as a special case – we cannot even dream of solving the SU(3) gauge theory exactly. The exact S-matrix of this theory is far more complicated than anything we can write down, much less calculate. Therefore it is necessary to find some sort of approximation scheme.

A good approximation scheme is probably possible only if there is an expansion parameter. What possible expansion parameter does QCD possess?

The ordinary coupling constant g is not really a free parameter in QCD, because in view of the renormalization group, it is absorbed into defining the scale of masses. This is one of the most important facts that we known about QCD, and it is, of course, the fact that makes the theory difficult. The quark bare masses are important parameters in describing heavy quark bound states, which we believe we understand comparatively well. But the up, down, and strange quark bare masses are very small, and in thinking about the conspicuous unsolved problems of QCD, the quark bare masses can be considered as negligibly small or zero.

* Research supported in part by the National Science Foundation under Grant No. PHY77-22864.

The theory that we would like to understand – the SU(3) gauge theory with very small or zero quark bare masses – has no obvious free parameter that could be used as an expansion parameter. It is, at least to appearances, a unique, zero parameter theory.

There being no obvious free parameter, we must find a free parameter that is not obvious, if we expect to make progress. In fact, as was originally pointed out by 't Hooft, [1] there is in QCD a not-so-obvious, hidden candidate for a possible expansion parameter.

't Hooft suggested that one should generalize QCD from three colors and an SU(3) gauge group to N colors and an SU(N) gauge group. The hope is that it may be possible to solve the theory in the large N limit, and that the $N = 3$ theory may be qualitatively and quantitatively close to the large N limit.

As will be discussed at length below, QCD simplifies as N becomes large, and there exists a systematic expansion in powers of $1/N$. In various ways, to be discussed later, this expansion is reminiscent of known phenomenology of hadron physics, indicating that an expansion in powers of $1/N$ may be a good approximation at $1/N = 1/3$.

The simplification that occurs in QCD for large N is, as we will see, very striking from a conceptual and qualitative point of view, but so far has not provided the basis for a quantitative approximation scheme. However, the qualitative results that are available seem to be quite sufficient encouragement to justify considering this subject seriously.

Actually, these qualitative results have in the past been limited to mesons and glue states (by a glue state is meant a color singlet, quarkless bound state of gluons). 't Hooft [1] and subsequent authors [2] (who wrote in some cases about two dimensions, but whose qualitative results, as we will see, are also valid in four dimensions) determined the qualitative nature of the large N limit for mesons and glue states in QCD. Perhaps the most interesting results are the following. In Yang-Mills theory at $N = \infty$ the mesons and glue states are free, stable, and non-interacting. Meson decay amplitudes are of order $1/\sqrt{N}$, and meson-meson elastic scattering amplitudes are of order $1/N$ (for glue states there are analogous results). Most strikingly, these elastic amplitudes are given (as in Regge phenomenology) by a sum of tree diagrams involving the exchange, not of quarks and gluons, but of physical mesons. Furthermore, Zweig's rule is exact at $N = \infty$, and at $N = \infty$ mesons are pure $q\bar{q}$ states (rather than, for instance, $qq\bar{q}\bar{q}$).

In sections two, three and four, the derivations of these and other results will be reviewed, together with a discussion of the phenomenological status of the $1/N$ expansion.

The purpose of this paper is to determine for baryons the qualitative nature of the large-N limit, in the same sense that the qualitative nature of this limit is already known for mesons and glue states.

2. Feynman diagrams for large N

In this section we will review the combinatorics of Feynman diagrams in the large-N limit, and in the next section we will discuss the physical results that can be deduced from this combinatorics. Most of this material can be found in refs. [1] and [2], and except for incidental details, none of it is novel. (For discussions of the $1/N$ expansion in tractable models, see ref. [3]).

In the $1/N$ expansion, one considers QCD with N colors and an SU(N) gauge group, in the large N limit. For large N there are many colors and therefore many possible intermediate states in Feynman diagrams, so that the sums over intermediate states give rise to large combinatoric factors. These combinatoric factors are responsible for the nature of the large N limit.

To be specific, the gluon field is an $N \times N$ matrix $A^i_{\mu j}$, which has N^2 components. (Actually, the matrix must be traceless, and so has only $N^2 - 1$ components, but for large N the difference between N^2 and $N^2 - 1$ can be neglected, and as 't Hooft showed, the constraint of tracelessness plays no role for large N.)

In contrast, the quark and antiquark fields q^i and $\bar{q}_i$ each have N components. Much of what follows reflects the fact that for large N there are many quark and antiquark states (N) but even more gluon states (N^2).

Let us consider now a typical low order Feynman diagram, such as the gluon contribution to the one-loop gluon vacuum polarization (fig. 1). It is fairly easy to see (and will be explained below) that even after the color quantum numbers of the initial and final states are specified, there are still N possibilities for the quantum numbers of the intermediate state gluons. As a result, this diagram receives a combinatoric factor of N.

On the other hand, there is also a factor of coupling at each of the two interaction vertices in fig. 1. If we want the one-loop gluon vacuum polarization to have a smooth limit for large N, we must choose the coupling constant to be $g/\sqrt{N}$, where g is to be held fixed as N becomes large. This is indicated in fig. 2 (and differs slightly from the convention followed in much of the previous literature). With this choice, the two vertex factors of $g/\sqrt{N}$ in fig. 1 combine with the combinatoric factor of N to give a smooth large N behavior: $(g/\sqrt{N})^2 N = g^2$, independent of N. Thus, the vanishing for large N of the coupling constant cancels the divergence of the combinatoric factor, to produce a smooth large N limit.

Since, however, we have normalized the coupling constant to be of order $1/\sqrt{N}$, it is clear that Feynman diagrams will have factors of $1/\sqrt{N}$ at each vertex. In order to survive as $N \to \infty$, a Feynman diagram must have combinatoric factors large enough to

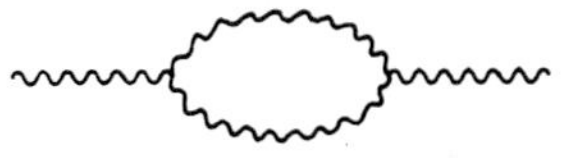

Fig. 1. The lowest order gluon vacuum polarization.

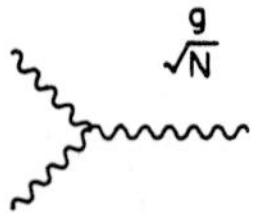

Fig. 2. The gluon-gluon-gluon interaction vertex.

compensate for the vertex factors. It turns out that a certain class of Feynman diagrams, the so-called planar diagrams, have combinatoric factors large enough to just cancel the vertex factors. All other diagrams have smaller combinatoric factors and vanish for large N. The large N limit is therefore given by the sum of the planar diagrams.

Let us now return and discuss the combinatoric analysis leading to this conclusion. It is made particularly easy by a technique introduced in ref. [1]. The gluon field $A^i_{\mu j}$ has one upper index like the quark field q^i and one lower index like the antiquark field $\bar{q}_i$. For keeping track of color quantum numbers (and for this purpose only) one may think of the gluon as a quark-antiquark combination, $A_\mu{}^i{}_j \sim q^i\bar{q}_j$. This suggests that, just as we frequently represent a quark or antiquark in a Feynman diagram as a single line with an arrow, the direction of the arrow distinguishing quark from antiquark, so we should represent the gluon as a double line, one line for the quark and one line for the antiquark of the q$\bar{\text{q}}$ pair which has the same SU(N) quantum numbers as the gluon (fig. 3).

Now, we must learn to write interaction vertices in the double line notation. The trilinear vertex of Yang-Mills theory, for example, is $\mathrm{Tr} A_\mu A_\nu \partial_\mu A_\nu$, which can be written more explicitly, in terms of indices, as

$$A_\mu{}^i{}_j A_\nu{}^j{}_k \partial_\mu A_\nu{}^k{}_i \,. \tag{1}$$

In terms of the double line notation, this can be written as in fig. 4. In other words, notice that in eq. (1) the lower index of the first gluon field is contracted with the upper index of the second gluon field; the lower index of the second gluon field is contracted with the upper index of the third; and the lower index of the third field is contracted with the upper index of the first. This is indicated schematically in fig. 4. in the fact that the outgoing (antiquark) index of each gluon field is contracted with the incoming (quark) index of the next gluon field (the next one in the

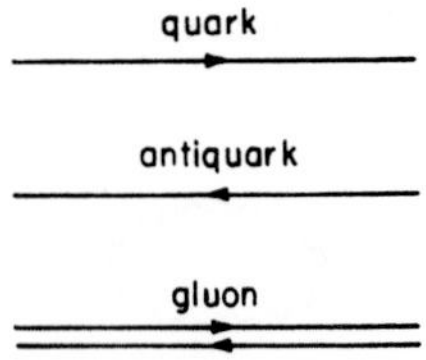

Fig. 3. The double-line notation for gluons.

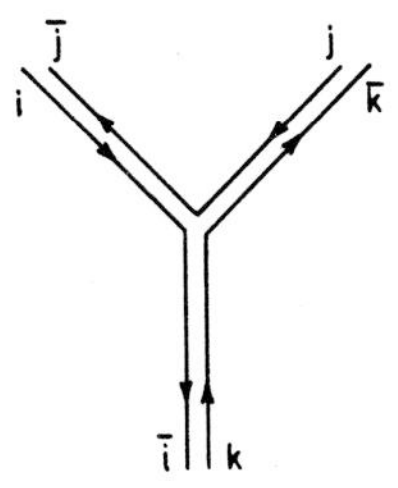

Fig. 4. The three-gluon vertex, in the double-line notation.

clockwise direction). Color conservation in fig. 4 is simply expressed in the fact that each color line that enters the diagram also leaves it.

Likewise, in fig. 5 the gluon vertex, $\bar{q}_i \gamma^\mu q^j A_\mu{}^i{}_j$, and the quartic gluon vertex, $\mathrm{Tr} A_\mu A_\nu A_\mu A_\nu = A_\mu{}^i{}_j A_\nu{}^j{}_k A_\mu{}^k{}_l A_\nu{}^l{}_i$, are drawn in the double line notation. (In the quartic gluon vertex there is a second term, with the last two Lorentz indices exchanged but the same color structure.)

We are now in a position to understand why in fig. 1 there is a combinatoric factor of N. In fig. 6 the diagram of fig. 1 has been redrawn in the double line language. In fig. 6 the color index lines at the edge of the diagram are contracted with those of the initial and final states. They carry quantum numbers (shown as i and $\bar{j}$ in the diagram) that are fixed once the initial and final states are specified.

However, at the center of the figure, there is a closed color line (shown in the figure as having color k) that is contracted only with itself. The value of k, the color running around this loop, is unspecified even when the initial and final states are given, and the sum over k gives a factor of N. This is the combinatoric factor of N associated with the diagram of figs. 1 and 6.

Using the double line notation, it is not difficult to determine whether a given diagram survives in the large N limit. For example, in fig. 7 there is drawn, both in the ordinary notation and in the double line notation, a two-loop contribution to the gluon propagator. This diagram has four interaction vertices, each contributing a factor of $1/\sqrt{N}$, but it has two closed color loops that are self-contracted, each contributing a factor of N. Altogether, the diagram is of order $(1/\sqrt{N})^4 N^2 = 1$, and so survives in the large N limit.

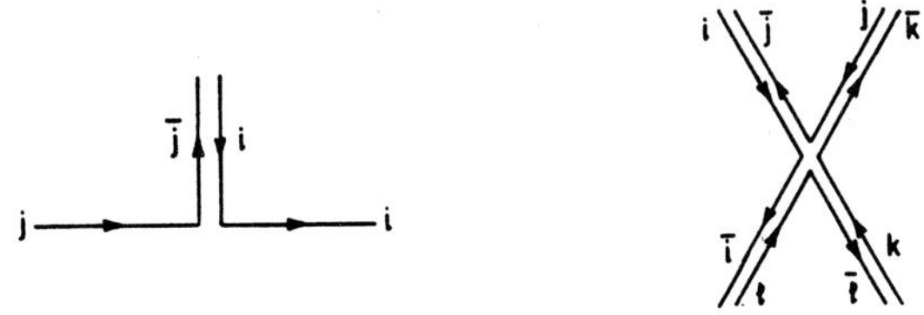

Fig. 5. The quark-quark-gluon and four-gluon vertices.

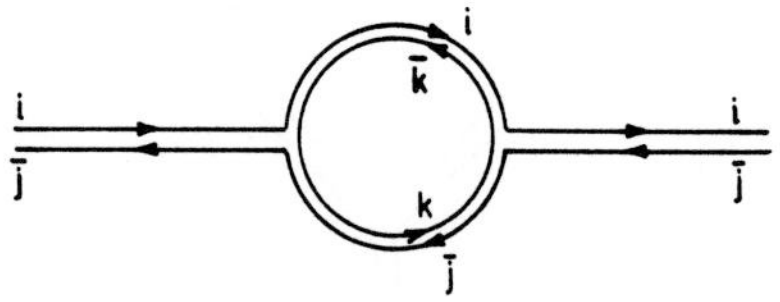

Fig. 6. The lowest-order gluon vacuum polarization.

Likewise, in fig. 8 is drawn a three-loop contribution to the gluon propagator. This diagram, with six interaction vertices and three-self-contracted loops, is of order $(1/\sqrt{N})^6 N^3$, so it too survives in the large N limit.

It is not true, however, that all diagrams survive in the large N limit. A typical diagram that does not survive is indicated in fig. 9. This diagram has six interaction vertices, but only one large and tangled closed color loop, for a single factor of N. The diagram is, therefore, of order $(1/\sqrt{N})^6 N = 1/N^2$, and vanishes like $1/N^2$ as N becomes large.

What distinguishes fig. 9 from the previous examples is that it is non-planar, that is, it is impossible to draw this diagram on the plane without line crossings (at points where there are no interaction vertices). In fig. 9 the crossing occurs at the center of the diagram. The diagrams of fig. 6, 7, and 8 are, by contrast, planar – they can be drawn on the plane.

By experimenting with simple examples, the reader should be able to see that non-planar Feynman diagrams always vanish at least like $1/N^2$ for large N.

On the other hand, planar diagrams made from an arbitrary number of gluon lines are non-vanishing in the large N limit. For example, by rewriting fig. 10 in the double line language, one should be able to see that this diagram is of order one for large N. With a little practice, one can see that adding an extra gluon line to a planar Feynman diagram in such a way that the diagram is still planar always creates two extra vertices and one extra closed loop. One thus obtains a factor $(1/\sqrt{N})^2 N = 1$, so that the dependence on N is unchanged.

So the first "selection rule" in the large N limit is that non-planar diagrams are suppressed.

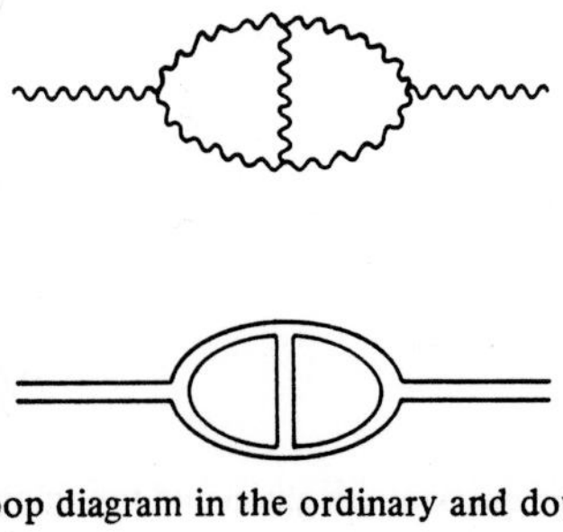

Fig. 7. The two-loop diagram in the ordinary and double-line notations.

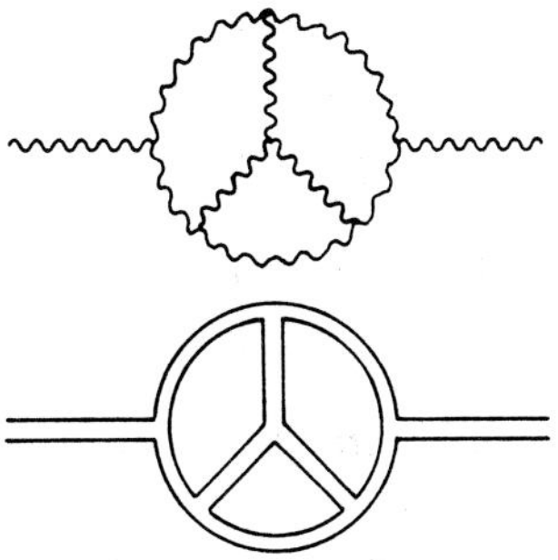

Fig. 8. A three-loop diagram, in the ordinary and double-line notations.

There is also a second "selection rule". It appears when one incorporates quarks and reflects the fact that for large N there are N^2 gluon states but only N quark states, so that diagrams with internal quark lines have fewer possible intermediate states and smaller combinatoric factors.

Consider the one-quark-loop contribution to the gluon propagator. This diagram is drawn in fig. 11, both in the usual notation and in the notation in which a gluon is represented as a double line. Because the quark propagator corresponds to a single color line, not two, the closed color line present in 6 is absent in 11. So fig. 11 has no large combinatoric factor, and its only dependence on N comes from factors of $1/\sqrt{N}$ at each of the two vertices. So 11 vanishes like $1/N$ for large N.

In summary, there are two selection rules for Feynman diagrams in the large N limit:

(i) Non-planar diagrams are suppressed by factors of $1/N^2$.

(ii) Internal quark loops are suppressed by factors of $1/N$.

The leading diagrams for large N are the planar diagrams with a minimum number of quark loops.

So far we have been discussing the gluon propagator, which is the two-point

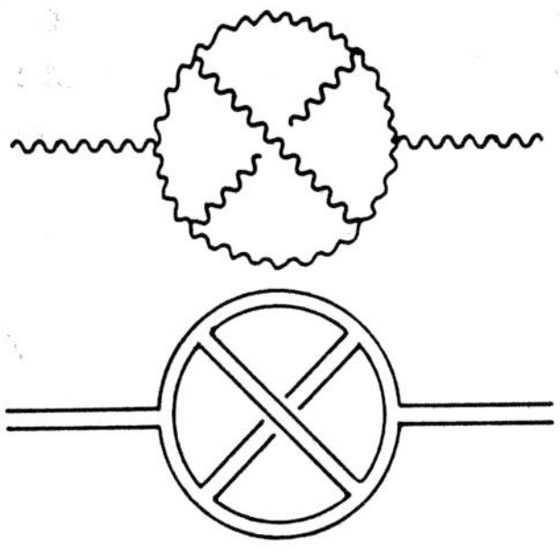

Fig. 9. A non-planar diagram.

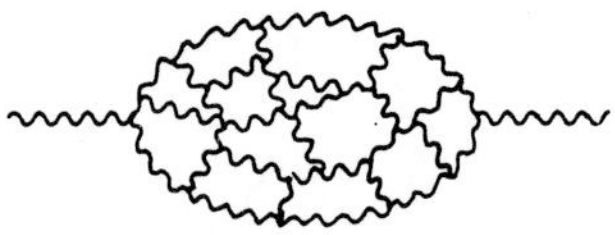

Fig. 10. A typical planar diagram with many loops.

function of the gluon field A_μ. For applications, however, we will want to consider matrix elements of gauge-invariant operators such as the quark bilinears $\bar{q}q$ or $\bar{q}\gamma^\mu q$. Let us, therefore, repeat the above analysis for matrix elements of quark bilinears.

Which diagram contribute, in the large N limit, to the two-point function of a quark bilinear J? In free field theory, we have only the one-loop diagram of fig. 12, and it is of order N, corresponding to a sum over the color of the quark running around the loop.

As the above discussion leads us to expect, arbitrary gluon insertions may be made without changing the dependence on N, as long as the planarity is preserved. For instance, the diagram of fig. 13 of order $(1/\sqrt{N})^2N^2 = N$, and that of fig. 14 is of order $(1/\sqrt{N})^6N^4 = N$, so both have the same dependence on N as the free field theory diagram.

However, planarity alone is not sufficient to ensure that a diagram contributing to matrix elements of quark bilinears will be of leading order in $1/N$. For instance, the diagram of fig. 15 is drawn in the plane, but by writing it in double line language one can see that it has only a single closed color loop and so is of order $(1/\sqrt{N})^4N = 1/N$ – down by two powers of N compared to the diagrams of figs. 12, 13, and 14.

The diagrams of fig. 15 differs from the previous examples in that there is at one place a gluon line (indicated in the figure by an arrow) at the edge of the diagram. The diagrams of figs. 12, 13, and 14 have, instead, only quarks at the edge. By considering simple examples, one can readily see that the appropriate generalization is that the dominant contributions to matrix elements of quark bilinears are planar diagrams with only quarks at the edge.

Furthermore, we learned in the discussion of the gluon propagator that for large N quark loops are suppressed, this being a reflection of the fact that there are N^2 gluon degrees of freedom but only N quark degrees of freedom. In considering quark bilinears, every diagram has at least one quark loop, since the quark bilinear

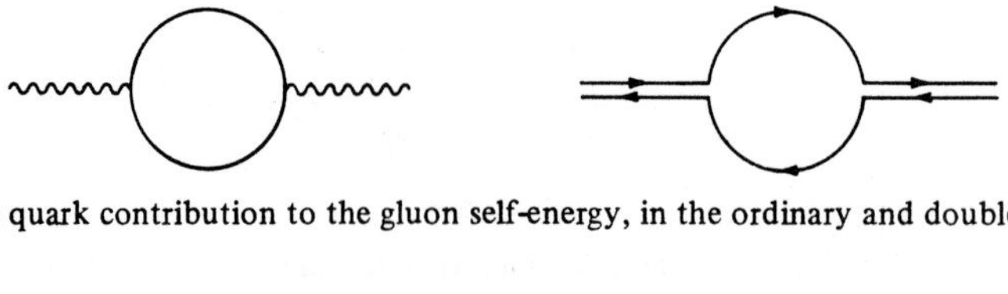

Fig. 11. The quark contribution to the gluon self-energy, in the ordinary and double-line notations.

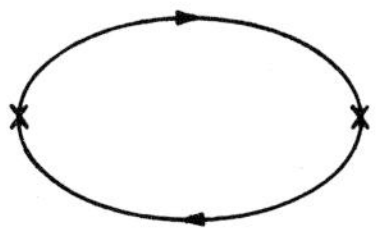

Fig. 12. The current correlation function in free field theory.

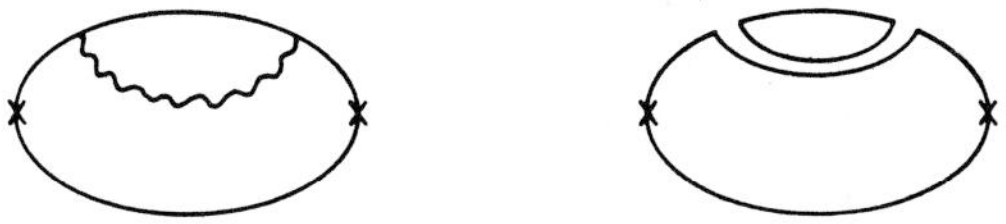

Fig. 13. A two-loop correction to the current correlation function.

Fig. 14. A four-loop, leading contribution to the current correlation function.

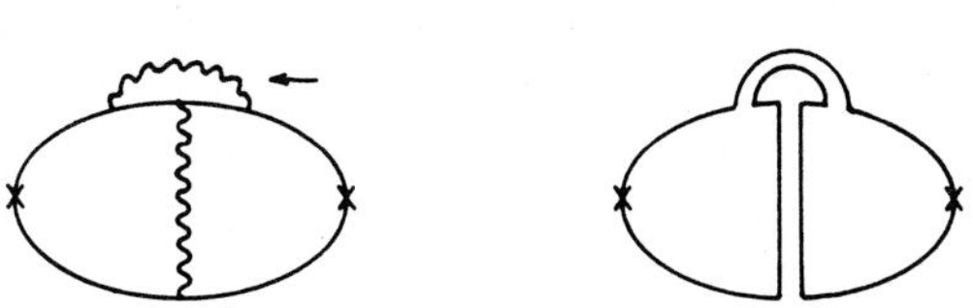

Fig. 15. A non-leading contribution to the current correlation function.

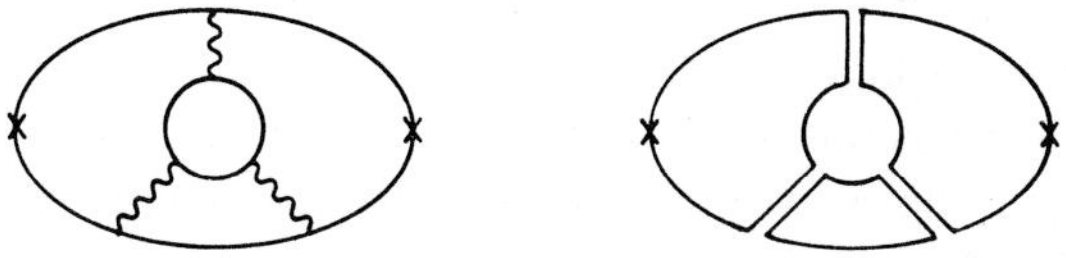

Fig. 16. A diagram with an internal quark loop.

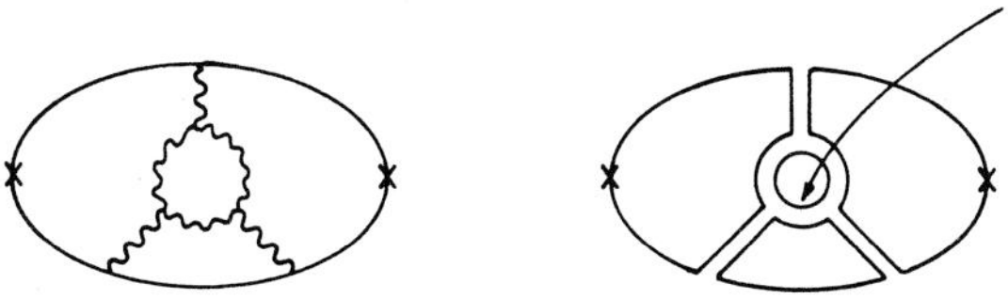

Fig. 17. The same diagram, with the extra quark loop replaced by a gluon.

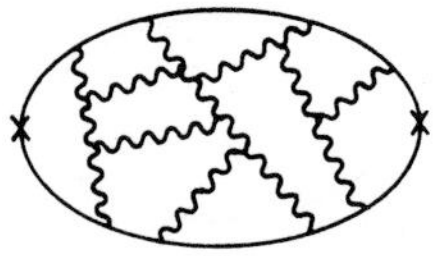

Fig. 18. A typical leading contribution to ⟨JJ⟩.

directly creates a $q\bar{q}$ pair, but we may guess that the dominant diagrams will be those with only the minimum possible number of quark loops, one. This is indeed true; for instance, the diagram 16 with an internal quark loop is of order $1/N$ compared to the similar diagram 17 with the internal quark loop replaced by a gluon loop. The extra factor of N in 17 arises from the closed color loop at the center of the diagram (indicated by an arrow).

Thus, the general statement is that the dominant diagrams for large N are the planar diagrams with only a single quark loop which runs at the edge of the diagram. A typical diagram of this sort is shown in fig. 18.

The summation of the leading diagrams is a very formidable task. We certainly do not know how to evaluate complicated planar diagrams like those of figs. 10 and 18. (Some indication that the problem is not hopeless is provided, however, by remarkable recent work of Brézin, Parisi, Zuber and Itzykson, and of Casartelli, Marchesini and Onofri, and by results of several other authors [4]).

One might have expected that, without being able to sum the planar diagrams, one could learn very little about the large N limit. But it is, perhaps surprisingly, possible to obtain a rather clear qualitative picture of the large N limit, simply from knowing that it is the planar diagrams that dominate. In the next section, this qualitative picture will be described.

3. Properties of mesons for large N

In this section we will review the qualitative picture of mesons (and glue states) in the large N limit that emerges from refs. [1] and [2].

It is necessary at the outset to make an important assumption. QCD is, apparently, a confining theory at $N = 3$, and we will assume that the confinement persists also at large N. In other words, while we are unable to sum the planar diagrams, we will assume that the planar diagrams sum up to give a confining theory. Our task is then to determine what can be deduced by combining the assumption of color confinement with the knowledge that, for large N, it is the planar diagrams that dominate. We will see that confinement and planarity combine rather efficiently, into an attractive picture, and this is, in fact, one of the main reasons to believe that confinement is present for large N.

We would like to show that under these assumptions, the large N theory has the following properties:

(i) Mesons for large N are free, stable, and non-interacting. Mesons masses have smooth limits for large N, and the number of meson states is infinite.

(ii) Meson decay amplitudes are of order $1/\sqrt{N}$; meson-meson elastic scattering amplitudes are of order $1/N$, and are given by a sum of tree diagrams (with local vertices and free field propagators) involving the exchange, not of quarks and gluons, but of physical mesons. More generally, meson physics in the large N limit is described by the tree diagrams of an effective local Lagrangian, with local vertices and local meson fields (but with masses and coupling constants that are, unfortunately, unknown).

(iii) Zweig's rule is exact at large N; singlet-octet mixing and mixing of mesons with glue states are suppressed, so that mesons come in nonets; and mesons for large N are pure $q\bar{q}$ states (no mixing with, for instance, $qq\bar{q}\bar{q}$).

The comments that follow are meant as an introduction to the subject and will not satisfy experts. Many statements will be made that would be valid if there were only a finite number of meson states, so that sums over intermediate meson states would be finite sums that would always converge. These statements need modification when there are an infinite number of mesons, so that, for example, sums over s-channel poles may not converge for all values of t. A more careful treatment of some of the questions that are avoided here has been given by Brower, Ellis, Schmidt, and Weis [2].

We will discuss mesons in QCD by considering matrix elements of operators that have the appropriate quantum numbers to create a meson. Such operators are local quark bilinears, like the scalar $q\bar{q}$ or the current $\bar{q}\gamma^\mu q$. We will denote as $J(x)$ the generic quark bilinear, and we will refer generically to these bilinears as "currents".

The first point we wish to establish is that the operator $J(x)$, acting on the vacuum, creates, in the large N limit, only one-meson states. It has vanishing matrix elements to create, for instance, a two-meson state, or a state consisting of a meson and a glue particle or of glue particles only.

It is equivalent to claim that the only singularities of the two-point function of J are one-meson poles. In other words, we claim that to lowest order in $1/N$,

$$\langle J(k)\, J(-k)\rangle = \sum_n \frac{a_n^2}{k^2 - m_n^2} \tag{2}$$

with no multiparticle cuts, and with the sum running over meson states only. Here m_n is the mass of the nth meson, and $a_n = \langle 0|J|n\rangle$ is the matrix element for J to create the nth meson from the vacuum.

Thus, we must show that if one cuts the leading contributions to the two-point function of J, the only intermediate states that one reveals are one-meson intermediate states.

If one looks at a typical diagram of the dominant sort – planar diagrams with quarks only at the edges – cut in a typical way, one immediately sees that the intermediate state always contains exactly one $q\bar{q}$ pair. This is so because the single quark loop in the diagram is always cut exactly twice, once at the top and once at the bottom of the diagram. For instance, in fig. 19 a typical way is shown to cut the typical diagram of fig. 18.

In a confining theory, the $q\bar{q}$ pair are always bound together into a meson, and the fact that there is always exactly one $q\bar{q}$ pair means that there is always exactly one meson. However, we still wish to show that the intermediate states in $\langle J(x)J(y)\rangle$ are precisely one-meson states, rather than consisting of one meson plus glue states. This requires a more careful study.

Let us examine more closely a planar diagram like 20, cut so as to reveal an intermediate state with one quark, one antiquark, and three gluons.

Because of color confinement, we expect the physical hadrons in this theory to be color singlets. What must be shown in this theory is that the quark, antiquark, and three gluons in this state form a single color singlet hadron (or rather, form a perturbative approximation to a single hadron). The alternative possibility would be that, for instance, the quark, antiquark and one gluon are coupled to total color zero and form a perturbative approximation to a meson, while the other two gluons are coupled to color zero and form a perturbative approximation to a color singlet glue state.

The question can be answered by looking at how the SU(N) indices run in fig. 20. The quark, antiquark, and three gluons are coupled together in the pattern

$$\bar{q}_l A^l{}_k A^k{}_j A^j{}_i q^i \,. \tag{3}$$

In other words, by inspection of the diagram one can see that the color of the antiquark is combined with one color index of the lowermost gluon, the other color index of the lowermost gluon is coupled with one index of the central gluon, the other index of the central gluon is coupled with one index of the upper gluon, and the other index of the upper gluon is coupled with the quark. This is indicated in eq. (3).

The pattern in eq. (3) has the property that while the five fields together are coupled to a color singlet, no smaller combination of them is separately a color singlet. For example, the first gluon field in (3) is in the adjoint representation $A^l{}_k$; the first two are coupled together, not to a singlet, but to the adjoint representa-

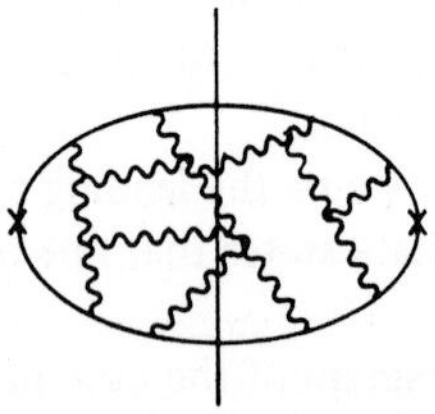

Fig. 19. The preceding diagram cut in a typical way.

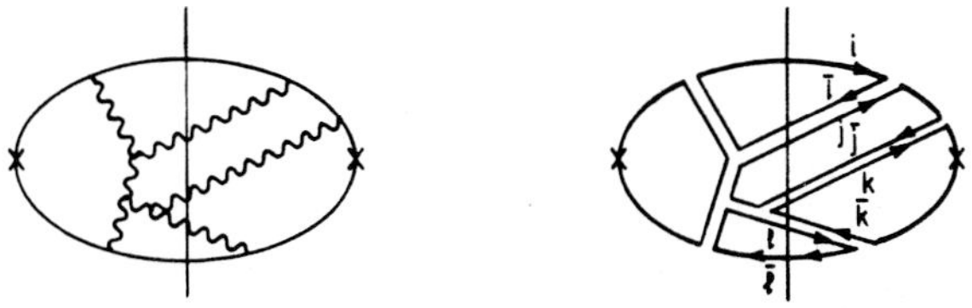

Fig. 20. A closer look at the intermediate states contributing to $\langle JJ\rangle$.

tion $A^l{}_k A^k{}_j$ and the three gluon fields together are coupled as $A^l{}_k A^k{}_j A^j{}_i$, also the adjoint representation. (Although $A^l{}_k A^k{}_j$ has a singlet piece, it is only one part in N^2.)

Because no group of fields in (3) other than all five fields together is in a color singlet state, it is impossible to split (3) into two or more color singlet pieces. Therefore, in a confining theory, the intermediate state in diagram 20 is a perturbative approximation to a single hadron; it cannot be split into two or more pieces representing two or more color singlet hadrons. By comparison, the group structure

$$\bar{q}_k A^k{}_l q^l A^j{}_m A^m{}_j \tag{4}$$

is a product of two color singlet operators, $\bar{q}_k A^k{}_l q^l$ and $A^j{}_m A^m{}_j$. An intermediate state with this structure could be interpreted as representing one meson $(\bar{q}_k A^k{}_l q^l)$ and one color singlet glue state $(A^j{}_m A^m{}_j)$.

But the intermediate states in the planar diagrams are always of type (3). If QCD is a confining theory, they are one-meson states. (Non-planar diagrams, on the other hand, generally contain intermediate states like (4).) From this we can draw several interesting conclusions. First of all, the meson masses have smooth limits for large N. In fact, as noted before, the fact that the intermediate states are one-particle states means that the two-point function of J can be written

$$\langle J(k)\, J(-k)\rangle = \sum_n \frac{a_n{}^2}{k^2 - m_n{}^2}\ . \tag{5}$$

The left-hand side has a smooth limit for large N – it is the sum of the planar diagrams, which each have the same dependence on N. So the right-hand side of (5) also has a smooth limit, which means that the meson masses have smooth limits, independent of N.

Second, the number of meson states is infinite. In fact, by asymptotic freedom the asymptotic behavior of the left-hand side of (5) is known; it behaves logarithmically for large k^2. If the sum on the right hand of (5) had only a finite number of terms, it would behave as $1/k^2$ for large k^2. A logarithmic behavior is only possible if the number of terms in the sum is infinite, and therefore, it must be infinite.

Third, we can already deduce that the meson states are stable at $N = \infty$. The one-particle poles in (5) must be on the real axis, since poles off the real axis would violate the spectral representation. The fact that the poles are on the real axis means that the mesons are stable. We cannot yet determine how rapidly the meson widths vanish as $N \to \infty$; that will be discussed later.

Finally, we can determine the dependence on N of the a_n, the matrix elements for the operator J to create a meson from the vacuum. In fact, the two-point function of J is of order N – the lowest order diagram 12 is of order N, and our counting showed that all planar diagrams are of the same order. The fact that (5) is of order N means that $a_n = \langle 0|J|n\rangle$ is of order $\sqrt{N}$, so the matrix element for a current to create a meson is of order $\sqrt{N}$.

Our results for the two-point function can be summarized in a way that will soon be useful by saying that $\langle J(x)J(y)\rangle$ is a sum of tree diagrams in which J creates, with amplitude a_n, a meson which propagates with a bare propagator $1/(k^2 - m_n{}^2)$. This is sketched in fig. 21.

To learn more, we must consider now the matrix element of a product of more than two J's. Let us consider, for instance, the three-point function $\langle J(p)J(q)J(r)\rangle$. We wish to establish the following result, which is a generalization of the result just stated for the two-point function. The three-point function is a sum of tree diagrams, with free field propagators and local vertices. These tree diagrams (diagram 22) may be of two types. Each current creates one meson, and the three mesons combine in a local meson-meson-meson vertex (first part of diagram 22), or one of the currents creates two mesons, each of which is absorbed by one current (second part of the diagram). The sum indicated in the diagram is a sum over which meson is propagating in which line.

To establish this is surprisingly simple. It is enough to establish that the only singularities of the amplitude in any channel are one-particle single poles. This being so, the amplitude is a sum of terms of the following sort. There may be simultaneous poles in each of the kinematical variables p^2, q^2, and r^2, a typical term being

$$\frac{A}{(p^2 - a^2)(q^2 - b^2)(r^2 - c^2)} . \tag{6}$$

Or there might be simultaneous poles in only two variables, a typical term being

$$\frac{B}{(p^2 - a^2)(q^2 - b^2)} . \tag{7}$$

In these expressions, A and B must be completely non-singular as functions of p, q, and r, and therefore, for the asymptotic behavior to be physically acceptable, A and B must be polynomials in the momenta.

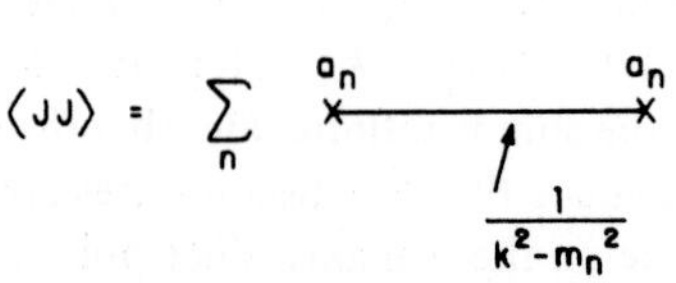

Fig. 21. $\langle JJ\rangle$ represented as a sum of one-meson poles.

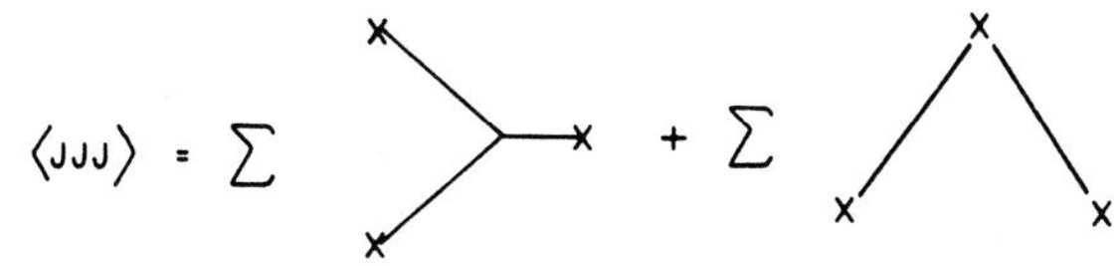

Fig. 22. ⟨JJJ⟩ as a sum of tree diagrams.

The polynomials can be interpreted as local interaction vertices, while the pole factors $1/(p^2 - a^2)$, $1/(q^2 - b^2)$, $1/(r^2 - c^2)$ can be interpreted as the propagators of mesons of masses a^2, b^2, and c^2. Thus, triple-pole terms can be interpreted as tree diagrams of the first type shown in 22, while double-pole terms correspond to tree diagrams of the second type.

We still must show that the only singularities in any channel are single poles, or equivalently, that the only intermediate states, when the amplitude is cut in any channel, are one-meson states. The argument for this is a near repetition of the argument considered previously.

In diagram 23 there is drawn a typical diagrams of the sort that dominates the three-point function ⟨JJJ⟩ for large N. The diagram has been cut in a typical way, to determine the singularities in the variable p^2. Exactly the same argument as before shows that the state revealed by cutting the diagram is (a perturbative approximation to) a one-meson state. Therefore, the singularities are one-meson poles. As was explained above, this is enough to ensure that ⟨JJJ⟩ is given by a sum of tree diagrams. From this, several interesting facts can be inferred.

First, we can now determine how narrow the meson states are. We already know that they are stable at $N = \infty$; now we can see that the amplitudes for two-body decays such as $A \to \mathrm{BC}$ are of order $1/\sqrt{N}$.

The argument is sketched in fig. 24. The three-point function is of order N, because in free field theory it is given by the one-loop diagram of fig. 24, which is of order N (coming from the sum over the quark color), and we know that the more elaborate planar diagrams have the same dependence on N that the free field theory has. But this amplitude has a term which is of the form $\langle 0|J|m\rangle^3 \Gamma_{mmm}$ – a product of three matrix elements for J to create mesons, times a trilinear meson-

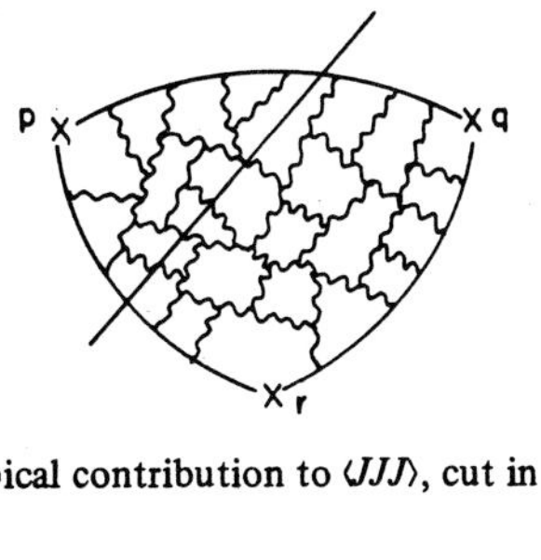

Fig. 23. A typical contribution to ⟨JJJ⟩, cut in a typical way.

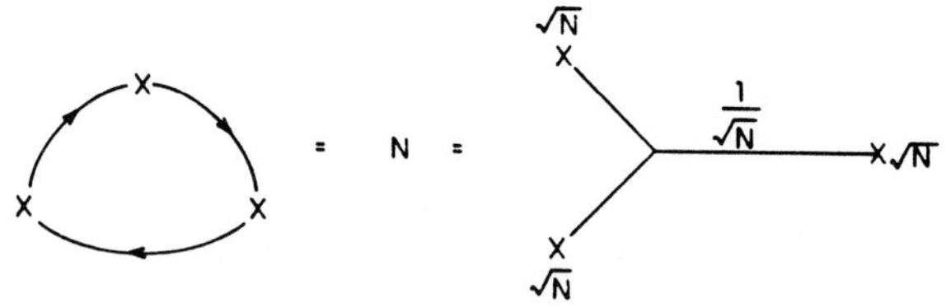

Fig. 24. N dependence of the three meson vertices.

meson-meson vertex. Since $\langle 0|J|m\rangle$ is already known to be of order $\sqrt{N}$, Γ_{mmm} must be of order $1/\sqrt{N}$.

By similar reasoning, we can see that the matrix element for J to create two mesons, $\langle 0|J|mm\rangle$, is of order 1. This follows (fig. 25) from the fact that $\langle JJJ\rangle$, of order N, has a term of the form $\langle 0|J|m\rangle^2\langle 0|J|mm\rangle$, while $\langle 0|J|m\rangle$ is of order $\sqrt{N}$.

As a final example, we would like to consider two-body meson scattering amplitudes. These can be studied by studying the four-point function of quark bilinears, $\langle J(x)J(y)J(z)J(w)\rangle$.

For this four-point function we will make a claim similar to the claim already made for the two- and three-point functions. Specifically, we will claim that it is given, in the large N limit, by a sum of tree diagrams involving the exchange of mesons, with free field propagators and local interaction vertices. These tree diagrams can take many forms, four of which are shown in fig. 26. Once again, the sum shown in the diagram is a sum over which mesons are propagating in which lines.

The argument is just the same as before. It is sufficient to show that the only intermediate states when the diagram is cut in any channel are one-particle states, so that the only singularities in any channel are one-particle poles. To argue this, we consider the leading (planar) contributions to the four-point function, cut so as to look for a singularity in any channel. As before, we may use the planarity to argue that the intermediate states are one-meson states, so that the singularities are single poles, and the amplitude can be written as a sum of tree diagrams.

Actually, there is a gap in the above reasoning; to fill it we must use crossing and unitarity. So far we have shown only that the four-point amplitudes are sums of terms with one-particle poles in the various channels (plus a possible contact term). The conclusion we wish to state is that the various pole terms are in fact the sum of

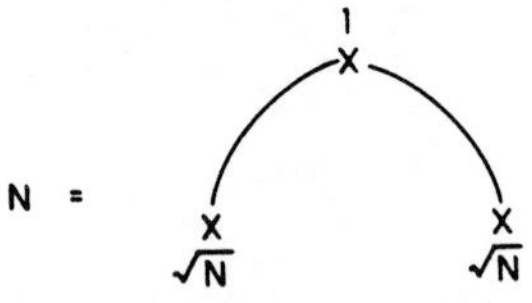

Fig. 25. N dependence of $\langle 0|J|mm\rangle$.

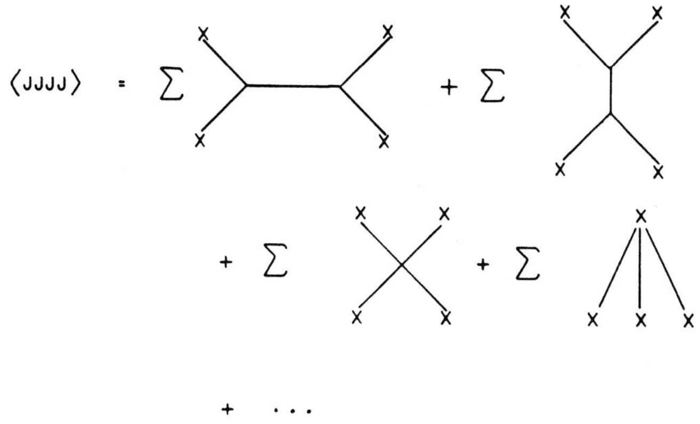

Fig. 26. ⟨JJJJ⟩ as a sum of tree diagrams.

all the tree diagrams that can be drawn from some local Lagrangian. To reach this conclusion we use crossing and unitarity, as follows. Crossing ensures that if a given pole term appears in one channel, it appears in all the crossed channels. Unitarity ensures that if a given pole term appears in, say, a three-point function, it must appear whenever that three-point function is possible as a subdiagram in the four-point function. Crossing and unitarity together ensure that if an amplitude is a sum of pole terms, it cannot be simply a miscellaneous sum of pole terms; they must be the pole terms coming from the tree approximation to some Lagrangian.

We can now determine the dependence on N of the vertices appearing in our tree diagrams.

The four-point function of quark bilinears is of order N, because (fig. 28) in free field theory it is given by the one-loop diagram, which is of order N, and the higher loop corrections do not change the dependence on N. This four-point function contains a term of the form $\langle 0|J|m\rangle^4 \Gamma_{mmm}{}^2$, four currents creating one meson each

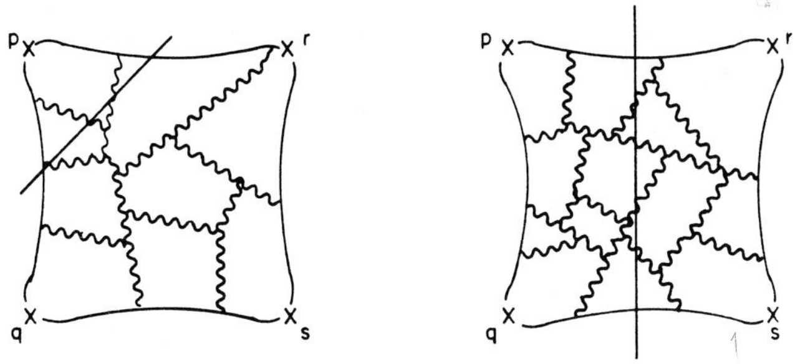

Fig. 27. A typical contribution to ⟨JJJJ⟩, cut to look for singularities in p^2 (first diagram) and in $(p + q)^2$ (second diagram).

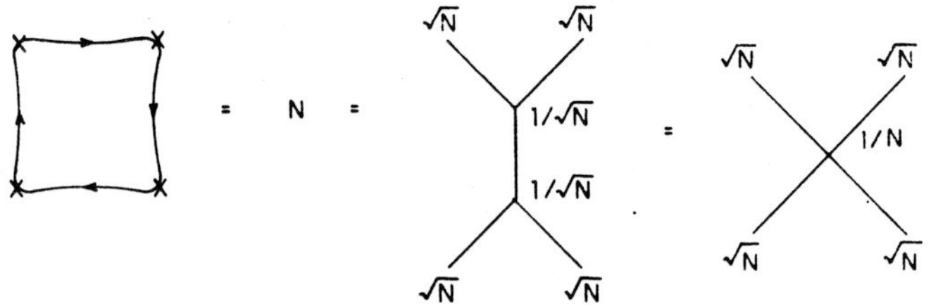

Fig. 28. The N dependence of the four meson vertices, and a consistency check for the three meson vertices.

from the vacuum, and two trilinear meson vertices. This is consistent with the knowledge we already have that $\langle 0|J|m\rangle$ is of order $\sqrt{N}$, and Γ_{mmm} is of order $1/\sqrt{N}$. The current four-point function also contains a term of the form $\langle 0|J|m\rangle^4\Gamma_{mmmm}$, four currents creating mesons from the vacuum, and a quartic meson vertex. From this we learn (diagram 28) that the quartic meson vertex is of order $1/N$.

Now we can make some qualitative comments about arbitrary two-body meson scattering amplitudes, AB → CD. Such amplitudes can be found from the current four-point function as the residue of a simultaneous pole in each external momentum. Of the various patterns of tree diagrams shown in diagram 26, the first three have simultaneous poles in each external momentum.

In particular, the amplitude for AB → CD is given by a sum of tree diagrams, since it is the residue of a particular pole in the current four-point function, and we know that the entire current four-point function is a sum of tree diagrams. These tree diagrams may involve meson exchange, with local vertices of order $1/\sqrt{N}$, and there may also be a local ABCD contact interaction, of order $1/N$. In the diagrams with meson exchange, one must sum over all possibilities for the meson exchanged. For example, in the reaction $\pi\pi \to \pi\pi$, the sum would include the ρ, the f, the ρ', the g, and every other meson that can couple to the $\pi\pi$ channel.

There are two important conclusions here. The first is that the scattering amplitude vanishes at $N = \infty$ – it is of order $1/N$ – so that at $N = \infty$ the mesons are free and non-interacting, and QCD at $N = \infty$ is a free field theory.

The second important conclusion is that the leading, $1/N$ terms in the scattering amplitudes are sums of tree diagrams with physical hadrons exchanged. The importance of this will be discussed in the next section.

Clearly, the preceding analysis can be extended to amplitudes with more than four external lines. One can show that arbitrary current n-point functions, and likewise n-body meson scattering amplitudes, are given by sums of tree diagrams with effective local meson fields and local vertices. We already know that three-meson vertices are of order $1/\sqrt{N}$ and four-meson vertices of order $1/N$. By a similar analysis one can show (diagram 30) that a local vertex with five mesons is of order

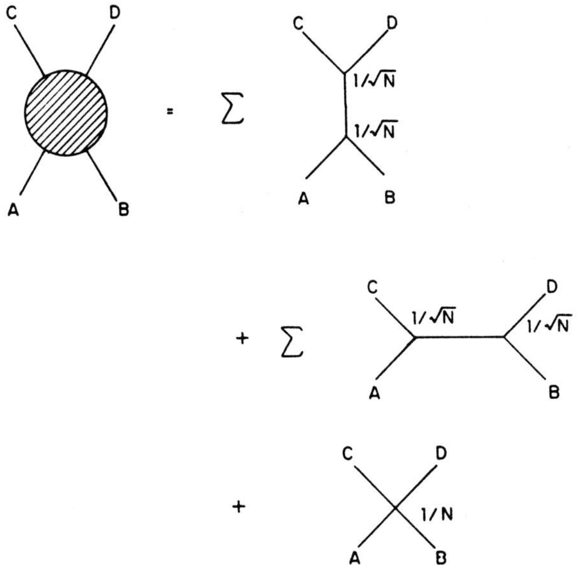

Fig. 29. Meson-meson scattering amplitudes as sums of tree diagrams.

$1/N^{3/2}$, and a local vertex with six mesons is of order $1/N^2$. In general a local vertex with k mesons is of order $1/N^{(k-2)/2}$.

It is also possible to extend this analysis to include glue states. To do so, one considers gauge invariant operators constructed from gluon fields, such as the Lorentz scalar Tr $G_{\mu\nu}G_{\mu\nu}$, or the pseudoscalar Tr $G_{\mu\nu}\tilde{G}_{\mu\nu}$.

By applying to these operators reasoning analogous to the reasoning that we have used for quark bilinears, one can derive for glue states – states created from the vacuum by gauge-invariant gluon operators – results analogous to the results that we have already derived for mesons.

In particular, at $N = \infty$ the glue states are free, stable, and non-interacting, and infinite in number. To lowest order in $1/N$, the glue states are decoupled from

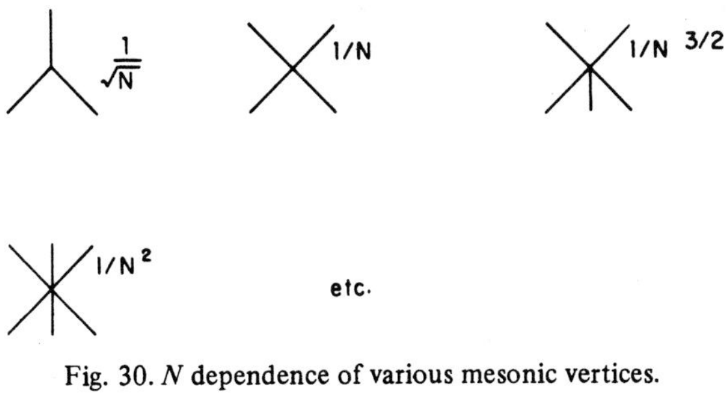

Fig. 30. N dependence of various mesonic vertices.

mesons. (This is why we have been able to discuss mesons so far without mentioning glue states.) The amplitude for a glue state to mix with a meson turns out to be of order $1/\sqrt{N}$. The amplitude for a glue state to decay to two glue states or to two mesons is of order $1/N$. The amplitudes for glue-glue or glue-meson elastic scattering are each of order $1/N^2$. Amplitudes with arbitrary numbers of mesons and glue states are given, to lowest order in $1/N$, by sums of tree diagrams. In these diagrams, the general local vertex with k mesons and l glue states is of order $N^{-l-k/2+1}$ (except when $k = 0$, when it is N^{-l+2}).

It is not as easy to include baryons in this picture as it is to include glue states. (For some interesting remarks on this question, from a viewpoint different from the one that will be followed here, see Durgut [6]. Another interesting approach is that of Veneziano and Rossi [13].) This problem will be the subject of this paper.

4. Attractiveness of the large N limit

In the last two sections we have examined some qualitative features of QCD in the large N limit. In this section, let us compare this qualitative picture with phenomenology, and discuss why one might expect an expansion in powers of $1/N$ to be a good approximation at $1/N = 1/3$. (For a review, see Veneziano, [2], and for a related, S-matrix theory discussion, see ref. [5].)

Several aspects of observed hadron phenomenology which have no other known explanation in QCD can be seen to be valid to lowest order in $1/N$. Insofar as this phenomenology is actually valid in the real world, its validity is a persuasive sign that the $1/N$ expansion is a good approximation to nature.

Some aspects of this phenomenology are the following:

(i) The suppression in hadronic physics of the $q\bar{q}$ sea; the fact that mesons are approximately pure $q\bar{q}$ states; the absence, or at least suppression, of $q\bar{q}q\bar{q}$ exotics.

(ii) Zweig's rule; the fact that mesons come in nonets of flavor SU(3); the decoupling of glue states.

(iii) The fact that multiparticle decays of unstable mesons are dominated by resonant two-body final states, when these are available.

(iv) Regge phenomenology; the success of a phenomenology that describes the strong interactions in terms of tree diagrams with exchange of physical hadrons.

Let us consider these points in turn.

We have seen that internal quark loops are suppressed by factors of $1/N$ in the large N limit; this reflected the fact that there are $O(N^2)$ gluon degrees of freedom, but only $O(N)$ quark degrees of freedom. Thus, at $N = \infty$ the $q\bar{q}$ "sea" is absent. The sea is certainly suppressed in phenomenology, and this is a success of the $1/N$ expansion.

A closely related statement is that $q\bar{q}q\bar{q}$ exotics are absent at $N = \infty$. One way to see this is that mesons at large N are non-interacting (the meson-meson interaction is of order $1/N$) so that at large N two mesons would not be bound together into an exotic.

A related way to see that exotics are absent at $N = \infty$ is to try to write down an operator with the right quantum numbers to create an exotic. The only gauge-invariant $q\bar{q}q\bar{q}$ operators are products of two gauge-invariant $q\bar{q}$ operators. Thus, one could consider creating an exotic from the vacuum with an operator $K(x) = \bar{q}_i q^i \bar{q}_j q^j(x)$. But to leading order in $1/N$ one finds that $\langle K(x)K(y)\rangle$ factorizes as $\langle\bar{q}q(x)\bar{q}q(y)\rangle^2$ so that, instead of an exotic, we have two freely propagating mesons.

(This discussion of exotics is really most applicable to hypothetical exotics that would couple strongly to a meson-meson pair. "Baryonium" states – exotics coupling strongly to a baryon-antibaryon pair – will be discussed later.)

Exotics are probably not entirely absent in the real world, but they are certainly suppressed – they are certainly not conspicuous in phenomenology. The only known field theoretic reason for this suppression is the $1/N$ expansion.

Another prediction of the $1/N$ expansion is Zweig's rule. For instance, a two body meson decay $A \to BC$ that violates Zweig's rule is suppressed by a factor of $1/N$ in the amplitude ($1/N^2$ in the branching rate) relative to a Zweig's rule conserving decay. This can be seen by counting powers of N in typical diagrams that violate Zweig's rule (left of fig. 31) and typical diagrams that conserve Zweig's rule (right of fig. 31).

In some situations, there exist special explanations of Zweig's rule. For example, in the decay of heavy quark systems one may invoke asymptotic freedom to explain Zweig's rule. However, there are many examples of situations in which Zweig's rule works, but the special explanations are not relevant. (For example: the suppression of the $\rho - \omega$ mass difference, and more generally, the fact, discussed below, that mesons come in nonets.) The only known theoretical explanation of Zweig's rule which is sufficiently general is the $1/N$ expansion.

As a special case of the above, we may ask why mesons are more accurately described as nonets of flavor SU(3) than as octets and singlets. If the u, d, and s quark masses were equal, then, to leading order in $1/N$, there would be exact singlet-octet degeneracy (because the diagrams that split singlets from octets involve $q\bar{q}$ annihilation, and are of order $1/N$).

Clearly related to Zweig's rule is the fact that to lowest order in $1/N$, glue states are decoupled from mesons. To lowest order in $1/N$, glue states are not produced in any reactions that are initiated by ordinary, quark-containing hadrons, or by weak or electromagnetic currents. This decoupling, to leading order in $1/N$, of the glue

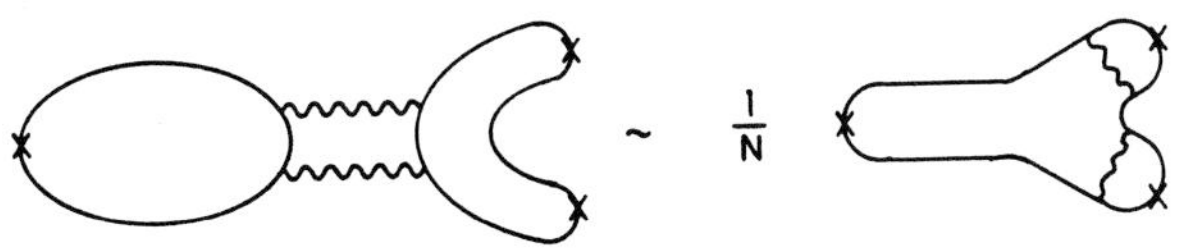

Fig. 31. Zweig's rule.

states, may be part of the reason why they are not conspicuous in phenomenology.

Next, we may consider the case of multi-body decays of unstable mesons. It is observed that such decays proceed mainly through resonant two-body states, when such are kinematically available. For example, the B, at 1237 MeV, decays to 4π, but the decay proceeds mainly as $B \to \omega\pi$, with subsequent $\omega \to 3\pi$. (It is generally believed that the tendency for decays to be two body dominated persists even when phase space is allowed for.)

From the point of view of the $1/N$ expansion, this is expected, because (recall fig. 30) a three meson coupling, like $B \to \omega\pi$, is of order $1/\sqrt{N}$, but the direct, non-resonant amplitude $B \to \pi\pi\pi\pi$ is a five meson coupling, of order $1/N^{3/2}$. Thus, $B \to \pi\pi\pi\pi$ is suppressed by a factor of $1/N$ in the amplitude, and $1/N^2$ in the probability, relative to $B \to \omega\pi$. Once again, from the point of view of field theory, the $1/N$ expansion is the only known exchanged in any given channel.

Finally, we should consider Regge phenomenology in relation to the $1/N$ expansion.

We have already seen that, to leading order in $1/N$, meson scattering amplitudes are given by sums of tree diagrams with exchange of physical mesons. The sum in question is always an infinite sum, because to lowest order in $1/N$ there are always an infinite number of mesons that could be exchanged in any given channel.

It is interesting that there exists a rather successful phenomenology, Regge phenomenology, in which the strong interactions are interpreted as an infinite sum of tree diagrams with hadron exchange. Of course, we are very far from deriving Regge phenomenology from the $1/N$ expansion, because we have not even shown that the infinite number of mesons of the large N limit lie on Regge trajectories. However, from the point of view of quantum chromodynamics, it does not seem likely that there is any sense other than the large N limit in which the strong interactions are given by an infinite sum of one meson exchanges. Consequently, the $1/N$ expansion is likely to be a major part of any eventual derivation of Regge phenomenology from quantum chromodynamics.

The preceding comments can be read in two ways. One may say that we have used the $1/N$ expansion to explain certain qualitative facts about the strong interactions. I personally prefer to reason in the opposite way, and to say that we may use certain qualitative facts about the strong interactions as diagnostic tests showing that the $1/N$ expansion is probably a good approximation to nature. In other words, assessing whether the $1/N$ expansion is likely to be a good approximation to nature is a very important matter from a theoretical point of view. It has a great bearing on how one would tend to think about the aspects of the strong interactions that can't be understood in perturbation theory. Without being able to actually sum the planar diagrams, one cannot assess this matter directly, from a theoretical point of view, and the best that we can do is to reason indirectly, using certain observed features of hadron physics as diagnostic tests showing that the $1/N$ expansion is probably a good approximation to nature.

Actually, there are some additional qualitative aspects of the strong interactions which can be compared with the $1/N$ expansion, and which are perhaps just as important as the points treated above, but were not included in the above list because they are more general and imprecise:

(i) the narrowness and very existence of resonances;

(ii) the general success of quark model spectroscopy;

(iii) the connection with the string model.

In some sense, resonances are narrow or they would not be noticeable. Thus, given that the strong interactions are strong and that the ρ decays strongly, into $\pi\pi$, one may aks why the ρ is narrow enough to be noticeable.

To make slightly stronger the claim that resonances are in some sense narrow, let us note that in Regge phenomenology it is believed that the straightness of Regge trajectories is related, by analyticity, to the narrowness of resonances. If there is any approximate sense in QCD in which the trajectories are linear, this must be an approximation in which the resonances are narrow.

It is thus an attractive feature of the $1/N$ expansion that resonances are narrow for large N, with widths of order $1/N$. This, together with an assumption that the $1/N$ expansion is a good approximation, explains why resonances are narrow enough to be noticeable, and offers a hope that at $N = \infty$ the Regge trajectories are linear.

Actually, in connection with unstable mesons like the ρ, some additional questions should be asked.

Let us compare an "old-fashioned" picture of the ρ, in which the ρ is considered as a bound state of two pions, to a "modern" picture, in which the ρ is considered as a bound state of $q\bar{q}$. Which picture will be more fruitful in QCD?

Contrary to a tempting assumption, it does *not* follow from confinement alone that the modern, $q\bar{q}$ picture is better. Confinement alone does not imply that the ρ couples less strongly to $\pi\pi$ than to $q\bar{q}$, and if this were not true, the older point of view about the ρ would be closer to the truth. For instance, confinement alone would allow the possibility that the π might be a tightly bound, nearly pointlike, $q\bar{q}$ state, and the ρ a loose, non-relativistic bound state (or resonance) of $\pi\pi$. (A two-dimensional confining model is known in which this occurs – the massive Schwinger model for large e/m [7].) Of course, nature is nothing like this. To explain why it is the modern $q\bar{q}$ picture of hadron resonances that is more fruitful, some ingredient other than confinement is needed. The $1/N$ expansion, in which the coupling of ρ to $\pi\pi$ is suppressed and the ρ couples only to $q\bar{q}$, is quite probably the needed ingredient.

Actually, it is not only the narrowness of resonances and the fact that they are best viewed as $q\bar{q}$ states that has to be explained. The very existence of resonances must be explained. For instance, confinement alone does not imply that the ρ or any state, stable or unstable, narrow or broad, would have to exist in the $J^{PC} = 1^{--}$ channel. Once one realizes that, in a confining theory, the ρ might have been mainly $\pi\pi$, it is easy to see how, in a confining theory, the ρ could fail to exist at all. Suppose that, in a world in which the ρ is mainly $\pi\pi$, one could, without losing confinement, vary some parameter (such as a quark mass, the gauge group, or a coupling of some scalar field) so that the $\pi\pi$ interaction would become a repulsion. Then the ρ could cease to exist even as a resonance.

In the large N limit, nothing of this sort occurs, and there are, on the contrary, an infinite number of states in each J^{PC} channel. This can be seen, along the lines of our previous discussion, by considering the two point function of a $\bar{q}q$ operator of given J^{PC}. To lowest order in $1/N$, only one meson intermediate states are possible, and an infinite number of such states must exist to satisfy the asymptotic freedom prediction for the high energy behavior.

Certainly, no derivation of the quark model spectroscopy from the $1/N$ expansion is known. But I believe that the preceding comments are significant hints that the success of the quark model spectroscopy is related to the fact that the $1/N$ expansion is a good approximation, and that the $1/N$ expansion will figure in any eventual derivation of the quark model spectroscopy from field theory.

The last phenomenological point to be mentioned here is quite important but very imprecise: the connection of the $1/N$ expansion with the string model. Starting with 't Hooft, many physicists have felt that the planar diagrams of the large N limit are related to the string model and dual theories. The idea that the $1/N$ expansion is related to a topological expansion in the motion of physical strings is also supported by arguments from the strong coupling expansion of lattice gauge theories [8].

The string model, which gives a natural and simple picture of how there could be a linear potential between quark and antiquark, is widely felt to be an attractive picture of how confinement comes about. From the point of view of QCD, the most reasonable way to understand how a string picture could be relevant to nature is to suppose that the string model is relevant to the large N limit, and the large N limit is a good approximation to nature.

This concludes our discussion of the connection of the $1/N$ expansion with phenomenology. There are, however, some arguments of a quite different sort which indicate that the $1/N$ expansion should be taken seriously. These are very general arguments of a theoretical sort.

First, the $1/N$ expansion is the tree approximation to the strong interactions. We have already seen that strong interaction scattering amplitudes are given, to lowest

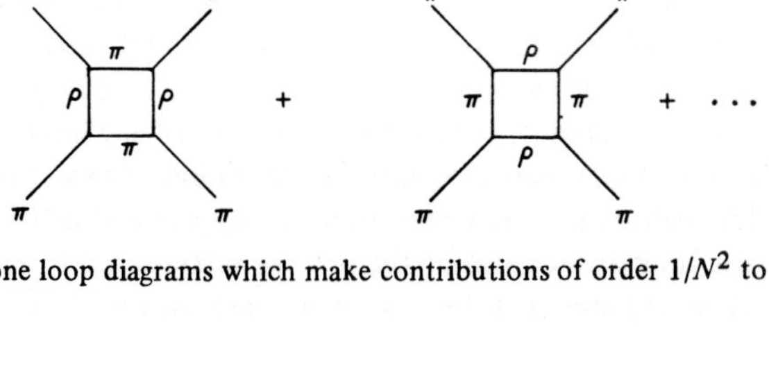

Fig. 32. Typical one loop diagrams which make contributions of order $1/N^2$ to meson-meson scattering.

order in $1/N$, by sums of tree diagrams with physical hadrons exchanged. It is likewise possible to show, by considering unitarity plus the diagrammatic counting rules of section two, that the higher order corrections in the $1/N$ expansion are sums of loop diagrams of hadrons. For instance, in fig. 32 are shown typical one loop diagrams, contributing corrections of order $1/N^2$ to the $\pi - \pi$ scattering amplitude.

Just as in any theory one understands the tree approximation before trying to consider loop diagrams, so in QCD we should understand the tree approximation – the large N limit – before trying to understand the loop diagrams (of arbitrarily high order) that must be included (although they may give only small contributions) at $N = 3$.

Put in a slightly different way, for QCD, the weak coupling regime, in terms of the physical degrees of freedom, is large N. It is for large N that the physical degrees of freedom, the mesons, are weakly coupled. As for any theory, so for QCD, we should understand the weak coupling regime as the first step in trying to understand the theory. For QCD, understanding the weak coupling regime means thinking about the $1/N$ expansion.

Second, the idea of the $1/N$ expansion is sometimes questioned on the grounds that $1/N = 1/3$ is not very small. What, really, is the status of this claim?

One can't really know, theoretically, how large N must be for the $1/N$ expansion to be a good approximation except by calculating the coefficients of some of the terms that are suppressed by powers of $1/N$. How small x must be for a series $\Sigma a_n x^n$ to be deominated by the first few terms depends entirely on how large are the coefficients a_n. If the coefficients are very small, $x = \frac{1}{3}$ can be considered a small number. If the coefficients are very large, $x = \frac{1}{3}$ would be a large number, from the standpoint of this series.

Without summing the planar diagrams to actually solve QCD in the large N limit, it is not possible to say with certainty on theoretical grounds whether $1/N = 1/3$ can be considered small in QCD. The best that we can do is to appeal to phenomenology. The purpose of the preceding discussion was to persuade the reader that there are significant phenomenological reasons to think that $1/N = 1/3$ is small enough for the $1/N$ expansion to be a good approximation in QCD.

The non-planar diagrams are actually of order $1/N^2$, not $1/N$. Even a skeptic might expect that $1/N^2 = 1/9$ is small enough to justify an expansion. Diagrams with internal quark loops are of order $1/N$, but are definitely known in phenomenology not to be important. A typical effect of these diagrams is the mixing of a $q\bar{q}$ meson with a $q\bar{q}q\bar{q}$ state, and such effects are definitely known, from the success of quark model spectroscopy, to be unimportant.

Just for the sake of comparison, let us ask why perturbation theory is successful in QED. It is not enough to say that "the electric charge is small." In fact, normalized in the usual way so that the interaction vertex is just $e\gamma^\mu$, the electric charge is approximately $e = 0.302$. Perturbation theory is a good approximation in QED because when one carries out perturbative expansions, one finds that the typical expansion parameter is really $e^2/4\pi$. If the typical parameter had turned out to be $4\pi e^2$, perturbation theory would not have been very successful for e as large as 0.302. And if we had not yet learned how to do perturbative calculations, we would have been unable to judge, just from the fact $e = 0.302$, whether an expansion in e^2 would be a good approximation.

If, for instance, as is perfectly possible, the characteristic parameter in the non-planar diagrams is really not $1/N^2$ but $1/4\pi N^2 = 1/113$, then non-planar diagrams in QCD are almost as tiny as electromagnetic corrections. While this is only an extreme possibility, there is no reason to be surprised that phenomenology seems to show that the $1/N$ expansion is a good approximation.

Another, very general, reason to believe that the $1/N$ expansion is an important approach is the following. It is very unlikely that we can solve exactly in four space-time dimensions any theory with a non-trivial S-matrix. A full interacting S-matrix in four dimensions would be far more complicated than anything we could write down, much less calculate. Most likely, we can expect to solve exactly a four-dimensional theory only in some limit in which the physical degrees of freedom decouple from each other, in which the S-matrix of the physical degrees of freedom (which may be complicated functions of the original fields) is one. Then solving the theory means only determining the masses, and it is at least conceivable that one could determine this much information, while it is not conceivable that we could determine all the information contained in a full interacting S-matrix. Moreover, we can probably realistically expect to understand a four-dimensional theory only by perturbing around a limit in which the physical degrees of freedom are decoupled.

In QCD the large N limit is such a limit, and it is probably the only such limit. Therefore the large N expansion should be taken seriously.

Finally, we should consider the large N expansion seriously because $1/N$ is a possible expansion parameter and is the only expansion parameter the theory is known to have.

In the last analysis, the strongest reason to think that the $1/N$ expansion will be quantitatively a good approximation is that it seems to be qualitatively correct. And experience in physics shows that qualitatively correct approximations tend to be quantitatively good, often for reasons that are understood only in hindsight.

5. Baryons in the large N limit: non-relativistic theory

In the previous sections we have reviewed the known results concerning the nature of the large N limit in QCD. One noticeable gap in this discussion is the failure to explain what baryons are like for large N. The purpose of the rest of this paper is to fill that gap.

In a theory with gauge group SU(N), a baryon is a completely antisymmetric state of N quarks. This is to be constrasted with mesons, which are made of $q\bar{q}$ regardless of what the gauge group is.

In studying mesons, one draws the same Feynman diagrams – the diagrams of the $q\bar{q}$ sector – for any gauge group. Only the combinatoric factors associated with the diagrams vary from group to group. To determine the large N limit, it is necessary to determine which diagrams have the largest combinatoric factors. This is the analysis that was first carried out by 't Hooft, and that has been reviewed in the preceding sections.

The problem of baryons is in a sense more subtle, because not only the combinatoric factors associated with the diagrams, but the diagrams themselves, depend on N. To determine the large N limit, we must determine the combined N dependence coming from the fact that a baryon contains N quarks, so that our Feynman diagrams have N quarks in them, as well the N dependence coming from any explicit factors of N that appear in the diagrams.

To appreciate the problems, one may look at the lowest order perturbative correction to the free propagation of an N quark state. This arises (fig. 33) from a diagram with one-gluon exchange between two of the quarks. Because the quark-gluon coupling constant is, as we have seen, of order $1/\sqrt{N}$, this diagram has an explicit factor of $1/N$ appearing at the vertices, so at first sight it appears small. But we must realize that the gluon may have been exchanged by any two of the N quarks in the baryon. There are $\frac{1}{2}N(N-1)$, which for large N is essentially $\frac{1}{2}N^2$, quark pairs in the baryon. So summing over which pair of quarks exchanged the gluon, the total contribution of diagrams like (33) is of order

$$\tfrac{1}{2}N^2\left(\frac{1}{N}\right) \sim N. \tag{8}$$

Thus, these contributions *grow* with N.

The task of determining the large N limit may seem unpromising, given that the lowest order diagram diverges as N becomes large. The situation becomes only worse when higher order diagrams are considered. A diagram (fig. 34) with two gluons exchanged has factors $(1/\sqrt{N})^4$ coming from the four vertices, but this is overwhelmed by the fact that the number of ways to choose the four quarks that exchanged the two gluons is of order N^4. The total contribution from diagrams of this kind is, therefore, of order

$$N^4\left(\frac{1}{\sqrt{N}}\right)^4 \sim N^2 \tag{9}$$

so that these diagrams diverge even more severely than the lowest order diagrams diverge.

With each diagram diverging more severely than the one before, one might despair of the existence of a large N limit. It turns out, however, that a simple large N limit exists, but that diagrams are not a very convenient way to study this limit.

Let us first try to get a heuristic understanding of the apparently divergent behavior of perturbation theory. In determining the large N limit of baryons, we would, first of all, like to determine how the baryon masses depend on N. In the

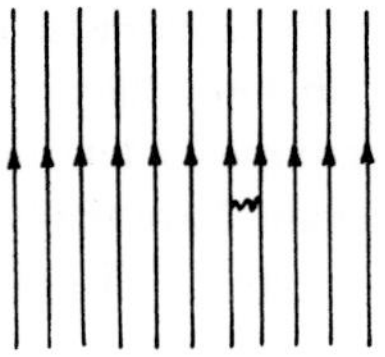

Fig. 33. The lowest order correction to the free propagation of N quarks.

simplest quark model picture, one might expect

$$\text{Baryon mass} = (\text{quark masses}) + (\text{quark kinetic energy}) + (\text{quark-quark potential energy})\,. \tag{10}$$

(The conclusion that follows is, as we will see, more general than this assumption.)

Since there are N quarks in the baryon, the quark masses contribute an amount NM to the baryon mass, where M is the quark mass. For the quark kinetic energy, we may guess that the kinetic energy of N quarks is N times the kinetic energy T of one quark. As for the potential energy, the interaction between one pair of quarks is of order $1/N$ – let us say it is $1/N$ times some V – but the total potential energy is a sum of all of the pair interactions, and there are $\frac{1}{2}N^2$ pairs. In sum, the baryon mass is

$$\begin{aligned} M_{\mathrm{B}} &= NM + NT + \tfrac{1}{2}N^2\left(\frac{1}{N}V\right) \\ &= N(M + T + \tfrac{1}{2}V)\,. \end{aligned} \tag{11}$$

The delicate point is that in the potential energy, the factors N^2 and $1/N$ combine to give a contribution of order N – of the same order of magnitude as all the other terms in the baryon mass. Thus, the entire baryon mass is of order N. We will see that this conclusion is much more general than the above derivation.

Once it is understood that the baryon mass is of order N – say $M_{\mathrm{B}} = N(f(g, M) + \mathrm{O}(1/N))$ for some function f that depends on g and M but not N – it is easy to understand the apparently divergent behavior of perturbation theory found in (8)

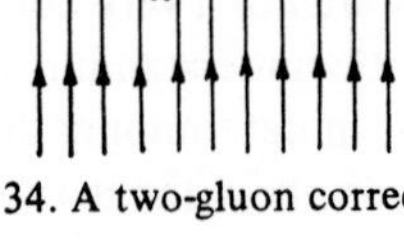

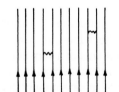

Fig. 34. A two-gluon correction.

and (9). Suppose, for instance, that we are trying to use diagrams like 33 and 34 to compute the propagation of an N quark state for a time t. The amplitude for such propagation is e^{-iM_Bt}, where M_B is the baryon mass. If in fact the baryon mass is N times some $f(g, M)$, the desired amplitude is $e^{-iNtf(g,M)}$. When such an amplitude is expanded in powers of g, each successive term will be more and more divergent in N. If, just to be definite, $f(g, M) = M(1 + g^2)$, then the expansion is

$$e^{-itNM(1+g^2)} = e^{-itNM}(1 - iMtNg^2 - \tfrac{1}{2}M^2t^2N^2g^4 + ...) \tag{12}$$

and we note that, just as in fig. 33 and 34, the terms of higher and higher order in g^2 are associated with higher and higher powers of N.

Thus the apparently bad behavior of perturbation theory does not mean that a large N limit for baryons does not exist, but only that the baryon mass is of order N. For determining the large N limit, diagrammatic methods are not very feasible, and we will use in this paper Hamiltonian and path integral methods.

Let us first consider the case that is conceptually simplest – baryons made from very heavy quarks.

If the quarks are very heavy, then, whether or not N is large, we may write a non-relativistic Schrödinger equation to describe the N quarks bound into a baryon. For quarks that are really very heavy, the potential in the Schrödinger equation can be taken to be a simple Coulomb potential. Thus, the Hamiltonian, including the quark bare masses, is

$$H = NM + \sum_i -\frac{\nabla_i^2}{2M} - \frac{g^2}{N}\sum_{i<j}\frac{1}{|x_i - x_j|} . \tag{13}$$

Note that the quark-quark interaction is of order $1/N$, and that in a state antisymmetric in the SU(N) indices, such as a baryon, the interaction is attractive, which explains the minus sign in the last term of (13).

One might feel that instead of a constant, g^2, in (13), we should write an effective coupling constant $\bar{g}^2(x)$. But once the effective coupling becomes small, it changes only very slowly. For the few orders of magnitude of distances that are significant in a baryon made from very heavy quarks, the effective coupling $\bar{g}^2(x)$ may be replaced by a constant, $\bar{g}^2(M)$, which has been called g^2 in (13).

The Hamiltonian (13) is not realistic, in the sense that even the b quark is probably not heavy enough for this Hamiltonian to give a valid description. The purpose in discussing the limiting case of baryons made from very heavy quarks is to derive some qualitative results that are valid also when the quarks are not so heavy.

Even for very heavy quarks, of course, the Coulomb potential is a good description only at short enough distances, and therefore the Hamiltonian (13) can be used only for baryons that are not too highly excited.

As a final preliminary, we must consider the question of statistics. Quarks, of course, are fermions, so that the wave function must be antisymmetric with respect to exchange of all of the coordinates of any pair of quarks. In a baryon, however, the wave function is antisymmetric with respect to the SU(N) indices (this was the original motivation for introducing color!), so we require complete symmetry in the other coordinates.

In most of this paper, we will consider baryons made from only a single flavor of quark. In other words, we will discuss the large N analogues of the Δ^{++} (made from up quarks) or the Ω^- (made from strange quarks), and postpone until later the discussion of baryon states that contain more than one flavor of quark.

Therefore, we must symmetrize with respect to space and spin. For very heavy quarks, the spin dependent forces are negligible. The low-lying wave functions of the Hamiltonian (13) are symmetric in space, so to describe the ground state baryons we will symmetrize in space, and also in spin. Having symmetrized in spin, we may forget about the spin, except for some comments later on the effects of including spin dependent forces when the quarks are not so heavy. Because we have symmetrized in space, we are, effectively, studying the Hamiltonian (13) for a system of bosons. We are, thus, studying the problem of N bosons with attractive Coulomb potentials of strength $1/N$.

How would one determine the large N limit of this problem?

What we must not do is to try to treat the last term in (13) as a perturbation. The tempting factor of $1/N$ is overwhelmed by the fact that there are $\frac{1}{2}N^2$ terms in the sum over quark pairs.

The large N limit is, instead, given by a sort of Hartree approximation. The logic behind this approxiatiom is as follows. For large N the interaction between any given pair of quarks is negligible – of order $1/N$. But the total potential experienced by any one quark is of order one, since any one quark interacts with N other quarks, each with strength $1/N$.

Thus, the total potential experienced by any one quark is of order one, but is a sum of many small, separately insignificant terms. As in statistical mechanics, when a quantity is a sum of many separately insignificant terms, the fluctuations around the mean value are very small. Thus, the potential experienced by one quark, apart from being of order one, can be regarded as a background, c-number potential – the fluctuations are negligible.

To find the ground state baryon, each quark should be placed in the ground state of the average potential that it experiences. By symmetry, the average potential is the same for each quark, so we should place each quark in the same ground state of the average potential.

In other words, the many-body wave function $\psi(x_1, x_2, ..., x_N)$ of the time independent Schrödinger equation should be written as a product

$$\psi(x_1, ..., x_N) = \prod_{i=1}^{N} \phi(x_i) \tag{14}$$

with each quark in the same properly normalized one particle wave function ϕ – the ground state of the average potential (which still must be determined). The ansatz (14), for reasons that we have discussed above in a heuristic way, becomes exact for large N; we will return to this issue later.

How is ϕ to be determined? The most straightforward method is to use the well-known variational principle associated with the time-independent Schrödinger equation. The exact many-body wave function $\psi(x_1, ..., x_N)$ makes stationary the variational functional $\langle\psi|H - E|\psi\rangle$, or equivalently $\langle\psi|H - N\epsilon|\psi\rangle$, where the total energy E has been written as $N\epsilon$, ϵ being the energy per quark. If we believe that the ansatz (14) becomes exact for large N, then instead of varying $\langle\psi|H - N\epsilon|\psi\rangle$ with respect to ψ, we may equivalently insert the ansatz (14) and vary only with respect to ϕ. In terms of ϕ (and requiring ϕ to be normalized to unity), $\langle\psi|H - N\epsilon|\psi\rangle$ becomes

$$NM + N\int \mathrm{d}^3x \frac{(\nabla\phi^*)(\nabla\phi)}{2M} + \tfrac{1}{2}N^2\left(\frac{g^2}{N}\right)\int\frac{\mathrm{d}^3x\,\mathrm{d}^3y\phi^*\phi(x)\,\phi^*\phi(y)}{|x-y|}$$
$$- N\epsilon\int \mathrm{d}^3x\phi^*\phi\,. \tag{15}$$

Notice that each term in (15) is proportional to N. The only delicate point is that in the potential energy term, N^2 and $1/N$ combine into a factor of N, as was suggested by some heuristic comments above. Moreover, N appears in (15) only as an overall factor which will not affect the equation obtained by varying with respect to ϕ. In other words, one may write (15) as

$$N\left[M + \int\frac{\mathrm{d}^3x\nabla\phi^*\nabla\phi}{2M} - \tfrac{1}{2}g^2\int\mathrm{d}^3x\,\mathrm{d}^3y\frac{\phi^*\phi(x)\,\phi^*\phi(y)}{|x-y|}\right.$$
$$\left. - \epsilon\int\mathrm{d}^3x\phi^*\phi(x)\right] \tag{16}$$

and the overall factor of N does not affect the variational equation, which turns out to be

$$-\frac{\nabla^2}{2M}\phi(x) - g^2\phi(x)\int\frac{\mathrm{d}^3y\phi^*\phi(y)}{|x-y|} = \epsilon\phi(x) \tag{17}$$

and is supplemented by the requirement $\int \mathrm{d}^3x\phi^*\phi(x) = 1$. (17) is the Hartree equation which determines the large N limit of the baryon wave function; the ground state baryon corresponds to the solution with the lowest value of ϵ.

Eq. (17) can apparently not be solved analytically, but it can be converted from an integro-differential equation into a differential equation by the following device. Dividing by ϕ, acting with the Laplacian ∇^2, and using the fact $\nabla^2(1/(x-y)) = -4\pi\delta(x-y)$, one finds that (17) yields

$$-\frac{1}{2M}\nabla^2\left(\frac{1}{\phi}\nabla^2\phi\right) + 4\pi g^2\phi^*\phi(x) = 0\,. \tag{18}$$

Thus, (17) can be converted into a fourth order differential equation. For the ground state baryon, one looks for a rotation invariant solution, $\phi(x) = \phi(r)$, where $r = \sqrt{x^2+y^2+z^2}$, and one finds for $\phi(r)$ the fourth order ordinary differential equation

$$-\frac{1}{2M}\left(\frac{d^2}{dr^2}+\frac{2}{r}\frac{d}{dr}\right)\left(\frac{1}{\phi}\left(\frac{d^2}{dr^2}+\frac{2d}{rdr}\right)\phi\right)+4\pi g^2\phi^*\phi=0\,. \tag{19}$$

But (19) apparently must be solved numerically.

Two important conclusions emerge from this discussion.

First, the baryon masses really are of order N. In fact, the computation makes clear that in wave functions like (14), each term in the Hamiltonian has an expectation value of order N. Therefore the total mass is of order N.

Second, while the baryon mass is of order N, the size and shape of the baryon have smooth limits as $N \to \infty$. (By the "shape," I means the shape of the charge profile as measured by electron scattering.) In fact, the charge density in the baryon receives identical contributions from each of the N quarks, and so is $N\phi^*\phi(x)$, where $\phi^*\phi(x)$ is the charge density due to one quark. Since N does not appear in the equation (17) that determines ϕ, ϕ does not depend on N. So apart from an overall factor of N, the charge distribution in the baryon is independent of N for large N, and the size and shape are dtermined by the N independent function $\phi^*\phi(x)$.

Two additional matters will be discussed here: the generalization from the time-independent to the time-dependent Schrödinger equation; and the question of understanding why it is that the ansatz (14) really does become exact as $N \to \infty$.

So far we have been discussing the time-independent Schrödinger equation $H\psi = E\psi$ in the N quark sector. For subsequent applications, however, we will need to consider the time-dependent Schrödinger equation. How would one determine the behavior, for large N, of the time-dependent Schrödinger equation in the N quark sector?

We must generalize the ansatz (14) to the time-dependent case. The appropriate generalization is to say that the time-dependent wave function, $\psi(x_1, ..., x_N; t)$, is a product of identical one-body wave functions, but these one-body wave functions depend on time:

$$\psi(x_1, ..., x_N; t) = \prod_i \phi(x_i, t)\,. \tag{20}$$

Moreover, ϕ is required to be normalized, $\langle\phi|\phi\rangle = 1$. According to the ansatz, which, like (14), becomes exact for large N, all N quarks are, at any given time, in the same one-body state, but the one-body state has a time dependence which must be calculated. The ansatz (20) is known as the time-dependent Hartree approximation; for some recent literature on this approximation see Kerman and Koonin, and Kerman and Jackiw [9].

What equation should the one-body wave function ϕ satisfy? The easiest way to determine the right equation is to consider the variational principle associated with the time-dependent Schrödinger equation. The time-dependent Schrödinger equation can be derived by varying the quantity

$$\int \mathrm{d}t \left\langle \psi \middle| H - i\frac{\partial}{\partial t} \middle| \psi \right\rangle \tag{21}$$

with respect to ψ. If for large N the ansatz (20) becomes exact, then instead of varying (21) with respect to ψ, one may insert the ansatz (20) and vary only with respect to ϕ. In this way one obtains the time-dependent Hartree equation:

$$-\frac{\nabla^2\phi(x, t)}{2M} - g^2\phi(x, t)\int \frac{\mathrm{d}y\phi^*(y, t)}{|x - y|} = i\frac{\partial}{\partial t}\phi(x, t) . \tag{22}$$

As one might have guessed, the time-dependent Hartree equation is simply the time-independent equation (17), but with the eigenvalue term $\epsilon\phi$ replaced by $\mathrm{i}\partial\phi/\partial t$.

Eq. (22) has the simple solution $\phi(x, t) = \phi_0(x)\,\mathrm{e}^{-i\epsilon t}$, where ϕ_0 is a solution of (17); this solution describes a baryon at rest. Eq. (22) also has Galilean-boosted solutions $\phi(x, t) = \phi_0(x - vt)\exp(iMv \cdot x)\exp(-i\epsilon t - \frac{1}{2}iMv^2t)$ which describe a baryon in a state of uniform motion with velocity v. However, (22) also has many solutions which are not simply built from solutions of the time-independent equation. Indeed, one may choose arbitrary initial data $\phi(x, 0)$ at time zero and integrate (22) to obtain a solution of (22) valid at all times. When inserted back into (20), this yields, to lowest order in $1/N$, a valid solution of the time-dependent Schrödinger equation. The significance of these solutions will be discussed later, in connection with an analysis of excited baryon states.

Finally, we must address the question of why the ansatzes (14) and (20) become exact as N becomes large.

One approach to seeing that (under suitable conditions) the Hartree approximation becomes exact when the number of particles is large is given in many textbooks. The Hartree wave function, given by (14) and (17), has been adjusted to be an exact eigenstate, not of the true Hamiltonian, but of a simplified Hamiltonian,

$$\hat{H} = NM + \sum_i -\frac{\nabla_i^2}{2M} + \sum_i V(x_i) , \tag{23}$$

where V is the average potential,

$$V(x) = -g^2 \int \frac{\mathrm{d}y\phi^*\phi(y)}{|x - y|} . \tag{24}$$

Since the Hartree wave function is an exact eigenstate of $\hat{H}$, one writes the true Hamiltonian H as $H = \hat{H} + (H - \hat{H})$, and attempts to treat the difference $(H - \hat{H})$ as a perturbation. Evaluating the effects of the perturbation $(H - \hat{H})$ by standard time-independent perturbation theory methods, one finds that these effects are small when N is large. One finds in fact, that there is a systematic expansion of the true answer in powers of $1/N$, the Hartree approximation being the first term and perturbation theory in $(H - \hat{H})$ supplying the corrections.

There is, however, another way to see that the Hartree approximation becomes exact for large N and is the first term in an expansion in powers of $1/N$. This alternative derivation, which many particle physicists might find more attractive, involves the use of path integrals, and is presented in a later section.

In general, for systems of many particles, the Hartree or Hartree-Fock approximation becomes exact in the limit of large particle number if the interactions are such that in this limit the system becomes dense. For example, an atom with many electrons is a dense system, and Hartree-Fock (or even the cruder Thomas-Fermi) becomes exact as $Z \to \infty$. By contrast, because of the saturating property of nuclear forces, a nucleus with many nucleons is large rather than dense, and Hartree-Fock does not become exact as one increases the number of nucleons. Baryons are in this respect similar to atoms rather than nuclei; we have seen that the size of a baryon is of order one as $N \to \infty$, and the density of order N. So Hartree is exact for baryons as $N \to \infty$, just as Thomas-Fermi is exact for atoms as $Z \to \infty$. The Hartree or Thomas-Fermi approximations become exact when the density is so large that each constituent is interacting with many others. In this limit, the interaction between any given pair is negligible, and the cumulative effect of many pair interactions can be treated statistically via the Hartree or Thomas-Fermi approximations.

Of course, all of the discussion so far is in the context of the non-relativistic limit of baryons made from heavy quarks. Later we will argue that baryons made from light quarks are still given, as $N \to \infty$, by a Hartree approximation: N quarks moving independently in a certain average potential. When the quarks are not heavy, however, a relativistic Hartree equation is needed. We will subsequently write the explicit relativistic Hartree equations intwo dimensions. In four dimensions we are unable to write the explicit relativistic equations (this is related to the inability to sum the planar diagrams for mesons). We will, however, claim that qualitative conclusions can be drawn without knowing the explicit relativistic Hartree equations for four dimensions, just as qualitative conclusions can be drawn for mesons without being able to sum the planar diagrams.

6. Scattering processes

In this section we will discuss, within the context of the non-relativistic theory described in the last section, scattering processes involving baryons – baryon-baryon, baryon-antibaryon, and baryon-meson scattering.

As was originally shown by 't Hooft, meson-meson scattering amplitudes (and also meson-glueball and glueball-glueball amplitudes) vanish at $N = \infty$. By contrast, we will see that scattering processes involving baryons have non-trivial large N limits. These large N limits are semiclassical in nature and, in the limit of heavy quarks, they are described by certain integro-differential equations that can be written down explicitly.

Considering first the case of baryon-baryon scattering, before attempting a mathematical treatment let us first ask heuristically how strong is the baryon-baryon interaction. It will turn out that the dominant baryon-baryon interaction comes, for large N, from the exchange (fig. 35) of a pair of constituents. One quark from each baryon jumps to the other baryon, with exchange of a gluon between the two quarks. The N dependence of such an amplitude can be determined as follows. There is a factor of N from choosing a quark from the first baryon, a factor of N from choosing a quark in the second baryon, and a factor of $1/N$ from the gluon couplings. Altogether, then, the amplitude for this process is of order $N^2(1/N) = N$.

Actually (fig. 36)) the two quarks could also have jumped places without exchanging a gluon. The diagram of fig. 36, which comes with a factor of (-1), is simply a diagrammatic way to express the fact that the quark wave functions in the first baryon must be orthogonal to the quark wave functions in the second baryon. In this case of gluonless quark interchange, the exchanged quarks must have the same SU(N) quantum numbers, so as to preserve the color neutrality of the two baryons. (In the previous case, this neutrality could be restored by exchange of a gluon.) As a result we may choose arbitrarily a quark from the first baryon – giving a factor of N – but the other quark that is exchanged must then be chosen to have the same quantum numbers as the first one. Thus, we obtain only a single factor of N from selecting the first quark, and the amplitude for gluonless quark interchange, like the amplitude for quark interchange accompanied by exchange of a gluon, is of order N.

At first sight, this result may appear disastrous. How can baryon-baryon scattering have a smooth large N limit, if the baryon-baryon force is growing in proportion to N? At this point we must remember that the baryon mass is also of order N. As a result, for given velocity the baryon kinetic energy $\frac{1}{2}M_B v^2$ is of order N. The fact that the baryon-baryon interaction energy is of order N is precisely what is needed in order for this interaction to be of the same order of magnitude as the kinetic energy. This makes possible a smooth and non-trivial large N limit for the scattering cross sections. Had the baryon-baryon interaction been of order one, it would have been negligible compared to the kinetic energy, and the scattering cross sections would have vanished at large N.

Roughly speaking, the situation can be described in the following way. Because

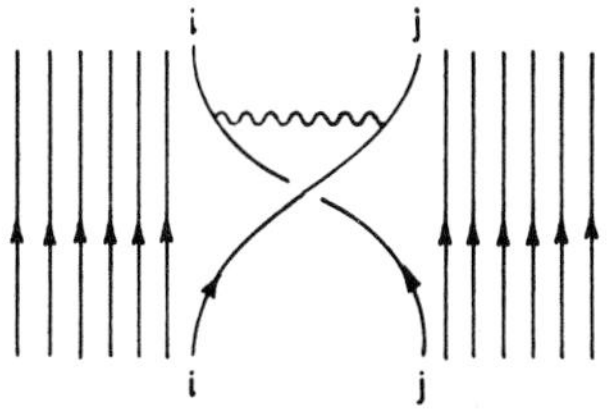

Fig. 35. Baryon-baryon scattering by constituent interchange, with gluon exchange.

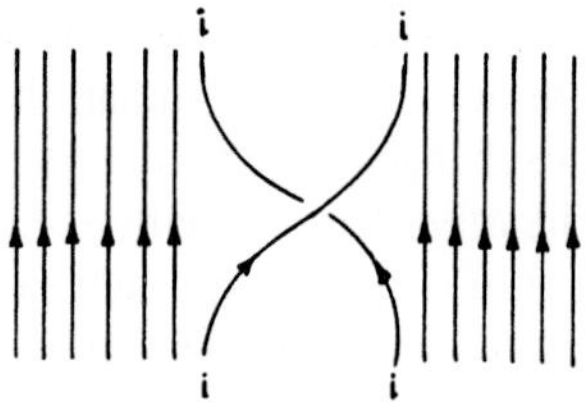

Fig. 36. Constituent interchange without gluon exchange.

the kinetic and interaction energy are each of order N, the total Hamiltonian H of the two-baryon sector can be written $H = N\hat{H}$, where $\hat{H}$ is a reduced Hamiltonian. The eigenvectors of $\hat{H}$, and therefore also the scattering cross sections, are not affected by the overall factor of N.

Proceeding now to a mathematical treatment of baryon-baryon scattering, we may reason as follows. To study a two-baryon process, we must study the Schrödinger equation in the sector with $2N$ quarks. Two baryons initially at rest will not remain at rest (they will attract or repel each other), and therefore it is not convenient to use the time-independent Schrödinger equation; we will consider instead the time-dependent Schrödinger equation in the sector with $2N$ quarks.

In the previous discussion of one-baryon problems, we placed all N quarks, with different colors, in the same space-spin wave function. Now that we have $2N$ quarks, and only N colors, the exclusion principle does not permit us to place all the quarks in the same space-spin wave function. The appropriate procedure (exact for large N) is to introduce a pair of time-dependent, space-spin wave functions $\phi_i(x, t), i = 1, 2$, and to place N quarks in ϕ_1 and the other N quarks in ϕ_2, antisymmetrizing with respect to which quarks have which colors and are in which of the ϕ_i. The ϕ_i are required to be orthonormal, $\langle\phi_i|\phi_j\rangle = \delta_{ij}$.

In other words, the many body wave function $\psi(x_1, \ldots x_{2N}, t)$ should be written

$$\psi(x_1 \ldots x_{2N}, t) = \sum_P (-1)^P \prod_{i=1}^{N} \phi_1(x_i, t) \prod_{j=1}^{N} \phi_2(x_j, t) \tag{25}$$

as a sum of products, with N quarks x_i, $i = 1 \ldots N$, in ϕ_1, and the other N quarks x_j, $j = 1 \ldots N$, in ϕ_2, and antisymmetrized with respect to which quarks are in the first group and which in the second group, and which colors they have.

To determine the time dependence of ϕ_1 and ϕ_2, one makes use of the usual time-dependent variational principle. Varying $\int dt \langle\psi|H - i(\partial/\partial t)|\psi\rangle$ with respect to ψ, one obtains the exact Schrödinger equation $i(\partial\psi/\partial t) = H\psi$. If one believes that the ansatz (25) is exact for large N, then one may insert this ansatz into the variational principle and vary only with respect to ϕ_1 and ϕ_2, obtaining in this way a pair of coupled equations for ϕ_1 and ϕ_2 which describe the large N limit of baryon-baryon scattering.

For example, in the case in which the two baryons have parallel spins, so that all quarks are in the same spin state and one may otherwise forget about spin, the variational equations for ϕ_1 and ϕ_2 turn out to be

$$i\frac{\partial}{\partial t}\phi_1(x, t) = -\frac{\nabla^2}{2M}\phi_1(x, t) - g^2\phi_1(x, t)\int\frac{dy\phi_1^*\phi_1(y, t)}{|x - y|}$$

$$- g^2\phi_2(x, t)\int\frac{dy\phi_2^*\phi_1(y, t)}{|x - y|} \tag{26}$$

and the same equation with ϕ_1 and ϕ_2 exchanged. Note that the factors of N have scaled out of the equation, as in our discussion of one-baryon problems.

Except for the last term, (26) coincides with the equation (22) derived previously to describe the free propagation of one baryon. Roughly speaking, ϕ_1 and ϕ_2 are each the wave function for one baryon; the separate propagation of the two baryons is described by the first few terms in (26), and the baryon-baryon interaction by the last term.

To describe baryon-baryon scattering, we would choose initial data in (26) such that in the far past, ϕ_1 and ϕ_2 are localized in different regions of space but heading for a collision. In this case, in the far past, the coupling term in (26) vanishes and the two baryons propagate freely. At a certain time, however, the two baryons collide, ϕ_1 and ϕ_2 overlap, and the coupling term is non-zero. The two baryons then scatter in the fashion described by (26).

Baryon-antibaryon scattering can be described in a similar way. The dominant process (fig. 37) is annihilation of a quark in the baryon with an antiquark in the antibaryon. The amplitude for this process contains a factor of N from choosing the quark, a factor of N from choosing the antiquark, and a factor of $1/N$ from the gluon couplings. It is therefore of order N. As before, the fact that the baryon-antibaryon interaction energy is of order N means that it is of the same order of magnitude as the kinetic energy; this is what is needed to have a smooth and nontrivial large N limit.

For a mathematical treatment, we consider the time dependent Schrödinger equation in the sector with N quarks and N antiquarks. We introduce a one-body time-dependent wave function $\phi(x, t)$ in which the quarks will be placed, and a one-

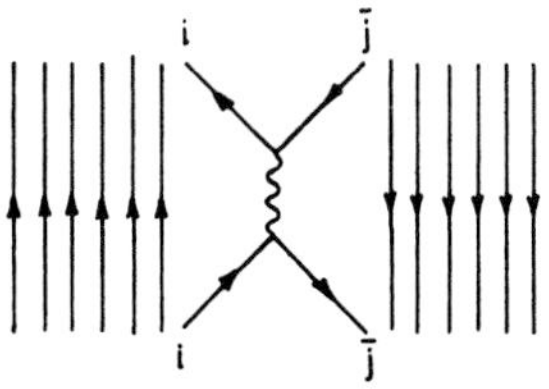

Fig. 37. Baryon-antibaryon scattering.

body time-dependent wave function $\omega(y, t)$ for the antiquarks. Then the many-body wave function $\psi(x_1 \dots x_N, y_1 \dots y_N, t)$ (the x_i are the quark and the y_j the antiquark coordinates) is written as a product

$$\psi(x_1 \dots x_N, y_1 \dots y_N, t) = \prod_i \phi(x_i, t) \prod_j \omega(y_j, t) . \tag{27}$$

To determine the time dependence of ϕ and ω, one inserts this ansatz in the usual variational principle $\int \mathrm{d}t \langle\psi| H - i(\partial/\partial t)|\psi\rangle$ and varies with respect to ϕ and ω, requiring $\langle\phi|\phi\rangle = \langle\omega|\omega\rangle = 1$. In this way, one obtains a pair of coupled equations for ϕ and ω. As usual, the factors of N cancel out of these equations, so that baryon-antibaryon scattering has a smooth large N limit. The actual variational equations are similar to (26) and will not be written here.

Turning finally to our last scattering channel, meson-baryon scattering, in this case the amplitude is again non-trivial even as $N \to \infty$, but it is given by a linear equation for motion of a meson in a background baryon field.

Let us again begin by asking qualitatively how the meson-baryon interaction depends on N. In this case, it turns out that a variety of processes contribute in the large N limit. One such contribution is sketched in fig. 38. The N dependence of this amplitude is as follows. We obtain a factor of N from choosing a quark in the baryon and a factor of $1/N$ from the gluon couplings. No factors of N come from the meson because, at any moment, the meson contains only one quark. So the amplitude is of order $N(1/N) = 1$.

The meson-baryon interaction, being of order one, is thus negligible compared to the baryon kinetic energy, which is of order N, and so it is too small to affect the motion of the baryon. To leading order in $1/N$, the baryon propagates freely, as if the meson were not present at all. The meson mass, on the other hand, is of order one. Consequently, the meson-baryon interaction is of the same order of magnitude as the meson kinetic energy, and is large enough to influence the motion of the meson. As a result, the meson is scattered by the baryon.

To translate this into a mathematical language, we consider the time-dependent Schrödinger equation in the sector with $N + 1$ quarks and one antiquark. We introduce a one-body space-spin wave function $\phi(x, t)$ representing the baryon, into which N of the quarks are put, and a two-body space-spin wave function $u(x, y, t)$ representing the meson, into which we put the extra quark and the antiquark. The wave functions ϕ and u are required to be normalized, and to satisfy an orthogonal-

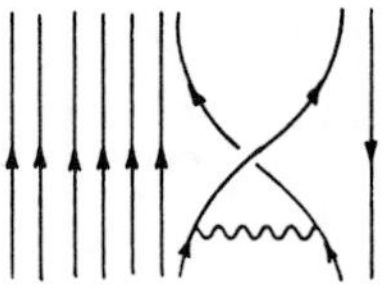

Fig. 38. Baryon-meson scattering.

ity condition $\int d^3x\phi^*(x)\,u(x, z)$ which states that the wave function of the quark in u is orthogonal to the wave functions of the quarks in ϕ. One then writes an ansatz for the full many-body wave function

$$\psi(x_1 \ldots x_{N+1}, y, t) = \Sigma(-1)^P \prod_{I=1}^{N} \phi(x_i, t)\,u(x', y, t) \tag{28}$$

with N quarks x_i in ϕ, and with the leftover quark x' and the antiquark y in u, and antisymmetrized with respect to the colors of the quarks in ϕ, and with respect to which quarks are in ϕ and which one in u. The time dependence of ϕ and u is then determined in the usual way by inserting the ansatz (28) in the variational principle $\int dt\langle\psi|H - i(\partial/\partial t)|\psi\rangle$.

It turns out that to lowest order in $1/N$, the resulting equation for ϕ is simply the one-baryon equation (22). The equation for ϕ is unaffected by the presence of u. The baryon propagates as if the meson were not there, as was suggested by the preceding heuristic discussion.

On the other hand, the equation for u is affected by the presence of ϕ. In the simplest case in which all quarks and the antiquarks have parallel spins, so that we may assign them the same spin wave function and then forget about spin, the equation for u turns out to be

$$\begin{aligned} i\frac{\partial}{\partial t}u(x, y, t) &= -\frac{1}{2M}\nabla_x^2 u(x, y, t) - \frac{1}{2M}\nabla_y^2 u(x, y, t) \\ &- g^2\frac{u(x, y, t)}{|x-y|} - g^2\phi(x)\int\frac{dz\phi^*(z, t)\,u(z, y, t)}{|x-z|} \\ &- g^2\phi(x)\int\frac{dz\phi^*(z, t)\,u(z, y, t)}{|z-y|}. \end{aligned} \tag{29}$$

As usual, the factors of N have canceled out, so that meson baryon scattering has a large N limit that is independent of N.

Notice that with the last two terms deleted, (29) would be simply the Schrödinger equation for a meson. The last two terms vanish when ϕ and u are localized in different regions of space, so that as long as the baryon is far away, u describes a freely propagating meson. But in general the last two terms do not vanish; when the meson and baryon collide, the meson is scattered by the baryon.

Notice also that (29) is linear in u. It is a linear equation for the scattering of a meson by a background baryon field.

An alternative method for deriving (29) and related equations will be presented later, in connection with a discussion of relativistic baryons in two dimensions.

7. Excited baryon states

So far we have been discussing baryons in their ground state. Let us now return to a discussion of excited baryon states.

We have seen that the ground state baryon can be described in terms of an average potential which is self-consistently determined. The quarks are placed in the ground state of this average potential, and the average potential is in turn determined by the motion of the quarks. Explicitly, in terms of the quark wave function ϕ which is the ground state solution of the time independent baryon equation (17), the average potential V is

$$V(x) = -g^2 \int \frac{\mathrm{d}y \phi^*\phi(y)}{|x-y|} . \tag{30}$$

It may now be fairly obvious how one should describe low-lying excitations of the baryon. Instead of placing all quarks in the ground state of V, one excites one (or more) of the quarks into excited states of motion in this potential. When one excites the baryon in this way, it is not necessary to recalculate the average potential V, for V is a cooperative effect which is determined by all N quarks together, and changing the motion of just one quark (or of a number insignificant compared to N) does not change the average potential, to lowest order in $1/N$.

We must, therefore, consider the one-body Schrödinger equation in the potential V:

$$-\frac{\nabla^2}{2M} \psi(x) + V(x)\, \psi(x) = \lambda \psi(x) . \tag{31}$$

If the excitation energies of this equation (measured relative to the ground state) are ϵ_k,, then the excited baryons will have masses $M_{B*} = M_B{}^0 + \Sigma n_k \epsilon_k$, where $M_B{}^0$ is the mass of the ground state baryon, and n_k is the number of quarks in the kth excited state. This formula is valid when the total number of quarks excited is small compared to N, so that for $N = 3$ it is valid, if at all, only when just one quark is excited.

Thus, as in the quark model, the low-lying excited baryons are single-body excitations. If N is very large, there exist also highly excited baryon states, which are best regarded as collective excitations of the baryon. We will now discuss some properties of these highly excited states. It will be clear from the discussion that most of the comments are relevant only when N is really very large, much larger than 3. The discussion of collective excitations will be included here mainly for its conceptual interest. In addition, it is not completely impossible that some trace of the collective excitations could exist even at $N = 3$.

Our previous formula $M_{B*} = M_B{}^0 + \Sigma n_k \epsilon_k$ is valid for any given number of excited quarks when N is sufficiently large. In other words, this formula is valid in the limit in which the number of excited quarks is kept fixed, while N is taken to infinity. One could, instead, consider a limit in which the *fraction* of quarks which are excited is kept fixed as N is taken to infinity. It is this limit that leads to highly excited baryon states, or collective excitations.

Imagine then that N is very large, and that we wish to describe a state in which not just one or two, but a non-zero fraction of all the quarks in the baryon are in an excited state. Thus, let p be a number between zero and one, and suppose that a fraction pN of the quarks are in an excited state, and only $(1-p)N$ are in the ground state.

With a non-zero fraction of all the quarks excited, we cannot simply use the average potential $V(x)$ determined by studying the ground state baryon; we must recalculate the average potential. The most straightforward way to deal with this situation is to consider the time-independent Schrödinger equation, and introduce a wave function $\phi_0(x)$ to represent $(1-p)N$ quarks in the ground state, and a second wave function $\phi_1(x)$ to represent pN quarks in an excited state. These wave functions are required to satisfy $\langle\phi_0|\phi_0\rangle = \langle\phi_1|\phi_1\rangle = 1$ and $\langle\phi_0|\phi_1\rangle = 0$. We then consider for the full wave function an ansatz

$$\psi(x_1 \dots x_N) = \Sigma(-1)^P \prod_{i=1}^{pN} \phi_1(x_i) \prod_{i=1}^{(1-p)N} \phi_0(x_j) , \tag{32}$$

with pN quarks in ϕ_1, $(1-p)N$ quarks in ϕ_0, and antisymmetrized with respect to color and with respect to which quarks are in which wave function. Inserting this ansatz into the usual variational principle $\langle\psi|H-E|\psi\rangle$, one obtains a pair of coupled non-linear equations for ϕ_1 and ϕ_0. These equations depend parametrically on the number p and could be solved, at least numerically, to get the wave function of an excited baryon state in which a fraction p of all the quarks are in an excited state.

Various generalizations of this procedure can be envisaged. For instance, one could look for baryon states in which a fraction p of the quarks are in one excited state, a fraction q are in a second excited state, and fractionally $(1-p-q)$ are in the ground state. This would be done by introducing three wave functions ϕ_0, ϕ_1, and ϕ_2, and repeating the above procedure.

However, there is a more comprehensive, and probably more instructive, way to study the highly excited baryon states. Let us return to an earlier equation, the time-dependent Hartree equation [eq. (22)] of the one-baryon sector. It is rewritten here for convenience:

$$-\frac{\nabla^2}{2M}\phi(x,t) - g^2\phi(x,t)\int\frac{\mathrm{d}^3y\phi^*\phi(y,t)}{|x-y|} = i\frac{\partial\phi(x,t)}{\partial t} . \tag{33}$$

As was mentioned previously, (33) has traveling wave solutions $\phi(x,t) = \phi_0(x-vt)\exp(iMv\cdot x)\exp(-i\epsilon t - \frac{1}{2}iMv^2 t)$ which are constructed in a simple way from solutions of the time-independent equation. However, (33) has many other solutions that are not of this form. In fact, one may choose arbitrary initial data $\phi(x, 0)$ at time zero; if these data are inserted in equation (33), the equation itself will determine $\partial\phi/\partial t$, and can be integrated to give a solution valid for all times.

For any solution $\phi(x, t)$ of (33), the product ansatz $\psi(x_1 \ldots x_N, t) = \Pi_i \phi(x_i, t)$ gives, to lowest order in $1/N$, a valid solution of the time dependent Schrödinger equation $i\partial\psi/\partial t = H\psi$. This can be checked by directly inserting the product ansatz and eq. (33) in the Schrödinger equation. We are therefore beset with an embarrassment of riches – for any function $\phi(x, 0)$ that is inserted in (33) at time zero, we get a solution of the time-dependent Schrödinger equation, valid to lowest order in $1/N$. What can we learn from these solutions?

These solutions describe baryon states excited in some way, but not excited in an energy eigenstate, since the time dependence determined by (33) from given initial data will generally not be harmonic. How can we determine from these time-dependent solutions the energy eigenstates, the energies of excited baryon states?

The correct procedure turns out to be the following. One looks for solutions of (33) that are periodic in time – not necessarily harmonic, but periodic. The solution may wiggle and shake, but it does so periodically. One then quantizes the periodic motions in the fashion of Dashen, Hasslacher and Neveu [10] requiring the action $\int_0^T dt\langle\psi|H - i(\partial/\partial t)|\psi\rangle$ in a period to be a multiple of 2π. This gives the energy eigenstates.

The DHN condition arises as follows. Given any periodic solution $\phi(x, t)$ of (33), we have we have a periodic solution $\psi(x_1 \ldots x_N, t) = \Pi_i \phi(x_i, t)$ of the time-dependent Schrödinger equation. The solution $\psi(x_1 \ldots x_N, t)$ is periodic but not harmonically varying – it does not correspond to an energy eigenstate. By the time translation invariance of the Schrödinger equation $\psi(x_1 \ldots x_N, t - t_0)$ is also a solution, for any t_0, and because the Schrödinger equation is linear, we may take linear combinations of the solutions corresponding to different t_0. Thus, we attempt to construct a harmonically varying, stationary state solution of the Schrödinger equation by writing $\hat{\psi}(x_1 \ldots x_N, t) = \int_0^T dt_0 \, e^{-it_0E} . \psi(x_1 \ldots x_N, t - t_0)$. The condition of constructive interference under which this procedure gives a non-zero stationary state solution of the Schrödinger equation is the DHN condition.

The coupled equations associated with (32) can be rederived in this approach by inserting the ansatz $\phi(x, t) = \phi_1(x)\, e^{-i\epsilon_1 t} + \phi_2(x)\, e^{-i\epsilon_2 t}$ in (33). This ansatz, in (33), will reproduce the equations that can be derived by inserting (32) in the time-independent variational principle $\langle\psi|H - E|\psi\rangle$.

The semi-classical quantization of periodic solutions of (33) has many applications. For example, we can look for solutions of (33) of the following form. The simplest solution of (33) is $\phi(x, t) = \phi_0(x)\, e^{-i\epsilon t}$, where $\phi_0(x)$ is a solution of the time-independent Hartree equation. One now expands around this solution, writing $\phi(x, t) = \phi_0(x)\, e^{-i\epsilon t} + \delta\phi(x, t)$. Assuming $\delta\phi$ to be small, one linearizes in $\delta\phi$, obtaining in this way a linear equation for low amplitude collective oscillations of the baryon.

Since $\phi_0(x)$, the ground state baryon wave function, is rotationally invariant, the linear equation for $\delta\phi$ has a conserved angular momentum, and we can look for solutions of definite partial wave. For instance, we can look for quadrupole solutions, $L = 2$. These describe phonons – baryonic analogues of the phonons of nuclear physics.

We have been assuming so far that all quarks are of the same flavor (as in the Δ^{++} or Ω^-). If we consider now baryons with two flavors (say up and down quarks), we will introduce separate wave functions for the two flavors of quarks, and the equation analogous to (33) will have solutions in which the up quarks will oscillate relative to the down quarks. These are baryon analogues of the giant dipole resonances of nuclear physics.

As a final example, we may consider rotational excitations of baryons. So far, we are assuming that the ground state baryon is spherically symmetric. This is a valid assumption when the quarks are very heavy and the spin-dependent forces are negligible. When the quarks are not so heavy and the spin-dependent forces must be included, one can no longer, in general, expect the ground state baryon to be spherically symmetric.

In fact, one flavor baryons such as the Δ^{++} or Ω^- have, in the ground state, all of the quark spins aligned. Such baryons therefore have spin $\frac{1}{2}N$ (spin $\frac{3}{2}$ for the actual Δ^{++} and Ω^-). Associated with this large total spin $\frac{1}{2}N$ is a preferred spin vector, the direction in which the baryon spin is pointing (which is a well-defined concept when the total spin is large). When spin orbit forces are included, the coupling of the spatial motion to the total spin vector will produce a ground state baryon wave function that is not rotationally symmetric. This is an analogue of the "deformed nuclei" of nuclear physics.

(The large N analogue of the nucleon, on the other hand, is probably not deformed. It is somewhat ambiguous how to generalize the nucleon to large N. The best procedure is probably to take N odd, $N = 2k + 1$ for large k, and consider a baryon with $k + 1$ up quarks and k down quarks, or vice-versa. The ground state of this system probably has spin $\frac{1}{2}$ for any k, because the spin-spin forces favor having the down quark spins antiparallel to the up quark spins. This small spin $\frac{1}{2}$ is too small to influence the orbital motion to lowest order in $1/N$. So the ground state "proton," unlike the large N analogues of the Δ^{++} and Ω^-, is probably rotationally symmetric.)

Once the ground state baryon is not rotationally symmetric, there will be solutions of (33) describing overall tumbling motions of the whole baryon. These are analogues of the rotational excitations of nuclear physics.

Thus, in the large N limit, many phenomena of nuclear physics have analogues for baryons. But baryon physics is much simpler than nuclear physics, because the time-dependent Hartree equation is exact for baryons when N is large; there is no such statement for nuclei.

8. Some additional phenomena

In this section we will consider several additional phenomena: nuclei and baryonium, deep inelastic scattering, and "forbidden processes."

8.1. Nuclei and baryonium

How would one describe nuclei – that is, bound states of several baryons – in the large N expansion?

Let us consider, for example, the simplest case of a nucleus made from two baryons. Such a nucleus is, microscopically, a bound state of $2N$ quarks.

We have already written an equation [eq. (26)] describing a system of $2N$ quarks. This equation was introduced in connection with a discussion of baryon-baryon scattering. Indeed (26) certainly has scattering solutions, since one can choose initial data such that in the far past ϕ_1 and ϕ_2 were localized in different regions of space and heading for a collision. Now, however, we are looking for bound states of the baryon-baryon system.

It is perfectly possible that (26) could have bound state solutions in addition to the scattering solutions. In fact, one may choose arbitrary initial values of ϕ_1 and ϕ_2 at time zero and use (26) to determine ϕ_1 and ϕ_2 at all times; this solution for ϕ_1 and ϕ_2, inserted back into the ansatz (25), gives, to lowest order in $1/N$, a valid solution of the time-dependent Schrödinger equation. It may be that, for suitable choices of the initial data, (26) yields bound solutions in which ϕ_1 and ϕ_2 do not separate from each other (they remain localized in the same region) even as the time goes to $+\infty$.

Even if such solutions do not exist in (26), they probably exist in the analogous equations in other spin-isospin channels. (Recall that (26) is written for the particularly simple case in which all quarks have the same spin and isospin; the generalizations of (26) to include more than one spin and isospin are, of course, more appropriate for real nuclei.)

Bound solutions of (26) correspond to bound solutions of the Schrödinger equation, but not to energy eigenstates. How would one find the energy eigenstates – the nuclear energy levels?

As in the previous discussion of highly excited baryon states, the correct prescription is to follow the procedure of Dashen, Hasslacher, and Neveu. One looks for solutions of (26) which are periodic in time – not necessarily with the harmonic $e^{-i\epsilon t}$ time dependence, but periodic. The solution may tumble and pulsate, but it does so periodically. One then quantizes the periodic motions in the manner of Dashen, Hasslacher, and Neveu, requiring that the action in a period be a multiple of 2π. One could thus, in principle, find the nuclear energy levels.

One may, likewise, wish to describe baryon-antibaryon bound states in the large-N limit. To accomplish this, one would write down the non-linear equations that follow from the ansatz (27) for the baryon-antibaryon system. One then would look for bound solutions of this equation, and would impose the semi-classical quantization condition on the bound solutions.

Finally, there has recently been a considerable amount of interest in possible "baryonium" states – color singlet states in the two-quark, two-antiquark system. A particularly interesting class of such states, which are expected to be narrow, are the states that are antisymmetric in the color of the two quarks and likewise antisymmetric in the color of the two antiquarks.

It seems that there is a way to generalize such states to large N so as to have a smooth large N limit. The appropriate generalization is to consider color singlet states with $N-1$ quarks and $N-1$ antiquarks, antisymmetric in the color of the quarks and in the color of the antiquarks.

Such "baryonium" states exist and become narrow as N becomes large. To describe them, one would consider an ansatz like (27) but with only $N-1$ quarks and $N-1$ antiquarks. Semiclassical quantization of bound state solutions of the corresponding Hartree equations would yield the baryonium states. (Such solutions definitely exist because in this channel the force is strong enough at large distances to ensure the existence of bound solutions.) The widths of the baryonium states vanish at $N=\infty$ because the Hartree approximation is exact at $N=\infty$, and the Hartree equations yield bona fide, stable bound states.

8.2. Deep inelastic scattering

Deep inelastic scattering from a baryon target, as from any other target in QCD, will show a scaling behavior at large Q^2. It is interesting, though, that in scattering from a baryon target, for large N, precocious scaling can be expected in the sense that scaling sets in while Q^2 is much *less* than the mass squared of the baryon. (This comment was suggested by K. Wilson.)

In fact, for large N the baryon mass is of order N, but scaling will set in at momentum transfer of order 1. For large N the baryon consists of N independently moving quarks. Deep inelastic scattering involves only one of those quarks; the other $N-1$ are spectators. Scaling occurs as soon as Q^2 is large compared to the parameters relevant to the motion of the one quark that participates in the scattering process. These parameters (the quark mass, the spatial extent of its wave function, etc.) are of order one. For scaling, Q^2 must be large compared to these parameters, but not compared to the baryon mass squared, which is of order N^2.

The following formal argument (suggested by M. Peskin) backs up the above conclusion. The corrections to scaling are controlled by the ratios of matrix elements of twist-four operators to those of twist-two operators. Both the twist-two and twist-four operators have matrix elements in a baryon state of order N. (This conclusion does not depend on the quark masses being large.) The ratios of the twist-four to twist-two matrix elements is therefore of order one, and to get scaling, Q^2 must be large compared to this ratio, but need not be large compared to the baryon mass squared.

8.3. "Forbidden processes"

Rather than leave the impression that *everything* has a smooth large N limit, let us now consider some examples of processes that have cross sections of order e^{-N}

– processes that are forbidden in every finite order of the $1/N$ expansion.

The simplest such process (fig. 39) is $e^+e^- \to B\bar{B}$. The virtual photon created by the electron and positron directly creates one quark-antiquark pair. But to produce $B\bar{B}$ we need N quark-antiquark pairs. If the probability to create one additional such pair is x, which is a number less than one, then the probability to create $N-1$ additional pairs is x^{N-1}, which vanishes exponentially for large N, like e^{-cN}, where $c = -\ln x$ is a positive number. So the cross section for $e^+e^- \to B\bar{B}$ vanishes like e^{-cN}.

One may also ask about the crossed process, $eB \to eB$. Since Yang-Mills theory has crossing symmetry for every N, it also has crossing symmetry order by order in $1/N$. Let us see how this works out.

In lepton-baryon elastic scattering there are, as N becomes large, two distinct kinematic regimes to consider. One may consider the regime in which the momentum transfer is kept fixed as $N \to \infty$. In this case the change in the baryon velocity is of order $1/N$, because the baryon mass that appears in the relationship $p = M_B v$ (or its relativistic generalization) is of order N. Alternatively, one may consider the case in which the change in the baryon velocity in the scattering process is of order one; then the momentum transfer is of order N.

In the first regime, fixed momentum transfer for large N, the baryon electromagnetic form factor is simply the Fourier transform of the baryon change density, which, as we noted in the discussion of eq. (17), is in the non-relativistic approximation $N\phi^*\phi(x)$. The Fourier transform of this function is N times a function that depends only on the momentum (and not on N), so in the fixed momentum transfer regime, the baryon form factor is of order N.

In the fixed velocity change regime, the situation is very different. We are interested in the matrix element of the current between a baryon of velocity v and one of velocity v'. The wave function of a quark in a baryon of velocity v is different from the wave function in a baryon of velocity v'. The current directly couples to one quark and changes the wave function of that quark, but the baryon has N quarks. In calculating the current matrix element $\langle B(v')|J_\mu|B(v)\rangle$ we encounter, for each of the $N-1$ quarks to which the current does not couple, an overlap integral between the wave function of a quark in a baryon of velocity v and the wave function in a baryon of velocity v'. Denoting this overlap integral as y, which is a num-

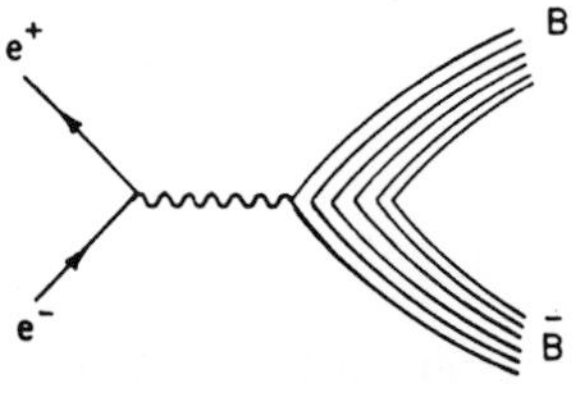

Fig. 39. $e^+e^- \to B\bar{B}$.

ber less than one, the current matrix element has a factor y^{N-1}, so the form factor in the regime of fixed velocity transfer vanishes exponentially for large N.

Under crossing, the physical annihilation process $e^+e^- \to B\bar{B}$ is mapped into $eB \to eB$ in the regime of fixed velocity transfer. Our results are entirely consistent with crossing symmetry – the two processes that cross into each other both vanish exponentially for large N.

For similar reasons, the cross sections for production or annihilation of a baryon-antibaryon pair in mesonic reactions vanish exponentially for large N. For instance, $MM \to B\bar{B}$ and its time reversed mate $B\bar{B} \to MM$ vanish exponentially. Again, this is because of the difficulty in creating or destroying N quarks pairs.

A more delicate case is the inclusive annihilation process $B\bar{B} \to$ any number of mesons. In the regime in which the baryon velocity is kept fixed as $N \to \infty$, this too vanishes exponentially. Heuristically, this is because of the large and complicated rearrangement needed to turn $B\bar{B}$ into mesons. Mathematically, we may note that the equation following from ansatz (27) is exact in the fixed velocity, large N regime. In the approximation of this equation, the N-quark, N-antiquark system propagates indefinitely as a baryon-antibaryon pair (rather than N mesons) if that is the initial situation. One could calculate to any finite order in $1/N$ (as will become clearer in the next section) by perturbing around (27). Finite orders of perturbation around (28) cannot change the baryon-antibaryon pair into N mesons, because the rearrangement from $B\bar{B}$ to N mesons is too large to be carried out in finite order. Therefore the inclusive annihilation cross section, at fixed center of mass velocity, for large N, is smaller than any power of $1/N$.

It should be stressed that this result depends on keeping the velocity fixed as $N \to \infty$. If the velocity is of order $1/N$, the result will be different. Baryon-antibaryon annihilation at rest is definitely not forbidden. It probably proceeds (for very large N!) mainly through the following cascade. The baryon-antibaryon pair with very small kinetic energy emit a single meson, forming a baryonium state ($N-1$ quarks, $N-1$ antiquarks) that is below the $B\bar{B}$ threshold. Once a state is formed that is below $B\bar{B}$ threshold, decay of the baryonium into mesons (and/or glueballs) is inevitable because no other channels for the final end products of the reaction are open. The baryonium state, being narrow, will decay slowly, but inevitably, eventually emitting a meson and leaving a state of $N-2$ quarks and $N-2$ antiquarks (if this channel is open; otherwise a different process will occur, but leading to a similar cascade). The state of $N-2$ quarks and $N-2$ antiquarks is also narrow, for large N, but it too would eventially decay, probably to a narrow state of $N-3$ quarks and $N-3$ antiquarks. And so on – after a long but inexhorable process, the $B\bar{B}$ pair has disappeared into mesons and glueballs.

9. Baryons made from light quarks

So far our attention has been restricted to non-relativistic phenomena. The Hartree equations that have been derived are valid only for baryons made from heavy quarks. The treatment of scattering processes by way of non-relativistic Hartree equations assumes that all of the particles have small velocities. Obviously, it is essential to know whether there are similar results for baryons made from light quarks, or for baryons interacting with relativistic velocities. This question will be addressed in this section.

Another matter that will be considered here is the question of justifying the Hartree approximation. The claim that the Hartree equations become exact for large N has not been fully justified in any of the discussion above. We will reconsider this question in the context of a path integral treatment of the baryon problem that is introduced below.

Concerning the quark mass, the basic claim that will be made here is that even when the quarks are not heavy, the large N limit of baryons is still a kind of Hartree approximation – N quarks moving independently in an average potential. Moreover, the qualitative results that we have derived concerning the N dependence of various quantities are still valid. For example, the baryon mass is of order N, the baryon size is of order one, and the scattering processes involving baryons have smooth and non-trivial large N limits, whether the quark masses are large or small.

However, when the quark masses are not large, one can no longer use the non-relativistic Hartree equations to calculate the average potential in which the quarks are moving. One would have to use relativistic Hartree equations. The appropriate relativistic Hartree equations are not known in four dimensions; the difficulties that prevent us from determining them are similar to the difficulties that prevent us from summing the planar diagrams to determine the meson spectrum.

How, without knowing the appropriate relativistic Hartree equations, can one be convinced that the large N limit for baryons really does consist of N quarks moving independently in an average potential?

We are dealing with a system of N objects, quarks, which, after antisymmetrization with respect to color, are effectively bosons. These N "bosons" have attractive interactions. Under very general conditions, a system of N bosons with attractive interactions has a Hartree-type large N limit, with the N particles all moving independently in an average field.

If the interactions are relativistic, then relativistic Hartree equations must be considered, to determine the correct average field. But the general picture of baryons presented in this paper, which depends only on the fact that all N quarks are moving in some average field, would still be valid.

For example, if baryons could be described simply in terms of two-body forces among the quarks, we could definitely expect a Hartree-type large N limit, because the quark-quark interaction is of order $1/N$, and a system of N particles with attractive two-body forces of order $1/N$ always has a Hartree type limit.

Thus, if we consider Feynman diagrams for the propagation of N quarks which are not extremely heavy, we encounter diagrams like that of fig. 40, in which two of the N quarks in a baryon interact through a process more complicated than simple Coulomb exchange. Such diagrams modify the quark-quark interaction, and

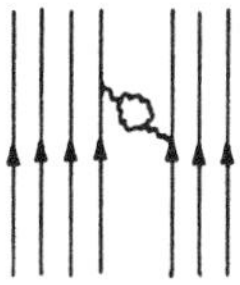

Fig. 40. A higher order correction to the quark-quark interaction.

the modified quark-quark interaction must be inserted into the Hartree equations, but otherwise the picture does not change.

Actually, we must consider also diagrams which do not simply modify the two-body forces, but introduce three-body forces, four-body forces, etc. However, it is easy to see that the dependence on N is such that the many-body forces simply induce, for large N, corrections to the average Hartree potential.

For example, in fig. 41 is shown a typical diagram describing a three-quark scattering process. This diagram introduces three-quark forces. The three-quark forces are of order $1/N^2$, because in fig. 41 there are four interaction vertices, each of order $1/\sqrt{N}$. The three-quark forces are negligible for very heavy quarks, but with not so heavy quarks, they must be included.

Three-quark forces of order $1/N^2$ lead to a picture very similar to the picture that comes from two-quark forces of strength $1/N$. The number of ways to pick three quarks in the baryon is of order N^3, so the three-quark forces contribute to the baryon mass a term of order $N^3(1/N^2)$ or N. Any one quark participates in N^2 of these three-quark combinations. Any one quark, in other words, interacts with N^2 quark pairs. The strength of interaction with any one of the quark pairs is of order $1/N^2$, so the total force on any one quark coming from the three-body forces is of order one. Being a sum of many small terms, this force can be regarded as an average background potential. In other words, the three-quark forces simply renormalize the average Hartree potential.

One can likewise study diagrams more complicated than those of figs. 40 and 41, and see that they can be interpreted, for large N, as modifications of the average potential.

The picture that emerges from this point of view is that for baryons made from light quarks – like those made from heavy quarks – the large N limit is of the Hartree nature. To write the appropriate relativistic Hartree equations in four

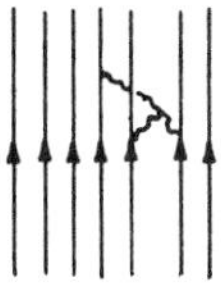

Fig. 41. A diagram that gives a three-quark interaction.

dimensions is, however, a task well beyond our present abilities.

In this situation, we are only able to draw qualitative conclusions about the large N limit for baryons. The qualitative conclusions are that, for instance, the baryon mass is of order N, the baryon size and shape are of order one, baryon S matrix elements have smooth and non-trivial large N limits, the cross section for baryon-antibaryon production by mesons or leptons is of order e^{-N}, etc.

The situation for baryons is actually quite similar to the situation for mesons. For mesons made from very heavy quarks the spectrum (of the low-lying states) would be given simply by the Schrödinger equation with a Coulomb potential. For mesons made from light quarks we know, from 't Hooft's work, that to obtain the large N limit we must sum the planar diagrams. This is beyond our present ability, so, for the time being, we can draw only qualitative conclusions (meson masses of order one, two-body decay amplitudes of order $1/\sqrt{N}$, three-body decay amplitudes of order $1/N$, etc.).

Two final topics will be considered in this section. The first is a comparison of the above remarks with expectations from the string model. The second is a consideration of two-dimensional quantum chromodynamics, in which it is possible to write down explicitly the relativistic Hartree equations that are relevant when the quarks are light. (In connection with the study of two-dimensional quantum chromodynamics, we will also find a more convincing answer to the question of showing that the Hartree approximation becomes exact for large N.)

Turning first to the strong model, there is a certain string picture according to which a baryon is a system of N quarks, each at the end of a string. The strings meet at a common junction. This is sketched in fig. 42.

Actually, this string picture is not necessarily very accurate in QCD. It is easy to suspect that the bag picture may be more accurate than the string picture, at least for low-lying baryons. Moreover, from our preceding discussion of baryon S matrix elements, we know that amplitudes involving baryons are *not* dual in the usual sense. The baryon-baryon scattering amplitude at $N = \infty$, for example, is *not* a sum of single poles.

The reason for considering the string model here is that it is useful to see how

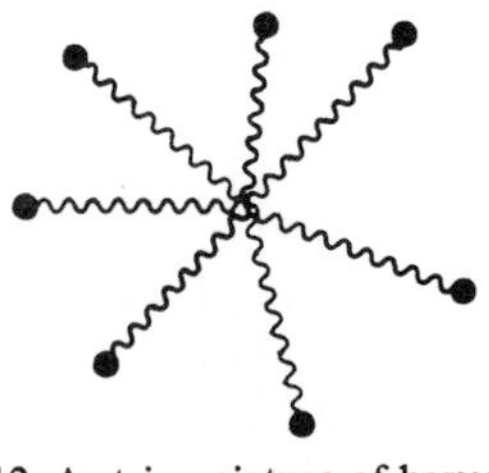

Fig. 42. A string picture of baryons.

the Hartree picture comes about for large N in a model very different from our previous discussion.

Indeed, the string model of the baryon simplifies dramatically for large N. The basic problem afflicting the string model of the baryon is that it is hard to take account of the motion of the junction. For large N, however, the motion of the junction can be neglected. The junction coupled to N quarks can be regarded as a heavy object, and its motion can be ignored, to lowest order in $1/N$. To put it differently, the junction is pulled on by N quarks, but simply from randomness, the pulls of the various quarks cancel each other to high accuracy, so that the junction does not move. (To state the same thing in a Fourier transformed version, the constraint of conservation of momentum at the junction can simply be ignored, to lowest order in $1/N$, because the total momentum transfer at the junction is the sum of the momentum transfer from each of the N quarks, and the N terms will always cancel each other, to high accuracy, just on statistical grounds, so that momentum is conserved automatically and the constraint need not be imposed.)

Thus, to lowest order in $1/N$, one should consider N quarks interacting with a fixed junction. The interaction with a fixed junction is a string model analogue of the Hartree potential. The motion of the junction should then be considered perturbatively in $1/N$.

So far, we have tacitly assumed that string-string collisions can be neglected. It is instructive to realize that this assumption is a poor one. It is true that the amplitude for a collision of two given strings is of order $1/N$, but there are N strings, and therefore N^2 possible pairs of strings that might collide. Altogether, in fact, the number of string-string interactions per unit time is of order N, and the string-string collisions must be considered even in the large N picture.

Since the density of strings at any given distance from the junction can be considered as static and non-fluctuating, the string-string collisions can be regarded, from the point of view of any one string, as providing an average background potential in which that string is moving.

Thus a Hartree-like picture emerges from a very different starting point, the string model.

Finally, let us turn our attention to two-dimensional quantum chromodynamics. It is possible in two dimensions to write the explicit relativistic Hartree equations that should be used for light quarks. This is somewhat analogous to the fact that in two dimensions, unlike four, it is possible to sum the planar diagrams.

Thus, let us consider the action of two-dimensional quantum chromodynamics with an arbitrary quark mass:

$$A = \int \mathrm{d}^2x \left[\bar{\psi}_i (i\partial\!\!\!/ - m)\, \psi^i - F_{\mu\nu}{}^i{}_j F_{\mu\nu}{}^j{}_i - \frac{g}{\sqrt{N}} \bar{\psi}^i \gamma_\mu \psi_j A_\mu{}^j{}_i \right] . \tag{34}$$

What makes two-dimensional quantum chromodynamics simple is that in appropriate gauges it is possible to explicitly eliminate the gluon fields. For example, in the Coulomb gauge $A_1 = 0$, the gauge field can be completely eliminated in favor of

Coulomb interactions:

$$A = \int dx\, dt \bar{\psi}_i (i\partial\!\!\!/ - m)\, \psi^i(x,t) - \frac{g^2}{N}\int dx\, dy\, dt \bar{\psi}_i \gamma^0 \psi^j(x,t)$$

$$\times\, \bar{\psi}_j \gamma_0 \psi^i(y,t) |x-y| \,. \tag{35}$$

Note the appearance of the linear Coulomb potential, $(g^2/N)|x-y|$.

We would like to analyze (35) by a path integral method, considering the integral $\int d\psi\, d\bar{\psi}\, e^{iA}$. Because of the fact that the action (35) is not quadratic in the ψ and $\bar{\psi}$ fields, a direct path integral treatment of (35) is difficult. It is therefore useful to introduce an action equivalent to (35) but quadratic in the Fermi fields.

As far as the SU(N) indices are concerned – ignoring the non-locality – the interaction term in (35) is of the type $\bar{\psi}_i \psi^j \bar{\psi}_j \psi^i$, or, permuting the various terms, it is of the type $\bar{\psi}_i \psi^i \bar{\psi}_j \psi^j$. Given a local $\bar{\psi}_i \psi^i \bar{\psi}_j \psi^j$ interaction, there is a fairly well-known trick [11] to introduce an equivalent Lagrangian that is quadratic in the Fermi fields. One introduces an auxiliary field σ into the theory and adds to the action a term $-(\sigma + \bar{\psi}\psi)^2$, so that $(\bar{\psi}\psi)^2$ is replaced by $(\bar{\psi}\psi)^2 - (\sigma + \bar{\psi}\psi)^2$. The introduction of σ and the addition of the extra term make no change in the theory, because by shifting σ by an amount $-\bar{\psi}\psi$, one could convert this Lagrangian to $(\bar{\psi}\psi)^2 - \sigma^2$, so that the new field σ is simply decoupled from the old fields. Instead of shifting, however, we may simply expand the square in $(\bar{\psi}\psi)^2 - (\sigma + \bar{\psi}\psi)^2$. The $(\bar{\psi}\psi)^2$ term cancels and we are left with $-\sigma^2 - 2\sigma\bar{\psi}\psi$, which must be equivalent to the original $(\bar{\psi}\psi)^2$ interaction.

Another way to see the equivalence is to take the Lagrangian $-\sigma^2 - 2\sigma\bar{\psi}\psi$ and eliminate σ by means of its equations of motion. Since there are no derivatives of σ in the Lagrangian, σ can be explicitly eliminated by using the equations of motion, and we return to the previous $(\bar{\psi}\psi)^2$ Lagrangian.

To find a Lagrangian quadratic in ψ and $\bar{\psi}$ and equivalent to (35) we will use a non-local version of this trick. We introduce a non-local field σ, dependent on two space coordinates and one time coordinate, which will roughly have the significance

$$\sigma(x,y,t) = \bar{\psi}_i(x,t)\, \psi^i(y,t) \,. \tag{36}$$

Actually, σ should be introduced as a matrix in Dirac space, $\sigma_{\alpha\beta}(x,y,t) = \bar{\psi}_{\alpha i}(x,t)\, \psi^i_\beta(y,t)$. We will, however, henceforth treat Dirac algebra in a cavalier way, not keeping track of the spinor indices.

In terms of σ defined in (36), we can write a new action equivalent to (35):

$$A = \int dx\, dt \bar{\psi}_i (i\partial\!\!\!/ - m)\, \psi^i(x,t) + \int dx\, dy\, dt \sigma(x,y,t)\, \sigma^*(x,y,t)$$

$$+ \frac{g}{\sqrt{N}} \Big(\int dx\, dy\, dt \bar{\psi}_i(x,t)\, \psi^i(y,t)\, \sigma(x,y,t) \sqrt{|x-y|} + \text{h.c.} \Big) \,. \tag{37}$$

The equivalence between (37) and (35) can be seen in several ways. One may eliminate σ from (37) by means of its equations of motion, arriving back at (35).

Equivalently, one may shift $\sigma(x, y, t)$ by an amount proportional to $(g/\sqrt{N})$ $\sqrt{|x-y|}\, \bar{\psi}_i(x, t)\, \psi^i(y, t)$, obtaining from (37) a decoupled σ and a ψ field interacting according to (35).

We now will treat (37) by a path integral method,

$$\int \mathrm{d}\psi\, \mathrm{d}\bar{\psi}\, \mathrm{d}\sigma\, \mathrm{d}\sigma^* \exp iA \ . \tag{38}$$

Since, however, we wish to study baryons, we must introduce a source that creates and destroys baryons. In fact, we will introduce a local operator $J(x)$ with the quantum numbers to create a baryon. $J(x)$ will be the product of all N colors of quark fields at the point x:

$$J(x) = \psi_1(x)\, \psi_2(x)\, \psi_3(x) \ldots \psi_N(x) \ . \tag{39}$$

Notice that because the quark fields anticommute, it is not necessary to antisymmetrize (39) in color. (39) defines a color singlet operator as it stands. (Since we are not keeping track of Dirac indices, the Dirac indices have not been written in (39). If one wishes to keep track of Dirac indices, one should choose in (39) the same Dirac component for each of the N quark fields. For instance, one may consider the positive chirality component of each quark field.)

To study baryons we will study the two-point function of the operator J:

$$\langle J(x)\, J^+(0)\rangle \ . \tag{40}$$

We will study this two-point function by studying the path integral formula

$$\int \mathrm{d}\psi\, \mathrm{d}\bar{\psi}\, \mathrm{d}\sigma\, \mathrm{d}\sigma^* \psi_1(x)\, \psi_2(x) \ldots \psi_N(x)\, \psi_1{}^+(0)\, \psi_2{}^+(0) \ldots \psi_N{}^+(0)$$
$$\times \exp i\Big[\int \mathrm{d}x\, \mathrm{d}t \bar{\psi}_i(i\partial\!\!\!/ - m)\, \psi^i + \int \mathrm{d}x\, \mathrm{d}y\, \mathrm{d}t \sigma(x, y, t)\, \sigma^*(x, y, t)$$
$$+ \frac{g}{\sqrt{N}} \int \mathrm{d}x\, \mathrm{d}y\, \mathrm{d}t(\sigma(x, y, t)\, \bar{\psi}_i(x, t)\, \psi^i(y, t) \sqrt{|x-y|} + \text{h.c.})\Big] \tag{41}$$

To deal with (41) one first integrates out the quark fields. This can be carried out explicitly, although only formally, because the quark fields appear only quadratically in the Lagrangian.

Upon doing so, we obtain in the exponential the trace of the logarithm of the quandratic operator to which the quarks are coupled. This quadratic operator is the free Dirac operator $(i\partial\!\!\!/ - m)$ plus an interaction with the σ field which we will write symbolically as $g\sigma/\sqrt{N}$ (from (41), we see that the interaction with σ is the operator whose kernel is $g\sigma(x, y, t) \sqrt{|x-y|}/\sqrt{N}$; we will write this symbolically as $g\sigma/\sqrt{N}$).

In addition to the "trace log" term appearing in the exponential, we will get, upon integrating the quarks out of (41), an extra term reflecting the insertion of the product $\psi_1(x)\, \psi_2(x) \ldots \psi_N(x)\, \psi_1{}^+(0)\, \psi_2{}^+(0) \ldots \psi_N{}^+(0)$ in (41). In fact, for each of the N species of quark, we will encounter a factor consisting of the quark propagator from 0 to x in the background field σ. We will write this propagator as $S(x, 0; g\sigma/\sqrt{N})$.

The result of integrating the quarks out of (41) is then

$$\int \mathrm{d}\sigma\, \mathrm{d}\sigma^* (S(x, 0; g\sigma/\sqrt{N}))^N \exp(N \operatorname{Tr} \ln(i\not\partial - m - g\sigma/\sqrt{N}) + i \int \mathrm{d}x\, \mathrm{d}y\, \mathrm{d}t \sigma^* \sigma(x, y, t)) , \tag{42}$$

which, on bringing the $(S(x, 0; g\sigma/\sqrt{N}))^N$ term into the exponential, can be rewritten

$$\int \mathrm{d}\sigma\, \mathrm{d}\sigma^* \exp(N \operatorname{Tr} \ln(i\not\partial - m - g\sigma/\sqrt{N}) + i \int \mathrm{d}x\, \mathrm{d}y\, \mathrm{d}t \sigma^* \sigma + N \ln S(x, 0; g\sigma/\sqrt{N})) . \tag{43}$$

The advantage of this procedure is that in (43) the only N dependence comes from the factors of N that are explicitly written. Because there is now only a single (non-local) field over which to integrate, there is no longer an implicit N dependence from having a number of fields that depends on N. Because the N dependence is explicit, it is comparatively easy to determine the large N limit.

The first step is to absorb a factor of $\sqrt{N}$ in the σ field, $\sigma/\sqrt{N} \to \sigma$. (43) can then be rewritten

$$\int \mathrm{d}\sigma\, \mathrm{d}\sigma^* \exp N(\operatorname{Tr} \ln(i\not\partial - m - g\sigma) + i \int \mathrm{d}x\, \mathrm{d}y\, \mathrm{d}t \sigma^* \sigma + \ln S(x, 0; g\sigma)) . \tag{44}$$

The important fact about (44) is that the only dependence on N comes from an overall factor of N multiplying the whole action. Thus, (44) is of the general form

$$\int \mathrm{d}\sigma\, \mathrm{d}\sigma^* \exp[iN\Gamma(\sigma, \sigma^*; x)] , \tag{45}$$

with a functional $\Gamma(\sigma, \sigma^*; x)$ that is written explicitly in (44). The functional Γ depends on the space-time point x because of the term $\ln S(x, 0; g\sigma)$ in (44).

The large N limit of an integral such as (44) or (45) – with N appearing only as an overall factor multiplying the entire action – can be calculated by stationary phase. As N becomes large, because of the factor of N in the exponent of (44) or (45), deviations from stationary phase are more and more strongly suppressed. At $N = \infty$ the stationary phase approximation becomes exact.

Thus, to evaluate (44) or (45) we first look for a "classical σ field," σ_{cl}, satisfying

$$\left(\frac{\partial \Gamma}{\delta \sigma}\right)_{\sigma = \sigma_{\mathrm{cl}}} = 0 . \tag{46}$$

To a first approximation, the integral (45) is just equal to $\exp iN\Gamma(\sigma_{\mathrm{cl}})$.

It is possible to proceed further and construct a systematic expansion in powers of $1/N$. We simply write $\sigma = \sigma_{\mathrm{cl}} + (1/\sqrt{N})\, \delta\sigma$ and expand the effective action in powers of $\delta\sigma$, near $\delta\sigma = 0$. The linear term in the expansion vanishes, because σ_{cl} is a stationary point of the effective action. The quadratic term in the expansion is of order one, because the overall factor of N in front of the action cancels two factors of $1/\sqrt{N}$ from expanding to quadratic order in $\delta\sigma$. The terms cubic and higher in $\delta\sigma$ are suppressed by powers of $1/\sqrt{N}$. Thus, the expansion is

$$
\begin{aligned}
N\Gamma(\sigma) &= N\Gamma(\sigma_{\rm cl} + \delta\sigma/\sqrt{N}) \\
&= N\Gamma(\sigma_{\rm cl}) + \int \mathrm{d}x\, \mathrm{d}y \left(\frac{\delta^2\Gamma}{\delta\sigma(x)\,\delta\sigma(y)}\right)_{\sigma_{\rm cl}} \delta\sigma(x)\,\delta\sigma(y) \\
&\quad + \frac{1}{\sqrt{N}} \int \mathrm{d}x\, \mathrm{d}y\, \mathrm{d}z \left(\frac{\delta^3\Gamma}{\delta\sigma(x)\,\delta\sigma(y)\,\delta\sigma(z)}\right)_{\sigma_{\rm cl}} \delta\sigma(x)\,\delta\sigma(y)\,\delta\sigma(z) \\
&\quad + \mathrm{O}(1/N)\,. \qquad (46)
\end{aligned}
$$

The integral that we must do is therefore

$$
\int \mathrm{d}\delta\sigma \exp[iN\Gamma(\sigma_{\rm cl})] \exp\left[i\int \frac{\delta^2\Gamma}{\delta\sigma(x)\,\delta\sigma(y)} \delta\sigma(x)\,\delta\sigma(y)\,\mathrm{d}x\,\mathrm{d}y + \mathrm{O}(1/\sqrt{N})\right] \qquad (47)
$$

and the answer is

$$
\frac{\exp[iN\Gamma(\sigma_{\rm cl})]}{\det\left(\dfrac{\delta^2\Gamma}{\delta\sigma(x)\,\delta\sigma(y)}\right)} (1 + \mathrm{O}(1/\sqrt{N}))\,, \qquad (48)
$$

where the terms of order $1/\sqrt{N}$ that figure on the right hand side of (48) could be calculated, in principle, by treating perturbatively the terms of order $1/\sqrt{N}$ that appear in (46). In this way one could obtain a systematic expansion in powers of $1/\sqrt{N}$.

What is the connection of this discussion with our previous, non-relativistic analysis?

It is possible to show that in the non-relativistic limit (very heavy quarks) (46) can be reduced to the non-relativistic Hartree equation (17). (46) is a relativistic generalization of (17) which describes baryons made from quarks that need not be heavy. The above derivation of (46) thus justifies our non-relativistic Hartree equation, as well as extending that equation to the relativistic case.

(The reduction of (46) to (17) in the heavy quark limit is not completely straightforward and involves the following. First, one takes the limit of x going to timelike infinity in (44) or (45) since to determine the baryon spectrum we want $\langle J(x)\,J^{+}(0)\rangle$ for large time. In this limit the term $\ln S(x, 0; g\sigma)$ in (44) plays only the role of defining a boundary condition. The resulting equation can be reduced to (17), with the identification $\sigma(x, y, t)/\sqrt{|x-y|} = \phi(x)\,\phi^{*}(y)$, ϕ being the wave function in (17).)

Our other formulas can also be rederived from this point of view. For instance,

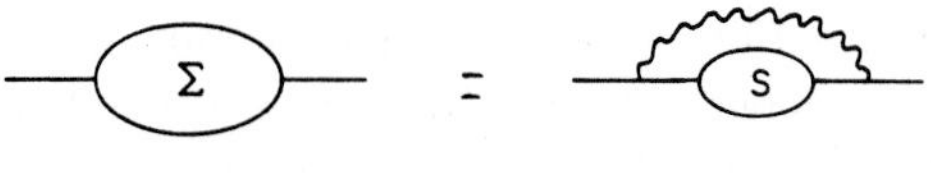

Fig. 43. 't Hooft's equation for the quark propagator.

meson-baryon scattering is especially easy to understand in this treatment. First, though, we must understand how to recover 't Hooft's results for mesons.

Let us recall that $\sigma(x, y, t)$ was introduced to represent $\overline{\psi}_i(x, t)\, \psi^i(y, t)$, which is the product of a quark and an antiquark field, at different points. σ therefore has the quantum numbers to create a meson.

The calculation leading from (41) to (44) could be carried out in the vacuum or meson sector, without the insertion of the baryon operators $J(x)$ and $J^{+}(0)$ in (41). In this way, we reach the same formula (41), but without the $N \ln S(x, 0; g\sigma)$ term that reflects the presence of the baryon. Again, one could carry out the $1/N$ expansion along the lines of equations (46) through (48). In this way, one is called upon first to solve the equation $\delta\Gamma/\delta\sigma = 0$ (with the baryon term omitted from Γ). And second, in order to do the Gaussian integral over $\delta\sigma$, one must invert the operator $\delta^2\Gamma/\delta\sigma(x)\,\delta\sigma(y)$.

The equation $\delta\Gamma/\delta\sigma = 0$ turns out to yield 't Hooft's equation for the quark self-energy, sketched in fig. 43. What about the equation for inverting the operator $\delta^2\Gamma/\delta\sigma\delta\sigma'$?

Because σ is a bilocal field with meson quantum numbers, in inverting the operator $\delta^2\Gamma/\delta\sigma\delta\sigma'$, we are solving for the propagator of a field with the quantum numbers to create or destroy a meson. In fact, the equation for inverting $\delta^2\Gamma/\delta\sigma\delta\sigma'$ turns out to be equivalent to 't Hooft's Bethe-Salpeter equation for mesons (fig. 44).

What happens now when one includes a baryon? We must now reinsert the baryon term $N \ln S(x, 0; g\sigma)$ in (41). We still must, in constructing the $1/N$ expansion, invert the operator $\delta^2\Gamma/\delta\sigma\delta\sigma'$. Now, however, this operator contains an extra term, the baryon term. The equation for inverting $\delta^2\Gamma/\delta\sigma\delta\sigma'$ is still a linear equation for the motion of a meson, but it is no longer 't Hooft's Bethe-Salpeter equation; the propagation of the meson is modified by the presence of the baryon.

Indeed, the equation for inverting $\delta^2\Gamma/\delta\sigma\delta\sigma'$ is a relativistic version of an equation that we have considered previously in the non-relativistic case – the linear equation (29) for motion of a meson in a background baryon field.

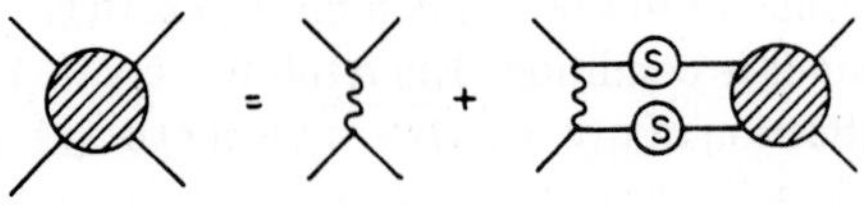

Fig. 44. 't Hooft's Bethe-Salpeter equation for mesons.

Apart from permitting a treatment of relativistic phenomena in two dimensions, the path integral method discussed here is a fairly satisfying demonstration that the Hartree approximation becomes exact for large N, and an efficient way to construct a systematic expansion in powers of $1/N$.

10. Conclusion: baryons as the "monopoles" of QCD

There is, actually, a simple way to summarize all of the results we have obtained – a single statement that encompasses the above conclusions.

From 't Hooft's work we know that mesons become free and non-interacting at $N = \infty$. The meson-meson couplings are, in fact, of order $1/N$.

For large N, we may regard QCD as a weakly coupled field theory of mesons. It is a theory of effective local meson fields with effective local interactions, of order $1/N$.

Weakly coupled field theories sometimes possess, apart from the usual particles, additional states whose masses diverge, for weak coupling, like the inverse of the coupling. Such states are solitons or Polyakov-'t Hooft monopoles [12].

DO such states exist in QCD? Are there states in QCD whose masses diverge in the weak coupling (large N) regime, and whose other properties can be understood by thinking of these states as QCD analogues of the Polyakov-'t Hooft monopoles?

The results of this paper can be understood by saying that baryons are such states.

Indeed, the baryon mass is of order N, which can be written as $1/(1/N)$. But $1/N$ is the "coupling constant" of the strong interactions, which characterizes the interaction among mesons. $1/N$ plays in QCD roughly the role that α plays in spontaneously broken gauge theories of the weak and electromagnetic interactions. The fact that the baryon mass is of order $1/(1/N)$ is analogous to the fact that the Polyakov-'t Hooft monopole mass is of order $1/\alpha$.

The baryon structure is determined for large N by solving the non-linear Hartree equations. N scales out of these equations, and therefore the baryon size and shape are independent of N for large N.

Likewise, the monopole structure is determined for small α by solving the classical equations. α scales out of these equations, and therefore the size and shape of the monopole are independent of α for small α.

This analogy extends also to the other processes we have considered.

For example, although for large N the mesons become non-interacting, S matrix elements involving baryons have a non-trivial large N limit. For example, baryon-baryon and baryon-antibaryon scattering can be calculated, for large N, by solving certain non-linear time-dependent Hartree equations. And meson-baryon scattering can be calculated by solving certain linear equations for the motion of a meson in a background baryon field. N scales out of all of these equations, so that all of these processes have non-trivial large N limits.

In a similar way, although for small α electrons and positrons become non-interacting, S matrix elements involving magnetic monopoles have a non-trivial small α limit. Monopole-monopole or monopole-antimonopole scattering can be calculated, for weak couplings, by solving the classical non-linear field equations in the monopole-monopole or monopole-antimonopole sector. And electron-monopole scattering can be understood by solving the linear equations for propagation of small disturbances (electrons) in a background monopole field. The coupling constant α scales out of all of these equations, so that all of these processes have small α limits.

The analogy can be extended to other processes. For instance, we saw that the cross section for $e^+e^- \to$ baryon-antibaryon is of order $\exp -N$ or $\exp[-1/(1/N)]$. This is analogous to the fact that the cross section for $e^+e^- \to$ monopole-antimonopole is of order $\exp(-1/\alpha)$. (The latter is so because, to any finite order in α, the outcome of an electron-positron collision can be calculated by evaluating Feynman diagrams. One does not see monopoles in the diagrams, and therefore the cross section for e^+e^- to produce a monopole-antimonopole pair is smaller than any power of α.)

In short, the analogy between QCD for large N and weakly coupled local field theories extends to baryons as well as to mesons and glueballs. The mesons and glueballs are the QCD analogues of the ordinary particles which become weakly coupled when the coupling is small. And the baryons are the QCD analogues of the solitons or magnetic monopoles, whose masses diverge like the inverse of the coupling.

Note added

A recent paper on some aspects of large N phenomenology not considered here is ref. [14].

References

[1] G. 't Hooft, Nucl. Phys. B72 (1974) 461; B75 (1974) 461.
[2] G. Veneziano, Review talk at the 1978 Tokyo meeting and references cited therein: Nucl. Phys B117 (1976) 519;
C.G. Callan, Jr., N. Coote and D.J. Gross, Phys. Rev. D13 (1976) 1649;
M.B. Einhorn, Phys. Rev. D14 (1976) 3451;
J.L. Cardy; Phys. Lett. 61B (1976) 293;
R.C. Brower, J. Ellis, M.G. Schmidt and J.H. Weis, Nucl. Phys. B128 (1977) 131, 175;
M.B. Einhorn, S. Nussinov and E. Rabinovici, Phys. Rev. D15 (1977) 2282;
S. Shei and H. Tsao, Nucl. Phys. B141 (1978) 445;
P. Tomaras, Harvard preprint HUTP-78/A054 (1978).
[3] K.G. Wilson, Phys. Rev. D7 (1973) 2911;
L. Dolan and R. Jackiw, Phys. Rev. D9 (1974) 3320;
S. Coleman, R. Jackiw and H.D. Politzer, Phys. Rev. D10 (1974) 2491;
H.J. Schnitzer, Phys. Rev. D10 (1974) 1800, 2042;

D. Gross and A. Neveu, Phys. Rev. D10 (1974) 3235;
L.F. Abbott, J.S. Kang, and H.J. Schnitzer, Phys. Rev. D13 (1975) 2212.
[4] E. Brézin, C. Itzykson, G. Parisi and J.B. Zuber, Comm. Math. Phys. 59 (1978) 35;
M. Casartelli, G. Marchesini and E. Onofri, Univ. of Parma preprint (February, 1979);
J. Koplik, A. Neveu and S. Nussinov, Nucl. Phys. B123 (1977) 109;
C. Thorne, Phys. Rev. D17 (1978) 1073;
R. Giles, L. McLerran and C.B. Thorne, Phys. Rev. D17 (1977) 2058;
R. Brower, R. Giles and C.B. Thorne, Phys. Rev. D18 (1978) 484.
[5] G.G. Chew and C. Rosenzweig, Phys. Reports 41 (1978) 265.
[6] M. Durgut, Nucl. Phys. B116 (1976) 223.
[7] S. Coleman, Ann. of Phys. 101 (1976) 239.
[8] K. Kikkawa and M. Sato, Phys. Rev. Lett. 38 (1977) 1309.
[9] A.K. Kerman and S.E. Koonin, Ann. of Phys. 100 (1976) 332;
A.K. Kerman and R. Jackiw, MIT preprint, (1979).
[10] R. Dashen, B. Hasslacher and A. Neveu, Phys. Rev. D12 (1975) 2443.
[11] D. Gross and A. Neveu, Phys. Rev. D10 (1974) 3235.
[12] L.D. Faddeev and V.E. Korepin, Phys. Reports 42 (1978) 1;
G. 't Hooft, Nucl. Phys. B79 (1974) 276;
A.M. Polyakov, JETP Sov. Phys. 41 (1976) 989.
[13] G. Rossi and G. Veneziano, Nucl. Phys. B123 (1977) 597.
[14] P. Aurenche and L. Gonzalez-Mestres, Z. Phys. C1 (1979) 307.

Nuclear Physics B223 (1983) 422–432
© North-Holland Publishing Company

GLOBAL ASPECTS OF CURRENT ALGEBRA

Edward WITTEN*

Joseph Henry Laboratories, Princeton University, Princeton, New Jersey 08544, USA

Received 4 March 1983

A new mathematical framework for the Wess-Zumino chiral effective action is described. It is shown that this action obeys an a priori quantization law, analogous to Dirac's quantization of magnetic change. It incorporates in current algebra both perturbative and non-perturbative anomalies.

The purpose of this paper is to clarify an old but relatively obscure aspect of current algebra: the Wess-Zumino effective lagrangian [1] which summarizes the effects of anomalies in current algebra. As we will see, this effective lagrangian has unexpected analogies to some 2 + 1 dimensional models discussed recently by Deser et al. [2] and to a recently noted SU(2) anomaly [3]. There also are connections with work of Balachandran et al. [4].

For definiteness we will consider a theory with $SU(3)_L \times SU(3)_R$ symmetry spontaneously broken down to the diagonal SU(3). We will ignore explicit symmetry-breaking perturbations, such as quark bare masses. With $SU(3)_L \times SU(3)_R$ broken to diagonal SU(3), the vacuum states of the theory are in one to one correspondence with points in the SU(3) manifold. Correspondingly, the low-energy dynamics can be conveniently described by introducing a field $U(x^\alpha)$ that transforms in a so-called non-linear realization of $SU(3)_L \times SU(3)_R$. For each space-time point x^α, $U(x^\alpha)$ is an element of SU(3): a 3×3 unitary matrix of determinant one. Under an $SU(3)_L \times SU(3)_R$ transformation by unitary matrices (A, B), U transforms as $U \to AUB^{-1}$.

The effective lagrangian for U must have $SU(3)_L \times SU(3)_R$ symmetry, and, to describe correctly the low-energy limit, it must have the smallest possible number of derivatives. The unique choice with only two derivatives is

$$\mathcal{L} = \tfrac{1}{16} F_\pi^2 \int d^4x \, \mathrm{Tr}\, \partial_\mu U \, \partial_\mu U^{-1}, \qquad (1)$$

* Supported in part by NSF Grant PHY80-19754.

where experiment indicates $F_\pi \simeq 190$ MeV. The perturbative expansion of U is

$$U = 1 + \frac{2i}{F_\pi} \sum_{a=1}^{8} \lambda^a \pi^a + \cdots, \tag{2}$$

where λ^a (normalized so $\operatorname{Tr} \lambda^a \lambda^b = 2\delta^{ab}$) are the SU(3) generators and π^a are the Goldstone boson fields.

This effective lagrangian is known to incorporate all relevant symmetries of QCD. All current algebra theorems governing the extreme low-energy limit of Goldstone boson S-matrix elements can be recovered from the tree approximation to it. What is less well known, perhaps, is that (1) possesses an extra discrete symmetry that is *not* a symmetry of QCD.

The lagrangian (1) is invariant under $U \leftrightarrow U^{\mathrm{T}}$. In terms of pions this is $\pi^0 \leftrightarrow \pi^0$, $\pi^+ \leftrightarrow \pi^-$; it is ordinary charge conjugation. (1) is also invariant under the naive parity operation $x \leftrightarrow -x$, $t \leftrightarrow t$, $U \leftrightarrow U$. We will call this P_0. And finally, (1) is invariant under $U \leftrightarrow U^{-1}$. Comparing with eq. (2), we see that this latter operation is equivalent to $\pi^a \leftrightarrow -\pi^a$, $a = 1, \dots, 8$. This is the operation that counts modulo two the number of bosons, N_{B}, so we will call it $(-1)^{N_{\mathrm{B}}}$.

Certainly, $(-1)^{N_{\mathrm{B}}}$ is not a symmetry of QCD. The problem is the following. QCD is parity invariant only if the Goldstone bosons are treated as pseudoscalars. The parity operation in QCD corresponds to $x \leftrightarrow -x$, $t \leftrightarrow t$, $U \leftrightarrow U^{-1}$. This is $P = P_0(-1)^{N_{\mathrm{B}}}$. QCD is invariant under P but not under P_0 or $(-1)^{N_{\mathrm{B}}}$ separately. The simplest process that respects all bona fide symmetries of QCD but violates P_0 and $(-1)^{N_{\mathrm{B}}}$ is $\mathrm{K}^+\mathrm{K}^- \to \pi^+\pi^0\pi^-$ (note that the ϕ meson decays to both $\mathrm{K}^+\mathrm{K}^-$ and $\pi^+\pi^0\pi^-$). It is natural to ask whether there is a simple way to add a higher-order term to (1) to obtain a lagrangian that obeys *only* the appropriate symmetries.

The Euler-Lagrangian equation derived from (1) can be written

$$\partial_\mu \left(\tfrac{1}{8} F_\pi^2 U^{-1} \partial_\mu U \right) = 0. \tag{3}$$

Let us try to add a suitable extra term to this equation. A Lorentz-invariant term that violates P_0 must contain the Levi-Civita symbol $\varepsilon_{\mu\nu\alpha\beta}$. In the spirit of current algebra, we wish a term with the smallest possible number of derivatives, since, in the low-energy limit, the derivatives of U are small. There is a unique P_0-violating term with only four derivatives. We can generalize (3) to

$$\partial_\mu \left(\tfrac{1}{8} F_\pi^2 U^{-1} \partial_\mu U \right) + \lambda \varepsilon^{\mu\nu\alpha\beta} U^{-1}(\partial_\mu U) U^{-1}(\partial_\nu U) U^{-1}(\partial_\alpha U) U^{-1}(\partial_\beta U) = 0, \tag{4}$$

λ being a constant. Although it violates P_0, (4) can be seen to respect $P = P_0(-1)^{N_{\mathrm{B}}}$.

Can eq. (4) be derived from a lagrangian? Here we find trouble. The only pseudoscalar of dimension four would seem to be $\varepsilon^{\mu\nu\alpha\beta} \operatorname{Tr} U^{-1}(\partial_\mu U) \cdot U^{-1}(\partial_\nu U) U^{-1} (\partial_\alpha U) U^{-1}(\partial_\beta U)$, but this vanishes, by antisymmetry of $\varepsilon^{\mu\nu\alpha\beta}$ and cyclic symmetry of the trace. Nevertheless, as we will see, there is a lagrangian.

Let us consider a simple problem of the same sort. Consider a particle of mass m constrained to move on an ordinary two-dimensional sphere of radius one. The lagrangian is $\mathfrak{L}=\frac{1}{2}m\int \mathrm{d}t\,\dot{x}_i^2$ and the equation of motion is $m\ddot{x}_i + mx_i(\Sigma_k \dot{x}_k^2)=0$; the constraint is $\Sigma x_i^2=1$. This system respects the symmetries $t \leftrightarrow -t$ and separately $x_i \leftrightarrow -x_i$. If we want an equation that is only invariant under the combined operation $t \leftrightarrow -t$, $x_i \leftrightarrow x_i$, the simplest choice is

$$m\ddot{x}_i + mx_i\left(\sum_k \dot{x}_k^2\right) = \alpha\varepsilon_{ijk}x_j\dot{x}_k, \tag{5}$$

where α is a constant. To derive this equation from a lagrangian is again troublesome. There is no obvious term whose variation equals the right-hand side (since $\varepsilon_{ijk}x_i x_j \dot{x}_k = 0$).

However, this problem has a well-known solution. The right-hand side of (5) can be understood as the Lorentz force for an electric charge interacting with a magnetic monopole located at the center of the sphere. Introducing a vector potential $\boldsymbol{A}$ such that $\nabla \times \boldsymbol{A} = \boldsymbol{x}/|x|^3$, the action for our problem is

$$I = \int \left(\tfrac{1}{2}m\dot{x}_i^2 + \alpha A_i \dot{x}_i\right)\mathrm{d}t. \tag{6}$$

This lagrangian is problematical because A_i contains a Dirac string and certainly does not respect the symmetries of our problem. To explore this quantum mechanically let us consider the simplest form of the Feynman path integral, $\mathrm{Tr}\,\mathrm{e}^{-\beta H} = \int \mathrm{d}x_i(t)\mathrm{e}^{-I}$. In e^{-I} the troublesome term is

$$\exp\left(i\alpha\int_\gamma A_i\,\mathrm{d}x^i\right), \tag{7}$$

where the integration goes over the particle orbit γ: a closed orbit if we discuss the simplest object $\mathrm{Tr}\,\mathrm{e}^{-\beta H}$.

By Gauss's law we can eliminate the vector potential from (7) in favor of the magnetic field. In fact, the closed orbit γ of fig. 1a is the boundary of a disc D, and by Gauss's law we can write (7) in terms of the magnetic flux through D:

$$\exp\left(i\alpha\int_\gamma A_i\,\mathrm{d}x^i\right) = \exp\left(i\alpha\int_\mathrm{D} F_{ij}\,\mathrm{d}\Sigma^{ij}\right). \tag{8}$$

The precise mathematical statement here is that since $\pi_1(\mathrm{S}^2)=0$, the circle γ in S^2 is the boundary of a disc D (or more exactly, a mapping γ of a circle into S^2 can be extended to a mapping of a disc into S^2).

The right-hand side of (8) is manifestly well defined, unlike the left-hand side, which suffers from a Dirac string. We could try to use the right-hand side of (8) in a Feynman path integral. There is only one problem: D isn't unique. The curve γ also bounds the disc D′ (fig. 1c). There is no consistent way to decide whether to choose

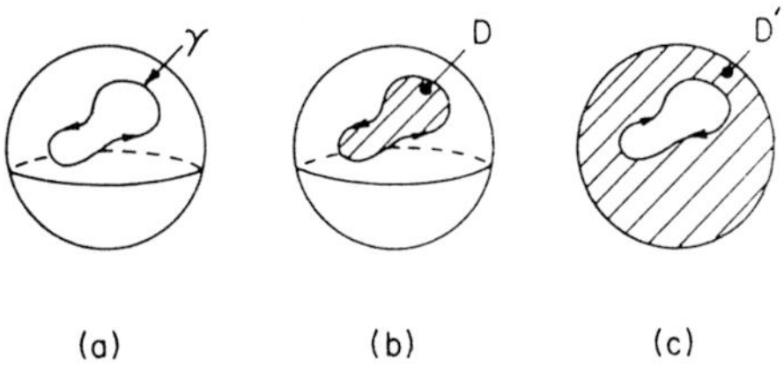

Fig. 1. A particle orbit γ on the two-sphere (part (a)) bounds the discs D (part (b)) and D′ (part (c)).

D or D′ (the curve γ could continuously be looped around the sphere or turned inside out). Working with D′ we would get

$$\exp\left(i\alpha\int_{\gamma}A_i\,\mathrm{d}x^i\right)=\exp\left(-i\alpha\int_{D'}F_{ij}\,\mathrm{d}\Sigma^{ij}\right). \tag{9}$$

where a crucial minus sign on the right-hand side of (9) appears because γ bounds D in a right-hand sense, but bounds D′ in a left-hand sense. If we are to introduce the right-hand side of (8) or (9) in a Feynman path integral, we must require that they be equal. This is equivalent to

$$1=\exp\left(i\alpha\int_{D+D'}F_{ij}\,\mathrm{d}\Sigma^{ij}\right). \tag{10}$$

Since D + D′ is the whole two sphere S^2, and $\int_{S^2}F_{ij}\,\mathrm{d}\Sigma^{ij}=4\pi$. (10) is obeyed if and only if α is an integer or half-integer. This is Dirac's quantization condition for the product of electric and magnetic charges.

Now let us return to our original problem. We imagine space-time to be a very large four-dimensional sphere M. A given non-linear sigma model field U is a mapping of M into the SU(3) manifold (fig. 2a). Since $\pi_4(\mathrm{SU}(3))=0$, the four-sphere in SU(3) defined by U(x) is the boundary of a five-dimensional disc Q.

By analogy with the previous problem, let us try to find some object that can be integrated over Q to define an action functional. On the SU(3) manifold there is a unique fifth rank antisymmetric tensor ω_{ijklm} that is invariant under $\mathrm{SU(3)_L}\times \mathrm{SU(3)_R}$*. Analogous to the right-hand side of eq. (8), we define

$$\Gamma=\int_Q \omega_{ijklm}\,\mathrm{d}\Sigma^{ijklm}. \tag{11}$$

* Let us first try to define ω at $U=1$; it can then be extended to the whole SU(3) manifold by an $\mathrm{SU(3)_L}\times\mathrm{SU(3)_R}$ transformation. At $U=1$, ω must be invariant under the diagonal subgroup of $\mathrm{SU(3)_L}\times\mathrm{SU(3)_R}$ that leaves fixed $U=1$. The tangent space to the SU(3) manifold at $U=1$ can be identified with the Lie algebra of SU(3). So ω, at $U=1$, defines a fifth-order antisymmetric invariant in the SU(3) Lie algebra. There is only one such invariant. Given five SU(3) generators A, B, C, D and E, the one such invariant is $\mathrm{Tr}\,ABCDE-\mathrm{Tr}\,BACDE\pm$ permutations. The $\mathrm{SU(3)_L}\times\mathrm{SU(3)_R}$ invariant ω so defined has zero curl ($\partial_i\omega_{jklmn}\pm$ permutations $=0$) and for this reason (11) is invariant under infinitesimal variations of Q; there arises only the topological problem discussed in the text.

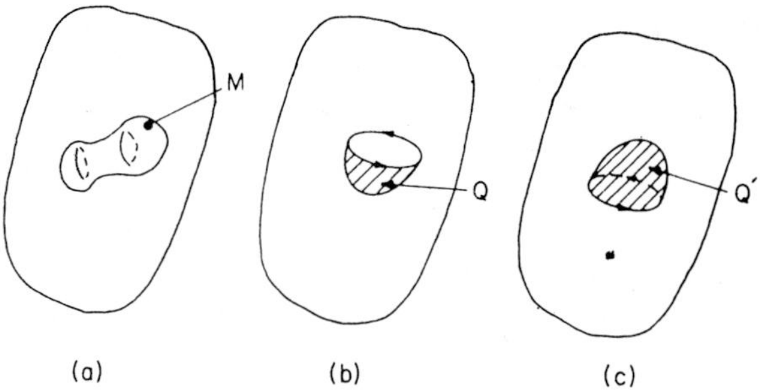

Fig. 2. Space-time, a four-sphere, is mapped into the SU(3) manifold. In part (a), space-time is symbolically denoted as a two sphere. In parts (b) and (c), space-time is reduced to a circle that bounds the discs Q and Q′. The SU(3) manifold is symbolized in these sketches by the interior of the oblong.

As before, we hope to include $\exp(i\Gamma)$ in a Feynman path integral. Again, the problem is that Q is not unique. Our four-sphere M is also the boundary of another five-disc Q′ (fig. 2c). If we let

$$\Gamma' = -\int_{Q'} \omega_{ijklm}\, \mathrm{d}\Sigma^{ijklm}, \tag{12}$$

(with, again, a minus sign because M bounds Q′ with opposite orientation) then we must require $\exp(i\Gamma) = \exp(i\Gamma')$ or equivalently $\int_{Q+Q'} \omega_{ijklm}\, \mathrm{d}\Sigma^{ijklm} = 2\pi \cdot \text{integer}$. Since Q + Q′ is a closed five-dimensional sphere, our requirement is

$$\int_{S} \omega_{ijklm}\, \mathrm{d}\Sigma^{ijklm} = 2\pi \cdot \text{integer},$$

for any five-sphere S in the SU(3) manifold.

We thus need the topological classification of mappings of the five-sphere into SU(3). Since $\pi_5(\mathrm{SU}(3)) = Z$, every five sphere in SU(3) is topologically a multiple of a basic five sphere S_0. We normalize ω so that

$$\int_{S_0} \omega_{ijklm}\, \mathrm{d}\Sigma^{ijklm} = 2\pi, \tag{13}$$

and then (with Γ in eq. (11)) we may work with the action

$$I = \tfrac{1}{16} F_\pi^2 \int \mathrm{d}^4x\, \mathrm{Tr}\, \partial_\mu U\, \partial_\mu U^{-1} + n\Gamma, \tag{14}$$

where n is an arbitrary integer. Γ is, in fact, the Wess-Zumino lagrangian. Only the a priori quantization of n is a new result.

The identification of S_0 and the proper normalization of ω is a subtle mathematical problem. The solution involves a factor of two from the Bott periodicity theorem. Without abstract notation, the result [5] can be stated as follows. Let y^i, $i = 1\dots5$ be coordinates for the disc Q. Then on Q (where we need it)

$$\mathrm{d}\Sigma^{ijklm}\,\omega_{ijklm} = -\frac{i}{240\pi^2}\mathrm{d}\Sigma^{ijklm}\left[\mathrm{Tr}\,U^{-1}\frac{\partial U}{\partial y^i}U^{-1}\frac{\partial U}{\partial y^j}U^{-1}\frac{\partial U}{\partial y^k}U^{-1}\frac{\partial U}{\partial y^l}U^{-1}\frac{\partial U}{\partial y^m}\right]. \tag{15}$$

The physical consequences of this can be made more transparent as follows. From eq. (2),

$$U^{-1}\partial_i U = \frac{2i}{F_\pi}\partial_i A + \mathrm{O}(A^2), \qquad \text{where } A = \Sigma\lambda^a\pi^a. \tag{16}$$

So

$$\omega_{ijklm}\mathrm{d}\Sigma^{ijklm} = \frac{2}{15\pi^2F_\pi^5}\mathrm{d}\Sigma^{ijklm}\mathrm{Tr}\,\partial_i A\,\partial_j A\,\partial_k A\,\partial_l A\,\partial_m A + \mathrm{O}(A^6)$$

$$= \frac{2}{15\pi^2F_\pi^5}\mathrm{d}\Sigma^{ijklm}\,\partial_i\left(\mathrm{Tr}\,A\,\partial_j A\,\partial_k A\,\partial_l A\,\partial_m A\right) + \mathrm{O}(A^6).$$

So $\int_Q\omega_{ijklm}\mathrm{d}\Sigma^{ijklm}$ is (to order A^5 and in fact also in higher orders) the integral of a total divergence which can be expressed by Stokes' theorem as an integral over the boundary of Q. By construction, this boundary is precisely space-time. We have, then,

$$n\Gamma = n\frac{2}{15\pi^2F_\pi^5}\int \mathrm{d}^4x\,\varepsilon^{\mu\nu\alpha\beta}\mathrm{Tr}\,A\,\partial_\mu A\,\partial_\nu A\,\partial_\alpha A\partial_\beta A + \text{higher order terms}. \tag{17}$$

In a hypothetical world of massless kaons and pions, this effective lagrangian rigorously describes the low-energy limit of $K^+K^- \to \pi^+\pi^0\pi^-$*. We reach the remarkable conclusion that in any theory with SU(3)×SU(3) broken to diagonal SU(3), the low-energy limit of the amplitude for this reaction must be (in units given in (17)) an integer.

What is the value of this integer in QCD? Were n to vanish, the practical interest of our discussion would be greatly reduced. It turns out that if N_c is the number of colors (three in the real world) then $n = N_c$. The simplest way to deduce this is a

* Our formula should agree for $n = 1$ with formulas of ref. [1], as later equations make clear. There appears to be a numerical error on p. 97 of ref. [1] ($\frac{1}{6}$ instead of $\frac{2}{15}$).

procedure that is of interest anyway, viz. coupling to electromagnetism, so as to describe the low-energy dynamics of Goldstone bosons and photons.

Let

$$Q = \begin{pmatrix} \frac{2}{3} & & \\ & -\frac{1}{3} & \\ & & -\frac{1}{3} \end{pmatrix}$$

be the usual electric charge matrix of quarks. The functional Γ is invariant under global charge rotations, $U \to U + i\varepsilon[Q, U]$, where ε is a constant. We wish to promote this to a local symmetry, $U \to U + i\varepsilon(x)[Q, U]$, where $\varepsilon(x)$ is an arbitrary function of x. It is necessary, of course, to introduce the photon field A_μ which transforms as $A_\mu \to A_\mu - (1/e)\partial_\mu \varepsilon$; e is the charge of the proton.

Usually a global symmetry can straightforwardly be gauged by replacing derivatives by covariant derivatives, $\partial_\mu \to D_\mu = \partial_\mu + ieA_\mu$. In the case at hand, Γ is not given as the integral of a manifestly $\mathrm{SU(3)_L} \times \mathrm{SU(3)_R}$ invariant expression, so the standard road to gauging global symmetries of Γ is not available. One can still resort to the trial and error Noether method, widely used in supergravity. Under a local charge rotation, one finds $\Gamma \to \Gamma - \int \mathrm{d}^4x\, \partial_\mu \varepsilon J^\mu$ where

$$J^\mu = \frac{1}{48\pi^2}\varepsilon^{\mu\nu\alpha\beta}\mathrm{Tr}\big[Q(\partial_\nu U U^{-1})(\partial_\alpha U U^{-1})(\partial_\beta U U^{-1}) + Q(U^{-1}\partial_\nu U)(U^{-1}\partial_\alpha U)(U^{-1}\partial_\beta U)\big], \tag{18}$$

is the extra term in the electromagnetic current required (from Noether's theorem) due to the addition of Γ to the lagrangian. The first step in the construction of an invariant lagrangian is to add the Noether coupling, $\Gamma \to \Gamma' = \Gamma - e\int \mathrm{d}^4x\, A_\mu J^\mu(x)$. This expression is still not gauge invariant, because J^μ is not, but by trial and error one finds that by adding an extra term one can form a gauge invariant functional

$$\tilde{\Gamma}(U, A_\mu) = \Gamma(U) - e\int \mathrm{d}^4x\, A_\mu J^\mu + \frac{ie^2}{24\pi^2}\int \mathrm{d}^4x\, \varepsilon^{\mu\nu\alpha\beta}(\partial_\mu A_\nu)A_\alpha \times \mathrm{Tr}\big[Q^2(\partial_\beta U)U^{-1} + Q^2U^{-1}(\partial_\beta U) + QUQU^{-1}(\partial_\beta U)U^{-1}\big]. \tag{19}$$

Our gauge invariant lagrangian will then be

$$\mathcal{L} = \tfrac{1}{16}F_\pi^2\int \mathrm{d}^4x\, \mathrm{Tr}\, D_\mu U D_\mu U^{-1} + n\tilde{\Gamma}. \tag{20}$$

What value of the integer n will reproduce QCD results?

Here we find a surprise. The last term in (18) has a piece that describes $\pi^0 \to \gamma\gamma$. Expanding U and integrating by parts, (18) has a piece

$$A = \frac{ne^2}{48\pi^2 F_\pi} \pi^0 \varepsilon^{\mu\nu\alpha\beta} F_{\mu\nu} F_{\alpha\beta} . \tag{21}$$

This agrees with the result from QCD triangle diagrams [6] if $n = N_c$, the number of colors. The Noether coupling $-eA_\mu J^\mu$ describes, among other things, a $\gamma\pi^+\pi^0\pi^-$ vertex

$$B = -\tfrac{2}{3} ie \frac{n}{\pi^2 F_\pi^3} \varepsilon^{\mu\nu\alpha\beta} A_\mu \, \partial_\nu \pi^+ \, \partial_\alpha \pi^- \, \partial_\beta \pi^0 . \tag{22}$$

Again this agrees with calculations [7] based on the QCD VAAA anomaly if $n = N_c$. The effective action $N_c \tilde{\Gamma}$ (first constructed in another way by Wess and Zumino) precisely describes all effects of QCD anomalies in low-energy processes with photons and Goldstone bosons.

It is interesting to try to gauge subgroups of $SU(3)_L \times SU(3)_R$ other than electromagnetism. One may have in mind, for instance, applications to the standard weak interaction model. In general, one may try to gauge an arbitrary subgroup H of $SU(3)_L \times SU(3)_R$, with generators K^σ, $\sigma = 1 \ldots r$. Each K^σ is a linear combination of generators T_L^σ and T_R^σ of $SU(3)_L$ and $SU(3)_R$, $K^\sigma = T_L^\sigma + T_R^\sigma$. (Either T_L^σ or T_R^σ may vanish for some values of σ.) For any space-time dependent functions $\varepsilon^\sigma(x)$, let $\varepsilon_L = \Sigma_\sigma T_L^\sigma \varepsilon^\sigma(x)$, $\varepsilon_R = \Sigma_\sigma T_R^\sigma \varepsilon^\sigma(x)$. We want an action with local invariance under $U \to U + i(\varepsilon_L(x) U - U \varepsilon_R(x))$.

Naturally, it is necessary to introduce gauge fields $A_\mu^\sigma(x)$, transforming as $A_\mu^\sigma(x) \to A_\mu^\sigma(x) - (1/e_\sigma)\, \partial_\mu \varepsilon^\sigma + f^{\sigma\tau\rho} \varepsilon^\tau A_\mu^\rho$ where e_σ is the coupling constant corresponding to the generator K^σ, and $f^{\sigma\tau\rho}$ are the structure constants of H. It is useful to define $A_{\mu L} = \Sigma_\sigma e_\sigma A_\mu^\sigma T_L^\sigma$, $A_\mu^R = \Sigma_\sigma e_\sigma A_\mu^\sigma T_R^\sigma$.

We have already seen that Γ incorporates the effects of anomalies, so it is not very surprising that a generalization of Γ that is gauge invariant under H exists only if H is a so-called anomaly-free subgroup of $SU(3)_L \times SU(3)_R$. Specifically, one finds that H can be gauged only if for each σ,

$$\mathrm{Tr}(T_L^\sigma)^3 = \mathrm{Tr}(T_R^\sigma)^3 , \tag{23}$$

which is the usual condition for cancellation of anomalies at the quark level.

If (23) is obeyed, a gauge invariant generalization of Γ can be constructed somewhat tediously by trial and error. It is useful to define $U_{\nu L} = (\partial_\nu U) U^{-1}$ and $U_{\nu R} = U^{-1} \partial_\nu U$. The gauge invariant functional then turns out to be

$$\tilde{\Gamma}(A_\mu, U) = \Gamma(U) + \frac{1}{48\pi^2} \int \mathrm{d}^4 x \, \varepsilon^{\mu\nu\alpha\beta} Z_{\mu\nu\alpha\beta} ,$$

where

$$
\begin{aligned}
Z_{\mu\nu\alpha\beta} = & -\mathrm{Tr}\left[A_{\mu\mathrm{L}}U_{\nu\mathrm{L}}U_{\alpha\mathrm{L}}U_{\beta\mathrm{L}} + (\mathrm{L}\to\mathrm{R})\right] \\
& + i\,\mathrm{Tr}\left[\left[(\partial_\mu A_{\nu\mathrm{L}})A_{\alpha\mathrm{L}} + A_{\mu\mathrm{L}}(\partial_\nu A_{\alpha\mathrm{L}})\right]U_{\beta\mathrm{L}} + (\mathrm{L}\to\mathrm{R})\right] \\
& + i\,\mathrm{Tr}\left[(\partial_\mu A_{\nu\mathrm{R}})U^{-1}A_{\alpha\mathrm{L}}\,\partial_\beta U + A_{\mu\mathrm{L}}U^{-1}(\partial_\nu A_{\alpha\mathrm{R}})\,\partial_\beta U\right] \\
& - \tfrac{1}{2}i\,\mathrm{Tr}\left(A_{\mu\mathrm{L}}U_{\nu\mathrm{L}}A_{\alpha\mathrm{L}}U_{\beta\mathrm{L}} - (\mathrm{L}\to\mathrm{R})\right) \\
& + i\,\mathrm{Tr}\left[A_{\mu\mathrm{L}}UA_{\nu\mathrm{R}}U^{-1}U_{\alpha\mathrm{L}}U_{\beta\mathrm{L}} - A_{\mu\mathrm{R}}U^{-1}A_{\nu\mathrm{L}}UU_{\alpha\mathrm{R}}U_{\beta\mathrm{R}}\right] \\
& - \mathrm{Tr}\Big[\left[(\partial_\mu A_{\nu\mathrm{R}})A_{\alpha\mathrm{R}} + A_{\mu\mathrm{R}}(\partial_\nu A_{\alpha\mathrm{R}})\right]U^{-1}A_{\beta\mathrm{L}}U \\
& \qquad - \left[(\partial_\mu A_{\nu\mathrm{L}})A_{\alpha\mathrm{L}} + A_{\mu\mathrm{L}}(\partial_\nu A_{\alpha\mathrm{L}})\right]UA_{\beta\mathrm{R}}U^{-1}\Big] \\
& \quad - \mathrm{Tr}\left[A_{\mu\mathrm{R}}U^{-1}A_{\nu\mathrm{L}}UA_{\alpha\mathrm{R}}U_{\beta\mathrm{R}} + A_{\mu\mathrm{L}}UA_{\nu\mathrm{R}}U^{-1}A_{\alpha\mathrm{L}}U_{\beta\mathrm{L}}\right] \\
& \quad - \mathrm{Tr}\left[A_{\mu\mathrm{L}}A_{\nu\mathrm{L}}U(\partial_\alpha A_{\beta\mathrm{R}})U^{-1} + A_{\mu\mathrm{R}}A_{\nu\mathrm{R}}U^{-1}(\partial_\alpha A_{\beta\mathrm{L}})U\right] \\
& \quad - i\,\mathrm{Tr}\Big[A_{\mu\mathrm{R}}A_{\nu\mathrm{R}}A_{\alpha\mathrm{R}}U^{-1}A_{\beta\mathrm{L}}U - A_{\mu\mathrm{L}}A_{\nu\mathrm{L}}A_{\alpha\mathrm{L}}UA_{\beta\mathrm{R}}U^{-1} \\
& \qquad + \tfrac{1}{2}A_{\mu\mathrm{L}}A_{\nu\mathrm{L}}UA_{\alpha\mathrm{R}}A_{\beta\mathrm{R}}U^{-1} + \tfrac{1}{2}A_{\mu\mathrm{R}}U^{-1}A_{\nu\mathrm{L}}UA_{\alpha\mathrm{R}}U^{-1}A_{\beta\mathrm{L}}U\Big]. \qquad (24)
\end{aligned}
$$

If eq. (22) for cancellation of anomalies is not obeyed, then the variation of $\tilde{\Gamma}$ under a gauge transformation does not vanish but is

$$
\begin{aligned}
\delta\tilde{\Gamma} = & -\frac{1}{24\pi^2}\int \mathrm{d}^4x\,\varepsilon^{\mu\nu\alpha\beta}\mathrm{Tr}\,\varepsilon_{\mathrm{L}}\left[(\partial_\mu A_{\nu\mathrm{L}})(\partial_\alpha A_{\beta\mathrm{L}}) - \tfrac{1}{2}i\,\partial_\mu(A_{\nu\mathrm{L}}A_{\alpha\mathrm{L}}A_{\beta\mathrm{L}})\right] \\
& - (\mathrm{L}\to\mathrm{R}), \qquad (25)
\end{aligned}
$$

in agreement with computations at the quark level [8] of the anomalous variation of the effective action under a gauge transformation.

Thus, Γ incorporates all information usually associated with triangle anomalies, including the restriction on what subgroups H of $\mathrm{SU(3)_L \times SU(3)_R}$ can be gauged. However, there is another potential obstruction to the ability to gauge a subgroup of $\mathrm{SU(3)_L \times SU(3)_R}$. This is the non-perturbative anomaly [3] associated with $\pi_4(\mathrm{H})$. Is this anomaly, as well, implicit in Γ? In fact, it is.

Let H be an SU(2) subgroup of $\mathrm{SU(3)_L}$, chosen so that an SU(2) matrix W is embedded in $\mathrm{SU(3)_L}$ as

$$
\hat{W} = \left(\begin{array}{cc|c} \multicolumn{2}{c|}{\multirow{2}{*}{W}} & 0 \\ & & 0 \\ \hline 0 & 0 & 1 \end{array}\right).
$$

This subgroup is free of triangle anomalies, so the functional $\tilde{\Gamma}$ of eq. (23) is invariant under infinitesimal local H transformations.

However, is $\tilde{\Gamma}$ invariant under H transformations that cannot be reached continuously? Since $\pi_4(\mathrm{SU}(2)) = \mathrm{Z}_2$, there is one non-trivial homotopy class of SU(2) gauge transformations. Let W be an SU(2) gauge transformation in this non-trivial class. Under $\hat{W}$, $\tilde{\Gamma}$ may at most be shifted by a constant, independent of U and A_μ, because $\delta\tilde{\Gamma}/\delta U$ and $\delta\tilde{\Gamma}/\delta A_\mu$ are gauge-covariant local functionals of U and A_μ. Also $\tilde{\Gamma}$ is invariant under $\hat{W}^2$, since $\hat{W}^2$ is equivalent to the identity in $\pi_4(\mathrm{SU}(2))$, and we know $\tilde{\Gamma}$ is invariant under topologically trivial gauge transformations. This does not quite mean that $\tilde{\Gamma}$ is invariant under W. Since $\tilde{\Gamma}$ is only defined modulo 2π, the fact that $\tilde{\Gamma}$ is invariant under W^2 leaves two possibilities for how $\tilde{\Gamma}$ behaves under W. It may be invariant, or it may be shifted by π.

To choose between these alternatives, it is enough to consider a special case. For instance, it suffices to evaluate $\Delta = \tilde{\Gamma}(U=1,\ A_\mu = 0) - \tilde{\Gamma}(U = \hat{W},\ A_\mu = ie^{-1}(\partial_\mu \hat{W})\hat{W}^{-1})$. It is not difficult to see that in this case the complicated terms involving $\varepsilon^{\mu\nu\alpha\beta}Z_{\mu\nu\alpha\beta}$ vanish, so in fact $\Delta = \Gamma(U=1) - \Gamma(U=\hat{W})$. A detailed calculation shows that

$$\Gamma(U=1) - \Gamma(U=\hat{W}) = \pi\,. \tag{26}$$

This calculation has some other interesting applications and will be described elsewhere [9].

The Feynman path integral, which contains a factor $\exp(iN_c\tilde{\Gamma})$, hence picks up under W a factor $\exp(iN_c\pi) = (-1)^{N_c}$. It is gauge invariant if N_c is even, but not if N_c is odd. This agrees with the determination of the SU(2) anomaly at the quark level [3]. For under H, the right-handed quarks are singlets. The left-handed quarks consist of one singlet and one doublet per color, so the number of doublets equals N_c. The argument of ref. [3] shows at the quark level that the effective action transforms under W as $(-1)^{N_c}$.

Finally, let us make the following remark, which apart from its intrinsic interest will be useful elsewhere [9]. Consider $\mathrm{SU}(3)_\mathrm{L} \times \mathrm{SU}(3)_\mathrm{R}$ currents defined at the quark level as

$$J^a_{\mu\mathrm{L}} = \bar{q}\lambda^a\gamma_\mu \tfrac{1}{2}(1-\gamma_5)q\,, \qquad J^a_{\mu\mathrm{R}} = \bar{q}\lambda^a\gamma_\mu \tfrac{1}{2}(1+\gamma_5)q\,. \tag{27}$$

By analogy with eq. (17), the proper sigma model description of these currents contains pieces

$$J^{\mu a}_{\mathrm{L}} = \frac{N_c}{48\pi^2}\varepsilon^{\mu\nu\alpha\beta}\mathrm{Tr}\,\lambda^a U_{\nu\mathrm{L}}U_{\alpha\mathrm{L}}U_{\beta\mathrm{L}}\,,$$

$$J^{\mu a}_{\mathrm{R}} = \frac{N_c}{48\pi^2}\varepsilon^{\mu\nu\alpha\beta}\mathrm{Tr}\,\lambda^a U_{\nu\mathrm{R}}U_{\alpha\mathrm{R}}U_{\beta\mathrm{R}}\,, \tag{28}$$

corresponding (via Noether's theorem) to the addition to the lagrangian of $N_c\Gamma$. In this discussion, the λ^a should be traceless SU(3) generators. However, let us try to construct an anomalous baryon number current in the same way. We define the baryon number of a quark (whether left-handed or right-handed) to be $1/N_c$, so that an ordinary baryon made from N_c quarks has baryon number one. Replacing λ^a by $1/N_c$, but including contributions of both left-handed and right-handed quarks, the anomalous baryon-number current would be

$$J^\mu = \frac{1}{24\pi^2}\varepsilon^{\mu\nu\alpha\beta}\mathrm{Tr}\, U^{-1}\,\partial_\nu U\, U^{-1}\,\partial_\alpha U\, U^{-1}\,\partial_\beta U . \tag{29}$$

One way to see that this is the proper, and properly normalized, formula is to consider gauging an arbitrary subgroup not of $SU(3)_L \times SU(3)_R$ but of $SU(3)_L \times SU(3)_R \times U(1)$, U(1) being baryon number. The gauging of U(1) is accomplished by adding a Noether coupling $-eJ^\mu B_\mu$ plus whatever higher-order terms may be required by gauge invariance. (B_μ is a U(1) gauge field which may be coupled as well to some $SU(3)_L \times SU(3)_R$ generator.) With J^μ defined in (29), this leads to a generalization of $\tilde{\Gamma}$ that properly reflects anomalous diagrams involving the baryon-number current (for instance, it properly incorporates the anomaly in the baryon number $SU(2)_L - SU(2)_L$ triangle that leads to baryon non-conservation by instantons in the standard weak interaction model). Eq. (29) may also be extracted from QCD by methods of Goldstone and Wilczek [10].

References

[1] J. Wess and B. Zumino, Phys. Lett. 37B (1971) 95

[2] S. Deser, R. Jackiw and S. Templeton, Phys. Rev. Lett. 48 (1982) 975; Ann. of Phys. 140 (1982) 372

[3] E. Witten, Phys. Lett. 117B (1982) 324

[4] A.P. Balachandran, V.P. Nair and C.G. Trahern, Syracuse University preprint SU-4217-205 (1981)

[5] R. Bott and R. Seeley, Comm. Math. Phys. 62 (1978) 235

[6] S.L. Adler, Phys. Rev. 177 (1969) 2426;
J.S. Bell and R. Jackiw, Nuovo Cim. 60 (1969) 147;
W.A. Bardeen, Phys. Rev. 184 (1969) 1848

[7] S.L. Adler and W.A. Bardeen, Phys. Rev. 182 (1969) 1517;
R. Aviv and A. Zee, Phys. Rev. D5 (1972) 2372
S.L. Adler, B.W. Lee, S.B. Treiman and A. Zee, Phys. Rev. D4 (1971) 3497

[8] D.J. Gross and R. Jackiw, Phys. Rev. D6 (1972) 477

[9] E. Witten, Nucl. Phys. B223 (1983) 433

[10] J. Goldstone and F. Wilczek, Phys. Rev. Lett. 47 (1981) 986

Nuclear Physics B223 (1983) 433–444
© North-Holland Publishing Company

CURRENT ALGEBRA, BARYONS, AND QUARK CONFINEMENT

Edward WITTEN*

Joseph Henry Laboratories, Princeton University, Princeton, New Jersey 08544, USA

Received 4 March 1983

It is shown that ordinary baryons can be understood as solitons in current algebra effective lagrangians. The formation of color flux tubes can also be seen in current algebra, under certain conditions.

The idea that in some sense the ordinary proton and neutron might be solitons in a non-linear sigma model has a long history. The first suggestion was made by Skyrme more than twenty years ago [1]. Finkelstein and Rubinstein showed that such objects could in principle be fermions [2], in a paper that probably represented the first use of what would now be called θ vacua in quantum field theory. A gauge invariant version was attempted by Faddeev [3]. Some relevant miracles are known to occur in two space-time dimensions [4]; there also exists a different mechanism by which solitons can be fermions [4].

It is known that in the large-N limit of quantum chromodynamics [5] meson interactions are governed by the tree approximation to an effective local field theory of mesons. Several years ago, it was pointed out [6] that baryons behave as if they were solitons in the effective large-N meson field theory. However, it was not clear in exactly what sense the baryons actually *are* solitons.

The first relevant papers mainly motivated by attempts to understand implications of QCD current algebra were recent papers by Balachandran et al. [7] and by Boguta [8].

We will always denote the number of colors as N and the number of light flavors as n. For definiteness we first consider the usual case $n = 3$. Nothing changes for $n > 3$. Some modifications for $n < 3$ are pointed out later. Except where stated otherwise, we discuss standard current algebra with global $\mathrm{SU}(n) \times \mathrm{SU}(n)$ spontaneously broken to diagonal $\mathrm{SU}(n)$, presumably as a result of an underlying $\mathrm{SU}(N)$ gauge interaction.

* Supported in part by NSF Grant PHY80-19754.

Standard current algebra can be described by a field $U(x)$ which (for each space-time point x) is a point in the SU(3) manifold. Ignoring quark bare masses, this field is governed by an effective action of the form

$$I = -\tfrac{1}{16}F_\pi^2 \int d^4x \, \mathrm{Tr}\, \partial_\mu U \partial_\mu U^{-1} + N\Gamma + \text{higher order terms}. \tag{1}$$

Here Γ is the Wess-Zumino term [9] which cannot be written as the integral of a manifestly SU(3)×SU(3) invariant density, and $F_\pi \simeq 190$ Mev. In quantum field theory the coefficient of Γ must a priori be an integer [10], and indeed we will see that the quantization of the soliton excitations of (1) is inconsistent (they obey neither bose nor fermi statistics) unless N is an integer.

Any finite energy configuration $U(x, y, z)$ must approach a constant at spatial infinity. This being so, any such configuration represents an element in the third homotopy group $\pi_3(\mathrm{SU(3)})$. Since $\pi_3(\mathrm{SU(3)}) \simeq \mathrm{Z}$, there are soliton excitations, and they obey an additive conservation law. Actually, higher-order terms in (1) are needed to stabilize the soliton solutions and prevent them from shrinking to zero size. We will see that such higher-order terms (which could be measured in principle by studying meson processes) must be present in the large-N limit of QCD and are related to the bag radius. Our remarks will not depend on the details of the higher-order terms.

A technical remark is in order. To study solitons, it is convenient to work with a euclidean space-time M of topology $S^3 \times S^1$. Here S^3 represents the spatial variables, and S^1 is a compactified euclidean time coordinate. A given non-linear sigma model field $U(x)$ defines a mapping of M into SU(3). We may think of M as the boundary of a five-dimensional manifold Q with topology $S^3 \times D$, D being a two-dimensional disc. Using the fact that $\pi_1(\mathrm{SU(3)}) = \pi_4(\mathrm{SU(3)}) = 0$, it can be shown that the mapping of M into SU(3) defined by $U(x)$ can be extended to a mapping from Q into SU(3). Then as in ref. [10] the functional Γ is defined by $\Gamma = \int_Q \omega$, where ω is the fifth-rank antisymmetric tensor on the SU(3) manifold defined in ref. [10]. By analogy with the discussion in ref. [10], Γ is well-defined modulo 2π. (It is essential here that because $\pi_2(\mathrm{SU(3)}) = 0$, the five-dimensional homology classes in $H_5(\mathrm{SU(3)})$ that can be represented by cycles with topology $S^3 \times S^2$ are precisely those that can be represented by cycles with topology S^5. There are closed five-surfaces S in SU(3) such that $\int_S \omega$ is an *odd* multiple of π, but they do not arise if space-time has topology $S^3 \times S^1$ and Q is taken to be $S^3 \times D$.)

Now let us discuss the quantum numbers of the current algebra soliton. First, let us calculate its baryon number (which was first demonstrated to be non-zero in ref. [7], where, however, different assumptions were made from those we will follow). In previous work [10] it was shown that the baryon-number current has an anomalous piece, related to the $N\Gamma$ term in eq. (1). If the baryon number of a quark is $1/N$, so that an ordinary baryon made from N quarks has baryon number 1, then the anomalous piece in the baryon number current B_μ was shown to be

$$B_\mu = \frac{\varepsilon_{\mu\nu\alpha\beta}}{24\pi^2} \mathrm{Tr}\left(U^{-1}\partial_\nu U\right)\left(U^{-1}\partial_\alpha U\right)\left(U^{-1}\partial_\beta U\right). \tag{2}$$

So the baryon number of a configuration is

$$B = \int \mathrm{d}^3x\, B_0 = \frac{1}{24\pi^2}\int \mathrm{d}^3x\, \varepsilon^{ijk}\mathrm{Tr}\left(U^{-1}\partial_i U\right)\left(U^{-1}\partial_j U\right)\left(U^{-1}\partial_k U\right). \tag{3}$$

The right-hand side of eq. (24) can be recognized as the properly normalized integral expression for the winding number in $\pi_3(\mathrm{SU}(3))$. In a soliton field the right-hand side of (3) equals one, so the soliton has baryon number one; it is a baryon. (In ref. [7] the baryon number of the soliton was computed using methods of Goldstone and Wilczek [11]. The result that the soliton has baryon number one would emerge in this framework if the elementary fermions are taken to be quarks.)

Now let us determine whether the soliton is a boson or a fermion. To this end, we compare the amplitude for two processes. In one process, a soliton sits at rest for a long time T. The amplitude is $\exp(-iMT)$, M being the soliton energy. In the second process, the soliton is adiabatically rotated through a 2π angle in the course of a long time T. The usual term in the lagrangian $L_0 = \frac{1}{16}F_\pi^2 \mathrm{Tr}\, \partial_\mu U\, \partial_\mu U^{-1}$ does not distinguish between the two processes, because the only piece in L_0 that contains time derivatives is quadratic in time derivatives, and the integral $\int \mathrm{d}t\, \mathrm{Tr}(\partial U/\partial t)(\partial U^{-1}/\partial t)$ vanishes in the limit of an adiabatic process. However, the anomalous term Γ is linear in time derivatives, and distinguishes between a soliton that sits at rest and a soliton that is adiabatically rotated. For a soliton at rest, $\Gamma = 0$. For a soliton that is adiabatically rotated through a 2π angle, a slightly laborious calculation explained at the end of this paper shows that $\Gamma = \pi$. So for a soliton that is adiabatically rotated by a 2π angle, the amplitude is not $\exp(-iMT)$ but $\exp(-iMT)\exp(iN\pi) = (-1)^N \exp(-iMT)$.

The factor $(-1)^N$ means that for odd N the soliton is a fermion; for even N it is a boson. This is uncannily reminiscent of the fact that an ordinary baryon contains N quarks and is a boson or a fermion depending on whether N is even or odd.

These results are unchanged if there are more than three light flavors of quarks. How do they hold up if there are only two light flavors? The field $U(x)$ is then an element of SU(2). Because $\pi_3(\mathrm{SU}(2)) = \mathrm{Z}$, there are still solitons. The baryon-number current has the same anomalous piece, and the soliton still has baryon number one. But in SU(2) current algebra, there is no Γ term, so how can we see that the soliton can be a fermion?

The answer was given long ago [2]. Although $\pi_4(\mathrm{SU}(3)) = 0$, $\pi_4(\mathrm{SU}(2)) = \mathrm{Z}_2$. With suitably compactified space-time, there are thus two topological classes of maps from space-time to SU(2). In the SU(2) non-linear sigma model, there are hence two

"θ-vacua": fields that represent the non-trivial class in $\pi_4(SU(2))$ may be weighted with a sign $+1$ or -1. An explicit field $U(x, y, z, t)$ which goes to 1 at space-time infinity and represents the non-trivial class in $\pi_4(SU(2))$ can (fig. 1) be described as follows (a variant of this description figures in recent work by Goldstone [12]). Start at $t \to -\infty$ with a constant, $U = 1$; moving forward in time, gradually create a soliton-anti-soliton pair and separate them; rotate the soliton through a 2π angle without touching the anti-soliton; bring together the soliton and anti-soliton and annihilate them. Weighting this field with a factor of -1, while a configuration without the 2π rotation of the soliton is homotopically trivial and gets a factor $+1$, corresponds to quantizing the soliton as a fermion. Thus, internally to $SU(2) \times SU(2)$ current algebra, one sees that the soliton can be a fermion. In $SU(3) \times SU(3)$ current algebra one finds the stronger result that the soliton *must* be a fermion if and only if N is odd.

Our results so far are consistent with the idea that quantization of the current algebra soliton describes ordinary nucleons. However, we have not established this. Perhaps there are ordinary baryons and exotic, topologically excited solitonic baryons. However, certain results will now be described which seem to directly show that the ordinary nucleons are the ground state of the soliton.

For simplicity, we will focus now on the case of only two flavors. Soliton states can be labeled by their spin and isospin quantum numbers, which we will call J and I, respectively. We will determine semiclassically what values of I and J are expected for solitons. A semiclassical description of current algebra solitons will be accurate quantitatively only in the limit of large N. (Since F_π^2 is proportional to N, N enters the effective lagrangian (1) as an overall multiplicative factor. Hence, N plays the role usually played by $1/\hbar$.) So we will check the results we find for solitons by comparing to the expected quantum numbers of baryons in the large-N limit.

Let us first determine the expected baryon quantum numbers. We make the usual assumption that the multi-quark wave function is symmetric in space and antisymmetric in color, and hence must have complete symmetry in spin and isospin. The spin-isospin group is $SU(2) \times SU(2) \sim O(4)$. A quark transforms as $(\frac{1}{2}, \frac{1}{2})$; this is the

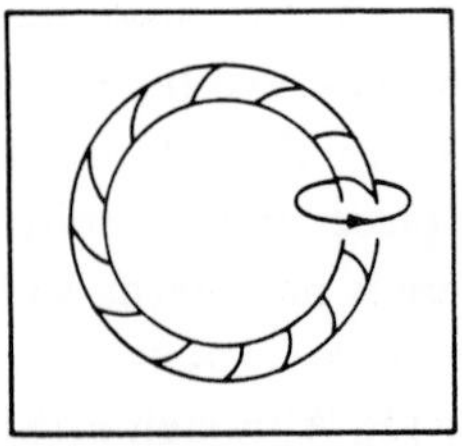

Fig. 1. A soliton-antisoliton pair is created from the vacuum; the soliton is rotated by a 2π angle; the pair is then annihilated. This represents the non-trivial homotopy class in $\pi_4(SU(2))$.

vector representation of O(4). We may represent a quark as ϕ_i, where $i = 1\dots4$ is a combined spin-isospin index labeling the O(4) four-vector.

We must form symmetric combinations of N vectors ϕ_i. As is well known, there is a quadratic invariant $\phi^2 = \sum_{i=1}^4 \phi_i^2$. One can also form symmetric traceless tensors of any rank $A^{(p)}_{i_1\dots i_p} = (\phi_{i_1}\phi_{i_2}\dots\phi_{i_p}$ − trace terms); this transforms as $(\frac{1}{2}p, \frac{1}{2}p)$ under SU(2) × SU(2). The general symmetric expression that we can make from N quarks is $(\phi^2)^k A^{(N-2k)}_{i_1\dots i_{N-2k}}$, where $0 \leqslant k \leqslant \frac{1}{2}N$. So the values of I and J that are possible are the following:

$$N \text{ even}, \qquad I = J = 0, 1, 2, 3, \dots,$$

$$N \text{ odd}, \qquad I = J = \tfrac{1}{2}, \tfrac{3}{2}, \tfrac{5}{2}, \tfrac{7}{2}, \dots. \tag{4}$$

For instance, in nature we have $N = 3$. The first two terms in the sequence indicated above are the nucleon, of $I = J = \frac{1}{2}$, and the delta, of $I = J = \frac{3}{2}$. If the number of colors were five or more, we would expect to see more terms in this series. Moreover, simple considerations involving color magnetic forces suggest that, as for $N = 3$, the mass of the baryons in this sequence is always an increasing function of I or J.

Now let us compare to what is expected in the soliton picture. (This question has been treated previously in ref. [7].) We do not know the effective action of which the soliton is a minimum, because we do not know what non-minimal terms must be added to eq. (1). We will make the simple assumption that the soliton field has the maximum possible symmetry. The soliton field cannot be invariant under I or J (or any component thereof), but it can be invariant under a diagonal subgroup $I + J$. This corresponds to an ansatz $U(x) = \exp[iF(r)]\boldsymbol{T}\cdot\boldsymbol{x}$, where $F(r) = 0$ at $r = 0$ and $F(r) \to \pi$ as $r \to \infty$.

Quantization of such a soliton is very similar to quantization of an isotropic rigid rotor. The hamiltonian of an isotropic rotor is invariant under an SU(2) × SU(2) group consisting of the rotations of body fixed and space fixed coordinates, respectively. We will refer to these symmetries as I and J, respectively. A given configuration of the rotor is invariant under a diagonal subgroup of SU(2) × SU(2). This is just analogous to our solitons, assuming the classical soliton solution is invariant under $I + J$.

The quantization of the isotropic rigid rotor is well known. If the rotor is quantized as a boson, it has $I = J = 0, 1, 2, \dots$. If it is quantized as a fermion, it has $I = J = \frac{1}{2}, \frac{3}{2}, \frac{5}{2}, \dots$. The agreement of these results with eq. (4) is hardly likely to be fortuitous.

In the case of three or more flavors, it may still be shown that the quantization of collective coordinates gives the expected flavor quantum numbers of baryons. The analysis is more complicated; the Wess-Zumino interaction plays a crucial role.

So far, we have assumed that the color gauge group is SU(N). Now let us discuss what would happen if the color group were O(N) or Sp(N). (By Sp(N) we will mean the group of $N \times N$ unitary matrices of quaternions; thus $\mathrm{Sp}(1) \simeq \mathrm{SU}(2)$.) We will see that also for these gauge groups, the topological properties of the current algebra theory correctly reproduce properties of the underlying gauge theory.

In an $\mathrm{O}(N)$ gauge theory, we assume that we have n multiplets of left-handed (Weyl) spinors in the fundamental N-dimensional representation of $\mathrm{O}(N)$. There is no distinction between quarks and antiquarks, because this representation is real. (If n is even, the theory is equivalent to a theory of $\frac{1}{2}n$ Dirac multiplets.) The anomaly free flavor symmetry group is $\mathrm{SU}(n)$. Simple considerations based on the most attractive channel idea suggest that the flavor symmetry will be spontaneously broken down to $\mathrm{O}(n)$, which is the maximal subgroup of $\mathrm{SU}(n)$ that permits all fermions to acquire mass. In this case the current algebra theory is based on a field that takes values in the quotient space $\mathrm{SU}(n)/\mathrm{O}(n)$.

In an $\mathrm{Sp}(N)$ gauge theory, we assume the fermion multiplets to be in the fundamental $2N$-dimensional representation of $\mathrm{Sp}(N)$. Since this representation is pseudoreal, there is again no distinction between quarks and antiquarks. In this theory the number of fermion multiplets must be even; otherwise, the $\mathrm{Sp}(N)$ gauge theory is inconsistent because of a non-perturbative anomaly [2] involving $\pi_4(\mathrm{Sp}(N))$. If there are $2n$ multiplets, the flavor symmetry is $\mathrm{SU}(2n)$. Simple arguments suggest that the $\mathrm{SU}(2n)$ flavor group is spontaneously broken to $\mathrm{Sp}(n)$, so that the current algebra theory is based on the quotient space $\mathrm{Su}(2n)/\mathrm{Sp}(n)$. This corresponds to symmetry breaking in the most attractive channel; $\mathrm{Sp}(n)$ is the largest unbroken symmetry that lets all quarks get mass.

In $\mathrm{O}(N)$, since there is no distinction between quarks and antiquarks, there is also no distinction between baryons and anti-baryons. A baryon can be formed from an antisymmetric combination of N quarks; $B = \varepsilon_{i_1 i_2 \ldots i_N} q^{i_1} q^{i_2} \ldots q^{i_N}$. But in $\mathrm{O}(N)$, a product of two epsilon symbols can be rewritten as a sum of products of N Kronecker deltas:

$$\varepsilon_{i_1 i_2 \ldots i_N} \varepsilon_{j_1 j_2 \ldots j_N} = \left(\delta_{i_1 j_1} \delta_{j_2 j_2} \ldots \delta_{i_N j_N} \pm \text{permutations} \right).$$

This means that in an $\mathrm{O}(N)$ gauge theory, two baryons can annihilate into N mesons.

On the other hand, in an $\mathrm{Sp}(N)$ gauge theory there are no baryons at all. The group $\mathrm{Sp}(N)$ can be defined as the subgroup of $\mathrm{SU}(2N)$ that leaves fixed an antisymmetric second rank tensor γ_{ij}. A meson made from two quarks of the same chirality can be described by the two quark operator $\gamma_{ij} q^i q^j$. In $\mathrm{Sp}(N)$ the epsilon symbol can be written as a sum of products of N γ's:

$$\varepsilon_{i_1 i_2 \ldots i_{2N}} = \left(\gamma_{i_1 i_2} \gamma_{i_3 i_4} \ldots \gamma_{i_{2N-1} i_{2N}} \pm \text{permutations} \right).$$

So in an $\mathrm{Sp}(N)$ gauge theory, a single would-be baryon can decay to N mesons.

Now let us discuss the physical phenomena that are related to the topological properties of our current algebra spaces SU(n)/O(n) and SU(n)/Sp(n). We recall from ref. (10) that the existence in QCD current algebra with at least three flavors of the Wess-Zumino interaction, with its a priori quantization law, is closely related to the fact that $\pi_5(\mathrm{SU}(n)) = \mathrm{Z}$, $n \geq 3$. The analogue is that

$$\pi_5(\mathrm{SU}(n)/\mathrm{O}(n)) = \mathrm{Z}, \qquad n \geq 3,$$
$$\pi_5(\mathrm{SU}(2n)/\mathrm{Sp}(n)) = \mathrm{Z}, \qquad n \geq 2. \tag{5}$$

So also the O(N) and Sp(N) gauge theories possess at the current algebra level an interaction like the Wess-Zumino term, provided the number of flavors is large enough. Built into the current algebra theories is the fact that in the underlying theory there is a parameter (the number of colors) which a priori must be an integer.

Now we come to the question of the existence of solitons. These are classified by the third homotopy group of the configuration space, and we have

$$\pi_3(\mathrm{SU}(n)/\mathrm{O}(n)) = \mathrm{Z}_2, \quad n \geq 4,$$
$$\pi_3(\mathrm{SU}(2n)/\mathrm{Sp}(n)) = 0, \qquad \text{any } n. \tag{6}$$

Thus, in the case of an O(N) gauge theory with at least four flavors, the current algebra theory admits solitons, but the number of solitons is conserved only modulo two. This agrees with the fact that in the O(N) gauge theory there are baryons which can annihilate in pairs. In current algebra corresponding to Sp(N) gauge theory there are no solitons, just as the Sp(N) gauge theory has no baryons.

For O(N) gauge theories with less than four light flavors we have

$$\pi_3(\mathrm{SU}(3)/\mathrm{O}(3)) = \mathrm{Z}_4,$$
$$\pi_3(\mathrm{SU}(2)/\mathrm{O}(2)) = \mathrm{Z}. \tag{7}$$

Thus, the spectrum of current algebra solitons seems richer than the expected spectrum of baryons in the underlying gauge theory. The following remark seems

TABLE 1
Some homotopy groups of certain homogeneous spaces

	SU(n)	SU(n)/O(n)	SU(2n)/Sp(n)
π_2	0	Z_2, $n \geq 3$	0
π_3	Z, all n	Z_2, $n \geq 4$	0
π_5	Z, $n \geq 3$	Z, $n \geq 3$	Z, $n \geq 3$

appropriate in this connection. It is only in the multi-color, large-N limit that a semiclassical description of current algebra solitons becomes accurate. Actually, large-N gauge theories are described by weakly interacting theories of mesons, but it is not only Goldstone bosons that enter; one has an infinite meson spectrum. Corresponding to the rich meson spectrum is an unknown and perhaps topologically complicated configuration space P of the large-N theory. Plausibly, baryons can always be realized as solitons in the large-N theory, even if all or almost all quark flavors are heavy. Perhaps $\pi_3(\mathrm{P})$ is always Z, Z_2, or O for SU(N), O(N), and Sp(N) gauge theories. The Goldstone boson space is only a small subspace of P and would not necessarily reflect the topology of P properly. Our results suggest that as the number of flavors increases, the Goldstone boson space becomes an increasingly good *topological* approximation to P. In this view, the extra solitons suggested by eq. (7) for O(N) gauge theories with two or three flavors become unstable when SU(2)/O(2) or SU(3)/O(3) is embedded in P.

One further physical question will be addressed here. Is color confinement implicit in current algebra?

Do current algebra theories in which the field U labels a point in SU(n), SU(n)/O(n), or SU($2n$)/SP(n) admit flux tubes? By a flux tube we mean a configuration $U(x, y, z)$ which is independent of z and possesses a non-trivial topology in the x-y plane. To ensure that the energy per unit length is finite, U must approach a constant as $x, y \to \infty$. The proper topological classification involves therefore the *second* homotopy group of the space in which U takes its values. In fact, we have

$$\begin{aligned} &\pi_2(\mathrm{SU}(n)) = 0, \\ &\pi_2(\mathrm{SU}(n)/\mathrm{O}(n)) = Z_2, \qquad n \geqslant 3, \\ &\pi_2(\mathrm{SU}(2n)/\mathrm{Sp}(n)) = 0. \end{aligned} \tag{8}$$

Thus, current algebra theories corresponding to underlying SU(N) and Sp(N) gauge theories do not admit flux tubes. The theories based on underlying O(N) gauge groups do admit flux tubes, but two such flux tubes can annihilate.

These facts have the following natural interpretation. Our current algebra theories correspond to underlying gauge theories with quarks in the fundamental representation of SU(N), O(N), or Sp(N). SU(N) or Sp(N) gauge theories with dynamical quarks cannot support flux tubes because arbitrary external sources can be screened by sources in the fundamental representation of the group. For O(N) gauge theories it is different. An external source in the spinor representation of O(N) cannot be screened by charges in the fundamental representation. But two spinors make a tensor, which can be screened. So the O(N) gauge theory with dynamical quarks supports only one type of color flux tube: the response to an external source in the spinor representation of O(N). It is very plausible that this color flux tube should be identified with the excitation that appears in current algebra because $\pi_2(\mathrm{SU}(n)/\mathrm{O}(n)) = \mathrm{Z}_2$.

The following fact supports this identification. The interaction between two sources in the spinor representation of O(N) is, in perturbation theory, N times as big as the interaction between two quarks. Defining the large-N limit in such a way that the interaction between two quarks is of order one, the interaction between two spinor charges is therefore of order N. This strongly suggests that the energy per unit length in the flux tube connecting two spinor charges is of order N. This is consistent with our current algebra identification; the whole current algebra effective lagrangian is of order N (since $F_\pi^2 \sim N$), so the energy per unit length of a current algebra flux tube is certainly of order N.

In conclusion, it still remains for us to establish the contention made earlier that the value of the Wess-Zumino functional Γ for a process consisting of a 2π rotation of a soliton is $\Gamma = \pi$.

First of all, the soliton field can be chosen to be of the form

$$V(x_i) = \left(\begin{array}{c|c} W(x_i) & 0 \\ \hline 0 & 1 \end{array} \right), \tag{9}$$

where the SU(2) matrix W is chosen to be invariant under a combined isospin rotation plus rotation of the spatial coordinate x_i. This being so, a 2π rotation of V in space is equivalent to a 2π rotation of V in isospin. Introducing a periodic time coordinate t which runs from 0 to 2π, the desired field in which a soliton is rotated by a 2π angle can be chosen to be

$$U(x_i, t) = \begin{pmatrix} e^{it/2} & & \\ & e^{-it/2} & \\ & & 1 \end{pmatrix} V(x_i) \begin{pmatrix} e^{-it/2} & & \\ & e^{it/2} & \\ & & 1 \end{pmatrix}. \tag{10}$$

Note that $U(x_i, t)$ is periodic in t with period 2π even though the individual exponentials $\exp(\pm \frac{1}{2} it)$ have period 4π. Because of the special form of V, we can equivalently write U in the much more convenient form

$$U(x_i, t) = \begin{pmatrix} 1 & & \\ & e^{-it} & \\ & & e^{it} \end{pmatrix} V(x_i) \begin{pmatrix} 1 & & \\ & e^{it} & \\ & & e^{-it} \end{pmatrix}. \tag{11}$$

This field $U(x_i, t)$ describes a soliton that is rotated by a 2π angle as t ranges from 0 to 2π. We wish to evaluate $\Gamma(U)$.

To this end we introduce a fifth parameter ρ $(0 \leqslant \rho \leqslant 1)$ so as to form a five-manifold of which space-time is the boundary; this five-manifold will have the topology of three-space times a disc. A convenient choice is to write

$$\tilde{U}(x_i, t, \rho) = A^{-1}(t, \rho) U(x_i, t) A(t, \rho), \tag{12}$$

where

$$A(t, \rho) = \begin{pmatrix} 1 & 0 & 0 \\ 0 & \rho e^{it} & \sqrt{1-\rho^2} \\ 0 & -\sqrt{1-\rho^2} & \rho e^{-it} \end{pmatrix}. \tag{13}$$

Note that at $\rho = 0$, $A(t, \rho)$ is independent of t. So we can think of ρ and t as polar coordinates for the plane, ρ being the radius and t the usual angular variable. Also $\tilde{U}(x_i, t, 1) = U(x_i, t)$ so the product of three space with the unit circle in the ρ-t plane can be identified with the original space-time. According to eq. (14) of ref. (10), what we must calculate is

$$\Gamma(U) = -\frac{i}{240\pi^2} \int_0^1 \mathrm{d}\rho \int_0^{2\pi} \mathrm{d}t \int \mathrm{d}^3x\, \varepsilon^{ijklm} \times \left[\mathrm{Tr}\, \tilde{U}^{-1} \partial_i \tilde{U}\, \tilde{U}^{-1} \partial_j \tilde{U}\, \tilde{U}^{-1} \partial_k \tilde{U}\, \tilde{U}^{-1} \partial_l \tilde{U}\, \tilde{U}^{-1} \partial_m \tilde{U} \right], \tag{14}$$

where i, j, k, l, and m may be ρ, t, x_1, x_2, or x_3. The integral can be done without undue difficulty (the fact that W is invariant under spatial rotations plus isospin is very useful), and one finds $\Gamma(U) = \pi$.

This calculation can also be used to fill in a gap in the discussion of ref. (10). In that paper, the following remark was made. Let $A(x, y, z, t)$ be a mapping from space-time into SU(2) that is in the non-trivial homotopy class in $\pi_4(\mathrm{SU}(2))$. Embed A in SU(3) in the trivial form

$$\hat{A} = \left(\begin{array}{cc|c} \multicolumn{2}{c|}{A} & \begin{matrix} 0 \\ 0 \end{matrix} \\ \hline 0 & 0 & 1 \end{array}\right).$$

Then $\Gamma(\hat{A}) = \pi$. In fact, as we have noted above, the non-trivial homotopy class in $\pi_4(\mathrm{SU}(2))$ differs from the trivial class by a 2π rotation of a soliton (which may be one member of a soliton-antisolition pair). The fact that $\Gamma = \pi$ for a 2π rotation of soliton means that $\Gamma = \pi$ for the non-trivial homotopy class in $\pi_4(\mathrm{SU}(2))$.

The following important fact deserves to be demonstrated explicitly. As before, let A be a mapping of space-time into SU(2) and let $\hat{A}$ be its embedding in SU(3). Then $\Gamma(\hat{A})$ depends only on the homotopy class of A in $\pi_4(\mathrm{SU}(2))$. In fact, suppose $\hat{A}$ is

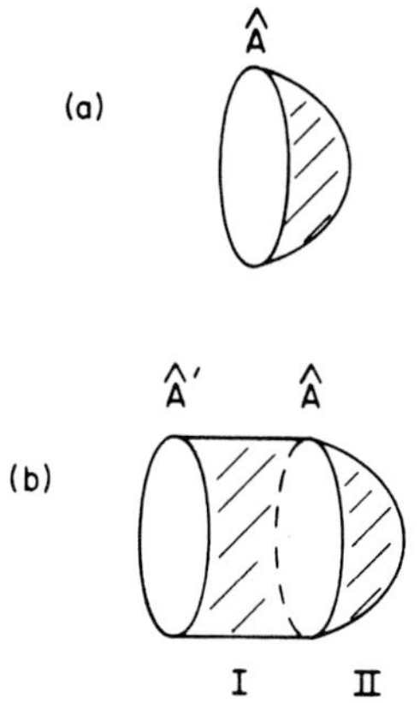

Fig. 2. A demonstration that Γ is a homotopy invariant for SU(2) mappings.

homotopic to $\hat{A}'$: Let us prove that $\Gamma(\hat{A}) = \Gamma(\hat{A}')$. To compute $\Gamma(\hat{A})$ we realize space-time as the boundary of a disc, extend $\hat{A}$ to be defined over that disc, and evaluate an appropriate integral (fig. 2a). To evaluate $\Gamma(\hat{A}')$ we again must extend $\hat{A}'$ to a disc. This can be done in a very convenient way (fig. 2b). Since $\hat{A}$ is homotopic to $\hat{A}'$, we first deform $\hat{A}'$ into $\hat{A}$ via matrices of the form $\left(\begin{array}{c|c} X & 0 \\ \hline 0 & 1 \end{array}\right)$ (matrices that are really SU(2) matrices embedded in SU(3)) and then we extend $\hat{A}$ over a disc as before. The integral contribution to $\Gamma(\hat{A}')$ from part I of fig. 2b vanishes because the fifth rank antisymmetric tensor that enters in defining Γ vanishes when restricted to any SU(2) subgroup of SU(3). The integral in part II of fig. 2b is the same as the integral in fig. 2a, so $\Gamma(\hat{A}) = \Gamma(\hat{A}')$.

The fact that Γ is a homotopy invariant for SU(2) mappings also means that Γ can be used to prove that $\pi_4(\mathrm{SU}(2))$ is non-trivial. Since Γ obviously is 0 for the trivial homotopy class in $\pi_4(\mathrm{SU}(2))$, while $\Gamma = \pi$ for a process containing a 2π rotation of a soliton, the latter process must represent a non-trivial element in $\pi_4(\mathrm{SU}(2))$. What cannot be proved so easily is that this is the only non-trivial element.

I would like to thank A.P. Balachandran and V.P. Nair for interesting me in current algebra solitons.

Note added in proof

Many physicists have asked how the soliton quantum numbers can be calculated if there are three flavors. Following is a sketch of how this question can be answered.

We assume that for $\mathrm{SU}(3) \times \mathrm{SU}(3)$ current algebra, the soliton solution is simply an SU(2) solution embedded in SU(3). Such a solution is invariant under combined spin-isospin transformations; and it is also invariant under hypercharge rotations.

There are now seven collective coordinates instead of three. They parametrize the coset space $X = SU(3)/U(1)$, where U(1) refers to right multiplication by hypercharge. Thus a point in X is an element U of SU(3) defined up to multiplication on the right by a hypercharge transformation. The space X has flavor SU(3) symmetry (left multiplication of U by an SU(3) matrix) and rotation SU(2) symmetry (right multiplication of U by an SU(3) matrix that commutes with hypercharge).

The crucial novelty of the three-flavor problem is that even when restricted to the space of collective coordinates, the Wess-Zumino term does not vanish. As usual, the quantization of collective coordinates involves the quantum mechanics of a particle moving on the manifold X, but in this case, the effect of the Wess-Zumino term is that the particle is moving under the influence of a simulated "magnetic field" on the X manifold. Moreover, this magnetic field is of the Dirac monopole type; it has string singularities which are unobservable if the Wess-Zumino coupling is properly quantized.

The wave functions of the collective coordinates are "monopole harmonics" on the X manifold with quantum numbers that depend on the "magnetic charge." For charge three (three colors) the lowest monopole harmonic is an SU(3) octet of spin $\frac{1}{2}$, and the next one is an SU(3) decuplet of spin $\frac{3}{2}$.

References

[1] T.H.R. Skyrme, Proc. Roy. Soc. A260 (1961) 127
[2] D. Finkelstein and J. Rubinstein, J. Math. Phys. 9 (1968) 1762
[3] L.D. Faddeev, Lett. Math. Phys. 1 (1976) 289
[4] S. Coleman, Phys. Rev. D11 (1975) 2088;
R. Jackiw and C. Rebbi, Phys. Rev. Lett. 36 (1976) 1116;
P. Hasenfratz and G. 't Hooft, Phys. Rev. Lett. 36 (1976) 1119
[5] G. 't Hooft, Nucl. Phys. B72 (1974) 461; B75 (1974) 461
[6] E. Witten, Nucl. Phys. B160 (1979) 57
[7] A.P. Balachandran, V.P. Nair, S.G. Rajeev and A. Stern, Phys. Rev. Lett. 49 (1982) 1124; Syracuse University preprint (1982)
[8] J. Boguta, Phys. Rev. Lett. 50 (1983) 148
[9] J. Wess and B. Zumino, Phys. Lett. 37B (1971) 95
[10] E. Witten, Nucl. Phys. B223 (1983) 422
[11] J. Goldstone and F. Wilczek, Phys. Rev. Lett. 47 (1981) 986
[12] J. Goldstone, private communication

Nuclear Physics B228 (1983) 552–566
© North-Holland Publishing Company

STATIC PROPERTIES OF NUCLEONS IN THE SKYRME MODEL

Gregory S. ADKINS[1]

Joseph Henry Laboratories, Princeton University, Princeton, New Jersey 08544, USA

Chiara R. NAPPI

The Institute for Advanced Study, Princeton, New Jersey 08540, USA

Edward WITTEN[1]

Joseph Henry Laboratories, Princeton University, Princeton, New Jersey 08544, USA

Received 20 June 1983

We compute static properties of baryons in an SU(2)×SU(2) chiral theory (the Skyrme model) whose solitons can be interpreted as the baryons of QCD. Our results are generally within about 30% of experimental values. We also derive some relations that hold generally in soliton models of baryons, and therefore, serve as tests of the $1/N$ expansion.

1. Introduction

Recent developments have provided partial confirmation of Skyrme's old idea [1] that baryons are solitons in the non-linear sigma model. We know that in the large-N limit, QCD becomes equivalent to an effective field theory of mesons [2]. Counting rules suggest [3] that baryons may emerge as solitons in this theory. Although we do not understand in detail the large-N theory of mesons, we know that at low energies this theory reduces to a non-linear sigma model of spontaneously broken chiral symmetry. Moreover, the solitons of the non-linear model have precisely the quantum numbers of QCD baryons [4] provided one includes the effects of the Wess–Zumino coupling [5, 6].

In this paper we will evaluate the static properties of nucleons such as masses, magnetic moments, and charge radii, in a soliton model. For simplicity we will restrict ourselves to the case of two flavors. One simplification in the SU(2) case is that the Wess–Zumino term vanishes. At a pedestrian level, for $U=1+iA+\mathrm{O}(A^2)$, the Wess–Zumino term is [5, 6]

$$n\Gamma = n\frac{2}{15\pi^2F_\pi^5}\int \mathrm{d}^4x\,\varepsilon^{\mu\nu\alpha\beta}\,\mathrm{Tr}\,[A\partial_\mu A\partial_\nu A\partial_\alpha A\partial_\beta A]+\text{higher orders}\,.$$

If $A=a_a\tau_a$, then

$$n\Gamma = n\frac{2}{15\pi^2F_\pi^5}\int \mathrm{d}^4x\,\varepsilon^{\mu\nu\alpha\beta}a_a\partial_\mu a_b\partial_\nu a_c\partial_\alpha a_d\partial_\beta a_e\,\mathrm{Tr}\,[\tau_a\tau_b\tau_c\tau_d\tau_e]\,.$$

[1] Supported in part by NSF grant no. PHY80-19754.

This term is completely antisymmetrical in the Lorentz indices, so it needs to be completely antisymmetrical in the isospin indices *b*, *c*, *d* and *e*. But that is impossible because there are only three independent SU(2) generators. More generally, the fifth-rank antisymmetric tensor ω discussed in [6] vanishes on the SU(2) group manifold. Nonetheless, the anomalous baryon-number current, which can be obtained from the WZ term or by the method of Goldstone and Wilczek [7], is still present in the two-flavor case.

Since the proper large-N effective theory is unknown, we will consider here a crude description in which the large-N theory is assumed to be a theory of pions only. In this context, it is necessary to add a non-minimal term to the non-linear sigma model to prevent the solitons from shrinking to zero-size. The simplest reasonable choice is the Skyrme model

$$L = \tfrac{1}{16}F_\pi^2 \,\mathrm{Tr}\,(\partial_\mu U \partial_\mu U^\dagger) + \frac{1}{32e^2}\mathrm{Tr}\,[(\partial_\mu U)U^\dagger, (\partial_\nu U)U^\dagger]^2 . \tag{1}$$

Here U is an SU(2) matrix, transforming as $U \to AUB^{-1}$ under chiral SU(2)× SU(2); $F_\pi = 186$ MeV is the pion decay constant; and the last term, which contains the dimensionless parameter *e*, was introduced by Skyrme to stabilize the solitons. It is the unique term with four derivatives which leads to a positive hamiltonian. (It is also the unique term with four derivatives that leads to a hamiltonian second order in time derivatives.) Although the Skyrme model is only a rough description, since it omits the other mesons and interactions that are present in the large-N limit of QCD, we regard it as a good model for testing the reasonableness of a soliton desciption of nucleons.

1. Kinematics

From the lagrangian (1) we find the soliton solution by using the Skyrme ansatz $U_0(x) = \exp[iF(r)\boldsymbol{\tau}\cdot\hat{\boldsymbol{x}}]$, where $F(r) = \pi$ at $r = 0$ and $F(r) \to 0$ as $r \to \infty$. If we substitute this ansatz in (1) we get the expression for the soliton mass:

$$M = 4\pi \int_0^\infty r^2 \left\{ \tfrac{1}{8}F_\pi^2 \left[\left(\frac{\partial F}{\partial r}\right)^2 + 2\frac{\sin^2 F}{r^2} \right] + \frac{1}{2e^2}\frac{\sin^2 F}{r^2}\left[\frac{\sin^2 F}{r^2} + 2\left(\frac{\partial F}{\partial r}\right)^2 \right] \right\} dr . \tag{2}$$

The variational equation from (2) is

$$(\tfrac{1}{4}\tilde{r}^2 + 2\sin^2 F)F'' + \tfrac{1}{2}\tilde{r}F' + \sin 2F\, F'^2 - \tfrac{1}{4}\sin 2F - \frac{\sin^2 F \sin 2F}{\tilde{r}^2} = 0 .$$

in terms of a dimensionless variable $\tilde{r} = eF_\pi r$. The behaviour of the numerical solution of eq. (3) is shown in fig. 1.

Now, if $U_0 = \exp[iF(r)\boldsymbol{\tau}\cdot\hat{\boldsymbol{x}}]$ is the soliton solution, then $U = AU_0A^{-1}$, where A is an arbitrary constant SU(2) matrix, is a finite-energy solution as well. A solution

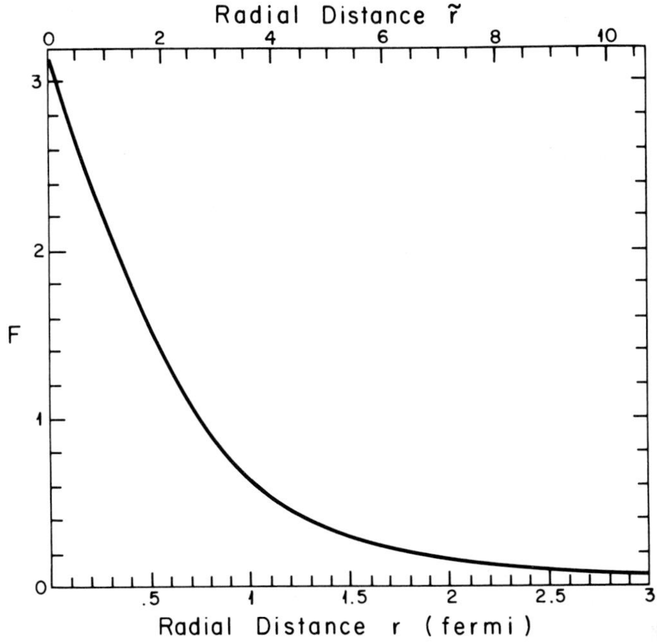

Fig. 1. Plot of F, the numerical solution of eq. (3). F appears in the Skyrme ansatz $U_0(x) = \exp[iF(r)\boldsymbol{\tau}\cdot\hat{\boldsymbol{x}}]$. The radial distance is measured in fm, and also in the dimensionless variable $\tilde{r} = eF_\pi r$.

of any given A is not an eigenstate of spin and isospin. We need to treat A as a quantum mechanical variable, as a collective coordinate. The simplest way to do this is to write the lagrangian and all physical observables in terms of a time dependent A. We substitute $U = A(t)U_0A^{-1}(t)$ in the lagrangian, where U_0 is the soliton solution and $A(t)$ is an arbitrary time-dependent SU(2) matrix. This procedure will allow us to write a hamiltonian which we will diagonalize. The eigenstates with the proper spin and isospin will correspond to the nucleon and delta.

So, substituting $U = A(t)U_0A^{-1}(t)$ in (1), after a lengthy calculation, we get

$$L = -M + \lambda\,\mathrm{Tr}\,[\partial_0 A \partial_0 A^{-1}]\,, \tag{4}$$

where M is defined in (2) and $\lambda = \frac{4}{6}\pi(1/e^3F_\pi)\Lambda$ with

$$\Lambda = \int \tilde{r}^2 \sin^2 F\left[1+4\left(F'^2+\frac{\sin^2 F}{\tilde{r}^2}\right)\right]\mathrm{d}\tilde{r}\,. \tag{5}$$

Numerically we find $\Lambda = 50.9$. The SU(2) matrix A can be written $A = a_0 + i\boldsymbol{a}\cdot\boldsymbol{\tau}$, with $a_0^2 + \boldsymbol{a}^2 = 1$. In terms of the a's (4) becomes

$$L = -M + 2\lambda \sum_{i=0}^{3} (\dot{a}_i)^2\,.$$

Introducing the conjugate momenta $\pi_i = \partial L/\partial \dot{a}_i = 4\lambda \dot{a}_i$, we can now write the hamiltonian

$$H = \pi_i \dot{a}_i - L = 4\lambda \dot{a}_i \dot{a}_i - L = M + 2\lambda \dot{a}_i \dot{a}_i = M + \frac{1}{8\lambda} \sum_i \pi_i^2 .$$

Performing the usual canonical quantization procedure $\pi_i = -i\partial/\partial a_i$ we get

$$H = M + \frac{1}{8\lambda} \sum_{i=0}^{3} \left(-\frac{\partial^2}{\partial a_i^2} \right), \tag{6}$$

with the constraint $\sum_{i=0}^{3} a_i^2 = 1$. Because of this constraint, the operator $\sum_{i=0}^{3} \partial^2/\partial a_i^2$ is to be interpreted as the laplacian ∇^2 on the three-sphere. The wave functions (by analogy with usual spherical harmonics) are traceless symmetric polynomials in the a_i. A typical example is $(a_0 + ia_1)^l$, with $-\nabla^2 (a_0 + ia_1)^l = l(l+2)(a_0 + ia_1)^l$. Such a wave function has spin and isospin equal to $\frac{1}{2}l$, as one may see by considering the spin and isospin operators

$$I_k = \tfrac{1}{2} i \left(a_0 \frac{\partial}{\partial a_k} - a_k \frac{\partial}{\partial a_0} - \varepsilon_{klm} a_l \frac{\partial}{\partial a_m} \right),$$

$$J_k = \tfrac{1}{2} i \left(a_k \frac{\partial}{\partial a_0} - a_0 \frac{\partial}{\partial a_k} - \varepsilon_{klm} a_l \frac{\partial}{\partial a_m} \right). \tag{7}$$

An important physical point must be addressed here. Since the nonlinear sigma model field is $U = AU_0A^{-1}$, A and $-A$ correspond to the same U. Naively, one might expect to insist that the wave function $\psi(A)$ obeys $\psi(A) = +\psi(-A)$. Actually, as discussed long ago by Finkelstein and Rubinstein [8], there are two consistent ways to quantize the soliton; one may require $\psi(A) = +\psi(-A)$ for all solitons, or one may require $\psi(A) = -\psi(-A)$ for all solitons. The former choice corresponds to quantizing the soliton as a boson. The latter choice corresponds to quantizing it as a fermion. We wish to follow the second road, of course, so our wave functions will be polynomials of *odd* degree in the a_i. So, the nucleons, of $I = J = \frac{1}{2}$, correspond to wave functions linear in a_i, while the deltas, of $I = J = \frac{3}{2}$, correspond to cubic functions. Wave functions of fifth order and higher correspond to highly excited states (masses$\geq$1730 MeV) which either are lost in the pion-nucleon continuum or else are artifacts of the model. The properly normalized wave functions for proton and neutron states of spin up or spin down along the z axis, and some of the Δ wave functions, are:

$$|p\uparrow\rangle = \frac{1}{\pi}(a_1 + ia_2), \qquad |p\downarrow\rangle = -\frac{i}{\pi}(a_0 - ia_3),$$

$$|n\uparrow\rangle = \frac{i}{\pi}(a_0 + ia_3), \qquad |n\downarrow\rangle = -\frac{1}{\pi}(a_1 - ia_2),$$

$$|\Delta^{++}, s_z = \tfrac{3}{2}\rangle = \frac{\sqrt{2}}{\pi}(a_1 + ia_2)^3,$$

$$|\Delta^{+}, s_z = \tfrac{1}{2}\rangle = -\frac{\sqrt{2}}{\pi}(a_1 + ia_2)(1 - 3(a_0^2 + a_3^2)). \tag{8}$$

Returning to eq. (6), the eigenvalues of the hamiltonian are $E = M + (1/8\lambda)l(l+2)$ where $l = 2J$. So, the nucleon and delta masses are given by

$$M_N = M + \frac{1}{2\lambda}\frac{3}{4}, \qquad M_\Delta = M + \frac{1}{2\lambda}\frac{15}{4}, \tag{9}$$

where M, obtained by evaluting (2) numerically, is given by $M = 36.5 F_\pi/e$ and $\lambda = \frac{4}{6}\pi(1/e^3 F_\pi)50.9$, as already said. We have found that the best procedure in dealing with this model is to adjust e and F_π to fit the nucleon and delta masses. The results are $e = 5.45$ and $F_\pi = 129$ MeV. Thus, on the basis of the values of the baryon masses, we require (or predict) in this model a value of F_π that is 30% lower than the experimental value of 186 MeV.

2. Currents, charge radii and magnetic moments

In order to compute weak and electromagnetic couplings of baryons, we need first to evaluate the currents in terms of collective coordinates. The Noether current associated with the V−A transformation $\delta U = iQU$ is

$$J^\mu_{V-A} = \tfrac{1}{8} i F_\pi^2 \,\mathrm{Tr}\,[(\partial^\mu U)U^\dagger Q] + \frac{i}{8e^2}\,\mathrm{Tr}\,\{[(\partial_\nu U)U^\dagger, Q][(\partial^\mu U)U^\dagger, (\partial^\nu U)U^\dagger]\}\,. \tag{10}$$

The V+A current is obtained by exchanging U with $U^\dagger$.

The anomalous baryon current is instead [7, 6]

$$B^\mu = \frac{\varepsilon^{\mu\nu\alpha\beta}}{24\pi^2}\,\mathrm{Tr}\,[(U^\dagger\partial_\nu U)(U^\dagger\partial_\alpha U)(U^\dagger\partial_\beta U)]\,, \tag{11}$$

where our notation is $\varepsilon_{0123} = -\varepsilon^{0123} = 1$.

If we substitute $U = A(t)U_0 A^{-1}(t)$ in (10), we get rather complicated expressions for the vector and axial currents V and A. The following angular integrals, which are much simpler, are adequate for our purposes:

$$\int \mathrm{d}\Omega\; V^{a,0} = \tfrac{1}{3} i 4\pi \Lambda'\,\mathrm{Tr}\,[(\partial_0 A)A^{-1}\tau_a]\,, \tag{12}$$

$$\int \mathrm{d}\Omega\; \boldsymbol{q}\cdot\boldsymbol{x} V^{a,i} = \tfrac{1}{3} i\pi\Lambda'\,\mathrm{Tr}\,(\boldsymbol{\tau}\cdot\boldsymbol{q}\tau_i A^{-1}\tau_a A)\,. \tag{13}$$

$$\int \mathrm{d}\Omega\; A^{a,i} = \tfrac{1}{3}\pi D'\,\mathrm{Tr}\,(\tau_i A^{-1}\tau_a A)\,, \tag{14}$$

where Λ' and D' are respectively

$$\Lambda' = \sin^2 F\left[F_\pi^2 + \frac{4}{e^2}\left(F'^2 + \frac{\sin^2 F}{r^2}\right)\right], \tag{15}$$

$$D' = F_\pi^2\left(F' + \frac{\sin 2F}{r}\right) + \frac{4}{e^2}\left(\frac{\sin 2F}{r}F'^2 + 2\frac{\sin^2 F}{r^2}F' + \frac{\sin^2 F\sin 2F}{r^3}\right). \tag{16}$$

In the computation of the above formulas from (10) we have neglected terms which are quadratic in time derivatives. In the semiclassical limit the solitons rotate slowly, so terms quadratic in time derivatives are higher order in the semiclassical approximation.

The expression in (7) for the isospin generator I_k can be derived from (12) by integrating over r, and replacing $\dot{a}_i$ by the canonical momentum.

From (11) we derive the baryon current and charge density

$$B^0 = -\frac{1}{2\pi^2}\frac{\sin^2 F}{r^2}F' , \tag{17}$$

$$B^i = i\frac{\varepsilon^{ijk}}{2\pi^2}\frac{\sin^2 F}{r}F'\hat{x}_k \operatorname{Tr}[(\partial_0 A^{-1})A\tau_j] . \tag{18}$$

The baryon charge per unit r is therefore

$$\rho_B(r) = 4\pi r^2 B^0(r) = -\frac{2}{\pi}\sin^2 F F' ,$$

and its integral $\int_0^\infty \rho_B(r)\,dr = 1$ gives the baryonic charge.

The isoscalar mean square radius is given by

$$\langle r^2\rangle_{I=0} = \int_0^\infty r^2\rho_B(r)\,dr = \frac{4.47}{e^2F_\pi^2} = 4.47\ (0.28)^2\ \mathrm{fm}^2 ,$$

and we get $\langle r^2\rangle_{I=0}^{1/2} = 0.59$ fm, while the corresponding experimental value is 0.72 fm.

From (12) and (15) we can compute the isovector charge density per unit r

$$\rho_{I=1}(r) = \frac{r^2\sin^2 F\,(F_\pi^2 + (4/e^2)(F'^2 + \sin^2 F/r^2))}{\int_0^\infty r^2\sin^2 F\,(F_\pi^2 + (4/e^2)(F'^2 + \sin^2 F/r^2))\,dr} ,$$

and finally derive the proton and neutron charge distributions which are plotted in fig. 2.

The isovector mean square charge radius $\int_0^\infty r^2\rho_{I=0}(r)\,dr$ is divergent, as expected in the chiral limit [9]. The introduction of quark masses in this model [10] will cure this problem, as it does in nature.

The definitions of isoscalar and isovector magnetic moments are respectively

$$\boldsymbol{\mu}_{I=0} = \tfrac{1}{2}\int \boldsymbol{r}\times\boldsymbol{B}\,d^3x , \tag{19}$$

$$\boldsymbol{\mu}_{I=1} = \tfrac{1}{2}\int \boldsymbol{r}\times\boldsymbol{V}^3\,d^3x . \tag{20}$$

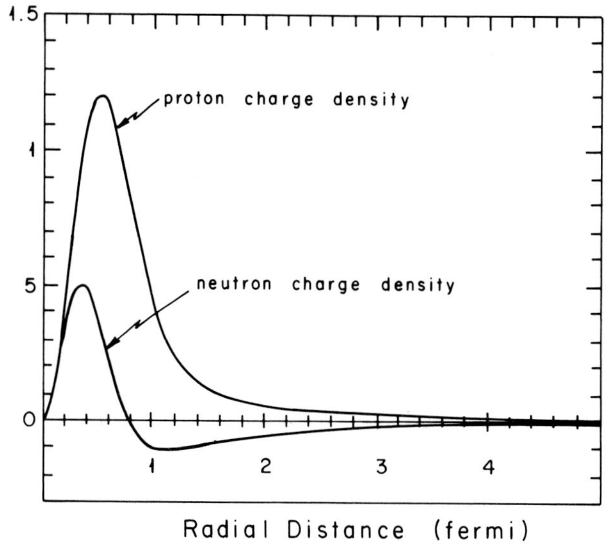

Fig. 2. Plot of the proton and neutron charge densities. These charge densities are given as functions of the radial distance r, and include a factor of $4\pi r^2$.

Therefore, from (18) the isoscalar magnetic moment density is

$$\rho_M^{I=0}(r) = \frac{r^2 F' \sin^2 F}{\displaystyle\int_0^\infty r^2 F' \sin^2 F \, \mathrm{d}r} .$$

The isoscalar magnetic mean radius is defined by

$$\langle r^2 \rangle_{M,I=0} = \int_0^\infty r^2 \rho_M^{I=0}(r) \, \mathrm{d}r .$$

We get $\langle r^2 \rangle_{M,I=0}^{1/2} = 0.92$ fm, against the experimental value of 0.81 fm.

The simplest way to extract the g factors is to calculate the expectation value of the magnetic moment operators in a proton state of spin up, using the forms given earlier for the wave functions. From (18) and (19) the isoscalar magnetic moment is

$$(\mu_{I=0})_i = \tfrac{1}{2} \int \mathrm{d}^3x \, \varepsilon_{lmi} x_l \langle \mathrm{p}\uparrow | B_m | \mathrm{p}\uparrow \rangle$$
$$= -\frac{1}{2} \frac{i}{2\pi^2} \int \mathrm{d}^3x \sin^2 F F' \hat{x}_l \hat{x}_k \varepsilon_{lmi} \varepsilon_{mjk} \langle \mathrm{p}\uparrow | \mathrm{Tr}\,[(\partial_0 A^{-1}) A \tau_j] | \mathrm{p}\uparrow \rangle .$$

It is easy to check that $\langle \mathrm{p}\uparrow | \mathrm{Tr}\,[(\partial_0 A^{-1}) A \tau_j] | \mathrm{p}\uparrow \rangle = -\delta_{j3} i / 2\lambda$. It follows that

$$(\boldsymbol{\mu}_{I=0})_3 = \frac{\langle \tilde{r}^2 \rangle_{I=0}}{\Lambda} \frac{e}{F_\pi} \frac{1}{4\pi} . \tag{21}$$

The g factor is defined by writing $\boldsymbol{\mu} = (g/4M)\boldsymbol{\sigma}$. The isoscalar g factor $g_{I=0} = g_p + g_n$ is 1.11 in this model (the experimental value is 1.76, instead).

In order to compute the isovector magnetic moment, we start from (13) and integrate in the radial variable. We get

$$\int d^3x\, \boldsymbol{q} \cdot \boldsymbol{x} V^{3,i} = \tfrac{1}{3} i\pi \frac{\Lambda}{F_\pi e^3} \operatorname{Tr}(\boldsymbol{\tau} \cdot \boldsymbol{q} \tau_i A^{-1} \tau_3 A),$$

with Λ given in (5). Now

$$\operatorname{Tr}(\boldsymbol{\tau} \cdot \boldsymbol{q} \tau_i A^{-1} \tau_j A) = i q_l \varepsilon_{\mathrm{lim}} \operatorname{Tr}(\tau_m A^{-1} \tau_j A).$$

A detailed calculation using the nucleon wave function given in (8) shows that for any nucleon states N and N′

$$\langle \mathrm{N}' | \operatorname{Tr}[\tau_i A^{-1} \tau_j A] | \mathrm{N} \rangle = -\tfrac{2}{3} \langle \mathrm{N}' | \sigma_i \tau_j | \mathrm{N} \rangle. \tag{22}$$

Therefore

$$\langle \mathrm{p}\uparrow | \int d^3x\, \boldsymbol{q} \cdot \boldsymbol{x} V_i^3 | \mathrm{p}\uparrow \rangle = -\tfrac{1}{3} q_l \pi \frac{\Lambda}{F_\pi e^3} \varepsilon_{li3}(-\tfrac{2}{3}),$$

$$\langle \mathrm{p}\uparrow | \int d^3x\, x_l V_i^3 | \mathrm{p}\uparrow \rangle = \tfrac{2}{9} \pi \frac{\Lambda}{F_\pi e^3} \varepsilon_{li3}.$$

In conclusion, from (20) we get

$$(\boldsymbol{\mu}_{I=1})_3 = \tfrac{2}{9} \pi \frac{\Lambda}{F_\pi e^3}. \tag{23}$$

The isovector g factor $g_{I=1} = g_p - g_n$ turns out to be 6.38 against the experimental value of 9.4. The magnetic moments for the proton and neutron, measured in terms of Bohr magneton, are $\mu_p = \frac{1}{2} g_p = 1.87$ and $\mu_n = \frac{1}{2} g_n = -1.31$ respectively. The ratio $|\mu_p/\mu_n|$ turns out to be 1.43 (see table 1), as opposed to 1.5 in the quark model and 1.46 experimentally.

TABLE 1

Quantity	Prediction	Experiment
M_N	input	939 MeV
M_Δ	input	1232 MeV
F_π	129 MeV	186 MeV
$\langle r^2 \rangle^{1/2}_{I=0}$	0.59 fm	0.72 fm
$\langle r^2 \rangle^{1/2}_{M,I=0}$	0.92 fm	0.81 fm
μ_p	1.87	2.79
μ_n	−1.31	−1.91
$\left\lvert \frac{\mu_p}{\mu_n} \right\rvert$	1.43	1.46
g_A	0.61	1.23
$g_{\pi NN}$	8.9	13.5
$g_{\pi N\Delta}$	13.2	20.3
$\mu_{N\Delta}$	2.3	3.3

3. Mass relations

It is interesting to form certain combinations of experimentally measured quantities from which the parameters of the Skyrme model cancel out. Combining our various formulas, one finds the following formula for the isoscalar g factor in terms of experimentally measured quantities

$$g_{I=0}=\tfrac{4}{9}\langle r^2\rangle_{I=0}M_{\rm N}(M_\Delta-M_{\rm N})\,. \tag{24}$$

This formula is very well satisfied experimentally. The left-hand side is 1.76 and the right-hand side is 1.66. We also find a formula for the isovector g factor from which the Skyrme model parameters cancel out:

$$g_{I=1}=\frac{2M_{\rm N}}{M_\Delta-M_{\rm N}}\,. \tag{25}$$

This relation is not so well satisfied experimentally, the left-hand side being 9.4 and the right-hand side 6.38.

Relations (24) and (25) are clearly much more general than the rest of our formulas. For instance, it is easy to see that they continue to hold if an arbitrary isospin conserving potential energy $V(U)$ is included in the model, the most obvious candidate being a term Tr U to simulate the effects of quark masses. It is natural to wonder exactly how broad is the range of validity of these formulas.

Consider the soliton before it begins to rotate as a spherically symmetric classical body with an energy density $T_{00}(r)$. (We will treat the soliton as a non-relativistic object and ignore the pressure T_{ij} relative to T_{00}. Actually the proper inclusion of T_{ij} does not modify the formulas.) If such a body begins to rotate with angular frequency $\boldsymbol{\omega}$, the velocity at position $\boldsymbol{x}$ is $\boldsymbol{v}(r)=\boldsymbol{\omega}\times\boldsymbol{x}$, and the momentum density is $T_{0i}(\boldsymbol{x})=T_{00}(r)\varepsilon_{ijk}\omega_j x_k$. The angular momentum of the spinning body is

$$\begin{aligned} J_i &= \int d^3x\,\varepsilon_{ijk}x_jT_{0k}(x) \\ &= \int d^3x\,(\omega_i r^2 - x_i\boldsymbol{x}\cdot\boldsymbol{\omega})T_{00} \\ &= \tfrac{2}{3}\omega_i \int d^3x\,T_{00}r^2\,. \end{aligned}$$

We have simply obtained the formula $\boldsymbol{J}=I\boldsymbol{\omega}$, where the moment of inertia is $I=\frac{2}{3}\int d^3x\,T_{00}r^2$. If the body begins to rotate its kinetic energy will be

$$\begin{aligned} T &= \tfrac{1}{2}\int d^3x\,T_{00}\boldsymbol{v}^2 \\ &= \tfrac{1}{2}\int d^3x\,T_{00}(\boldsymbol{\omega}\times\boldsymbol{x})^2 \\ &= \tfrac{1}{3}\boldsymbol{\omega}^2\int d^3x\,T_{00}r^2 = \frac{\boldsymbol{J}^2}{2I}\,. \end{aligned}$$

For the nucleon, $\boldsymbol{J}^2=\frac{3}{4}\hbar^2$; for the Δ, $\boldsymbol{J}^2=\frac{15}{4}\hbar^2$. Interpreting the mass difference between the delta and nucleon as a consequence of the rotational kinetic energy, we find for the moment of inertia $I=\frac{3}{2}(M_\Delta-M_{\rm N})^{-1}$. The rotational frequency of the nucleon is hence $\boldsymbol{\omega}=\boldsymbol{J}/I=\frac{2}{3}(M_\Delta-M_{\rm N})\boldsymbol{J}$.

The soliton before it begins to spin has some isoscalar charge density $\rho_{I=0}(r)$, but the isoscalar current density vanishes for a soliton at rest because of spherical symmetry and current conservation (or because of time reversal invariance). A rotating soliton has the current density $\boldsymbol{B}=\rho_{I=0}\boldsymbol{v}=\rho_{I=0}(\boldsymbol{\omega}\times\boldsymbol{x})$. So the magnetic moment of the rotating soliton is

$$\boldsymbol{\mu}_{I=0}=\frac{1}{2}\int \mathrm{d}^3x\,\boldsymbol{x}\times\boldsymbol{B}=\frac{1}{2}\int \mathrm{d}^3x\,\rho_{I=0}\boldsymbol{x}\times(\boldsymbol{\omega}\times\boldsymbol{x})$$

$$=\frac{1}{3}\boldsymbol{\omega}\int \mathrm{d}^3x\,\rho_{I=0}r^2=\frac{1}{3}\omega\langle r^2\rangle_{I=0}\,. \tag{26}$$

Combining this with $\boldsymbol{\omega}=\frac{2}{3}(M_\Delta-M_{\rm N})\boldsymbol{J}$ and with the definition $\boldsymbol{\mu}=(g/2M)\boldsymbol{J}$ of the g factor, we find the result (24) for the isoscalar magnetic moment of the nucleon.

Now, to what extent is this result general? The relations $\boldsymbol{J}=I\boldsymbol{\omega}$, $T=\boldsymbol{J}^2/2I$ are completely general formulas for the angular momentum and kinetic energy of a slowly rotating body. (These formulas hold even when the hamiltonian after elimination of non-propagating degrees of freedom is non-local.) The nucleon and delta are slowly rotating bodies in the large-N limit, with I of order N and $\boldsymbol{\omega}$ of order $1/N$. The formula $\boldsymbol{\omega}=\frac{2}{3}(M_\Delta-M_{\rm N})\boldsymbol{J}$ is a rigorous formula for the rotational frequency of the nucleon or delta in the large-N limit or in any semi-classical soliton description.

Unfortunately, the formula $\boldsymbol{B}=\rho_{I=0}(\omega\times\boldsymbol{x})$ is not a completely general formula for the current density induced in a static object when it begins to rotate. This formula holds for a *macroscopic* body, but whether it holds for a *microscopic* body such as a soliton depends on how the current and charge densities are constructed from the elementary fields. Likewise the formula (26) is not a completely general formula for the magnetic moment of a rotating sphere. In general there may be a non-locality in the relation between the charge density and the induced current; this non-locality spoils the relation $\boldsymbol{\mu}_{I=0}=\frac{1}{3}\boldsymbol{\omega}\langle r^2\rangle_{I=0}$.

If the baryon current is given by Skyrme's formula (11) then we will have (26), and the successful relation (24) will hold regardless of the choice of the chiral model lagrangian. However, a realistic description of nature requires additions to the Skyrme current. For instance, the $J=1$, $I=0$ ω meson is observed to couple to the isoscalar current. This suggests the addition to the current of an extra term $\Delta B_\mu=\alpha\partial_\nu\omega_{\mu\nu}$, where $\omega_{\mu\nu}=\partial_\mu\omega_\nu-\partial_\nu\omega_\mu$. With this addition to the current the relation (26) no longer holds for a rotating soliton, and with it eq. (24) is lost.

Thus, the successful formula (24) depends on the definition of the baryon current but not on the choice of the lagrangian. It can likewise be shown that eq. (25) holds as long as the lagrangian only involves spinless fields and their first derivatives, but can be modified by including higher derivatives or fields of higher spin.

4. Axial coupling and the Goldberger-Treiman relation

To evaluate the axial coupling g_A we calculate the integral $\int d^3x\, A_i^a(x)$ in a soliton state. The relation of this integral with the axial coupling is slightly subtle. The standard definition of the axial current matrix element is

$$\langle N'(p_2)|A_\mu^a(0)|N(p_1)\rangle = \bar{u}(p_2)\tau^a(\gamma_\mu\gamma_5 g_A(q^2)+q_\mu\gamma_5 h_A(q^2))u(p_1)\,. \tag{27}$$

Current conservation implies $2mg_A(q^2)+q^2h_A(q^2)=0$. In the non-relativistic limit, for the spatial components of the current, (27) becomes

$$\langle N'(p_2)|A_i^a(0)|N(p_1)\rangle = g_A(q^2)\left(\delta_{ij}-\frac{q_iq_j}{|q|^2}\right)\langle N'|\sigma_j\tau^a|N\rangle\,. \tag{28}$$

The $1/|q|^2$ singularity in (28) reflects, of course, the pion pole. The $\boldsymbol{q}\to 0$ limit of (28) is ambiguous. Taking the limit in a symmetric way, replacing q_iq_j by $\frac{1}{3}\delta_{ij}|q|^2$, the right-hand side of (28) becomes $\frac{2}{3}g_A\langle N'|\sigma_i\tau^a|N\rangle$ in the limit $q\to 0$; here $g_A = g_A(0)$ is the usual axial coupling constant.

Corresponding to this subtlety, the integral $\int d^3x\, A_i^a(x)$ in a soliton state is not absolutely convergent. Performing first the angular integral and then the radial integral corresponds to the symmetric limit just described. With this prescription for the integral we find

$$\int d^3x\, A_i^a(x) = \frac{\pi}{3e^2} D\, \mathrm{Tr}[\tau_i A^{-1}\tau_a A]\,, \tag{29}$$

where

$$D = \int_0^\infty d\tilde{r}\, \tilde{r}^2\left[\left(F'+\frac{\sin 2F}{\tilde{r}}\right)+4\left(\frac{\sin 2F}{\tilde{r}}(F')^2+\frac{2\sin^2 F}{\tilde{r}^2}F'+\frac{\sin^2 F\sin 2F}{\tilde{r}^3}\right)\right].$$

Numerically we find $D=-17.2$. As we have discussed before (22) $\mathrm{Tr}[\tau_i A^{-1}\tau_a A]$, evaluated in a nucleon state, equals $-\frac{2}{3}\langle\sigma_i\tau_a\rangle$. Setting (29) equal to $\frac{2}{3}g_A$ (corresponding to the symmetric $\boldsymbol{q}\to 0$ limit of (28)) we get

$$g_A = \tfrac{3}{2}(-\tfrac{2}{3})\frac{\pi}{3e^2}D = 0.61\,, \tag{30}$$

which unfortunately is not in good agreement with the experimental value $g_A = 1.23$. Although the Adler–Weisberger sum rule, which is a consequence of chiral symmetry, is surely obeyed in the Skyrme model, we do not know how it works out.

There is another useful way to compute g_A, which links it to the long-distance behaviour of the soliton solution $F(r)$, and turns out to be particularly useful for proving the Goldberger-Treiman relation.

The requirement of current conservation $\partial_\mu A^\mu = 0$ reduces to $\partial_i A^i = 0$ in the static approximation. Therefore the volume integral of the axial current can be computed as a surface integral by using the divergence theorem, as follows:

$$\int d^3x\, A_i^a = \int d^3x\, \partial_j(x_i A_j^a) = \int_S x_i A_j^a \hat{x}_j\, dS. \tag{31}$$

The definition of the axial current from (10) is

$$A_i^a = \tfrac{1}{8} i F_\pi^2 \operatorname{Tr}[(\partial_i U_0) U_0^\dagger + U_0^\dagger \partial_i U_0) A^{-1} \tau^a A] + \text{higher derivatives}, \tag{32}$$

where $U_0 = \cos F + i \sin(F) \boldsymbol{\tau} \cdot \hat{\boldsymbol{x}}$ is the soliton solution. At large distances $F(r)$ goes like B/r^2 where B can be extracted from the computer solution and is $B = B'/e^2 F_\pi^2$ with $B' = 8.6$ Therefore at large distances

$$U_0 = 1 + i\frac{B}{r^2} \boldsymbol{\tau} \cdot \hat{\boldsymbol{x}},$$

$$\partial^i U_0 = -i\frac{B}{r^3}(\tau_i - 3\boldsymbol{\tau} \cdot \hat{\boldsymbol{x}} \hat{x}^i).$$

It follows from (32) that the current to be used in formula (31) is

$$A_i^a = \tfrac{1}{4} F_\pi^2 \frac{B}{r^3} [(\tau_i - 3\boldsymbol{\tau} \cdot \hat{\boldsymbol{x}} \hat{x}_i) A^{-1} \tau_a A] + \cdots. \tag{33}$$

Therefore from (31) we obtain

$$\int d^3x\, A_i^a = -F_\pi^3 B \tfrac{2}{3}\pi \operatorname{Tr}[\tau_i A^{-1} \tau^a A].$$

From (22) and the definition of g_A we get therefore

$$g_A = \tfrac{3}{2} F_\pi^2 B \tfrac{2}{3} \pi \tfrac{2}{3} = 2B' \frac{\pi}{3e^2} = 0.61, \tag{34}$$

as before. Eqs. (30) and (34) imply a relation between D and B'. Indeed, by using (3) one can integrate D with the result $D = -2B'$.

Finally, let us check the Goldberger-Treiman relation in this model. The old fashioned lagrangian for pions π coupled to nucleons ψ is

$$\mathscr{L} = \tfrac{1}{2}(\partial_\mu \pi^a)^2 + i g_{\pi NN} \pi^a \bar{\psi} \gamma_5 \tau^a \psi.$$

The non-relativistic reduction of the coupling term is $(g_{\pi NN}/2M_N)\partial_i \pi^a \bar{\psi} \sigma_i \tau^a \psi$. From this form one can find the large-distance behaviour of the expectation value of the pion field in a nucleon state

$$\langle \pi^a(x) \rangle = -\frac{g_{\pi NN}}{8\pi M_N} \frac{x_i}{r^3} \langle \sigma_i \tau^a \rangle. \tag{35}$$

On the other hand, we can find the expectation value of the pion field at great distances from a soliton by studying the asymptotic behaviour of the soliton solution. The small fluctuations of U around its vacuum expectation value are related to the pion field by

$$U = 1 + 2i\frac{\boldsymbol{\tau}\cdot\boldsymbol{\pi}}{F_\pi} + \cdots .$$

With $U = AU_0A^{-1}$ and $U_0 = 1 + i(B/r^2)\boldsymbol{\tau}\cdot\hat{\boldsymbol{x}}\ldots$, we find the large-distance behaviour of the pion field:

$$\pi^a = \tfrac{1}{4}BF_\pi \frac{x_i}{r^3}\,\mathrm{Tr}\,[\tau_i A^{-1}\tau^a A] .$$

By using (22) and (34)

$$\langle \pi^a \rangle = -\tfrac{1}{6}BF_\pi \frac{x_i}{r^3}\langle \sigma_i \tau^a \rangle = -\frac{g_\mathrm{A}}{F_\pi}\frac{1}{4\pi}\langle \sigma_i \tau^a \rangle . \tag{36}$$

So comparing (35) and (36) we finally get the Goldberger–Treiman relation

$$g_\mathrm{A} = \frac{F_\pi g_{\pi\mathrm{NN}}}{2M_\mathrm{N}} .$$

The predicted value of $g_{\pi\mathrm{NN}}$ is 8.9 compared with the experimental value of 13.5.

5. Decays of the Δ

In this section, we will calculate the amplitudes for the decay processes $\Delta \to \mathrm{N}\pi$ and $\Delta \to \mathrm{N}\gamma$. The decay $\Delta \to \mathrm{N}\gamma$ is related by a simple quark model argument [11] to the nucleon magnetic moment. A similar quark model argument [12] relates the amplitude $\Delta \to \mathrm{N}\pi$ to the pion-nucleon coupling. For a review of the quark model relations, see [13]. We will see that the $1/N$ expansion makes predictions for Δ decays analogous to the predictions of the quark model. These predictions are model-independent in the sense that they hold for any soliton model of baryons and serve as quantitative tests of the $1/N$ expansion. The Skyrme model will not enter in this section except in the concluding paragraph.

In the large-N limit the Δ and the nucleon are nearly degenerate, so the decays $\Delta \to \mathrm{N}\pi$ and $\Delta \to \mathrm{N}\gamma$ involve soft pions and photons. Also, the nucleon and the Δ are described by the same classical soliton solution with different but known wave functions for the collective coordinates (8). Hence the coupling of the soft pion or photon in Δ decay can be computed in terms of the static coupling of pions or photons to nucleons.

In view of chiral symmetry, the pion couplings to baryons can be expressed as derivative couplings. For soft pions, the coupling will involve mainly the first derivative of the pion field $\partial_i \pi^a$, multiplied by some operator $\mathcal{O}_i^a$ acting on the collective coordinates. In $\mathcal{O}_i^a$ time derivatives of A can be neglected (since the nucleon rotates slowly in the large-N limit) so $\mathcal{O}_i^a$ must be a function of A only. The only function of A that transforms properly under spin and isospin ($\mathcal{O}_i^a$ must have $I=J=1$) is $\mathrm{Tr}\,[\tau_i A^{-1}\tau_a A]$. So in the large-$N$ limit, irrespective of other details, the coupling of soft pions to baryons is of the form

$$\mathcal{L}_\pi = \delta\partial_i\pi^a\,\mathrm{Tr}\,[\tau_i A^{-1}\tau_a A]\,, \tag{37}$$

for some δ.

The pion-nucleon coupling is related to δ by evaluating the matrix element of (37) between initial and final nucleon states. We have already done this, in effect, in computing $g_{\pi NN}$ in the Skyrme model, and the relation is $\delta=\frac{3}{4}g_{\pi NN}/M_N$, M_N being the nucleon mass. On the other hand, we can describe the hadronic decay of the Δ by taking the matrix element of (37) between an initial Δ and a final nucleon.

Let us define a coupling $g_{\pi N\Delta}$ as follows (it is called $\mathcal{M}_{\uparrow\uparrow}$ in [12]). For a decay $\Delta^{++}(s_z=\frac{3}{2})\to p(s_z=\frac{1}{2})+\pi^+$, we define the amplitude to be $g_{\pi N\Delta}(k_x+ik_y)/2M_N$, where $\boldsymbol{k}$ is the c.m. momentum of the pion. Evaluating the matrix element of (37), we find $g_{\pi N\Delta}=\frac{3}{2}g_{\pi NN}$. The quark model relation of [12] is instead $g_{\pi N\Delta}=\frac{6}{5}g_{\pi NN}$. The relation $g_{\pi N\Delta}=\frac{3}{2}g_{\pi NN}$, which follows from the $1/N$ expansion without other assumptions, is in excellent agreement with experiment. With the experimental value $g_{\pi NN}=13.5$, it gives a value of 125 MeV for the width of the Δ; the experimental value is about 120 MeV.

A similar analysis can be made for the electromagnetic decay of the Δ. The decay $\Delta\to N\gamma$ violates isospin, so it involves only the isovector part of the electromagnetic current. The isovector coupling of the magnetic field $\boldsymbol{B}$ to baryons must be of the form $\boldsymbol{B}\cdot\boldsymbol{\mu}$, where $\boldsymbol{\mu}$ is an operator acting on the collective coordinates of baryons. $\boldsymbol{\mu}$, the isovector magnetic moment operator of baryons, must be the third component of an isovector. Neglecting time derivatives, the only possibility is $\mu_i=\alpha\,\mathrm{Tr}\,[\tau_i A^{-1}\tau_3 A]$ where α is some constant. So the magnetic coupling to baryons is

$$\mathcal{L}_{\mathrm{mag}}=\boldsymbol{B}\cdot\boldsymbol{\mu}=\alpha B_i\,\mathrm{Tr}\,[\tau_i A^{-1}\tau_3 A]\,. \tag{38}$$

A relation of this form holds in any soliton description of baryons; only the value of α is model dependent. The value of α determines the isovector part of the nucleon magnetic moment. The relation is obtained by calculating the matrix element of (38) between initial and final nucleon states; the calculation is essentially the one we have already performed in deriving eq. (23). Writing the proton and neutron magnetic moments as $\boldsymbol{\mu}_p=\mu_p\boldsymbol{\sigma}$, $\boldsymbol{\mu}_n=\mu_n\boldsymbol{\sigma}$, the relation is $\alpha=\frac{3}{4}(\mu_p-\mu_n)$.

We can now calculate the amplitude for $\Delta\to N\gamma$ by evaluating the matrix element of (38) between the initial Δ and final nucleon. Let us define a transition moment $\mu_{N\Delta}$ by the formula $\mu_{N\Delta}=\langle p, s_z=\frac{1}{2}|\mu_z|\Delta^+, s_z=\frac{1}{2}\rangle$ where $\mu_z=\frac{3}{4}(\mu_p-\mu_n)\,\mathrm{Tr}\,[\tau_3 A^{-1}\tau_3 A]$ is the z component of the baryon magnetic moment operator. Using wave functions in (8) we find $\mu_{N\Delta}=\sqrt{\frac{1}{2}}\,(\mu_p-\mu_n)$. This agrees very well with the experimental value $\mu_{N\Delta}=(0.70\pm0.01)\;(\mu_p-\mu_n)$. The quark model

(11) gives $\mu_{N\Delta}=\frac{2}{5}\sqrt{2}(\mu_p-\mu_n)=0.57(\mu_p-\mu_n)$ (this relation is often written $\mu_{N\Delta}=\frac{2}{3}\sqrt{2}\,\mu_p$; we are using here the quark model prediction $\mu_n=-\frac{2}{3}\mu_p$).

The model-independent tests of the $1/N$ expansion $g_{\pi N\Delta}=\frac{3}{2}g_{\pi NN}$ and $\mu_{N\Delta}=\sqrt{\frac{1}{2}}\,(\mu_p-\mu_n)$ work very well (perhaps fortuitously so) if one takes $g_{\pi NN}$ and $\mu_p-\mu_n$ from experiment. The Skyrme model, however, is less successful. Since the Skyrme model values of $g_{\pi NN}$, μ_p, and μ_n are all about 30% too small, the predictions for $\mu_{N\Delta}$ and $g_{\pi N\Delta}$ are too low (see table 1) by a similar margin.

C.R.N. acknowledges useful conversations with S. Gupta.

References

[1] T.H.R. Skyrme, Proc. Roy. Soc. A260 (1961) 127
[2] G. 't Hooft, Nucl. Phys. B72 (1974) 461; B75 (1974) 461
[3] E. Witten, Nucl. Phys. B160 (1979) 57
[4] E. Witten, Nucl. Phys. B223 (1983) 433
[5] J. Wess and B. Zumino, Phys. Lett. 37B (1971) 95
[6] E. Witten, Nucl. Phys. B223 (1983) 422
[7] J. Goldstone and F. Wilczek, Phys. Rev. Lett. 47 (1981) 986
[8] D. Finkelstein and J. Rubinstein, J. Math. Phys. 9 (1968) 1762
[9] M.A.B. Bet and A. Zepeda, Phys. Rev. D6 (1972) 2912
[10] G. Adkins, C.R. Nappi, The Skyrme model with pion masses, IAS preprint 83
[11] M.A. Beg, B.W. Lee, and A. Pais, Phys. Rev. Lett. 13 (1964) 514
[12] C. Becchi and G. Morpurgo, Phys. Rev. 149 (1966) 1284;
R. van Royen and V.F. Weisskopf, Nuovo Cim. 50 (1966); 51 (1967) 583
[13] J.J.J. Kokkedee, The quark model (W.A. Benjamin, 1969)

STABILIZATION OF CHIRAL SOLITONS VIA VECTOR MESONS

Gregory S. ADKINS
Department of Physics, Franklin and Marshall College, Lancaster, PA 17604, USA

and

Chiara R. NAPPI [1]
Joseph Henry Laboratories, Princeton University, Princeton, NJ 08544, USA

Received 23 December 1983

We introduce vector mesons in a chiral lagrangian. The ω meson stabilizes the chiral solitons. We compute the static properties of baryons in this model.

1. Introduction. The idea that baryons are solitons of a low-energy effective lagrangian for QCD has recently attracted interest [1–12]. If the low-energy effective lagrangian is the non-linear σ model, then the solitons are topologically stable, given that $\pi_3(\mathrm{SU}(N)) = Z$. However they are not energetically stable, unless higher order derivative terms are added to the lagrangian, in order to avoid Derrick's theorem. The simplest suitable higher order derivative term is the Skyrme term [13].

The elementary fields of the non-linear σ model are the Goldstone boson fields – the pions, kaons, and eta. The soliton excitations provide the baryons, whose masses are in the GeV range. It seems therefore reasonable to include in the theory also vector mesons, like the ω and the ρ, whose masses are lower than the masses of the baryons. This is what we do in this paper.

The interesting outcome is that the introduction of vector mesons by itself stabilizes the soliton, without any further need of the Skyrme term. The model we analyze in detail contains the omega vector meson only. We calculate the static properties of nucleons in such a model. The predictions, reported in table 1, are again generally within 30% of experimental values, although slightly improved with respect to the Skyrme model [4,5].

2. The model. To describe a non-linear σ-model coupled to an ω vector meson field, we assume a lagrangian of the form

$$L = -\tfrac{1}{4}(\partial_\mu\omega_\nu - \partial_\nu\omega_\mu)(\partial^\mu\omega^\nu - \partial^\nu\omega^\mu) + \tfrac{1}{2}m_\omega^2\omega_\mu\omega^\mu + \beta\omega_\mu B^\mu + \tfrac{1}{16}F_\pi^2 \operatorname{tr}[(\partial_\mu U)(\partial^\mu U^\dagger)] + \tfrac{1}{8}F_\pi^2 m_\pi^2(\operatorname{tr} U - 2). \quad (1)$$

Above U is an SU(2) matrix related to the pion field in the usual way

$$U = \exp[(2\mathrm{i}/F_\pi)\boldsymbol{\tau}\cdot\boldsymbol{\pi}] ,$$

ω_μ is the omega field, B^μ the baryonic current [3]

$$B^\mu = (\epsilon^{\mu\nu\alpha\beta}/24\pi^2)\operatorname{tr}[(U^\dagger\partial_\nu U)(U^\dagger\partial_\alpha U)(U^\dagger\partial_\beta U)] . \quad (2)$$

Moreover, m_ω is the mass of the omega, m_π the mass of the pion and F_π the pion decay constant (in our units F_π = 186 MeV). The term in (1) containing tr U is an explicit chiral symmetry breaking term [5].

The interaction piece of the lagrangian $\beta\omega_\mu B^\mu$ describes the coupling of the omega to three pions. In fact

[1] Research supported in part by the National Science Foundation under grant No. PYH80-19754.

0.370-2693/84/$ 03.00 © Elsevier Science Publishers B.V.
(North-Holland Physics Publishing Division)

$$U^{\dagger}\partial_{\mu}U = (2\mathrm{i}/F_{\pi})\boldsymbol{\tau}\cdot\partial_{\mu}\boldsymbol{\pi} + \mathrm{O}(\boldsymbol{\pi}^2)\,. \tag{3}$$

The constant β can be related to the $\omega \to \pi^+\pi^-\pi^0$ decay rate. Our computation gives

$$\Gamma_{\omega\to 3\pi} = (\beta^2 m_{\omega}/2^{10}\times 3\times\pi^7)\times(m_{\omega}/F_{\pi})^6\times 0.030\,.$$

The experimental value of $\Gamma_{\omega\to 3\pi}$ (= 9.07 MeV) would then assign to β a value $\beta = 25.4$. However, in the real world the decay $\omega\to 3\pi$ is enhanced by the resonance $\omega\to\rho+\pi$, which we do not take in account in our coupling. Therefore we expect the value 25.4 to be only an upper bound on β. Our approach will be to leave β as a free parameter, and determine it by fitting the properties of the nucleons.

We remark that we have not included in (1) any higher order derivative term to stabilize the soliton solution. Therefore in absence of the ω field the solitons of the non-linear σ model in (1) would have zero size.

As usual, we make the ansatz for the soliton solution [13]

$$U(x) = U_0(x) = \exp\,[\mathrm{i}\boldsymbol{\tau}\cdot\hat{\boldsymbol{x}}F(r)]\,,$$

with the boundary conditions $F(0) = \pi$, $F(\infty) = 0$. With this choice of U

$$B^0 = -(1/2\pi^2)[(\sin^2 F)/r^2]F'\,, \qquad B_i = 0\,. \tag{4}$$

So there is no "source term" for the spatial part of ω_{μ}, and we might as well take $\omega_i = 0$. Finally we assume $\omega_0 = \omega(r)$, with $\omega(0)$ finite and $\omega(\infty) = 0$.

Therefore the lagrangian (1) simplifies as follows

$$L[U_0,\omega] \equiv -M = \tfrac{1}{2}(\omega')^2 + \tfrac{1}{2}m_{\omega}^2\omega^2 - (\beta\omega/2\pi^2)[(\sin^2 F)/r^2]F' - \tfrac{1}{8}F_{\pi}^2[F'^2 + 2(\sin^2 F)/r^2] - \tfrac{1}{4}F_{\pi}^2 m_{\pi}^2(1-\cos F)\,. \tag{5}$$

From (5) we have numerically found $F(r)$ and $\omega(r)$ by employing relaxation techniques. In figs. 1 and 2 we plot the solutions $F(r)$ and $\omega(r)$ for various values of the ratio $\bar{\beta} = \beta m_{\omega}/2\pi^2 F_{\pi}$.

So the presence of the ω stabilizes the soliton. This can be made more transparent if we eliminate ω from the action by using the variational equation

$$[-(\mathrm{d}/\mathrm{d}r)(r^2\,\mathrm{d}/\mathrm{d}r) + r^2 m_{\omega}^2]\,\omega(r) = -\beta r^2 B^0(r)\,. \tag{6}$$

The solution to this equation can be written as

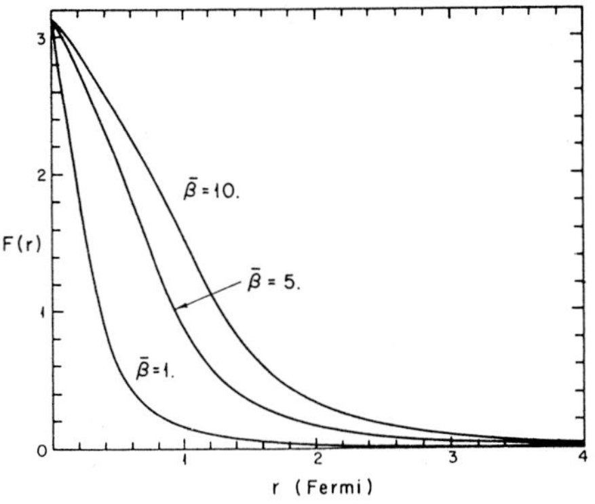

Fig. 1. Plot of the function $F(r)$ which appears in the baryon ansatz $U_0)x) = \exp\,[\mathrm{i}F(r)\boldsymbol{\tau}\cdot\hat{\boldsymbol{x}}]$ for various values of $\bar{\beta} = \beta m_{\omega}/2\pi^2 F_{\pi}$.

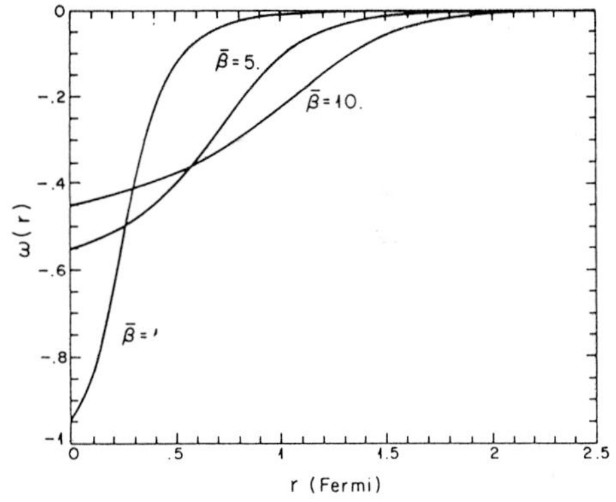

Fig. 2. The omega field $\omega(r)$ for various values of $\bar{\beta} = \beta m_{\omega}/2\pi^2 F_{\pi}$.

$$\omega(r) = \int_0^\infty dr'\, G(r,r')\,[-\beta r'^2 B^0(r')]\,, \tag{7}$$

where $G(r, r')$ is the radial Green's function

$$G(r,r') = (2m_\omega r r')^{-1}\,\{\exp(-m_\omega|r-r'|) - \exp[-m_\omega(r+r')]\}\,. \tag{8}$$

In terms of the above Green's function the action reads

$$S = -\int d^3x\,\{\tfrac{1}{8}F_\pi^2[(F')^2 + 2(\sin^2 F)/r^2] + \tfrac{1}{4}F_\pi^2 m_\pi^2(1-\cos F)\}$$

$$-\frac{4\pi}{2}\int_0^\infty dr\,dr'\,\beta r^2 B^0(r)\,G(r,r')\,[\beta r'^2 B^0(r')]\ ;\,. \tag{9}$$

with $B^0(r)$ as in (4). One can check that the second piece in (9) diverges for small ϵ as $1/\epsilon$, where ϵ is the size of the soliton. So the action can be finite only if the soliton has non-zero size. Actually one could have suspected from the beginning that the coupling $\omega^\mu B_\mu$ could stabilize the soliton simply on dimensional grounds, given that this term scales like $1/r$.

3. *Static properties of nucleons*. As in ref. [4] we make a time dependent SU(2) rotation of our static soliton solution

$$U = A(t)\,U_0(x)\,A^{-1}(t)\,.$$

Then

$$L[U,\omega] = -M + I\,\mathrm{tr}\,[\partial_0 A\,\partial_0 A^\dagger]\,, \tag{10}$$

with M as in (5) and I the moment of inertia of the soliton. By rescaling $\omega = F_\pi\tilde{\omega}$ and $r = t/m_\omega$, M and I can be written as

$$M = (4\pi F_\pi^2/m_\omega)\tilde{M}\,, \qquad I = (2\pi F_\pi^2/3m_\omega^3)\tilde{I}\,, \tag{11}$$

where $\tilde{M}$ and $\tilde{I}$ are now dimensionless numbers, fully determined by the solutions $F(t)$ and $\tilde{\omega}(t)$,

$$\tilde{M} = \int_0^\infty dt\,[-\tfrac{1}{2}t^2\tilde{\omega}'^2 - \tfrac{1}{2}t^2\tilde{\omega}^2 + \bar{\beta}\,\tilde{\omega}(\sin^2 F)F' + \tfrac{1}{8}t^2F'^2 + \tfrac{1}{4}(\sin^2 F) + \tfrac{1}{4}(m_\pi^2/m_\omega^2)t^2(1-\cos F)]\,, \tag{12}$$

$$\tilde{I} = \int_0^\infty t^2(\sin^2 F)\,dt + 2\bar{\beta}^2\int_0^\infty\!\!\int_0^\infty dt\,dt'\,\{[\sin^2 F(t)]\,F'(t)/t\}\{[\sin^2 F(t')]\,F'(t')/t'\}$$

$$\times\{\exp[-(t+t')](1+t+t'+t't) - \exp(-|t-t'|)(1+|t-t'|-t't)\}\,. \tag{13}$$

Above, the first term in $\tilde{I}$ is the contribution to the moment of inertia coming from the non-linear σ-model piece in the lagrangian and the second is the one coming from the term $\beta\omega_\mu B^\mu$. This second contribution arises because when the soliton is rotating we cannot assume anymore $\omega_i = 0$. Instead ω_i will satisfy an equation analogous to (7), with B_i replacing B_0 and with the appropriate Green's function. From the formula for the energy [4] $E = M + J^2/2I$, we derive the mass differences

$$m_\Delta - m_N = 3/2I = 9m_\omega^3/4\pi F_\pi^2\tilde{I}\,, \quad 5m_N - m_\Delta = 4M = 16\pi(F_\pi^2/m_\omega)\tilde{M}\,, \tag{14, 15}$$

and the mass ratio

$$\widetilde{M}/\widetilde{I} = (5m_N - m_\Delta)(m_\Delta - m_N)/36m_\omega^2 . \tag{16}$$

We will choose $\overline{\beta}$ so to fit the experimental value of the mass ratio in (16).

In order to compute other static properties of nucleons we need to evaluate the baryonic current (2) and the vector and axial currents in terms of the collective coordinates $A(t)$, as in ref. [4]. The vector and axial currents will now get contributions from the coupling of $\beta\omega^\mu B_\mu$.

The following angular integrals of the vector and axial currents are adequate for our purposes:

$$\int d\Omega_r\, V^{0,a} = \tfrac{4}{3}\pi i \{F_\pi^2 [\sin^2 F(r)] + 2(\beta/2\pi^2)^2 \{[\sin^2 F(r)]/r^2\} F'(r)\}$$

$$\times \int_0^\infty dr' [\sin^2 F(r')] F'(r') A(r,r')\, \mathrm{tr}\,[(\partial_0 A)A^\dagger \tau^a] , \tag{17}$$

where

$$A(r,r') = (rr' m_\omega)^{-1} \{\exp(-m_\omega |r-r'|)(rr' - |r-r'|/m_\omega - 1/m_\omega^2) + \exp[-m_\omega(r+r')]\,[rr' + (r+r')/m_\omega + 1/m_\omega^2]\}, \tag{18}$$

$$\int d\Omega_x\, \boldsymbol{q}\cdot\boldsymbol{x}\, V^{i,a} = \tfrac{4}{6}\pi\, [\tfrac{1}{2}F_\pi^2 + (\beta\omega/\pi^2)F'](\sin^2 F)\epsilon^{ilm} q^l\, \mathrm{tr}\,[\tau^m A^\dagger \tau^a A] , \tag{19}$$

$$\int d\Omega\, A^{i,a} = \tfrac{4}{3}\pi \{\tfrac{1}{4}F_\pi^2 \{F' + 2(\sin F)(\cos F)/r + (\beta\omega/2\pi^2)\{[2(\sin F)(\cos F)/r]F' + (\sin^2 F)/r^2\}\}\, \mathrm{tr}\,[\tau^i A^\dagger \tau^a A] . \tag{20}$$

From (2) we get the usual expressions for the components of the baryon current [4]

$$B^0 = -(1/2\pi^2)[(\sin^2 F)/r^2]F' , \quad B^i = i(\epsilon^{ijk}/2\pi^2)[(\sin^2 F)/r]F' \hat{x}_k\, \mathrm{tr}\,[(\partial_0 A^\dagger)A\tau_j] . \tag{21}$$

We have taken as imput the masses m_ω, m_π, and m_N and chosen the value of $\overline{\beta} = \beta m_\omega/2\pi^2 F_\pi$ to fit the mass of the delta, i.e. the mass formula (16). The value of $\overline{\beta}$ appropriate for the purpose is $\overline{\beta} = 5.0$. The value of F_π predicted through (14) or (15) is then $F_\pi = 124$ MeV. Therefore the parameter β turns out to be $\beta = 15.6$, compatible with the experimental upper bound $\beta \leqslant 25.4$.

In table 1 we have reported our predictions for this model and compared them with the predictions of the Skyrme model with massive pions [5]. For the details of the derivations we refer the reaser to ref. [4]. The quan-

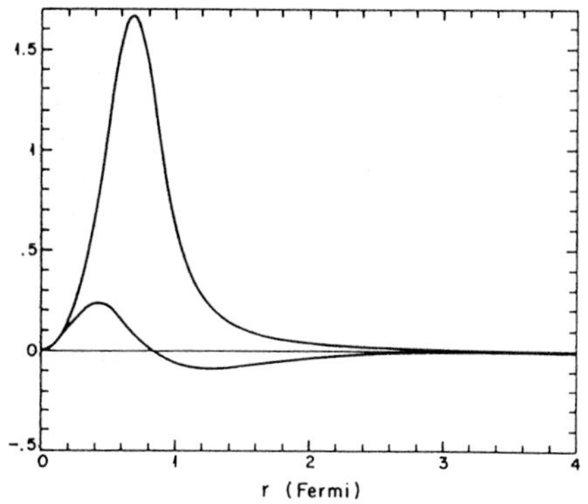

Fig. 3. Proton and neutron charge density as functions of the radial distance r for $\overline{\beta} = 5$.

Table 1

Quantity	Prediction in this model ($\bar{\beta}$ = 5.0)	Prediction in the Skyrme model [5]	Experiment
m_N (MeV)	input	input	938.9
m_π (MeV)	input	input	138
m_ω (MeV)	input	–	782.4
m_Δ (MeV)	input	input	1232
F_π (MeV)	124	108	186
$\langle r^2\rangle^{1/2}_{I=0}$ (fm)	0.74	0.68	0.72
$\langle r^2\rangle^{1/2}_{I=1}$ (fm)	1.06	1.04	0.88
$\langle r^2\rangle^{1/2}_{M,I=0}$ (fm)	0.92	0.95	0.81
$\langle r^2\rangle^{1/2}_{M,I=1}$ (fm)	1.02	1.04	0.80
μ_p	2.34	1.97	2.79
μ_n	−1.46	−1.24	−1.91
g_A	0.82	0.65	1.23
$g_{\pi NN}$	13.0	11.9	13.5
σ (MeV)	53	49	36±20
$g_{\pi N\Delta}$	19.5	17.8	20.3
$\mu_{N\Delta}$	2.7	2.3	3.3

tities $g_{\pi N\Delta}$ and $\mu_{N\Delta}$ have been calculated by using the model independent relations of ref. [4]. The last quantity in table 1, σ, is the pion–nucleon sigma term.

The results in this model seem to be slightly better than those of the Skyrme model. Still, however, we get very unsatisfying values of F_π and g_A.

4. *Final remarks.* The model we have described is obviously incomplete. To make it more realistic, one should include the ρ meson. This should be done in a chirally invariant way and in a way that obeys Zweig's rule, which is a property of the $1/N$ expansion.

A remark is in order here. Now that we have introduced vector mesons in a chiral lagrangian, probably there is no justification in defining the baryon current in terms of pion fields only. One would expect it to have contributions from the vector mesons too [4]. A plausible addition to the current is the divergenceless function $\partial_\nu\omega_{\mu\nu}$, which does not affect current conservation and definition of baryon number. Then the new baryon current would read

$$\hat{B}_\mu = B_\mu + \lambda\partial_\mu\omega_{\mu\nu}\,, \tag{22}$$

for an unknown parameter λ.

This addition will induce a change in our predictions for the isoscalar electrical charge radius $\langle r^2\rangle^{1/2}_{I=0}$ and for the isoscalar magnetic charge radius $\langle r^2\rangle^{1/2}_{M,I=0}$. One can easily check that it will not affect the isoscalar magnetic moment, at least for slowly rotating solitons (i.e. as long as we ignore time derivatives of ω_i).

The changes in the charge radii depend of course on the unknown parameter λ. If one believes the vector meson dominance hypothesis, then λ can be determined to be $\lambda = -1/\beta$.

In fact from the equations of motion

$$\partial_\nu\omega_{\mu\nu} = m_\omega^2\omega_\mu + \beta B_\mu\,,$$

one derives

$$\hat{B}_\mu = (1+\beta\lambda)B_\mu + m_\omega^2\lambda\omega_\mu\,.$$

If $\hat{B}_\mu$ has to be proportional to ω_μ, as in the vector meson dominance hypothesis, then $\lambda = -1/\beta$, i.e.

$$\hat{B}_\mu = -(m_\omega^2/\beta)\omega_\mu \,. \tag{23}$$

One can compute analytically the changes in $\langle r^2 \rangle_{I=0}$ and $\langle r^2 \rangle_{M,I=0}$ by using (23) and (7), with the following results

$$\delta \langle r^2 \rangle_{I=0} = (6/m_\omega^2) = 0.38 \text{ fm}^2 \,, \quad \delta \langle r^2 \rangle_{M,I=0} = (10/m_\omega^2) = 0.63 \text{ fm}^2 \,.$$

Consequently, the new values for the charge radii will be $\langle r^2 \rangle_{I=0}^{1/2} = 0.96$ and $\langle r^2 \rangle_{M,I=0}^{1/2} = 1.22$. The other predictions in table 1 instead remain unchanged.

We thank E. Witten for valuable discussions. We are indebted to V. Kaplunovsky for a very useful remark.

References

[1] N.K. Pak and H.C. Tze, Ann. Phys. 117 (1979) 164;
J. Gibson and H.C. Tze, Nucl. Phys. B183 (1981) 524.
[2] A.P. Balachandran, V.P. Nair, S.G. Rajeev and A. Stern, Phys. Rev. Lett. 49 (1982) 1124; Phys. Rev. D27 (1983) 1153.
[3] E. Witten, Nucl. Phys.B 223 (1983) 422, 433.
[4] G. Adkins, C. Nappi and E. Witten, Nucl. Phys. B228 (1983) 552.
[5] G. Adkins and C. Nappi, Nucl. Phys. B233 (1984) 109.
[6] M. Rho, A.S. Goldhaber and G.E. Brown, Phys. Rev. Lett. 51 (1983) 747;
A.D. Jackson and M. Rho, Phys. Rev. Lett. 51 (1983) 751.
[7] A. Chodos and C.B. Thorn, Phys. Rev. D12 (1975) 2733.
[8] J. Goldstone and R.L. Jaffe, MIT-CTP-1100 (1983).
[9] U.-G. Meissner, Ruhr-Universität Bochum preprint (1983).
[10] M.C. Birse and M.K. Banerjee, University of Maryland PP No. 83-201.
[11] E. Guadagnini, Princeton University preprint (1983).
[12] E. D'Hoker and E. Farhi, MIT preprint CTP No. 1111 (1983).
[13] T.H.R. Skyrme, Proc. R. Soc. A 260 (1961) 127.

PHYSICS REPORTS (Review Section of Physics Letters) 142, Nos. 1 & 2 (1986) North-Holland, Amsterdam

THE SKYRME MODEL*

I. ZAHED and G.E. BROWN

Physics Department, State University of New York at Stony Brook, Stony Brook, New York 11794, U.S.A.

Received March 1986

Contents:

1. Introduction 3
2. The SU(2) Skyrme model 5
 2.1. Chiral symmetry in the context of QCD 5
 2.2. The non-linear σ-model 7
 2.3. Topology of the non-linear σ-model 9
 2.4. The Skyrme model 11
 2.5. The skyrmion 14
 2.6. Soft-pion theorems and the Skyrme model 18
3. The Wess–Zumino term 20
 3.1. QCD chiral anomalies and effective chiral models 20
 3.2. Analogy with the U(1) monopole 21
 3.3. Topological quantization of the Wess–Zumino term 24
 3.4. The Wess–Zumino term in the context of QCD 26
 3.5. Anomalous baryon current 30
 3.6. Spin statistics of the skyrmion 31
4. Gauged Wess–Zumino term and non-Abelian chiral anomalies 33
 4.1. The infinite-hotel story 34
 4.2. Topological character of the ABJ anomaly 36
 4.3. Non-Abelian anomalies 39
5. Vector meson dominance and the Skyrme model 43
 5.1. Vector mesons and non-Abelian anomalies 43
 5.2. ω-stabilized skyrmions 45
 5.3. The Skyrme term from ρ-mesons 48
 5.4. The skyrmion with vector mesons 51
 5.5. Sakurai's wisdom in the Skyrme model 53
6. The QCD generating functional with passive gluons at low energy 56
 6.1. Z_{QCD} and the low-energy collective modes 56
 6.2. The Wess–Zumino term from fermions 59
 6.3. A Skyrme-like Lagrangian from fermions 61
7. Phenomenological aspects of SU(2) hedgehog skyrmions 64
 7.1. Nucleon and Δ-isobar masses 64
 7.2. Axial coupling g_A 67
 7.3. Charge radii and magnetic moments 69
 7.4. Low-lying resonances in the Skyrme model 71
 7.5. Soft-pion corrections to the masses 76
 7.6. The skyrmion–skyrmion interaction 78
 7.7. Status of the many-skyrmion problem 80
 7.8. Skyrmion crystal 82
 7.9. Exotic skyrmions 83
8. The Skyrme model and boson exchange 83
 8.1. The role of bosons in the Skyrme model 83
 8.2. The attractive interaction between nucleons 84
 8.3. Short-distance cutoffs in boson exchange models and the finite size of the skyrmion 84
9. Discussion and conclusions 85
 9.1. The Cheshire cat philosophy and the chiral bag model 85
 9.2. Conclusions 86
Appendices 87
 A. Baryon number for general textures 87
 B. Weinberg scaling argument 88
 C. Asymptotic behavior of the pion field 89
 D. Additivity of the baryon number 90
 E. $SU(2)_L \otimes SU(2)_R$ Noether currents 91
 F. Evaluation of $\Gamma_{WZ}(2\pi)$ 95
 G. Quantization of the collective coordinates 96
References 99
Note added in proof 101

* Supported in part by the US Department of Energy under Contract No. DE-AC02-76ER13001 with the State University of New York.

Abstract:

We review the recent developments on the Skyrme model in the context of QCD, and analyze their relevance to low-energy phenomenology. The fundamentals of chiral symmetry and PCAC are presented, and their importance in effective chiral models of the Skyrme type discussed. The nature and properties of skyrmions are thoroughly investigated, with particular stress on the basic role of the Wess–Zumino term. The conventional Skyrme model is extended to the low-lying vector meson resonances, and the rudiments of vector meson dominance are elucidated. A detailed account of the static and dynamical properties of nucleons and Δ-isobars is presented. The relevance of the Skyrme model to the nuclear many-body problem is outlined and its importance for boson exchange models stressed.

1. Introduction

Although many believe that quantum chromodynamics (QCD) is the theory of strong interactions, our quantitative understanding of this theory in the long-wavelength approximation still remains a challenging problem. Aside from tedious but promising lattice gauge calculations, concepts such as confinement and chiral symmetry breaking are only qualitatively understood. Most of our present low-energy predictions stem from models believed to approximate QCD. These range from conventional quark models to effective chiral models.

In QCD, the absence of an explicit expansion parameter such as the fine structure constant α in QED, rules out from the word go any perturbative scheme upon which most of our present understanding of physics is based. Massless QCD is in fact a parameter free theory. The quark–gluon coupling g or equivalently the QCD cutoff Λ plays the role of an overall scale that can be factorized out of the starting dynamics.

Some years ago 't Hooft [1] and later on Witten [2], proposed to generalize QCD from SU(3) to SU(N_c) gauge theory, and use $1/N_c$ as an implicit expansion parameter. Assuming confinement, 't Hooft has shown that the large-N_c version of QCD is a weakly interacting phase of mesons and glue balls that decouple to leading order. Indeed, simple power counting based on the fact that g^2 scales like $1/N_c$, shows that the correlation function of any bilocal color singlet operator $J(x)$, e.g. $J(x) = \bar{q}q$, $\mathrm{TR}[F^{\mu\nu}F_{\mu\nu}]$, etc., is dominated by planar diagrams with exclusive glue insertion as shown in fig. 1a. Non-planar diagrams and quark insertions are suppressed by powers of $1/N_c$. Insisting on crossing and unitarity, Witten has shown that to leading order in N_c, the scattering amplitudes in QCD are sums of tree diagrams involving the exchange of physical $q\bar{q}$ resonances. In this spirit a 3-meson vertex scales like $1/\sqrt{N_c}$, a 4-meson vertex like $1/N_c$, etc., as illustrated in fig. 1c. Remarkably, large-N_c QCD seems to reduce smoothly to an effective field theory of non-interacting mesons to leading order. The latter

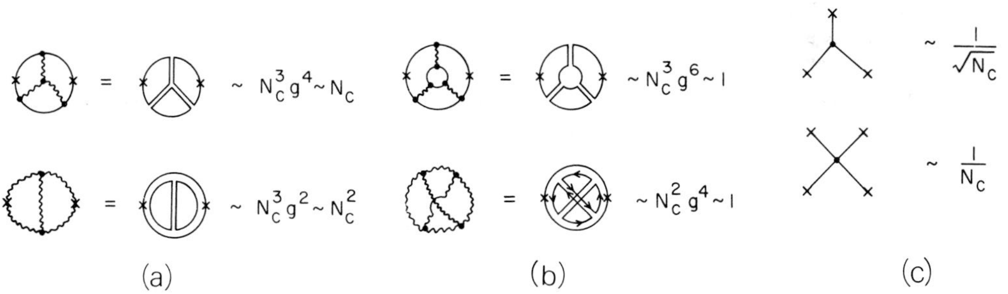

Fig. 1. (a) Dominant planar diagrams in the large-N_c world. In (b) are examples of non-leading diagrams that involve quark-loop and non-planar gluon insertions; (c) illustrates the large-N_c behavior of the 3- and 4-point functions. The straight lines stand for meson resonances.

involves infinitely many elementary fields, leads naturally to the suppression of exotics, and turns out to be consistent with Regge phenomenology and Zweig selection rules.

Using large N_c diagrammatics, Witten has argued that to leading order, baryons emerge as solitons in this weakly coupled phase of mesons. A baryon can be thought of as a bound state of N_c quarks interacting via one-gluon exchange to leading order. If M_B denotes the ground state energy, then,

$$M_B = N_c m_q + N_c T_q + \tfrac{1}{2}[N_c(N_c - 1)]g^2 V = O(N_c) ,$$

since g^2 scales like $1/N_c$. In other words, M_B is inversely proportional to $1/N_c$ which is the strength of the 4-meson coupling in the large N_c limit. This proportionality to the inverse of the coupling constant is characteristic of a soliton solution, and underlines a non-perturbative effect. Similarly, one deduces that the baryon–baryon interaction through one-gluon exchange is strong and scales like N_c, i.e. $BB = \frac{1}{2}N_c(N_c - 1)g^2 V \equiv O(N_c)$, while the baryon–meson coupling is weak and scales like 1, i.e. $BM = N_c V g^2 = O(1)$.

Aside from color, QCD involves different quark flavors, u, d, s, c, etc. The low-energy properties of QCD are dominated by the three light quarks u, d and s. The masses of the u- and d-quarks are small compared to the QCD cutoff Λ. If these masses are neglected, the QCD Lagrangian becomes invariant under $SU(2)_L \otimes SU(2)_R$ chiral transformations.* The absence of parity doublets in the hadronic spectra suggests that this symmetry is spontaneously broken to $SU(2)_V$ via the Nambu–Goldstone mechanism with the appearance of 3 massless pseudoscalar excitations, the pions.

In the absence of a quantitative understanding of non-perturbative QCD, a convenient alternative at low energy is provided by effective chiral descriptions of which the non-linear σ-model constitutes the starting point. While QCD is undoubtedly a fundamental theory of hadrons, chiral field models are specifically designed approximations to hadron dynamics suited for a low-energy treatment. In general their quantum range is limited by both short- and long-distance singularities. Despite the unrenormalizable character of such models, the relative success of chiral phenomenology has led one to believe that a systematic quantum pattern is worthwhile investigating.

The non-linear character of chiral field models leads naturally to non-trivial field configurations (solitons) as advocated by the large N_c wisdom, a rather unique scenario in strong interaction physics. Through solitons one can already describe bound states at tree level without resorting to the quantum mechanism of the Bethe–Salpeter equation. Geometry conspires with dynamics to give rise to stable and extended field configurations whose properties are similar to the ones of baryons. In short:

(i) They carry a conserved topological charge.
(ii) They are heavy and interact strongly.
(iii) They yield a rich quantum sector.

These are properties that are shared by baryons in the large N_c limit.

Twenty years ago, well before the advent of concepts such as chiral symmetry and QCD, Skyrme [3] in a pioneering series of work proposed to use the non-linear σ-model [4] to describe the light pseudoscalar mesons in which the non-strange baryons emerge as stable field configurations with a non-trivial geometrical structure. In retrospect, he was the first to provide via the soliton scenario a genuine mechanism to generate baryons as bound states of weakly interacting mesons already at the classical level. For more than a decade, Skyrme's work was ignored except for the attempts of

* Actually $SU(2)_L \otimes SU(2)_R \otimes U(1)_V$ as we will show later on.

Finkelstein and Rubinstein [5] that brought more insights to the geometrical character of Skyrme's solitons and Williams' proof [6] that shed light on their spinor structure.

In the past decade however, the situation has changed considerably. There has been significant progress in the understanding of non-perturbative phenomena such as instantons, monopoles, dyons, etc., in the general context of Yang–Mills theories. Recently, Skyrme's non-linear version of current algebra has been revived in the context of QCD under the impetus of Balachandran at Syracuse, and Witten at Princeton. It is the purpose of this article to report about some of these major developments.

Section 2 contains a review of the essential aspects of QCD relevant for a low-energy description of hadrons. This discussion leads naturally to the concept of chiral symmetry, a central issue in current algebra. It is shown that the non-linear σ-model embodies the essence of chiral symmetry, and constitutes the underlying script for the Skyrme model. The Riemannian character of its coset space allows for non-trivial vacuum configurations: skyrmions. In section 3, Witten's recent reinterpretation of the Wess–Zumino term is presented, and its relevance to QCD discussed. Skyrme's original conjecture identifying the topological current with the baryon current is confirmed. The topological quantization of the Wess–Zumino term is shown to provide a subtle relationship between N_c and the spin character of the skyrmions. In section 4, we stress the important relationship between the Wess–Zumino term and the non-Abelian chiral anomalies, using the powerful arguments of Stora and Zumino. In section 5, the old concept of vector meson dominance is examined in the context of the Skyrme model. The phenomenological aspects of the non-Abelian anomaly are discussed using the low-lying vector meson resonances. Their relevance to the skyrmion is analyzed. In section 6, we give a qualitative account on how a Skyrme-like Lagrangian can be obtained by integrating out the light quark components of QCD, assuming 'passive' gluons. In section 7, we summarize most of the SU(2) phenomenology associated with hedgehog skyrmions. In section 8, we underline the relevance of the Skyrme model for conventional boson exchange models. The last section summarizes our prejudices.

2. The SU(2) Skyrme model

2.1. Chiral symmetry in the context of QCD

In QCD, the structure of baryons and mesons is determined by an $SU(3)_C$ gauge theory of quarks and gluons described by the following Lagrangian*

$$\mathcal{L} = -\tfrac{1}{4}\mathrm{TR}[F^{\mu\nu}F_{\mu\nu}] + \bar{q}(\mathrm{i}\not{D} - m)q\,, \tag{1}$$

where the quark field q_i^a is in the fundamental representation of both $SU(3)_c$ and $SU(L)$ flavor, i.e. $a = 1, 2, 3$ and $i = 1, 2, \ldots L$, $F^a_{\mu\nu}$ is the field strength of the gauged gluons taken in the adjoint representation of $SU(3)_c$, D_μ is the usual covariant derivative

$$D_\mu = \partial_\mu + \mathrm{i}gA^a_\mu\lambda^a\,, \tag{2}$$

and m_{ij} is a mass matrix in flavor space. In general, $m_{ij} = m_i\delta_{ij}$ where the m_i are the current quark masses. Due to the non-Abelian character of QCD, the QCD vacuum is paramagnetic and exhibits

* Instantons effects are neglected, and consequently the vacuum angle θ is set to zero.

strong infrared forces. At short distances color anti-screening takes place and leads to an effective coupling constant that goes to zero, hence a free quark–gluon phase (asymptotic freedom) [7]. At large distances the coupling constant grows and one expects only color singlet states to show up in the asymptotic channels (color confinement) [7].

The low-energy behavior of QCD mostly relevant for nuclear physics is dominated by the u- and d-quarks whose current masses m_u and m_d (~10 MeV) are small compared to the QCD cutoff Λ (~200 MeV). The latter sets the scale for the confinement mechanism.* In the massless limit the two-flavor QCD Lagrangian is invariant under independent left × right transformations in flavor space as given by

$$q'_R = Rq_R\,, \qquad q'_L = Lq_L\,, \tag{3}$$

where R and L are U(2) matrices and $q_{R,L}$ are the usual right and left quark fields defined through

$$q_{R,L} = \tfrac{1}{2}(1 \pm \gamma_5)q\,. \tag{4}$$

In the chiral limit, both left- and right-vector currents are conserved classically by Noether's theorem,

$$\partial^\mu\left(\bar{q}\gamma_\mu\left(\frac{1-\gamma_5}{2}\right)\frac{\tau^i}{2}\,q\right) = 0\,, \qquad \partial^\mu\left(\bar{q}\gamma_\mu\left(\frac{1+\gamma_5}{2}\right)\frac{\tau^i}{2}\,q\right) = 0\,, \qquad i = 0,1,2,3\,, \tag{5}$$

where the τ^i are the generators of U(2) in the fundamental representation, i.e., $\tau^0 = 1$ while τ^1, τ^2, τ^3 are ordinary Pauli matrices. The U(2) conserved charges are

$$\begin{aligned} \tilde{Q}^i_L &= \int_{R^3} dx\, q^+\left(\frac{1-\gamma_5}{2}\right)\frac{\tau^i}{2}\,q = \int_{R^3} dx\, q^+_L\,\frac{\tau^i}{2}\,q_L\,,\\ \tilde{Q}^i_R &= \int_{R^3} dx\, q^+\left(\frac{1+\gamma_5}{2}\right)\frac{\tau^i}{2}\,q = \int_{R^3} dx\, q^+_R\,\frac{\tau^i}{2}\,q_R\,. \end{aligned} \tag{6}$$

Specifically, (1) is invariant under the continuous group of transformations

$$U(2)_L \otimes U(2)_R \equiv SU(2)_L \otimes SU(2)_R \otimes U(1)_L \otimes U(1)_R\,, \tag{7}$$

with $L = \exp(iQ_L)$ and $R = \exp(iQ_R)$. Vector transformations are generated by $Q_V = \frac{1}{2}(Q_L + Q_R)$, while axial-vector transformations by $Q_A = \frac{1}{2}(Q_R - Q_L)$. The $U(1)_A$ symmetry is explicitly broken at the quantum level by the well-known Adler–Bell–Jackiw anomaly [8], i.e.

$$\partial^\mu(\bar{q}\gamma_\mu\gamma_5 q) = \frac{g^2}{16\pi^2}\,\varepsilon^{\mu\nu\alpha\beta}\,\mathrm{TR}[F_{\mu\nu}F_{\alpha\beta}]\,, \tag{8}$$

leaving us with a global $SU(2)_L \times SU(2)_R \times U(1)_V$ invariance.

* See the brief discussion in section 9 on the chiral bag model, where it is argued that the confinement region of the nucleon is small because of the compression of the quark core by the pion cloud.

The absence of parity doublets in the physical spectrum with a pronounced isospin symmetry suggests that $SU(2)_L \times SU(2)_R$ is spontaneously broken to $SU(2)_V$ via the Nambu–Goldstone mechanism, with the appearance of 3 massless pseudoscalar excitations: $\pi^0, \pi^\pm$. In other words, the QCD ground state carries axial charge

$$Q_V^i|0\rangle = 0\,, \qquad Q_A^i|0\rangle \neq 0\,. \tag{9}$$

As a result, pions can decay into the vacuum.

2.2. The non-linear σ-model

The essence of chiral symmetry and the relative success of current algebra lies in the fact that the vacuum state in QCD breaks spontaneously chiral symmetry. If we denote by $|\pi^i(p)\rangle$ a pion state of momentum p, and by $A_\mu^i(x)$ the axial-vector current, then eq. (9) can be used to define

$$\langle 0|A_\mu^i(x)|\pi^j(p)\rangle = \mathrm{i} f_\pi p_\mu\, \mathrm{e}^{ipx}\delta^{ij}\,, \tag{10}$$

where $f_\pi = 93$ MeV is the observed pion decay constant. From eq. (10) follows the Goldberger–Treiman relation

$$g_A(0) = g_{\pi NN} f_\pi / m_N\,, \tag{11}$$

where $g_A = 1.34$ is the nucleon axial form factor, $g_{\pi NN} = 13.5$ is the πNN-coupling and $m_N = 938$ MeV the nucleon mass. Equations (10) and (11) play a central role in low-energy phenomenology.

Due to the non-perturbative character of QCD in the long-wavelength approximation, very little is known about f_π and g_A from first principles. There is no doubt that ultimately, lattice gauge calculations will provide a quantitative understanding of these low-energy parameters. Meanwhile, the large N_c limit as advocated by 't Hooft [1] and Witten [2] is suggestive of an effective mesonic description involving the dominant chiral degrees of freedom, π^0, $\pi^\pm$, as a substitute to QCD at low energy. Although $N_c = 3$ and not infinity, one hopes that this approach will provide the relevant starting lines for discussing low-energy phenomenology. In this spirit, the non-linear σ-model provides a pertinent script for chiral symmetry breaking, consistent with soft-pion threshold theorems. If we denote by $\sigma(x)$ the scalar meson field and by $\boldsymbol{\pi}$ the pseudoscalar pion field, then the resulting dynamics, described by

$$\mathscr{L} = \tfrac{1}{2}(\partial_\mu \sigma)^2 + \tfrac{1}{2}(\partial_\mu \boldsymbol{\pi})^2\,, \qquad \sigma^2 + \boldsymbol{\pi}^2 = f_\pi^2\,, \tag{12}$$

is manifestly chiral invariant since $\binom{\sigma}{\pi}$ corresponds to the $(1,0)$ representation of $SU(2)_L \otimes SU(2)_R \sim SO(4)$. Let $\boldsymbol{\alpha}$ and $\boldsymbol{\beta}$ be the parameters associated to vector and axial-vector transformations respectively, i.e.

$$Q_R = \tfrac{1}{2}(\boldsymbol{\beta} + \boldsymbol{\alpha})\cdot\boldsymbol{\tau}\,, \qquad Q_L = \tfrac{1}{2}(\boldsymbol{\beta} - \boldsymbol{\alpha})\cdot\boldsymbol{\tau}\,, \tag{13}$$

then

$$\begin{aligned} &\delta_V \sigma = 0\,, && \delta_A \sigma = -\boldsymbol{\beta}\cdot\boldsymbol{\pi}\,, \\ &\delta_V \boldsymbol{\pi} = -(\boldsymbol{\alpha}\times\boldsymbol{\pi})\,, && \delta_A \boldsymbol{\pi} = +\boldsymbol{\beta}\sigma\,, \end{aligned} \tag{14}$$

Fig. 2. (a) Profile of the scalar potential in the chiral symmetric linear σ-model. (b) Illustration of the effect of a chiral breaking term (tilted Mexican hat).

which clearly shows that axial transformations mix parity. The classically conserved Noether's currents associated to the continuous transformations (14) read

$$V^i_\mu = \varepsilon^{ijk}\pi^j\partial_\mu\pi^k\,, \qquad A^i_\mu = \sigma\partial_\mu\pi^i - \pi^i\partial_\mu\sigma\,, \tag{15}$$

from which one deduces that

$$\langle 0|A^i_\mu(x)|\pi^k(p)\rangle = \langle 0|\sigma|0\rangle \mathrm{i}p_\mu\, \mathrm{e}^{\mathrm{i}px}\delta^{ik}\,, \tag{16}$$

where we have used the fact that the vacuum expectation value of $\partial_\mu\sigma$ vanishes, along with the following pion field normalization:*

$$\langle 0|\pi^i(x)|\pi^k(p)\rangle = \delta^{ik}\,\mathrm{e}^{\mathrm{i}px}\,. \tag{17}$$

In the trivial vacuum the non-linear condition (12) translates into $\langle 0|\sigma|0\rangle = f_\pi$, which is the expected limit if one uses the linear σ-model with an infinitely heavy scalar particle ($m_\sigma \to \infty$). To account for the small but non-vanishing mass of the pion field in nature, one adds an explicit chiral breaking term in the form $\mathscr{L}_\mathrm{B} = -c\sigma$. In the trivial vacuum, the pion can be understood as fluctuations of the σ-field along the valley of the tilted Mexican hat of fig. 2b,

$$\begin{aligned}\mathscr{L}_\mathrm{B} &= -c\sigma = -c\sqrt{f_\pi^2 - \pi^2} \\ &= -cf_\pi + \frac{c}{2}\frac{\pi^2}{f_\pi} + \mathrm{O}\left(\frac{1}{f_\pi^3}\right).\end{aligned} \tag{18}$$

This yields $c = m_\pi^2 f_\pi$ along with the correct PCAC relation

$$\partial^\mu A^i_\mu = m_\pi^2 f_\pi \pi^i\,. \tag{19}$$

Already at this stage, we can see that the non-linear σ-model satisfies the basic low-energy requirements solely on the basis of chiral symmetry. In fact it does more as Skyrme noted long ago; it embodies an underlying topological structure that yields non-perturbative field configurations reminiscent of classical baryons.

* Strictly speaking this relation holds for the renormalized fields acting on asymptotic states.

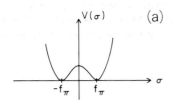

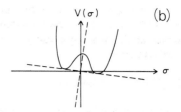

2.3. Topology of the non-linear σ-model

To grasp the geometrical intricacies of the non-linear σ-model, it is instructive to recast it in the Sugawara [3] form. For that, define the unitary 2×2 quaternion field $U(x)$

$$U(x)=\frac{1}{f_\pi}[\sigma+\mathrm{i}\boldsymbol{\tau}\cdot\boldsymbol{\pi}] \tag{20}$$

that transforms as the $(\frac{1}{2},\frac{1}{2})$ representation of $SU(2)_L\times SU(2)_R$,

$$U(x)\to\exp(\mathrm{i}Q_L)\,U\exp(-\mathrm{i}Q_R)\,, \tag{21}$$

in agreement with eqs. (14). In the quark picture, the analogue of U^{ij} is the complex 2×2 matrix $\bar{q}^i[(1-\gamma_5)/2]q^j$ corresponding to pseudoscalar mesons. $U(x)\in SU(2)$ whose group manifold is isomorphic to S^3. The latter is one of the four remarkable spheres (S^0,S^1,S^3,S^7) that are characterized by left × right parallelism by virtue of Adams' lemma [10]. The left and right currents on S^3 are defined to be

$$R_\mu=U\partial_\mu U^+\to\exp(\mathrm{i}Q_L)\,R_\mu\exp(\mathrm{i}Q_L)\,,\qquad L_\mu=U^+\partial_\mu U\to\exp(\mathrm{i}Q_R)L_\mu\exp(-\mathrm{i}Q_R)\,, \tag{22}$$

which show that $R_\mu(L_\mu)$ is invariant under right (left) chiral transformations. Since $\det U=1$, it follows that

$$\partial_\mu\det U=\partial_\mu\exp\mathrm{TR}\ln U=\mathrm{TR}[L_\mu]=\mathrm{TR}[R_\mu]=0\,. \tag{23}$$

It is instructive to note that for a weak pion field L_μ and R_μ reduce to

$$L_\mu\sim -R_\mu\sim\frac{\mathrm{i}}{f_\pi}\boldsymbol{\tau}\cdot\partial_\mu\boldsymbol{\pi}\,. \tag{24}$$

At any fixed time, the 2×2 field $U(\boldsymbol{x})$ defines a map from the three-dimensional space R^3 onto the group manifold S^3, with the natural boundary condition that $U(\boldsymbol{x})$ goes to the trivial vacuum, $\langle 0|\sigma|0\rangle=f_\pi$, for asymptotically large distances. This insures that the energy of the corresponding field configuration is finite,

$$U(|\boldsymbol{x}|\to\infty)=\mathbb{1}\,, \tag{25}$$

and implies that R^3 is compactified to S^3 as all points at infinity in R^3 are mapped into one fixed point in S^3. The set of static maps subject to (25),

$$U(\boldsymbol{x}):\quad S^3\to S^3\,,$$

is known to be non-trivial. In other words, at a given time, it is possible to split the set of all maps into homotopically distinct classes not continuously deformable into each other. These classes are called homotopy or Chern–Pontryagin classes. In our case, they constitute the third homotopy group

$\pi_3(S^3) \sim Z$, where Z is the additive group of integers. It is these integers that are referred to as winding numbers of the mapping $U(x)$. Since a continuous evolution in time can be understood as a homotopy transformation, the corresponding winding numbers are conserved by definition independently of the details of the underlying dynamics.

In order to construct an explicit form of the topological charge*

$$B^0: \pi_3(S^3) \rightarrow Z , \tag{26}$$

in the non-linear σ-model, it is convenient to use the vector representation (1, 0) of $SU(2)_L \otimes SU(2)_R$,

$$\phi^0 = \sigma/f_\pi , \qquad \phi^i = \pi^i/f_\pi . \tag{27}$$

An elementary surface element in the group manifold is characterized by

$$d^3\Sigma = \varepsilon^{ijkl}\phi^i\partial_1\phi^j\partial_2\phi^k\partial_3\phi^l \, dx^1 \, dx^2 \, dx^3 , \tag{28}$$

where (x^1, x^2, x^3) are the corresponding coordinates on R^3 obtained by stereographic projection from S^3 as illustrated in fig. 3a. Equation (28) is just the Jacobian associated to the affine transformation: $S^3 \rightarrow R^3$. Hence, the normalized topological density reads

$$B^0 = + \frac{1}{3! A_3} * \varepsilon^{ijkl}\varepsilon^{\nu\alpha\beta}\phi^i\partial_\nu\phi^j\partial_\alpha\phi^k\partial_\beta\phi^l , \tag{29}$$

where $A_3 = 2\pi^2$ is the surface of S^3 in R^4. To rewrite eq. (29) in terms of (22), notice that for a weak pion field, i.e., $\phi^0 \sim 1/f_\pi$ and $\phi^i \sim \pi^i/f_\pi$, we have

$$\begin{aligned} B^0 &= \frac{(-i)^3}{24\pi^2} \varepsilon^{\nu\alpha\beta} \mathrm{TR}\left[\partial_\nu\left(\frac{i\boldsymbol{\tau}\cdot\boldsymbol{\pi}}{f_\pi}\right)\partial_\alpha\left(\frac{i\boldsymbol{\tau}\cdot\boldsymbol{\pi}}{f_\pi}\right)\partial_\beta\left(\frac{i\boldsymbol{\tau}\cdot\boldsymbol{\pi}}{f_\pi}\right)\right] + O\left(\frac{1}{f_\pi^4}\right) \\ &= \frac{i}{24\pi^2} \varepsilon^{0\nu\alpha\beta} \mathrm{TR}[L_\nu L_\alpha L_\beta] . \end{aligned} \tag{30}$$

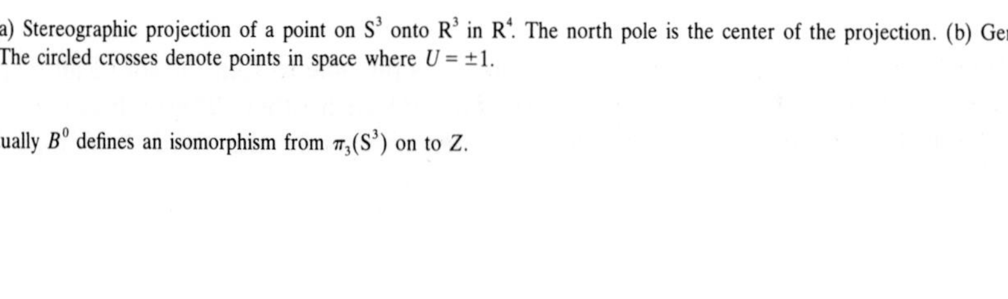

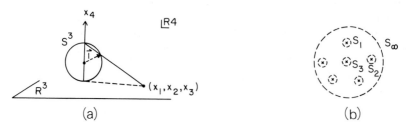

Fig. 3. (a) Stereographic projection of a point on S^3 onto R^3 in R^4. The north pole is the center of the projection. (b) General skyrmion-type texture. The circled crosses denote points in space where $U = \pm 1$.

* Actually B^0 defines an isomorphism from $\pi_3(S^3)$ on to Z.

The last equation follows from chiral invariance. Notice that B does not vanish if and only if the 3 pion degrees of freedom, π^0, $\pi^{\pm}$, are excited. It is rather obvious from the above construction that

$$\int_{R^3} B^0 \, dx = \text{winding number} . \tag{31}$$

The covariant current associated to the topological charge (30) is given by

$$B^{\mu} = + \frac{i\varepsilon^{\mu\nu\alpha\beta}}{24\pi^2} \, \mathrm{TR}[L_{\nu} L_{\alpha} L_{\beta}] , \tag{32}$$

and is conserved almost everywhere in R^3 independently of the equations of motion, i.e. $\partial^{\mu} B_{\mu} = 0$. Topological singularities of the pion field correspond to $U(\boldsymbol{x})$ taking on the values ± 1, and the total topological charge may be expressed in terms of properties of these singularities [11]. This is done by surrounding points, lines or surfaces where $U = \pm 1$ by surfaces S_{β} as shown in fig. 3b, and then calculating the flow of topological charge across these surfaces when the pion field is increased adiabatically from zero to its final value. From appendix A, we have

$$\begin{aligned} B &= \int_{R^3} d\boldsymbol{x} \, B^0(\boldsymbol{x}) \\ &= \frac{1}{4\pi^2} \sum_{\beta=1}^{N+1} \int_{S_{\beta}} d^2 S_{\beta} \left[\theta - \frac{\sin 2\theta}{2} \right] \hat{\theta} \cdot (\partial_1 \hat{\theta} \times \partial_2 \hat{\theta}) , \end{aligned} \tag{33}$$

where

$$U(x) = \exp(i\boldsymbol{\tau} \cdot \boldsymbol{\theta}(x)) , \qquad \boldsymbol{\theta}(\boldsymbol{x}) = \theta(\boldsymbol{x}) \hat{\theta}(\boldsymbol{x}) \quad \text{and} \quad \hat{\theta}^2(\boldsymbol{x}) = 1 . \tag{34}$$

The indices 1 and 2 in (33) refer the two comoving coordinates on S_{β} as detailed in appendix A. The sum in (33) runs over all surfaces $S_1, S_2, \ldots S_N$ excluding the source singularities with $S_{N+1} = S_{\infty}$ being the surface at infinity.

The topological charge (30) described above bears some similarities with the monopole and instanton charges. Like the latter, it follows from the underlying Riemannian structure of the group manifold, and is conserved regardless of the equations of motion. Its form is obtained by explicit construction of the isomorphism: $\pi_3(S^3) \rightarrow Z$. There are however, some essential differences as noticed by Rubakov and Sanyuk [12]. B^{μ} is a vector current and not an axial current as in the case of instantons and monopoles. The topological charge has no dual charge, and can be localized arbitrarily in space since (30) is not a total divergence.

2.4. The Skyrme model

So far, the non-linear σ-model has provided a rather economical framework for discussing low-energy phenomenology. Its finite-energy configuration space exhibits a non-trivial topological structure due to the underlying Riemannian geometry of the coset space. As a consequence, there exist static and finite field configurations other than the trivial vacuum, that are characterized by conserved topological charges. Unfortunately, these configurations are not energetically stable in 3-space by Derrick's theorem [13]. Indeed, if $U(\boldsymbol{x})$ is a static field configuration solution to the Euler–Lagrange equations associated to

$$\mathscr{L} = \tfrac{1}{4} f_\pi^2 \, \mathrm{TR}[L_\mu L^\mu] , \tag{35}$$

which is the Sugawara's form of (12), then

$$E = \int \mathrm{d}^D x \, \frac{f_\pi^2}{4} \, \mathrm{TR}[\partial^i U^+ \partial^i U] . \tag{36}$$

A simple rescaling of $U(\boldsymbol{x})$ in space, i.e. $U(\boldsymbol{x}) \to U(\lambda \boldsymbol{x})$, yields

$$E_\lambda = \lambda^{2-D} E . \tag{37}$$

Equation (37) clearly shows that for $D = 3$, the energetically favorable configurations have zero energy. In other words, the finite-energy solutions of the non-linear σ-model are unstable against scale transformations.

To avoid this collapse, Skyrme added by hand to (35) a quartic term in the currents L_μ (Skyrme term),

$$\mathscr{L}_\mathrm{S} = -\tfrac{1}{4} f_\pi^2 \, \mathrm{TR}[L^\mu L_\mu] + \tfrac{1}{4} \varepsilon^2 \, \mathrm{TR}[L_\mu , L_\nu]^2 . \tag{38}$$

Here ε is a dimensionless parameter that characterizes the size of the finite energy configurations. Since the Skyrme term scales like r^{-4} in three-dimensional space it will prevent the Skyrme solitons from collapsing to zero size. Indeed, a rescaling in space of any finite-energy solution to (38) translates to a rescaling in the ground state energy in the form

$$E_\lambda = \lambda^{2-D} E_{(2)} + \lambda^{4-D} E_{(4)} . \tag{39}$$

It is straightforward to show that E_λ exhibits a true minimum for $D \geq 3$, i.e.,

$$\text{(i)} \quad \mathrm{d}E_\lambda / \mathrm{d}\lambda = 0 \quad \to \quad E_{(2)} / E_{(4)} = -(D-4)/(D-2) , \tag{40a}$$

$$\text{(ii)} \quad \mathrm{d}^2 E_\lambda / \mathrm{d}\lambda^2 > 0 \quad \to \quad 2(D-2) E_{(2)} > 0 . \tag{40b}$$

In particular for $D = 3$ eq. (40a) shows that $E_{(2)} = E_{(4)}$, as of course expected from the virial theorem. Due to the underlying geometry, the energy is bounded from below by the topological charge. Indeed, for a static configuration

$$E = \int \mathrm{d}\boldsymbol{x} \{ -\tfrac{1}{4} f_\pi^2 \, \mathrm{TR}[L_i^2] - \tfrac{1}{4} \varepsilon^2 \, \mathrm{TR}[L_i , L_j]^2 \} . \tag{41}$$

A lower bound to (41) can be obtained using the Cauchy–Schwarz inequality,

$$E = -\frac{f_\pi^2}{4}\int \mathrm{d}\boldsymbol{x}\, \mathrm{TR}\Big[L_i^2 + \frac{\varepsilon^2}{f_\pi^2}(\sqrt{2}\varepsilon_{ijk}L_jL_k)^2\Big] \geq +\frac{f_\pi^2}{4}\int \mathrm{d}\boldsymbol{x} \Big|\mathrm{TR}\Big(\frac{2\sqrt{2}\varepsilon}{f_\pi}\,\varepsilon_{ijk}L_iL_jL_k\Big)\Big| , \tag{42}$$

where we have used the fact that the L_i's are anti-Hermitian. In terms of the topological charge B, this becomes

$$E \geq 12\sqrt{2}\pi^2\varepsilon f_\pi |B| . \tag{43}$$

Equation (43) is sometimes referred to as the Bogomol'ny bound. Since there are no self-dual chiral fields,* the energy is strictly larger than the estimate (43). For skyrmions the Bogomol'ny bound cannot be saturated, another difference from instantons. Equation (43) illustrates in a striking way the mechanism by which geometry induces local stability at the classical level.

The Skyrme term can be understood as a higher-order correction to the non-linear σ-model when cast in the general framework of an effective chiral description as advocated by Weinberg [15]. To lowest order in the pion momentum p_μ, this term exhibits the quantum numbers of a massive ρ-meson exchange in the D-wave ππ-channel, i.e.

$$\tfrac{1}{4}\varepsilon^2\,\mathrm{TR}[L_i, L_j]^2 \sim \frac{\varepsilon^2}{f_\pi^4}(\partial_i\boldsymbol{\pi}\times\partial_j\boldsymbol{\pi})^2 , \tag{44}$$

suggesting that the ad hoc parameter ε can be obtained from the ππ-data. We will comment more on this point later on. While the quadratic term (35) in the non-linear σ-model is unique to order $O(p^2)$, the Skyrme term is not to order $O(p^4)$. Indeed, under the general assumptions of locality, Lorentz-invariance, chiral symmetry, parity and G-parity, there are 3 independent terms to order $O(p^4)$, i.e.

$$L_{(4)} = \alpha\,\mathrm{TR}[L_\mu, L_\nu]^2 + \beta\,\mathrm{TR}[L_\mu, L_\nu]_+^2 + \gamma\,\mathrm{TR}[\partial_\mu L_\mu]^2 . \tag{45}$$

in the Weinberg expansion [15]. Other combinations can be eliminated by the Maurer–Cartan equation for the left-current on S^3,

$$\partial_\mu L_\nu - \partial_\nu L_\mu + [L_\mu, L_\nu] = 0 , \tag{46}$$

which follows trivially from the definitions (22). To this order, the Skyrme term is the unique term with four derivatives that leads to a Hamiltonian second order in time derivatives. This is important because there are problems with stability of the classical solution once higher-order terms in the time derivatives are included. We will briefly discuss the question of non-Skyrme four-derivative terms in section 8.

* The self-dual chiral field in the Skyrme model that would saturate the Bogomol'ny bound, i.e.

$$L_i = (\varepsilon\sqrt{2}/f_\pi)\varepsilon_{ijk}L_jL_k ,$$

is incompatible with the Maurer–Cartan equation (46).

2.5. The skyrmion

The Skyrme model embodies the $SO(4) \sim SU(2)_L \otimes SU(2)_R$ non-linear σ-model to leading order, breaks spontaneously $SU(2)_L \otimes SU(2)_R$ and fulfills current algebra requirements. The finite field configurations of non-trivial topology are made stable by the Skyrme term. From now on, these configurations will be referred to as Skyrme solitons or skyrmions. By soliton we will mean: classical, static, stable and finite-energy field configurations in weakly interacting non-linear field theories of mesons characterized by a degenerate vacuum state. In general, solitons are heavy objects with exactly conserved topological charges. The soliton–soliton interaction is strong while the soliton–meson interaction is weak. Once quantized, they exhibit a rich spectrum. Skyrmions, as it will become clear throughout the course of this work, enjoy many of these properties.

In his pioneering work two decades ago, Skyrme [3] believed that the field configurations of winding number one ($B=1$) in his model were fermions. He conjectured that the topological current (32) should be identified with the baryon current, suggesting that skyrmions were classical baryons. This remarkable statement in prehistoric times, was only confirmed recently in the context of QCD, as we will discuss it later on.

The classical equations of motion associated to (38) follow from the Euler–Lagrange equations subject to the unitarity constraint $U^+U=1$, or alternatively from the variations of the action functional associated to (38) to leading order in the left × right fluctuations. Indeed,

$$S[\tilde{U}(x;t)] = \int dx\{\tfrac{1}{4}f_\pi^2 \, \mathrm{TR}[L_\mu L^\mu] + \tfrac{1}{4}\varepsilon^2 \, \mathrm{TR}[L_\mu, L_\nu]^2\}, \tag{47}$$

where $\tilde{U}(x,t)$ is defined through

$$\tilde{U}(x,t) = U(x,t)\exp[i\phi_R(x,t)]. \tag{48}$$

The first-order variation in the action functional (47) reads

$$\delta^{(1)}S = i\frac{f_\pi^2}{2}\int dx \, \mathrm{TR}\Big[\partial^\mu\phi_R\Big(L_\mu - 2\frac{\varepsilon^2}{f_\pi^2}[L_\nu,[L_\mu, L_\nu]]\Big)\Big]. \tag{49}$$

Integration by parts yields the following equation of motion for the left currents L_μ

$$\partial^\mu L_\mu - 2\frac{\varepsilon^2}{f_\pi^2}\partial^\mu[L_\nu,[L_\mu, L_\nu]] = 0. \tag{50}$$

Similar equations can be obtained for R_μ, and follows from (50) through the substitution $L_\mu \to R_\mu$. Since the self-duality condition is incompatible with the Maurer–Cartan equation as indicated above, the Bogomol'ny construction to simplify (50) via the saturation mechanism (a well-known procedure for instantons) cannot be used in this case to obtain analytical solutions. A numerical treatment of (50) is required.

The above equations are highly non-linear and so far can only be handled under the assumption of maximal symmetry as suggested by Skyrme's static ansatz [3],

$$U(x) = \exp(i\boldsymbol{\tau}\cdot\hat{r}F(r)) \\ = \cos F(r) + i\boldsymbol{\tau}\cdot\hat{r}\sin F(r)\,. \tag{51}$$

This is the so-called hedgehog configuration in which the pion field is radial both in space and isospace. Equation (51) follows from the fact that in a given topological sector, the maximal compact symmetry group of the configuration space is

$$\mathrm{diag}(\mathrm{SU}(2)_\mathrm{L}\otimes\mathrm{SU}(2)_\mathrm{R}) \sim \mathrm{diag}(\mathrm{SO}(3)_I\otimes\mathrm{SO}((3)_J))\,, \tag{52}$$

where $\mathrm{SO}(3)_J$ and $\mathrm{SO}(3)_I$ refer to the orthogonal group of rotations in space and isospace respectively. In the ansatz (51), isospin (I) and angular momentum (J) are correlated in a way that neither of them is a good quantum number, but their sum is

$$\boldsymbol{K} = \boldsymbol{J} + \boldsymbol{I} \equiv (\boldsymbol{L}+\boldsymbol{S}) + \boldsymbol{I}\,. \tag{53}$$

$U(\boldsymbol{x})$ is left invariant under rotations in K-space, i.e.

$$[\boldsymbol{K}, U(\boldsymbol{x})] = i\sin F\Big(\Big[\Big(\boldsymbol{r}\times\frac{\boldsymbol{\nabla}}{i}\Big), \boldsymbol{\tau}\cdot\hat{r}\Big] + [\boldsymbol{\tau}/2, \boldsymbol{\tau}\cdot\hat{r}]\Big) \\ = i\sin F(-i(\boldsymbol{\tau}\times\hat{r}) - i(\hat{r}\times\boldsymbol{\tau})) \equiv 0\,, \tag{54}$$

suggesting that hedgehog skyrmions are scalar in K-space ($K=0$). Since parity is defined through*

$$\hat{\pi}_{\mathrm{op}} U(\boldsymbol{x}, t)\hat{\pi}_{\mathrm{op}}^{-1} = U^{+}(-\boldsymbol{x}, t) \tag{55}$$

one concludes that the ansatz (51) is parity invariant. Consequently, hedgehog skyrmions carry $K^{\pi} = 0^{+}$ and can be viewed as an admixture of states with $I = J$ and positive parity.

In terms of (51), eq. (50) simplifies into a radial equation for $F(r)$

$$F'' + \frac{2}{r}F' - \frac{\sin 2F}{r^2} + 8\frac{\varepsilon^2}{f_\pi^2}\left[\frac{\sin 2F\sin^2 F}{r^4} - \frac{F'^2\sin 2F}{r^2} - \frac{2F''\sin^2 F}{r^2}\right], \tag{56}$$

and the solution with unit winding number corresponds to

$$F(r=0) = \pi\,, \qquad F(r\to\infty) = 0\,.$$

* This operation corresponds to the usual transformation of the fermion fields under parity. Indeed, recalling the quark analogy $U^{ij}(x) \leftrightarrow \bar{q}^i[(1-\gamma_5)/2]q^j$ and the parity operation for spinor fields $\hat{\pi}_{\mathrm{op}} q(\boldsymbol{x}, t) = \gamma^0 q(-\boldsymbol{x}, t)$, we obtain

$$\hat{\pi}_{\mathrm{op}} U(\boldsymbol{x}, t)\hat{\pi}_{\mathrm{op}}^{-1} \equiv \bar{q}(-\boldsymbol{x}, t)\gamma^0[(1-\gamma_5)/2]\gamma^0 q(-\boldsymbol{x}, t) \\ = (q^{+}(-\boldsymbol{x}, t)[(1+\gamma_5)/2]\gamma^0 q(-\boldsymbol{x}, t))^{+} \equiv U^{+}(-\boldsymbol{x}, t)\,.$$

The Hermitian conjugation reflects the fact that pions are pseudoscalar.

Indeed, the topological density (30) for a hedgehog configuration becomes

$$B^0(r) = \frac{1}{2\pi^2} \sin^2 F \frac{F'}{r^2}, \tag{57}$$

leading to a total charge of the form

$$B = \frac{1}{\pi} (F(0) - F(\infty)) + \frac{1}{2\pi} (\sin 2F(\infty) - \sin 2F(0)) .$$

This configuration, provided that it exists, corresponds to a $B = 1$ classical baryon by Skyrme's conjecture.

Numerical solutions to the above boundary value problem have been obtained by Jackson et al. [20] and Adkins et al. [21]. Notice that close to the origin the chiral angle $F(r)$ has a simple linear behavior of the form $F(r \to 0) = n\pi - \alpha r$, while at large distance it falls off like a power law, $F(r \to \infty) = K^2/r^2$. Jackson et al. [20] used the empirical value of $f_\pi = 93$ MeV and insisted on having the correct axial coupling $g_A = 1.33$ by choosing $\varepsilon^2 = 0.00552$, so that asymptotically

$$F(r \to \infty) = \frac{16\varepsilon^2}{f_\pi^2} \frac{1.078}{r^2} . \tag{58a}$$

Alternatively, Adkins et al. [21] used $f_\pi = 64.5$ MeV and $\varepsilon^2 = 0.00421$ to insure the correct nucleon and Δ-mass (for more details see section 8). In their case, one has asymptotically

$$F(r \to \infty) = \frac{16\varepsilon^2}{f_\pi^2} \frac{1.081}{r^2} . \tag{58b}$$

For the parameters of refs. [21–22a] the behavior of the chiral angle $F(r)$ is displayed in fig. 4. Although there is no analytical proof for the global stability of the hedgehog configuration with unit baryon

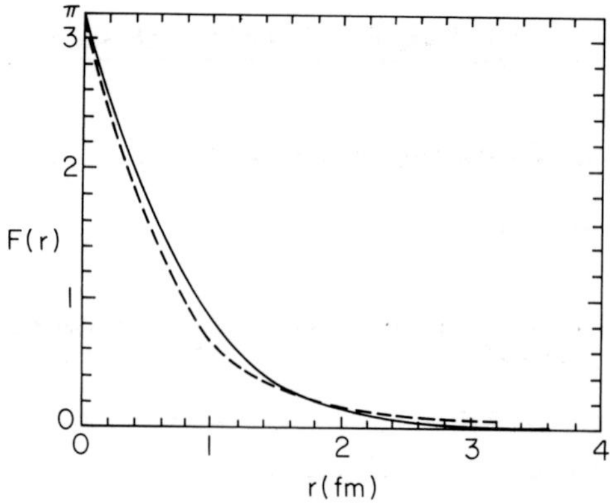

Fig. 4. Profile of the chiral angle $F(r)$ for a hedgehog configuration solution to (56) and subject to the boundary conditions (57). The continuous curve represents $F(r)$ for $m_\pi \neq 0$, as discussed in ref. [22a]. The dashed curve refers to the chiral case $m_\pi = 0$ of ref. [21].

number, the recent search for low-lying resonances in this model has shown that it is locally stable. The stability equations yield a positive definite spectrum.

To clarify the relationship between g_A and ε, it is instructive to notice that while asymptotically $F(r) \to K^2/r^2$, the pion field $\pi^a(\boldsymbol{r})$ has the following form

$$\pi^a(\boldsymbol{r}) = -\frac{f_\pi}{3} \times \frac{K^2}{r^2}\, \hat{r}^i \langle \sigma^i \tau^a \rangle\,, \qquad r \to \infty\,, \tag{59a}$$

which involves the expectation value in a hedgehog configuration. Old-fashioned soft-pion nucleon effective Lagrangians of the type described below, shows that the asymptotic behavior of the expectation value of the pion field in a nucleon state is (for more details see appendix C)

$$\pi^a(\boldsymbol{r}) = -\left(\frac{g_{\pi NN}}{8\pi m_N}\right) \times \frac{\hat{r}^i}{r^2} \langle \sigma^i \tau^a \rangle\,, \qquad r \to \infty\,, \tag{59b}$$

where $g_{\pi NN}$ is the πNN-coupling and m_N the nucleon mass. Using the Goldberger–Treiman relation and comparing (59a) and (59b), we obtain

$$g_A = \tfrac{8}{3}\pi K^2 f_\pi^2\,, \tag{60}$$

which shows that the large-distance behavior of the skyrmion is unambiguously related to g_A in the chiral limit. In terms of Skyrme's parameters, $K^2 = (16\varepsilon^2/f_\pi^2)k^2$, so that

$$\varepsilon^2 = \frac{3}{128\pi}\left(\frac{g_A}{k^2}\right). \tag{61}$$

Using the asymptotics (58), it is straightforward to check that for the parameters used by Adkins et al. $g_A = 0.61$, while for the ones used by Jackson et al. $g_A = 0.80$. Jackson et al. have argued that these values of the axial coupling correspond to $N_c \to \infty$, and that there are substantial finite N_c corrections. Using simple book-keeping arguments they have argued that for a given number of color N_c, the axial coupling g_A^N is related to g_A through

$$g_A^N = [(N_c + 2)/N_c]\, g_A\,.$$

This would be the dependence of g_A on N_c in the constituent quark model. For $N_c \to \infty$, $g_A^N = g_A$, while for $N_c = 3$, $g_A^N = \frac{5}{3} g_A$. In the case of Adkins et al. $g_A^N = 1.01$, while for Jackson et al. $g_A^N = 1.33$ in agreement with the empirical value. There may well be other N_c^{-1} corrections, however, so the systematics behind this factor are not entirely clear.

The mass of the hedgehog configuration can be obtained from (41), using the ansatz (51),

$$M_s = 4\pi \int_0^\infty r^2\, dr\, \frac{f_\pi^2}{2}\left[F'^2 + \frac{2\sin^2 F}{r^2}\right] + 4\varepsilon^2\, \frac{\sin^2 F}{r^2}\left[2F'^2 + \frac{\sin^2 F}{r^2}\right]. \tag{62a}$$

Using the dimensionless variable $x = (f_\pi/\varepsilon\sqrt{2})r$, together with the virial result discussed above, we obtain

Table 1
Results of Jackson and Rho (J.R.), and Adkins, Nappi and Witten (A.N.W.) for the hedgehog skyrmion. The (*) refers to input parameters

	ε^2	f_π [MeV]	g_A^H	M_H [MeV]	r_H [fm]
J.R.	0.00552	93.0*	0.80*	1425	0.48
A.N.W.	0.00421	64.5	0.61	863*	0.59

$$M_s = 4\pi\sqrt{2}\varepsilon f_\pi \int_0^\infty x^2\left[\left(\frac{dF}{dx}\right)^2 + 2\frac{\sin^2 F}{x^2}\right] dx\,. \tag{62b}$$

In terms of the Jackson et al. [20] parameters the $B = 1$ configuration has $M_s = 1.425$ GeV. The rms radius of this configuration is just the isoscalar radius constructed using (57),

$$r_{\rm rms} = \frac{\varepsilon}{\pi f_\pi}\left(-\int_0^\infty dx\, x^2 \sin^2 F\, F'\right)^{1/2}, \tag{63}$$

which is of the order of 0.48 fm for $B = 1$. For the purpose of completeness we have quoted the results of Adkins et al. [21] in table 1 along with the Jackson et al. [20] results referred to in this section. On the overall the results are satisfactory. The addition of a pion mass $m_\pi = 139$ MeV via a chiral breaking term into (38), does not yield substantial modifications [22].

2.6. Soft-pion theorems and the Skyrme model

So far, Skyrme's conjecture has provided us with a simple picture of baryons as static solitons in an effective chiral model of pseudoscalar mesons. The energy and size of the $B = 1$ hedgehog skyrmion do not seem to be unreasonable given the properties of nucleons and nucleon isobars. Our next step is to analyze the structure of the corresponding soft-pion–baryon Lagrangian in the light of soft-pion threshold theorems and current algebra. It is very likely that Skyrme's scenario will go through current algebra requirements since it embodies the non-linear σ-model and respects chiral symmetry.

Consider chiral fluctuations about the hedgehog skyrmion in the $B = 1$ sector,

$$U(\boldsymbol{x}, t) = e^{iQ} U_0(\boldsymbol{x})\, e^{iQ}\,, \tag{64}$$

where $Q \equiv (1/2f_\pi)\pi^a\tau^a$ is Lie-algebra valued, and $\pi^a(\boldsymbol{x}, t)$ is a perturbative pion field satisfying: $\pi^a(\boldsymbol{x}, t) \to 0$ as $|\boldsymbol{x}| \to 0, \infty$. These conditions insure that $\pi(\boldsymbol{x}, t)$ is geometrically trivial, so that the total topological charge is solely carried by $U_0(\boldsymbol{x})$. In terms of (64), we have

$$\begin{aligned}\tilde{L}_\mu &= U^+\partial_\mu U \\ &= \xi^+[U_0^+ l_\mu U_0 + L_\mu + r_\mu]\xi\,,\end{aligned}$$

where $\xi = e^{iQ}$ and $l_\mu = \xi^+\partial_\mu\xi = \xi^+ r_\mu \xi$. Expanding ξ and retaining only the leading contributions, yields l_μ and r_μ in the form

$$l_\mu = \mathrm{i}\partial_\mu Q + \tfrac{1}{2}[Q, \partial_\mu Q] + \mathrm{O}(Q^3)\ , \tag{65a}$$

$$r_\mu = \mathrm{i}\partial_\mu Q - \tfrac{1}{2}[Q, \partial_\mu Q] + \mathrm{O}(Q^3)\,. \tag{65b}$$

To deduce the correct soft-pion skyrmion Lagrangian we will neglect terms of order $\mathrm{O}(Q^3)$ along with hard-pion processes. The latter are terms quadratic in Q^2 that involve large momentum transfer. With this in mind, the substitution of eqs. (64–65) into (38) yields to leading order

$$\begin{aligned}\mathcal{L}_{\text{soft}} = \mathcal{L}_S + \mathcal{L}_\pi + \frac{1}{2f_\pi}\partial_\mu \boldsymbol{\pi}\cdot \mathrm{TR}\Big[\mathrm{i}\boldsymbol{\tau}\cdot\Big(-\frac{f_\pi^2}{2}L_\mu + \varepsilon^2[L_\nu,[L_\mu, L_\nu]] + (L\to R)\Big)\Big]\\ - \frac{1}{4f_\pi^2}(\boldsymbol{\pi}\times\partial_\mu\boldsymbol{\pi})\cdot\mathrm{TR}\Big[\mathrm{i}\boldsymbol{\tau}\cdot\Big(-\frac{f_\pi^2}{2}L_\mu + \varepsilon^2[L_\nu,[L_\mu, L_\nu]] - (L\to R)\Big)\Big]\,,\end{aligned} \tag{66}$$

where $\mathcal{L}_\pi$ is the free Lagrangian for massless pions, i.e. $\mathcal{L}_\pi = \frac{1}{2}(\partial_\mu\boldsymbol{\pi})^2$, $L_\mu = U_0^+\,\partial_\mu U_0$, etc. The Noether curents associated to the global $\mathrm{SU(2)_V}\otimes\mathrm{SU(2)_A}$ invariance of (38) are given by (for details see appendix D)

$$\boldsymbol{V}_\mu = \mathrm{TR}\Big[\mathrm{i}\boldsymbol{\tau}\Big(-\frac{f_\pi^2}{2}L_\mu + \varepsilon^2[L_\nu,[L_\mu, L_\nu]] + (L\to R)\Big)\Big]\,, \tag{67a}$$

$$\boldsymbol{A}_\mu = \mathrm{TR}\Big[\mathrm{i}\boldsymbol{\tau}\Big(-\frac{f_\pi^2}{2}L_\mu + \varepsilon^2[L_\nu,[L_\mu, L_\nu]] - (L\to R)\Big)\Big]\,. \tag{67b}$$

In terms of which the soft-pion Lagrangian (66) reads

$$\mathcal{L}_{\text{soft}} = \mathcal{L}_{\text{s}} + \tfrac{1}{2}(\partial_\mu\boldsymbol{\pi})(\partial^\mu\boldsymbol{\pi}) + \frac{1}{2f_\pi}(\partial_\mu\boldsymbol{\pi})\cdot\boldsymbol{A}^\mu + \frac{1}{4f_\pi^2}(\partial_\mu\boldsymbol{\pi}\times\boldsymbol{\pi})\cdot\boldsymbol{V}^\mu\,. \tag{68}$$

As noted by Schnitzer [23], eq. (68) has the appropriate form for a soft-pion skyrmion Lagrangian. Indeed in the chiral limit ($m_\pi = 0$), Weinberg's soft-pion nucleon Lagrangian reads [24]

$$\begin{aligned}\mathcal{L}_{\mathrm{W}} = \mathcal{L}_{\mathrm{N}} + \tfrac{1}{2}(\partial_\mu\boldsymbol{\pi})(\partial^\mu\boldsymbol{\pi}) + \Big(\frac{g_{\pi\mathrm{NN}}}{2M_{\mathrm{N}}}\Big)(\partial_\mu\boldsymbol{\pi})\cdot(\bar{N}\gamma^\mu\gamma_5\boldsymbol{\tau}N)\\ + \Big(\frac{G_{\mathrm{V}}}{G_{\mathrm{A}}}\frac{g_{\pi\mathrm{NN}}}{2M_{\mathrm{N}}}\Big)^2(\partial_\mu\boldsymbol{\pi}\times\boldsymbol{\pi})\cdot(\bar{N}\gamma^\mu\boldsymbol{\tau}N) + \cdots\,,\end{aligned} \tag{69}$$

where N is the nucleon wavefunction, and $\mathcal{L}_{\mathrm{N}}$ the free nucleon Lagrangian. Comparing (68) with (69) gives

$$g_{\mathrm{A}} = G_{\mathrm{A}}/G_{\mathrm{V}} = g_{\pi\mathrm{NN}}f_\pi/M_{\mathrm{N}}\,, \tag{70}$$

which is the Goldberger–Treiman relation (11). As expected, the addition of the Skyrme term does not alter the soft-pion threshold theorems satisfied already by the non-linear σ-model. The Skyrme term modifies only the nucleon structure as shown in (67). We will show later on how one can recover part of Sakurai's vector meson dominance, by gauging the skyrmion in the presence of the low-energy QCD anomalies. For that we need to introduce and discuss the fundamentals of the Wess–Zumino term.

3. The Wess–Zumino term

3.1. QCD chiral anomalies and effective chiral models

Consider the three-flavor QCD Lagrangian (1) in the chiral limit, i.e. $m_u = m_d = m_s = 0$. In this case (1) is globally invariant under $U(3)_L \times U(3)_R$, and by Noether's theorem there are nine conserved vector and axial-vector currents at the classical level. Because of the Adler–Bell–Jackiw anomaly the $U(1)_A$ symmetry is explicitly broken at the quantum level. So aside from the anomaly, it is believed that chiral symmetry is spontaneously broken through

$$U(3)_L \otimes U(3)_R/U(1)_A \equiv SU(3)_L \otimes SU(3)_R \otimes U(1)_V \to SU(3)_V \otimes U(1)_V , \tag{71}$$

with the appearance of eight massless Goldstone bosons, the pseudoscalar octet mesons: π^0, $\pi^{\pm}$, η, K^0, $\bar{K}^0$, $K^{\pm}$. In the spirit of the large N_c limit, the dynamics of the massless pseudoscalar mesons is dictated by a non-linear σ-model Lagrangian such as (35) to leading order, where $U(x)$ is now an SU(3)-valued field of the form

$$U(\boldsymbol{x}, t) = \exp \mathrm{i}\left[\lambda^a \frac{\pi^a(\boldsymbol{x}, t)}{f_\pi} \right] . \tag{72}$$

In (72), the λ's are the ordinary Gell-Mann matrices, normalized such that $\mathrm{TR}[\lambda^a \lambda^b] = 2\delta^{ab}$. The explicit form for the pseudoscalar octet meson field is

$$\lambda^a \pi^a(x, t) \equiv \sqrt{2} \begin{bmatrix} \pi^0/\sqrt{2} + \eta/\sqrt{6} & \pi^+ & K^+ \\ \pi^- & -\pi^0/\sqrt{2} + \eta/\sqrt{6} & K^0 \\ K^- & \bar{K}^0 & -\sqrt{\tfrac{2}{3}}\eta \end{bmatrix} \tag{73}$$

with $\pi^{\pm} = (\pi_x \pm \mathrm{i}\pi_y)/\sqrt{2}$. Under $U(3)_L \times U(3)_R$, $U(x, t)$ transforms as follows:

$$\exp(\mathrm{i}Q_L) U(\boldsymbol{x}, t) \exp(-\mathrm{i}Q_R) ,$$

where $Q_{L,R}$ are now the fundamental generators of U(3). Under parity, $U(\boldsymbol{x}, t)$ transforms according to (55), underlying the pseudoscalar character of the octet mesons, i.e.

$$\hat{\pi}_{op} \pi^a(\boldsymbol{x}, t) \hat{\pi}_{op}^{-1} = -\pi^a(-x; t) , \tag{74}$$

where $\hat{\pi}_{op}$ is the parity operator.

Witten [25] observed that while the non-linear σ-model (35) is invariant under global $U(3)_L \times U(3)_R$ and even under the QCD parity (55), it exhibits two discrete symmetries which are redundant with QCD, namely,

$$\text{(i)}\quad U(\boldsymbol{x}, t) \to U(-\boldsymbol{x}; t) , \qquad \text{(ii)}\quad U(\boldsymbol{x}, t) \to U^+(\boldsymbol{x}; t) . \tag{75}$$

The latters forbid anomalous processes of the type

$$\mathrm{K}^+\mathrm{K}^- \to \pi^+\pi^0\pi^- , \qquad \eta\pi^0 \to \pi^+\pi^-\pi^0 , \tag{76}$$

in which an even number of pseudoscalar mesons decay into an odd number, and vice versa. The processes (76) are mediated by anomalies in QCD, and turn out to be kinematically excluded by the non-linear σ-model which of course, does not account for the anomalous Ward identities. As a remedy, Witten [25] proposed to modify the classical equations of motion in the non-linear σ-model by adding explicitly a $\mathrm{U}(3)_\mathrm{L} \otimes \mathrm{U}(3)_\mathrm{R}$ invariant that breaks (i) and (ii) separately, while preserving their combination, i.e. the QCD parity operation (55). To break explicitly (i) while maintaining Lorentz-invariance requires the totally anti-symmetric Levi–Cevita tensor

$$\tfrac{1}{2} f_\pi^2\, \partial^\mu L_\mu + \lambda \varepsilon^{\mu\nu\alpha\beta} L_\mu L_\nu L_\alpha L_\beta = 0 . \tag{77}$$

Under $\boldsymbol{x} \to -\boldsymbol{x}$, we have $\varepsilon^{\mu\nu\alpha\beta} \to -\varepsilon_{\mu\nu\alpha\beta}$, $\partial^\mu \to \partial_\mu$, and $L^\mu \to L_\mu$, hence

$$\tfrac{1}{2} f_\pi^2 \partial_\mu L^\mu - \lambda \varepsilon_{\mu\nu\alpha\beta} L^\mu L^\nu L^\alpha L^\beta = 0 . \tag{78}$$

Under $\pi^a \to -\pi^a$, i.e. $U(x) \to U^+(x)$, $L_\mu \to R_\mu = -UL_\mu U^+$. Since

$$\partial^\mu R_\mu + U\partial^\mu L_\mu U^+ = 0 , \tag{79}$$

we obtain

$$\tfrac{1}{2} f_\pi^2 \partial^\mu L_\mu + \lambda \varepsilon^{\mu\nu\alpha\beta} L_\mu L_\nu L_\alpha L_\beta = 0 . \tag{80}$$

In other words, the redundant symmetries are lifted while their combination (QCD parity) is not.

To proceed to a quantum description starting from the classical field equations (77) we need the corresponding action functional. The answer to this question is non-trivial, since the obvious candidate for the added term

$$\varepsilon^{\mu\nu\alpha\beta}\, \mathrm{TR}[L_\mu L_\nu L_\alpha L_\beta] \tag{81}$$

vanishes identically in (3 + 1) dimensions due to the cyclic property of the trace. In fact, as we will show next, the pertinent action functional involves the Wess–Zumino term [26] of current algebra,* and turns out to be non-local** in (3 + 1) dimensions.

3.2. Analogy with the U(1) monopole

The solution of the problem raised by Witten [25] is suggested by the solution of the much simpler problem of an electron of mass m and charge e moving on the surface of a unit sphere surrounding a Dirac magnetic monopole [27] of charge q.

First, consider the constrained motion of the electron with magnetic field $\boldsymbol{B} = 0$. The appropriate dynamics is described by the following Lagrangian

* In a canonical formulation, this discussion will follow through anomalous commutators.

** By this we mean that it cannot be written in a closed form in (3 + 1) dimensions. It is however local in (4 + 1) dimensions.

$$\mathcal{L} = \tfrac{1}{2} m\dot{\boldsymbol{r}}^2 + \lambda(\boldsymbol{r}^2 - 1)\,, \tag{82}$$

where λ is the Lagrange multiplier constraining the motion to the unit sphere. Varying $\mathcal{L}$ yields

$$m\ddot{\boldsymbol{r}} = 2\lambda \boldsymbol{r}\,, \tag{83a}$$

hence

$$\lambda = \frac{m}{2}\frac{\ddot{\boldsymbol{r}}\cdot\boldsymbol{r}}{\boldsymbol{r}^2} \equiv \frac{m}{2}\,\ddot{\boldsymbol{r}}\cdot\boldsymbol{r} = -\frac{m}{2}\,\dot{\boldsymbol{r}}^2\,, \tag{83b}$$

where we have used the constraint equation $\boldsymbol{r}^2 = 1$, and have differentiated this twice in order to get the last equality. In the presence of the magnetic field $\boldsymbol{B} = g\hat{r}/r^2 \equiv g\hat{r}$ due to the magnetic monopole, we have

$$\begin{aligned} m\ddot{r}_i + m(\dot{\boldsymbol{r}}^2)r_i &= e(\dot{\boldsymbol{r}} \times \boldsymbol{B})_i \\ &= eg\varepsilon_{ijk}\dot{r}_j r_k\,. \end{aligned} \tag{84}$$

the Lorentz force has been put in by hand into the constrained equation (83a). There is no obvious term whose variation equals the right-hand side, since $\varepsilon_{ijk}\dot{r}_i r_j r_k = 0$. (Note the analogous situation in (81).)

The solution to this problem is well known. One introduces a vector potential such that $\boldsymbol{B} = g\boldsymbol{\nabla} \times \boldsymbol{A}$. The action is then

$$S = S_0 + S_{\mathrm{A}}\,, \tag{85a}$$

where

$$S_0 = \int \mathrm{d}t\,(m\dot{\boldsymbol{r}}^2/2 + \lambda(\boldsymbol{r}^2 - 1))\,, \qquad S_{\mathrm{A}} = \int \mathrm{d}t\, eg\dot{\boldsymbol{r}}\boldsymbol{A} = eg\int \boldsymbol{A}\cdot\mathrm{d}\boldsymbol{r}\,. \tag{85b}$$

Although (85) yields (84) as the Euler–Lagrange equations at the classical level, it is not a well-defined expression. Indeed, since $\boldsymbol{\nabla}\cdot\boldsymbol{B} \neq 0$ while $\boldsymbol{B} = \boldsymbol{\nabla}\times\boldsymbol{A}$, it is clear that $\boldsymbol{A}$ is singular in R^3. In fact, as noted by Dirac [27], $\boldsymbol{A}$ has a string of singularities as illustrated in fig. 5a. Moreover, the action functional (85) is not gauge-invariant. Under a gauge transformation

$${}^{\phi}\boldsymbol{A} = \boldsymbol{A} - \boldsymbol{\nabla}\phi\,, \tag{86}$$

the action changes by end-point terms

$${}^{\phi}S = S - eg\int \mathrm{d}\boldsymbol{r}\cdot\boldsymbol{\nabla}\phi = S - eg\int \mathrm{d}\phi\,. \tag{87}$$

The fact that S is not gauge-invariant makes no difference in classical physics which involves only the equation of motion (84). (The latter is manifestly gauge-invariant.) But the action does enter into quantum mechanics, as can be seen when it is cast in Feynman form

Fig. 5. (a) Schematic description of a U(1) Dirac monopole. The wiggly line refers to the string of singularities in the vector potential. (b) Illustration of the constrained motion of a charged particle on S^2 with a Dirac monopole at the origin. By Stoke's theorem line integrals can be reduced to surface integrals, provided that the string in (a) lies outside the concerned surface (undashed area).

$$Z[T] = \mathrm{TR}[\mathrm{e}^{-\mathrm{i}HT}] = \int_{r(T)=r(0)} \mathrm{d}[r(t)] \exp\left\{\frac{\mathrm{i}}{\hbar} S[\boldsymbol{r}(t)]\right\} . \tag{88}$$

The requirement of gauge invariance in the quantum theory can still be met, provided the change in S is an integral multiple of $\hbar$. The action in (88) involves only closed paths, since $\boldsymbol{r}(T) = \boldsymbol{r}(0)$, as shown in fig. 5b. By Stoke's theorem, the line integral of the vector potential around γ can be converted into an integral of the magnetic field over the surface bounded by γ (call this surface the disc D_2^+ as shown in fig. 5b, for instance), provided that the integrand is non-singular. But a vector potential $\boldsymbol{A}$ which gives rise to a magnetic monopole field is necessarily singular because of the presence of a Dirac string, the position of which is gauge-dependent. If the integral is cast in the disc D_2^+, the string can be rendered harmless by being relegated to the other surface D_2^-, hence

$$eg \int_{\gamma} \boldsymbol{A} \cdot \mathrm{d}\boldsymbol{r} = e \int_{D_2^+} \mathrm{d}\boldsymbol{\Sigma} \cdot \boldsymbol{B} . \tag{89}$$

But we must get equivalent physics by putting the string in D_2^+, carrying out the integral over D_2^-. By doing so, a minus sign comes in because γ bounds D_2^+ in a right-hand sense, and D_2^- in the left-hand sense. Introducing either into a Feynman path integral expression must be equivalent, hence

$$\exp\left[\frac{ie}{\hbar}\left(\int_{D_2^+} \mathrm{d}\boldsymbol{\Sigma} \cdot \boldsymbol{B}\right)\right] = \exp\left[-\frac{ie}{\hbar}\left(\int_{D_2^-} \mathrm{d}\boldsymbol{\Sigma} \cdot \boldsymbol{B}\right)\right], \tag{90}$$

leading to

$$\exp\left[\frac{ie}{\hbar} \int_{D_2^+ \cup D_2^-} \mathrm{d}\boldsymbol{\Sigma} \cdot \boldsymbol{B}\right] = 1 , \tag{91}$$

thus,

$$4\pi eg/\hbar = 2\pi n , \qquad n \in Z , \tag{92}$$

giving the conventional Dirac quantization of the U(1) monopole, $eg = n/2$, where n is an integer.

The same quantization rule can be deduced by requiring that $Z[T]$ in (88) is gauge invariant, i.e. $\delta_\phi Z = 0$. This condition implies that $\exp(\mathrm{i}^\phi S/\hbar)$ is single-valued, otherwise $Z[T]$ will vanish by destructive interference while integrating over the space of all closed paths which is infinitely connected by virtue of (87). Notice that for closed paths [28]

$$^{\phi}S = S - eg\int \mathrm{d}\phi = S - 2\pi egB\,, \tag{93}$$

where B is the properly normalized Chern–Pontryagin index associated to the ray representation of the gauge group U(1). Since $\pi_1(\mathrm{U}(1)) \sim Z$, B is topologically quantized and one obtains the Dirac quantization rule (92) through the requirement of single-valuedness.

3.3. Topological quantization of the Wess–Zumino term

The analogy of the preceding example with the $\mathrm{SU(3)_L} \otimes \mathrm{SU(3)_R}$ non-linear σ-model is striking if we notice that for $\boldsymbol{B} = 0$, the constrained equation on S^2 is invariant under $\boldsymbol{r} \to -\boldsymbol{r}$ and $t \to -t$ separately. The additional Lorentz force created by the magnetic monopole preserves only the combination $\boldsymbol{r} \to -\boldsymbol{r}$ and $t \to -t$. The Lorentz force in (84) is the analogue of the anomaly term in (77), while the geometrical analogue of the one-dimensional closed path S^1 on S^2 is a four-dimensional quasi-sphere $\mathrm{S}^3 \times \mathrm{S}^1$on $\mathrm{S}^3 \times \mathrm{S}^2$.

To elucidate these statements, it is best to work in Euclidean space with a compactified time direction,* i.e. $\mathrm{R}^4 = \mathrm{R}^3 \times \mathrm{R}^1 \to \mathrm{R}^3 \times \mathrm{S}^1$. Finite-field configurations yield a compactification of R^3 into S^3, and endow space-time with the topology of a quasi-sphere, $\mathrm{S}^3 \times \mathrm{S}^1$. The latter can be thought of as the boundary of a five-dimensional manifold D_5

$$\mathrm{D}_5^+ = \mathrm{S}^3 \times \mathrm{S}^1 \times [0,1]\,, \qquad \mathrm{D}_5^- = \mathrm{S}^3 \times \mathrm{S}^1 \times [-1,0]\,, \tag{94}$$

where we have used an obvious decomposition of S^2. The SU(3) field $U(x)$ acts as a mapping from $\mathrm{S}^3 \times \mathrm{S}^1$ onto SU(3) whose group manifold is isomorphic to $\mathrm{S}^5 \times \mathrm{S}^3$ by Bott's theorem [29]. In analogy with the U(1) monopole where the action associated to the Lorentz force was a U(1) invariant on the boundaries $\mathrm{D}_2^{\pm}$, the action functional corresponding to the anomaly term in (77) should be sought as an $\mathrm{SU(3)_L} \times \mathrm{SU(3)_R}$ invariant on $\mathrm{D}_5^{\pm}$. To achieve this, the SU(3) map $U(x)$ from $\mathrm{S}^3 \times \mathrm{S}^1$ onto SU(3) should be extended** to a map $U(x;s)$ from $\mathrm{D}_5^{\pm}$ onto SU(3). Since

$$(\mathrm{S}^3 \times \mathrm{S}^2, \mathrm{S}^5 \times \mathrm{S}^3) \sim (\mathrm{S}^5, \mathrm{S}^5) \sim \pi_5(\mathrm{S}^5) \sim Z\,,$$

then by De-Rham's theorem there must exist a topologically invariant and closed 5-form ω_5^0 on S^5, such that

$$\int_{\mathrm{S}^5} \omega_5^0 = \int_{\mathrm{S}^5} \mathrm{d}^5x\, Q_5^0 = 2\pi\,,$$

Q_5^0 is just the Chern–Pontryagin density associated to $\pi_5(\mathrm{S}^5) \sim Z$. To construct an explicit form of the pertinent isomorphism: $\pi^5(\mathrm{S}^5) \to Z$, we can use a straightforward generalization of the pedestrian

* This condition is reminiscent of $r(T) = r(0)$ in the functional form of the U(1) case.

** This extension is legitimate since $\pi_1(\mathrm{SU}(3)) = 0$.

construction that led to eq. (30). In particular, we have*

$$Q_5^0 = \frac{-\mathrm{i}}{240\pi^2}\,\varepsilon^{0\mu\alpha\beta\gamma\delta}\,\mathrm{TR}[L_\mu L_\alpha L_\beta L_\gamma L_\delta]\,. \tag{95}$$

Its corresponding closed 5-form ω_5^0 is exact. Indeed, if we define the 1-form $\alpha = L_\mu\, \mathrm{d}x^\mu$, then

$$\omega_5^0 = \frac{-\mathrm{i}}{240\pi^2}\,\mathrm{TR}[\alpha^5]\,, \tag{96}$$

where the wedge product has been omitted for convenience. If we notice that $\mathrm{d}\alpha + \alpha^2 = 0$, then ω_5^0 is closed since

$$\mathrm{d}\omega_5^0 = \frac{-\mathrm{i}}{48\pi^2}\,\mathrm{TR}[\mathrm{d}\alpha^5] = \frac{+\mathrm{i}}{48\pi^2}\,\mathrm{TR}[\alpha^6] = 0\,,$$

hence, locally exact by Poincaré's lemma.

In analogy with the monopole action (89), the Wess-Zumino Lagrangian associated to the anomaly term in (77) can be cast in the form

$$\Gamma_{\mathrm{WZ}} = +\lambda \int_{\mathrm{D}_5^+} \omega_5^0 = -\lambda \int_{\mathrm{D}_5^-} \omega_5^0 \qquad \lambda \in \mathrm{R}\,, \tag{97}$$

where $\mathrm{D}_5^\pm$ are the complementary discs defined in (94). To summarize:

(i) Γ_{WZ} is $\mathrm{SU}(3)_\mathrm{L} \otimes \mathrm{SU}(3)_\mathrm{R}$ invariant.
(ii) Γ_{WZ} is topologically invariant, since ω_5^0 is closed.
(iii) Γ_{WZ} depends only on the space-time boundary $\partial\mathrm{D}_5 = \mathrm{S}^3 \times \mathrm{S}^1$, since ω_5^0 is locally exact.

In terms of (97), the modified non-linear σ-model action is

$$S_\pm = -\frac{f_\pi^2}{4}\int \mathrm{d}^4x\,\mathrm{TR}[L^\mu L_\mu] \pm \frac{(-\mathrm{i})\lambda}{240\pi^2}\int_{\mathrm{D}_5^\pm} \mathrm{d}^5x\,\varepsilon^{\mu\nu\beta\gamma}\,\mathrm{TR}[L_\mu L_\nu L_\alpha L_\beta L_\gamma]\,, \tag{98}$$

whose constrained variations yield the equation of motion for the left × right currents** on S^3

$$-\frac{f_\pi^2}{2}\,\partial^\mu L_\mu + \frac{(-\mathrm{i}\lambda)}{48\pi^2}\,\varepsilon^{\nu\alpha\beta\gamma} L_\nu L_\alpha L_\beta L_\gamma = 0\,. \tag{99}$$

in agreement with (77). Witten [25] has argued that (97) is the unique choice that yields (99).

The contour ambiguity in (98) can be resolved if one requires the generating functional, and hence

* The normalization in (95) is obtained by first using a polar parametrization of S^5 which yields $2\pi/5!A_5$, with $A_5 = \pi^3$ being the surface of S^5, and then making the usual substitution, i.e. $\phi^0 = 1$, $\partial_\mu\phi^k \sim -\mathrm{i}L_\mu^k$, $k = 1, 2, 3, 4, 5$ for any subset of SU(3).

** The sign ambiguity in (98) disappears from the equation of motion (99) after using Stoke's theorem.

the exponential factors $\exp(\mathrm{i}S_\pm)$, to be contour independent. This is fulfilled if and only if $\exp(\mathrm{i}S_+) = \exp(\mathrm{i}S_-)$, i.e.

$$\lambda\Big(\int_{D_5^+} \omega_5^0 + \int_{D_5^-} \omega_5^0\Big) = \lambda \int_{D_5^+ \cup D_5^-} \omega_5^0 = \lambda \int_{S^3\times S^2} \omega_5^0 \equiv 2\pi\lambda \equiv 2\pi n\,, \qquad \lambda = n = \text{integer}\,. \tag{100}$$

This shows that λ is topologically quantized. Equation (100) is the analogue of the Dirac quantization condition (92). When analyzed in the context of QCD, this quantization is of fundamental relevance to the skyrmion.

Finally, notice that (98) can be expressed in a non-local form in ordinary space-time. Indeed, to leading order in the pseudoscalar fields $\phi(x) = \lambda^a\pi^a/f_\pi$, the Wess–Zumino term in (98) becomes

$$\begin{aligned}\Gamma_{\mathrm{WZ},\pm} &= \pm\frac{n}{240\pi^2}\int_{D_5^\pm} \mathrm{d}x^5\ \varepsilon^{\mu\nu\alpha\beta\gamma}\mathrm{TR}[\partial_\mu\phi\,\partial_\nu\phi\,\partial_\alpha\phi\,\partial_\beta\phi\,\partial_\gamma\phi] + \cdots \\ &= +\frac{n}{240\pi^2}\int_{\partial D_5 = S^3\times S^1} \mathrm{d}\Sigma_\mu\,\varepsilon^{\mu\nu\alpha\beta\gamma}\,\mathrm{TR}[\phi\,\partial_\nu\phi\,\partial_\alpha\phi\,\partial_\beta\phi\,\partial_\gamma\phi] + \cdots\,,\end{aligned} \tag{101}$$

where we have used Stoke's theorem. The expression (101) was originally investigated by Wess and Zumino [26] in the context of effective Lagrangians.

3.4. *The Wess–Zumino term in the context of QCD*

The phenomenological implications of the topological quantization of the Wess–Zumino term are of immediate relevance to low-energy observables such as $\pi^0 \to 2\gamma$, $K^+K^- \to \pi^+\pi^0\pi^-$, etc. At low energy, it provides the canonical link between QCD and chiral effective descriptions based on the non-linear σ-model. To illustrate these points and emphasize the phenomenological aspects of the preceding derivation, we will discuss anomalous electromagnetic interactions of pseudoscalar octet mesons as described by the gauged version of (98).

Let the electric charge matrix Q of SU(3) quarks, i.e.

$$Q = \mathrm{diag}(\tfrac{2}{3}, -\tfrac{1}{3}, -\tfrac{1}{3})\,, \tag{102}$$

be the generator of the unbroken $U(1)_v$ group that leaves (98) invariant under rigid rotations. To describe electromagnetic interactions it is necessary to promote $U(1)_v$ global to $U_v(1)$ local. The canonical pattern is through minimal substitution, i.e. $\partial_\mu \to \partial_\mu - \mathrm{i}A_\mu$, where the $U(1)_v$ gauge field transforms according to $(\varepsilon \to \varepsilon(x))$

$${}^{\varepsilon}A_\mu = A_\mu - \mathrm{i}\partial_\mu\varepsilon(x) \qquad (e = 1)\,. \tag{103}$$

While this works for the first term in (98), it does not for the Wess–Zumino term since the latter is not a local expression in four-dimensional space-time. A systematic way to gauge (98) is by trial and error [25, 31], of which the substitution $\partial_\mu \to \partial_\mu - \mathrm{i}A_\mu$ is a practical recipe when the requirement of locality is met. To achieve this it is convenient to work in terms of differential forms to get rid of the cumbersome tensorial structure of the expressions involved. For that, define the 1-forms

$$\alpha = U^{+}\partial_{\mu} U \, dx^{\mu} = U^{+} \, dU \equiv -dU^{+}U = -U\beta U^{+} ,$$
$$\beta = U\partial_{\mu} U^{+} \, dx^{\mu} = U \, dU^{+} = -dUU^{+} = -U^{+}\alpha U . \quad (104)$$

It is easy to show using Poincaré lemma ($d^2 = 0$) that the related 2-forms α^2 and β^2 are exact and satisfy*

$$d\alpha + \alpha^2 = 0 \quad \text{and} \quad d\beta + \beta^2 = 0 . \quad (105)$$

Notice that under $U(x) \rightarrow \exp[i\varepsilon(x)Q] \, U \, \exp[-i\varepsilon(x)Q]$, α and β transforms according to

$$\delta_Q \alpha = i\varepsilon[Q, \alpha] + i \, d\varepsilon \, U^{+}QU - i \, d\varepsilon Q , \qquad \delta_Q \beta = i\varepsilon[Q, \beta] + i \, d\varepsilon \, UQU^{+} - i \, d\varepsilon \, Q , \quad (106)$$

since $\delta_Q \, d = d\delta_Q$. In terms of (104), the Wess–Zumino term simplifies into [31]

$$\Gamma_{WZ} = N \int_{D_5^+} TR[\alpha^5] = N \int_{D_5^+} TR[\beta^5] , \quad (107)$$

with $N = -in/240\pi^2$. Under a U(1) gauge transformation we have*

$$\begin{aligned} \delta_Q \Gamma_{WZ} &= 5N \int_{D_5^+} TR[\delta_Q \alpha \alpha^4] \\ &= 5N \int_{D_5^+} d\varepsilon \, TR[-iQ\alpha^4 + iQU\alpha^4 U^{+}] \\ &= -i5N \int_{D_5^+} d\varepsilon \, TR[Q(\alpha^4 - \beta^4)] , \end{aligned} \quad (108)$$

where we have used the anticommuting character of the 1-forms, and kept only the terms proportional to $d\varepsilon$ since they ultimately yield to the gauge-currents as in the standard method of Gell–Mann and Levy. Notice that (108) can be made local using Stoke's theorem, i.e.

$$\begin{aligned} \delta_Q \Gamma_{WZ} &= i5N \int_{D_5^+} d\varepsilon \, TR[Q \, d(\alpha^3 - \beta^3)] \\ &= -i5N \int_{D_5^+} d(d\varepsilon \, TR[Q(\alpha^3 - \beta^3)]) \\ &= -i5N \int_{M_4} d\varepsilon \, TR[Q(\alpha^3 - \beta^3)] \end{aligned} \quad (109)$$

* Again, we have omitted the wedge product $\wedge$ for convenience. recall that $\alpha^2 = \alpha \wedge \alpha$ is a skew-symmetric form of rank 2, etc.

where $\partial D_5^+ = M_4$ is the Euclidean space-time boundary with the proper compactification for solitons. To compensate for the variation (109), we can introduce a term linear in the gauge field such that

$$\Gamma^{(1)}_{WZ} = \Gamma_{WZ} - 5N \int_{M_4} A\, \mathrm{TR}[Q(\alpha^3 - \beta^3)] . \tag{110}$$

Since $\delta_Q A = -\mathrm{i}\, \mathrm{d}\varepsilon$, we obtain

$$\begin{aligned} \delta_Q \Gamma^{(1)}_{WZ} &= -5N \int_{M_4} A\, \mathrm{TR}[Q\delta_Q(\alpha^3 - \beta^3)] \\ &= -5\mathrm{i}N \int_{M_4} A\, \mathrm{d}\varepsilon\, \mathrm{TR}[2Q^2(\beta^2 - \alpha^2) - Q\, \mathrm{d}U\, Q\, \mathrm{d}U^+ + Q\, \mathrm{d}U^+\, Q\, \mathrm{d}U] \\ &= -10\mathrm{i}N \int_{M_4} A\, \mathrm{d}\varepsilon\, \mathrm{TR}[Q^2(\beta^2 - \alpha^2) - Q\, \mathrm{d}U\, Q\, \mathrm{d}U^+] . \end{aligned} \tag{111}$$

Using (105) together with the following identity [31]:

$$Q\, \mathrm{d}U\, Q\, \mathrm{d}U^+ = \mathrm{d}(aQUQ\, \mathrm{d}U^+ - bQ\, \mathrm{d}U\, QU^+) \qquad a + b = 1 , \tag{112}$$

We conclude that for the parity invariant choice [31] $a = b = \frac{1}{2}$, we have

$$\begin{aligned} \delta_Q \Gamma^{(1)}_{WZ} &= +10\mathrm{i}N \int_{M_4} \mathrm{d}\varepsilon\, A\, \mathrm{d}\, \mathrm{TR}[Q^2(\alpha - \beta) + \tfrac{1}{2}(Q\, \mathrm{d}U\, QU^+ - QUQ\, \mathrm{d}U^+)] \\ &= +10\mathrm{i}N \int_{M_4} \mathrm{d}\varepsilon\, \mathrm{d}A\, \mathrm{TR}[Q^2(\alpha - \beta) + \tfrac{1}{2}(Q\, \mathrm{d}U\, QU^+ - QUQ\, \mathrm{d}U^+)] . \end{aligned} \tag{113}$$

To compensate for the gauge variation of $\Gamma^{(1)}_{WZ}$ we proceed as for Γ_{WZ} and define

$$\Gamma^{(2)}_{WZ} = \Gamma^{(1)}_{WZ} + 10N \int_{M_4} A\, \mathrm{d}A\, \mathrm{TR}[Q^2(\alpha - \beta) + \tfrac{1}{2}(Q\, \mathrm{d}U\, QU^+ - QUQ\, \mathrm{d}U^+)] , \tag{114}$$

which is finally U(1) gauge-invariant, i.e.

$$\delta\Gamma^{(2)}_{WZ} = 10N \int_{M_4} A\, \mathrm{d}A\, (\mathrm{i}\, \mathrm{d}\varepsilon)\, \mathrm{TR}[Q^2(U^+QU - UQU^+) + (Q^2UQU^+ - QUQ^2U^+)] = 0 . \tag{115}$$

So, in the presence of photons the gauged Wess–Zumino action reads [31]

$$\begin{aligned} \Gamma_{WZ}[U, A] = N \int_{D_5^+} \mathrm{TR}[\alpha^5] - 5N \int_{M_4} A\, \mathrm{TR}[Q(\alpha^3 - \beta^3)] \\ + 10N \int_{M_4} A\, \mathrm{d}A\, \mathrm{TR}[Q^2(\alpha - \beta) + \tfrac{1}{2}(Q\, \mathrm{d}U\, QU^+ - QUQ\, \mathrm{d}U^+)] , \end{aligned} \tag{116}$$

where $N = -in/240\pi^2$. The Noether current associated to a rigid $U(1)_V$ rotation can be read off (116) in the form

$$J^\mu = \frac{in}{48\pi^2}\,\varepsilon^{\mu\nu\alpha\beta}\,\mathrm{TR}[Q(L_\nu L_\alpha L_\beta - R_\nu R_\alpha R_\beta)]\,. \tag{117}$$

The expression (116) is expected to contain information about all QCD anomalies involving photons and pseudoscalar octet mesons at low energy, such as $\pi^0 \to 2\gamma$, $\gamma \to \pi^+\pi^0\pi^-$, etc. To see this, let us expand the last term in (116) in powers of $1/f_\pi$. Since

$$\alpha = U^+\,\mathrm{d}U = \frac{\mathrm{i}}{f_\pi}\,\mathrm{d}\pi + \mathrm{O}\Big(\frac{1}{f_\pi^2}\Big)\,, \qquad \beta = U\,\mathrm{d}U^+ = -\frac{\mathrm{i}}{f_\pi}\,\mathrm{d}\pi = \mathrm{O}\Big(\frac{1}{f_\pi^2}\Big)\,, \tag{118}$$

and

$$\int_{M_4} A\,\mathrm{d}A\,\mathrm{TR}[Q^2(\alpha - \beta)] = +\frac{2\mathrm{i}}{f_\pi}\int_{M_4} \mathrm{d}A\,\mathrm{d}A\,\mathrm{TR}[Q^2\pi] + \mathrm{O}\Big(\frac{1}{f_\pi^2}\Big)\,, \tag{119a}$$

$$\int_{M_4} A\,\mathrm{d}A\,\mathrm{TR}[\tfrac{1}{2}Q\,\mathrm{d}U\,QU^+] = +\frac{\mathrm{i}}{2f_\pi}\int_{M_4} \mathrm{d}A\,\mathrm{d}A\,\mathrm{TR}[Q^2\pi] + \mathrm{O}\Big(\frac{1}{f_\pi^2}\Big)\,, \tag{119b}$$

$$\int_{M_4} A\,\mathrm{d}A\,\mathrm{TR}[\tfrac{1}{2}QUQ\,\mathrm{d}U^+] = -\frac{\mathrm{i}}{2f_\pi^2}\int_{M_4} \mathrm{d}A\,\mathrm{d}A\,\mathrm{TR}[Q^2\pi] + \mathrm{O}\Big(\frac{1}{f_\pi^2}\Big)\,, \tag{119c}$$

one can extract from (116) the $\pi^0 \to 2\gamma$ vertex in the form [25] ($e = 1$)

$$\begin{aligned} v_{\pi^0\to 2\gamma} &= -\frac{n}{24\pi^2 f_\pi}\,\pi^0\varepsilon^{\mu\nu\alpha\beta}\partial_\mu A_\nu \partial_\alpha A_\beta \\ &= -\frac{n}{96\pi^2 f_\pi}\,\pi^0\varepsilon^{\mu\nu\alpha\beta}F_{\mu\nu}F_{\alpha\beta}\,, \end{aligned} \tag{120}$$

which coincides precisely with the $\pi^0 \to 2\gamma$ vertex obtained from the triangle-anomaly in the quark language as shown in fig. 6a, if we were to identify n with the number of colors in QCD, i.e. $n \equiv N_c$. The second term in (116), gives to leading order

$$\int_{M_4} A\,\mathrm{TR}[Q(\alpha^3 - \beta^3)] = 2\Big(\frac{\mathrm{i}}{f_\pi}\Big)^3 \int_{M_4} A\,\mathrm{TR}[Q(\mathrm{d}\pi)^3] + \mathrm{O}\Big(\frac{1}{f_\pi^4}\Big)\,. \tag{121}$$

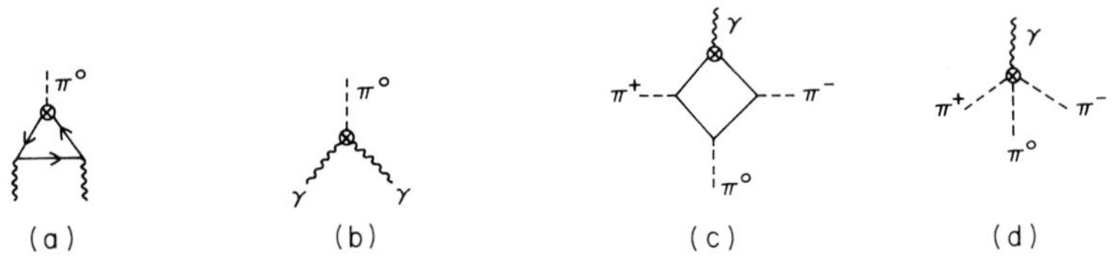

Fig. 6. (a), (c) Illustrations of the triangle and box anomaly respectively, with constituent chiral fermions looping around; (b) and (d) are their analogue in effective chiral models.

The contribution to the trace in (121) due to $\pi^0, \pi^{\pm}$ mesons is

$$\begin{aligned}\mathrm{TR}[Q(\mathrm{d}\pi)^3]_{\pi^0,\pi^\pm} &= \mathrm{d}\pi^0\,\mathrm{d}\pi^+\,\mathrm{d}\pi^-\,\mathrm{TR}[Q\lambda_0\lambda_+\lambda_-] + \mathrm{d}\pi^0\,\mathrm{d}\pi^-\,\mathrm{d}\pi^+\,\mathrm{TR}[Q\lambda_0\lambda_-\lambda_+]\\ &\quad + \mathrm{d}\pi^+\,\mathrm{d}\pi^0\,\mathrm{d}\pi^-\,\mathrm{TR}[Q\lambda_+\lambda_0\lambda_-] + \mathrm{d}\pi^+\,\mathrm{d}\pi^-\,\mathrm{d}\pi^0\,\mathrm{TR}[Q\lambda_+\lambda_-\lambda_0]\\ &\quad + \mathrm{d}\pi^-\,\mathrm{d}\pi^0\,\mathrm{d}\pi^+\,\mathrm{TR}[Q\lambda_-\lambda_0\lambda_+] + \mathrm{d}\pi^-\,\mathrm{d}\pi^+\,\mathrm{d}\pi^0\,\mathrm{TR}[Q\lambda_-\lambda_+\lambda_0]\\ &= +2\mathrm{i}\,\mathrm{d}\pi^+\,\mathrm{d}\pi^-\,\mathrm{d}\pi^0\,, \end{aligned} \tag{122}$$

where we have used the antisymmetry of the 1-forms, i.e. $\mathrm{d}\pi^+\,\mathrm{d}\pi^- + \mathrm{d}\pi^-\,\mathrm{d}\pi^+ = 0$, etc. In terms of (121) and (122) the $\gamma\to\pi^+\pi^0\pi^-$ vertex reads [25] ($e = 1$)

$$v_{\gamma\to 3\pi} = \frac{in}{12\pi^2 f_\pi^3}\,\varepsilon^{\mu\nu\alpha\beta} A_\mu \partial_\nu \pi^+ \partial_\alpha \pi^- \partial_\beta \pi^0\,, \tag{123}$$

which coincides with the VAAA vertex obtained from the box anomaly in the quark language as shown in fig. 6b, with again $n = N_c$. This extraordinary relationship between n and N_c is deep and important.

3.5. Anomalous baryon current

Skyrme's remarkable conjecture that skyrmions are low-energy baryons can be by now rigorously phrased in the light of Witten's reanalysis of the Wess–Zumino term. Although there were several previous arguments [32] demonstrating this relation, Witten's proof is simple and compelling.

The baryon current in the context of current algebra as described by (98) can be obtained as the conserved Noether current associated to a global $U(1)_V$ invariance. At the quark level, the baryon operator Q is given by

$$Q = 1/N_c\,\mathrm{diag}(1,1,1)\,. \tag{124}$$

The gauging of $U(1)_V$ in this case is entirely analogous to the gauging of $U(1)_{em}$ discussed above. If we denote by b_μ the corresponding U(1) gauge field and notice that (124) is diagonal in flavor space, then the contribution of the Wess–Zumino terms follows from (128),

$$\Gamma_{WZ}[U,b] = N\int_{D_5^+} \mathrm{TR}[\alpha^5] - 5\,\frac{N}{N_c}\int_{M_4} b\,\mathrm{TR}[\alpha^3 - \beta^3]\,. \tag{125}$$

The baryon current B_μ can be read off the Noether coupling $-b_\mu B^\mu$ in (125),

$$B^\mu = \left(\frac{n}{N_c}\right)\frac{\mathrm{i}}{48\pi^2}\,\varepsilon^{\mu\nu\alpha\beta}\,\mathrm{TR}[L_\nu L_\alpha L_\beta - R_\nu R_\alpha R_\beta]\,. \tag{126}$$

Assuming that $n = N_c$ as suggested by the chiral anomalies in QCD, we obtain the conserved Noether current in the form*

* The entire contribution to the baryon current is due to the Wess–Zumino term in chiral models. Since the latter follows from the QCD anomalies, the origin of the baryon current is anomalous in chiral effective descriptions.

$$B^{\mu} = \frac{\mathrm{i}}{24\pi^2}\, \varepsilon^{\mu\nu\alpha\beta}\, \mathrm{TR}[L_\nu L_\alpha L_\beta]\,, \tag{127}$$

which coincides precisely with the topological current of the non-linear σ-model (32). This shows that the $U(1)_V$ Noether current associated to the baryon charge in current algebra language, is the topological current associated to the pertinent Chern–Pontryagin index. This result was first established by Balachandran et al. [32] using methods of Goldstone and Wilczek [33].

3.6. *Spin statistics of skyrmions*

The question on how spin one-half fermions arise from field theories involving exclusively integer-spin mesons is one of the most intriguing concepts behind the proposal to describe spinor fields out of solitonic configurations. This question has been originally addressed by Finkelstein and Rubinstein [5]. Recently Witten [25] has discussed a more appealing procedure based on the Wess–Zumino term. Some subtle ambiguities raised by this proof have been recently discussed by Balachadran et al. [34] and Jaffe et al. [35] following arguments first used by Guadagnini [36].

To illustrate Witten's [25] proof, one has to single out time explicitly. For that consider an SU(3) skyrmion in (3 + 1)-dimensional space-time with topology $S^3 \times S^1$. To leading order in $\hbar$, the vacuum-to-vacuum amplitude of a skyrmion at rest is given by

$$\langle S(T) \,|\, S(0)\rangle = \exp(-\mathrm{i}TM/\hbar)(1 + O(\hbar))\,, \tag{128}$$

where M is the skyrmion mass. Now, rotate the skyrmion over 2π along S^1, infinitely slowly. Quantum mechanics tells us that the skyrmion wavefunction acquires a phase factor $\exp(-\mathrm{i}2\pi J/\hbar)$, where J is the skyrmion spin. In other words,

$$\langle S(T) \,|\, S(0)\rangle_{2\pi} = \exp(-\mathrm{i}TH/\hbar)\exp(\mathrm{i}2\pi J/\hbar)\,(1 + O(\hbar))\,, \tag{129}$$

which determines J up to an integer. The phase factors in (129) are given by the classical action of an adiabatically rotated skyrmion. To determine the latter, consider an SU(2) hedgehog $U_H(\boldsymbol{x})$ embedded in SU(3), i.e.

$$U(\boldsymbol{x}) = \begin{pmatrix} U_H(\boldsymbol{x}) & 0 \\ 0 & 1 \end{pmatrix}, \tag{130}$$

where $U_H(\boldsymbol{x})$ is given by (51). Because of the hedgehog character of $U_H(x)$, rotations are equivalent to isorotations,

$$U(\boldsymbol{x}, t) = \exp(\mathrm{i}\lambda_3 t/2)\, U(\boldsymbol{x}) \exp(-\mathrm{i}\lambda_3 t/2) = U(R_3(t)\boldsymbol{x})\,. \tag{131}$$

Explicitly, $U(\boldsymbol{x}, t)$ reads

$$U(\boldsymbol{x};t)=\begin{pmatrix}e^{it/2}&0&0\\0&e^{-it/2}&0\\0&0&0\end{pmatrix}U(\boldsymbol{x})\begin{pmatrix}e^{-it/2}&0&0\\0&e^{it/2}&0\\0&0&1\end{pmatrix}$$

$$=\begin{pmatrix}1&0&0\\0&e^{-it}&0\\0&0&e^{-it/2}\end{pmatrix}U(\boldsymbol{x})\begin{pmatrix}1&0&0\\0&e^{it}&0\\0&0&e^{it/2}\end{pmatrix}$$

$$=\begin{pmatrix}1&0&0\\0&e^{-it}&0\\0&0&e^{it}\end{pmatrix}U(\boldsymbol{x})\begin{pmatrix}1&0&0\\0&e^{it}&0\\0&0&e^{-it}\end{pmatrix}, \tag{132}$$

where we have used the fact that (130) is block-diagonal. In the adiabatic limit, only the Wess–Zumino term feels the rotation in time since it involves a single time derivative through $L_0 = U^+\partial_0 U$. Any other combination of L's involves at least two time derivatives, and yields a higher order correction in $\hbar$, hence

$$S_{\rm cl} = -M_s T + \left(\frac{-iN_c}{240\pi^2}\right)\int_{D_5^+} {\rm TR}[\alpha^5]\,. \tag{133}$$

Since ${\rm TR}[\alpha^5]$ is a closed form on D_5^+, only the global aspects of the integral in (132) are relevant. It is therefore sufficient to evaluate it using normalized topological manifolds. For this, extend $U(x,t)$ to $U(x,t,\rho)$ defined on the five-dimensional manifold D_5^+ as given by (94). Since (132) is topologically invariant, we can choose any regular and continuous extension of $U(x,t)$ to D_5^+ with the only constraint that on the space-time boundary $U(x,t,\rho=1) = U(x,t)$. From appendix F we have

$$\frac{i}{240\pi^2}\int_{D_5^+} {\rm TR}[\alpha^5] = +\frac{i}{240\pi^2}\int_{D_5^-} {\rm TR}[\alpha^5] = \pi\,,$$

in terms of which the vacuum-to-vacuum amplitude for the adiabatically rotated SU(3) skyrmion is

$$\langle S(T)\,|\,S(0)\rangle_{2\pi} = \exp(-iMT/\hbar)\exp(iN_c\pi)(1+O(\hbar))\,. \tag{134}$$

Comparison with (129) shows that for the three-flavor case considered, the skyrmion is a fermion if N_c is odd, and a boson otherwise. The generalization to an arbitrary number of flavors ($N_f > 3$) is straightforward.

The two-flavor case is slightly more subtle since the Wess–Zumino term vanishes identically because $\pi^5({\rm SU}(2)) = 0$. However the spin statistics of SU(2) skyrmions can be clarified using general arguments. For that, consider the process in which (i) a skyrmion–anti-skyrmion pair is created at $t=-\infty$, (ii) the pair is separated at large distance, (iii) the skyrmion is adiabatically rotated through 2π in the pair, and finally (iv) the pair is annihilated at $t=+\infty$, as illustrated in fig. 7b. The SU(2) field associated to this process satisfies

$$U(|x|\to\infty) = \text{vacuum} = \mathbb{1} \tag{135}$$

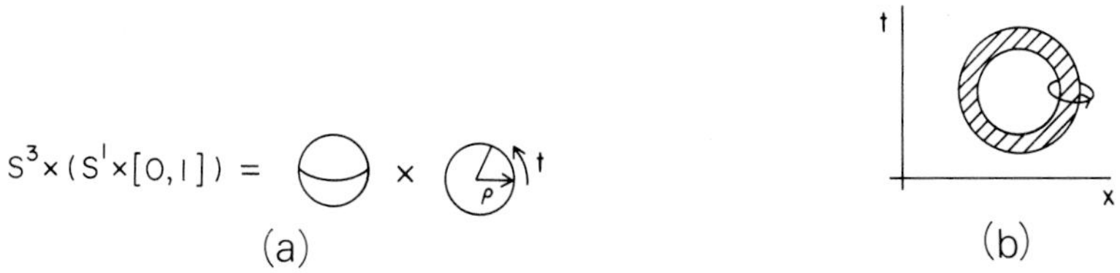

Fig. 7. (a) Simple illustration of the five-dimensional manifold used to rewrite the Wess–Zumino term in a closed form. S^3 is the compactified space, S^1 is the periodic Minkowski time and [0, 1] is the parameter space that characterizes the homotopy extension. (b) Intuitive demonstration that $\pi_4(SU(2)) = Z_2$.

where $x \equiv (t; \boldsymbol{x})$. Equation (135) corresponds to a compactification of R^4 into S^4. As such, $U(x)$ describes the set of Hopf maps from $S^4 \to SU(2)$, which are characterized by two homotopy classes since,

$$\pi_4(SU(2)) \sim \pi_4(S^3) \sim Z_2 \,, \tag{136}$$

where $Z_2 = \{+1; -1\}$ is the center of SU(2).* In other words, the adiabatically rotated pair can be characterized by either a trivial map (+1), in which case the constituents of the pair are bosons, or by a non-trivial map (−1), in which case the constituents of the pair are fermions.

So far, SU(2) skyrmions can be either fermions or bosons for a given N_c. To resolve the ambiguity, one can embed the SU(2) skyrmion into SU(3), and use the fact that the Wess–Zumino term does not vanish in this case. Using arguments similar to Witten [25], one can conclude that the spin of SU(2) skyrmions is half-integer if N_c is odd, and integer otherwise. The consecutive quantum ambiguities raised by the embedding can be avoided by restricting the discussion to the Hilbert sector of good triality [34, 35].

The above arguments showing that the skyrmion can be quantized as a fermion are somewhat formal. This matter should be clearer in section 7 where the nucleon wavefunctions are explicitly constructed (see eq. (301) and following).

4. Gauged Wess–Zumino term and non-Abelian chiral anomalies

Anomalies are playing an increasingly important role in our present understanding of quantum field theories with chiral fermions. Their perturbative discovery a decade ago by Adler, Bell and Jackiw [8] through the Abelian anomaly has provided the key explanation of the anomalous $\pi^0 \to 2\gamma$ decay in the context of chiral models with fermions. Since then, they have been signaled in various domains, ranging from solid state physics to supergravity. Their importance can hardly be exaggerated. They play a key role in the renormalizability condition in gauge theories, and the unitarity of the S-matrix; they are at the basis of the GIM mechanism [39] in lepto-quark families; they are at the origin of the fermion

* Recently, Wilczek and Zee [38] have investigated the spin statistics of static solitons in the SO(3) non-linear σ-model in (2 + 1) dimensions. They have found that while non-trivial maps are characterized by $\pi_2(SO(3)) \sim \pi_2(SU(2)/Z_2)) \sim Z$, their statistics is described by $\pi_3(SO(3)) \sim Z$ (Hopf mapping). As a result, the soliton wavefunction is found to be multivalued with periodicity π/Z (any number!). The latter is the first signal of an anomalous statistic.

number violation in the Callan–Rubakov effect [40]; they account for the missing baryon number in chiral bag models [41, 42]; etc.

In general, anomalies arise in quantum field theories involving γ^5-coupling.* When a global chiral invariance is promoted to a local invariance, the resulting vector and axial-vector currents are conserved classically by Noether's theorem. At the quantum level one finds in general, that some of these conservation laws are spoiled by anomalies. The necessity to use gauge invariant rather than chiral invariant regulators to monitor the short-distance singularities of fermions interacting with gauge fields, destroys the starting left × right symmetry and leads to anomalous terms in the Ward identities for fermions. In a Pauli–Villars regularization this effect is tantamount to introducing massive off-shell "fermions" that break chiral symmetry, while in dimensional regularization it is inherent to the difficulty to extend γ_5 to higher dimensions. By now, several methods have been proposed and used to evaluate these anomalies, based on either perturbative [44] or non-perturbative schemes [45]. The most transparent method so far has been discussed by Nielsen and Ninomiya [46]. It describes the Adler–Bell–Jackiw anomaly as an asymmetric production of left × right Weyl fermions in the Dirac sea, interacting with gauge fields. The most powerful method to discuss anomalies in general has been recently devised by Stora [47], and Zumino et al. [48]. It is a geometrical formulation based on the topological character of the anomaly. It is this method that we seek to present in this section to relate the Wess–Zumino term to the non-Abelian chiral anomalies in current algebra effective Lagrangians. Our strategy is as follows: we will first discuss the physical mechanism behind the anomaly, using the Abelian axial anomaly as a probe. Then we will clarify the asymmetry in terms of the Atiyah–Patodi–Singer [49] index theorem. The latter opens the door for a geometrical interpretation of the Abelian anomaly as the second Chern characteristic class in even-dimensional space–time. The topological character of the Chern classes yields naturally the Chern–Simons secondary classes in odd-dimensional space-time that are related to non-Abelian anomalies. The gauged Wess–Zumino action is none but the integrated non-Abelian anomaly subject to Bardeen's boundary condition [44].

4.1. *The infinite-hotel story*

The physical origin of the Abelian axial anomaly comes from a well-established quantum mechanical phenomenon associated with the chirality of fermions. To illustrate this, consider free massless fermions in $(1+1)$ dimensions,

$$\mathrm{i}\not{\partial}\psi(x) = 0\,. \tag{137}$$

Using light cone variables, i.e. $x_\pm = x_0 \pm x_1$ and $\not{\partial}_\pm = \partial_0 \pm \not{\partial}_1$, in the Weyl representation** with $\psi(x) = \binom{\psi_+}{\psi_-}$, yields

$$\mathrm{i}\not{\partial}_\pm \psi_\mp = 0\,. \tag{138}$$

This shows that the upper and lower component of $\psi(x)$ decouple, and that $\psi_+ \equiv \psi_+(x_+)$ (right-handed

* or self-dual stress-tensors with $\varepsilon_{\mu\nu\ldots}$-couplings in higher dimensions. This new class of anomalies has been discussed recently by Alvarez-Gaumé and Witten in the context of supergravity [43].

** In the Weyl representation

$$\gamma^0 = \gamma_0 = \sigma_x\,, \qquad \gamma^1 = -\gamma_1 = \mathrm{i}\sigma_y\,, \qquad \gamma^5 = \gamma_0\gamma^1 = -\sigma_z\,.$$

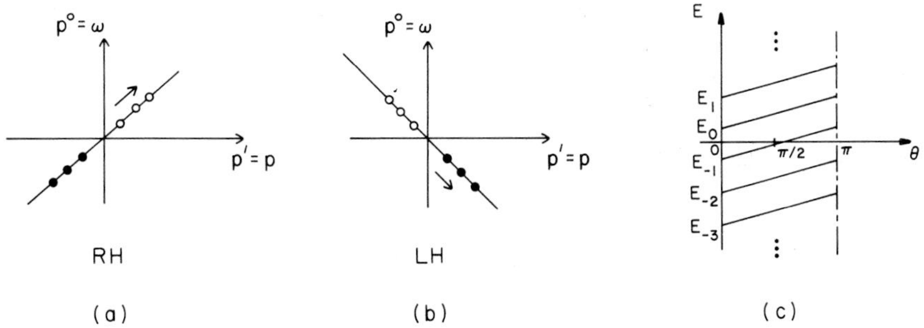

Fig. 8. (a) Illustrates the dispersion relation for right-handed (RH) Weyl fermions in the vacuum; (b) is the same as (a) for left-handed (LH) Weyl fermions in the vacuum; (a) and (b) are at the origin of the BJL anomaly; (c) illustrates the spectral flow in topological bag models that is at the origin of the baryon anomaly.

movers) while $\psi_- = \psi(x_-)$ (left-handed movers), with $\frac{1}{2}(1 \pm \gamma_5)\psi_\pm = \pm\psi_\pm$. Figure 8a illustrates the dispersive spectrum of free Weyl–fermions. At the classical level, left and right fermions decouple and their respective charges are separately conserved. If we denote by $Q_+(Q_-)$ the total number of RH(LH) fermions in the vacuum, by Q the fermion charge and by Q_5 the axial charge,* then

$$Q_+ = \int dx^1 \langle \psi_+^+ \psi_+ \rangle\,, \qquad Q_- = \int dx^1\, \langle \psi_-^+ \psi_- \rangle\,, \tag{139}$$

and

$$\dot{Q} = \dot{Q}_+ + \dot{Q}_- = 0\,, \qquad \dot{Q}_5 = \dot{Q}_- - \dot{Q}_+ = 0\,. \tag{140}$$

In the presence of an electromagnetic field, (137) reads**

$$(\mathrm{i}\not{\partial} - eA)\psi(x) = 0\,, \tag{141}$$

which still does not mix ψ_+ and ψ_-. The classical counterpart of (141) is given by Newton's equation for a charged particle moving along an electric field $E^1 = \dot{A}^1$ (temporal gauge, $A^0 = 0$) pointing in the x^1 direction,

$$dp^1/dt = -eF^{01} = +eF_{01} = eE\,, \tag{142}$$

the spectral flow associated to (141) is described by (142) as illustrated in fig. 8b.

Naively one would think that since there are "no particles" in the properly regularized vacuum state, and since LH and RH particles decouple through (141), the chiral charges $Q_\pm$ are separately conserved, and (140) still hold. Although this is true classically, it does not hold in a quantum description involving the Dirac sea (infinite hotel). Indeed, using figs. 8a and 8b one can easily see that as time goes on there

* Since the vacuum has infinite charge, these expressions are only meaningful when properly regularized and renormalized.

** In (1 + 1) dimensions the EM field strength $F_{\mu\nu}$ involves one independent field. Indeed, $B = 0$ and $E_1 = F_{01} = \partial_0 A_1 - \partial_1 A_0$ points in the x^1 direction.

is a production of right movers and anti-left movers. To characterize this production, consider a periodic universe of length L, i.e. $\psi(x^1 + L) = \psi(x^1)$, so that

$$L/2\pi = \text{right moving states/unit momentum}\,,$$
$$L/2\pi = \text{left moving states/unit momentum}\,. \tag{143}$$

Since $\dot{p}^1 = eE$, one concludes that the rate of right movers that pop out of the Dirac sea is: $\dot{Q}_+ = (L/2\pi)eE$, while the rate of left movers that disappear is: $Q_- = -(L/2\pi)eE$. Therefore,

$$\dot{Q} = \dot{Q}_+ + \dot{Q}_- = 0\,, \qquad \dot{Q}_5 = \dot{Q}_+ - \dot{Q}_- = (L/\pi)eE\,, \tag{144}$$

which clearly shows that although the fermion charge Q is conserved, the axial charge is not, leading to the conventional ABJ-axial anomaly.* Using (144) we have

$$Q_5 = \int_0^L \mathrm{d}x^1\, \langle \psi_-^+\psi_- - \psi_+^+\psi_+ \rangle = \int_0^L \mathrm{d}x^1\, \langle \bar{\psi}\gamma^0\gamma_5\psi \rangle = +\int_0^L \mathrm{d}x^1 \langle J_5^0 \rangle\,, \tag{145}$$

where J_5^0 is the time component of the axial current. Using the periodic character of space, we obtain

$$\dot{Q}_5 = \int_0^L \mathrm{d}x^1\, \langle \partial^0 J_5^0 \rangle \equiv \int_0^L \mathrm{d}x^1\, \langle \partial^\mu J_{5\mu} \rangle = \int_0^L \mathrm{d}x^1\, eE/\pi\,, \tag{146}$$

where we have used a trivial surface term together with (144). Assuming that the up and down pumping in the infinite hotel can be done locally, one deduces

$$\partial^\mu J_{5\mu}(x) = \frac{eE(x)}{\pi} \equiv -\frac{e}{2\pi}\,\varepsilon^{\mu\nu}F_{\mu\nu}(x)\,, \tag{147}$$

which is the local form of the ABJ-axial anomaly in $(1+1)$ dimensions.

4.2. *Topological character of the ABJ anomaly*

To exhibit the topological character of (147), notice that since the Weyl spinors are normalized within $[0, 1]$, we have**

$$Q_5 = \sum_- 1 - \sum_+ 1\,, \tag{148}$$

which is the Atiyah–Patodi–Singer index [49]. Using the anomaly equation (147), and Wick rotating to Euclidean space ($x^0 = -\mathrm{i}x^2$), yields

* At this stage, it is instructive to note that while the vector coupling described above generates an axial anomaly, i.e. $\dot{Q} = 0$ and $\dot{Q}_5 \neq 0$, an axial coupling would generate a vector anomaly, i.e. $\dot{Q} \neq 0$ and $\dot{Q}_5 = 0$. The latter is at the origin of the missing baryon number in topological bag models. Figure 8c illustrates the spectral flow in chiral bag models that leads to $\dot{Q} \neq 0$ in the vacuum.

** Equation (148) is conditionally convergent and needs regularization.

$$Q_5 = \int dx^0\, dx^1 \left(\frac{-e}{2\pi} \varepsilon^{\mu\nu} F_{\mu\nu} \right)$$

$$= \int_E dx^1\, dx^2 \left(\frac{ie}{2\pi} \varepsilon^{\mu\nu} F_{\mu\nu} \right). \tag{149}$$

Using the standard 2-form notation for the field strength, i.e. $F = \frac{1}{2} F_{\mu\nu}\, dx^\mu\, dx^\nu$, on a 2-manifold M_2 with boundary, gives

$$Q_5 = 2e \int_{D_2} \frac{iF}{2\pi} = 2e \int_{D_2} Ch_1(F), \tag{150}$$

where $Ch_1(F)$ is the first Chern characteristic class associated to the Abelian Chern character

$$Ch(F) = e^{iF/2\pi} = \sum_{n=0}^{\infty} \frac{1}{n!} \left(\frac{iF}{2\pi} \right)^n = \sum_{n=0}^{\infty} Ch_n F, \tag{151}$$

$Ch_1(F)$ is closed and exact. Indeed for an Abelian gauge group the Bianchy identity is trivially given by

$$dF = d^2 A = 0. \tag{152}$$

In the Abelian case $F = dA$ is globally exact, and one can define

$$dF = dA = dQ_1(A), \tag{153}$$

with $Q_1(A) = A$ being the second Chern–Simons characteristic class in one dimension. By Stoke's theorem we have

$$Q_5 = \frac{ie}{\pi} \int_{M_2} dA = \frac{ie}{\pi} \int_{S^1} A \tag{154}$$

where $S^1 = \partial M_2$ is the boundary of M_2. Equation (154) involves only the gauge field A on the boundary and does not depend on its detailed distribution in M_2. Pure gauge-configurations correspond to

$$A = -ie^{-1} g^+\, dg, \tag{155}$$

in terms of which (154) reads

$$\tfrac{1}{2} Q_5 = \frac{1}{2\pi} \int_{S^1} g^+\, dg, \tag{156}$$

with $g(x) \in U(1)$. This is none but the Chern–Pontryagin index associated to $\pi_1(S^1)$ as encountered in

(93). Notice that in the pure gauge sector, the Chern–Simons character reduces to the Chern–Pontryagin index.

This simple exercise in $(1+1)$ dimensions shows already a pattern that is going to prevail in higher dimensions, and for non-Abelian gauge groups as well. The Abelian anomaly in $2n$ dimensions is related to the nth characteristic class of the pertinent Chern character via the Atiyah–Patodi–Singer index. Since the Chern characters are closed forms, the index can be determined in terms of the Chern–Simons forms in $2n-1$ dimensions by Stoke's theorem. For pure gauge fields, the Chern–Simons characters are none but the Chern–Pontryagin indices or winding numbers of the pertinent mapping,

$$g(x)\colon \partial \mathrm{M}_{2n} \rightarrow G\,, \tag{157}$$

where $g(x)$ is an element of the gauge group G assumed semi-simple for convenience.

This sequence of relationships is amenable to immediate generalization. Indeed, in $2n$ dimensions the generic relation for the Abelian anomaly is locally given by

$$\partial^{\mu} J_{5\mu}(x) = a(A(x))\,, \tag{158}$$

where a stands for the anomaly term. Using Stoke's theorem on a $2n$ manifold with a spherical boundary yields

$$Q_5 = \tfrac{1}{2} \int_{\partial \mathrm{M}_{2n}} \mathrm{d}^{2n-1}x \langle [\bar{\psi}, \gamma_0\gamma_5\psi] \rangle = \tfrac{1}{2} \int_{\mathrm{M}_{2n}} \mathrm{d}^{2n}x\, a(x)\,, \tag{159}$$

which generalizes (149) to $2n$ dimensions. By analogy with (150), we will assume that the index (158) is given by the nth Chern characteristic class in $2n$ dimensions,* i.e.

$$\tfrac{1}{2} \int_{\partial \mathrm{M}_{2n}} \mathrm{d}^{2n-1}x \langle [\bar{\psi}, \gamma_0\gamma_5\psi] \rangle = \int_{\mathrm{M}_{2n}} \mathrm{Ch}_n(F)\,, \tag{160}$$

where we have defined

$$\mathrm{Ch}_n(F) = \frac{1}{n!}\, \mathrm{TR}\left(\frac{\mathrm{i}F}{2\pi}\right)^n . \tag{161}$$

Equation (159) together with eq. (160) yield the expected form for the Abelian anomaly in $2n$ dimensions, namely,

$$a(x) = \frac{2}{n!} \left(\frac{\mathrm{i}}{2\pi}\right)^n \frac{1}{2^n}\, \varepsilon^{\mu_1\mu_2\cdots\mu_{2n}}\, \mathrm{TR}[F_{\mu_1\mu_2} \cdots F_{\mu_{2n-1}\mu_{2n}}]\,. \tag{162}$$

* The gauge coupling is lumped into the definition of F for convenience.

4.3. Non-Abelian anomalies

In the preceding discussion, chiral symmetry is a global symmetry of the world, and the Abelian anomaly is generated by external gauge-coupling to fermions. If chiral symmetry is promoted to a local symmetry through the introduction of external chiral gauge fields coupled to fermions, the resulting theory exhibits non-Abelian anomalies. The origin of these anomalies in the conservation equation of chiral currents dates back to Bardeen [44]. In comparison with (158), the generic relation for the non-Abelian anomaly is that the covariant divergence of the axial-isovector current $J^a_{5\mu}$ does not vanish,*

$$D^\mu J^a_{5\mu}(x) = B^a(A(x)) \qquad a = 1, 2, \dots N_F^2 - 1\,, \tag{163}$$

where $B^a(A)$ stands for the non-Abelian anomaly term in Bardeen's form. The main departure from the Abelian case is that the gauge field $A(x)$ transforms under chiral transformations.

For a convenient discussion of the chiral gauge transformations, let us agree to define the Lie-valued 1-form

$$A = \mathrm{i}gA^a_\mu T^a\, \mathrm{d}x^\mu = A_\mu\, \mathrm{d}x^\mu\,, \tag{164}$$

whose corresponding field strength 2-form is given by

$$\begin{aligned} F &= \mathrm{i}g(\tfrac{1}{2}F^a_{\mu\nu}T^a\, \mathrm{d}x^\mu \wedge \mathrm{d}x^\nu) \\ &= \mathrm{i}g\{\tfrac{1}{2}(\partial_\mu A^a_\nu - \partial_\nu A^a_\mu + \mathrm{i}g[A_\mu ; A_\nu]^a)\}\, \mathrm{d}x^\mu \wedge \mathrm{d}x^\nu \\ &= \mathrm{d}A + A^2\,. \end{aligned} \tag{165}$$

Under local infinitesimal left × right chiral transformations,

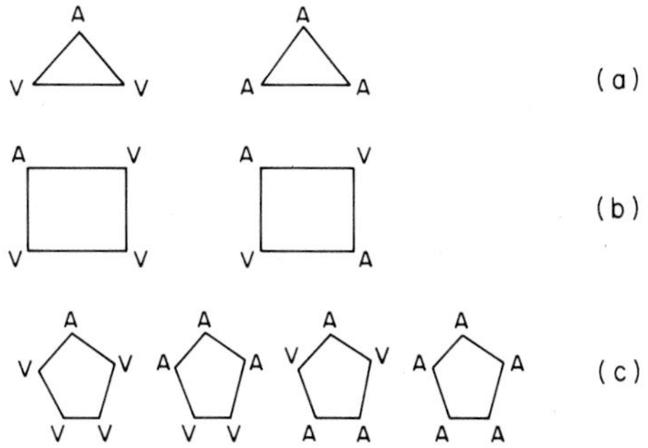

Fig. 9. (a) Represents the triangle anomalies (AVV, AAA), in both Abelian and non-Abelian gauge theories with constituent fermions; (b) and (c) give respectively the box (AVVV, VAAA) and the pentagonal (AVVVV, AAAVV, VAVAA, AAAAA) anomalies in non-Abelian gauge theories with constituent fermions.

* Choosing to have the anomaly in the axial current rather than the vector current, or any of their linear combination is conventional.

$$\Lambda_{\rm L} = \exp({\rm i}Q_{\rm L}) = 1 + l\,; \qquad l^+ = -l\,; \qquad {}^l A_{\rm L} = \Lambda_{\rm L}(A_{\rm L} + d)\Lambda_{\rm L}^+\,,$$
$$\Lambda_{\rm R} = \exp({\rm i}Q_{\rm R}) = 1 + r\,; \qquad r^+ = -r\,; \qquad {}^r A_{\rm R} = \Lambda_{\rm R}(A_{\rm R} + d)\Lambda_{\rm R}^+\,, \tag{166}$$

the gauge-fields transform as follows:

$$\delta_l A_{\rm L} = -[A_{\rm L}, l] - {\rm d}l = -[A_{\rm L} + d, l] = -D_{\rm L} l\,,$$
$$\delta_r A_{\rm R} = -[A_{\rm R}, r] - {\rm d}r = -[A_{\rm R} + d, t] = -D_{\rm R} r\,, \tag{167}$$

and their corresponding field strengths as follows:

$${}^l F_{\rm L} = \Lambda_{\rm L} F_{\rm L} \Lambda_{\rm L}^+ = {\rm d}\,{}^l A_{\rm L} + {}^l A_{\rm L}^2\,, \qquad {}^r F_{\rm R} = \Lambda_{\rm R} F_{\rm R} \Lambda_{\rm R}^+ = {\rm d}\,{}^r A_{\rm R} + {}^r A_{\rm R}^2\,, \tag{168}$$

$D_{\rm L,R}$ are the left, right covariant derivatives 1-form, and the gauge-fields $\Lambda_{\rm L,R}$ are Lie-valued 0-form. In our case, the Bianchi identity reads

$${\rm D}F \equiv [A, F] + {\rm d}F = 0\,, \tag{169}$$

and follows from the fact that F is gauge covariant.

Now, consider the third-Chern character in 6 dimensions,

$${\rm Ch}_3(F) = \frac{1}{3!}\left(\frac{\rm i}{2\pi}\right)^3 {\rm TR}[F^3] \tag{170}$$

which is gauge-invariant by inspection and closed*,

$$\begin{aligned} {\rm dCh}_3(F) &= \frac{1}{3!}\left(\frac{\rm i}{2\pi}\right)^3 \times 3\,{\rm TR}[{\rm d}F\,F^2] \\ &= \frac{1}{2!}\left(\frac{\rm i}{2\pi}\right)^3 \times {\rm TR}[({\rm d}F + [A, F])F^2] \\ &= \frac{1}{2!}\left(\frac{\rm i}{2\pi}\right)^3 \times {\rm TR}[{\rm D}F\,F^2] = 0\,. \end{aligned} \tag{171}$$

The latter follows from the Bianchi identity (169) and the cyclic property of the trace. By Poincare's lemma, ${\rm Ch}_3(F)$ is locally exact. In other words, there exists a local 5-form $Q_5^0(A)$ such that

$${\rm Ch}_3(F) = {\rm d}Q_5^0(A)\,, \tag{172}$$

$Q_5^0(A)$ is the Chern–Simons secondary character. Since ${\rm Ch}_3(F)$ is gauge invariant, then

$$\delta_\Lambda {\rm Ch}_3(F) = {\rm d}\delta_\Lambda Q_5^0(A) = 0\,. \tag{173}$$

* For any Chern class: $\delta_\Lambda {\rm Ch}_n(F) = 0$ and ${\rm dCh}_n(F) = 0$.

Again by Poincaré's lemma $\delta_\Lambda Q_5^0(A)$ is locally exact. If we denote by $Q_4^1(A;\Lambda)$ the 4-form linear in Λ, then*

$$\delta_\Lambda Q_5^0(A) = \mathrm{d}Q_4^1(A;\Lambda)\,. \tag{174}$$

Now, consider the generic form of the non-Abelian anomaly (163), and denote by $\Gamma(A)$ its corresponding action functional. By the Noether theorem we know that functional variations involving axial gauge-transformations** $\Lambda = 1+\gamma \equiv 1+\gamma^a T^a$, yield the anomaly condition

$$\delta_\gamma \Gamma(A) = \int \mathrm{d}^4x\, \mathrm{TR}[\gamma D^\mu J_{5\mu}] = \int \mathrm{d}^4x\, \gamma^a(x) B^a(A)\,. \tag{175}$$

To solve (175) we will proceed in two steps. First, notice that due to the Lie character of the gauge group $\Gamma(A)$ satisfies

$$\delta_\gamma\delta_\xi \Gamma(A) - \delta_\xi\delta_\gamma \Gamma(A) = \int \mathrm{d}^4x\, \mathrm{TR}([\gamma,\xi]D^\mu J_{5\mu}) = \delta_{[\gamma,\xi]}\Gamma(A)\,, \tag{176}$$

which is known as the Wess–Zumino consistency condition. Second, Kaymackalan et al. [31] have shown that in the presence of pseudoscalar fields $U(x)$, the anomaly equation in Bardeen's form is saturated if

$$\Gamma(A; U=1) = 0\,, \tag{177}$$

eqs. (176) and (177) are sufficient to solve for the anomalous action functional (175). Indeed, a general solution to (176) is given by the integrated form of the Chern character (170),

$$\Gamma(A) = \int_{\mathrm{M}_6} \mathrm{Ch}_3(F) = \int_{\mathrm{M}_5} Q_5^0(A)\,, \tag{178}$$

since

$$\begin{aligned}(\delta_\gamma\delta_\xi - \delta_\xi\delta_\gamma)\Gamma(A) &= \int_{\mathrm{M}_5} (\delta_{\gamma\xi}\gamma_{\xi\gamma}) Q_5^0(A)\\ &= \int_{\mathrm{M}_4} (Q_4^1(A;\gamma\xi) - Q_4^1(A;\xi\gamma))\\ &= \int_{\mathrm{M}_4} Q_4^1(A;[\gamma,\xi])\\ &= \delta_{[\gamma,\xi]}\Gamma(A)\,,\end{aligned} \tag{179}$$

* In the BRS formalism [50], (174) constitutes the first descent equation in the so-called Russian formula. The subset of gauge transformations enlarged with ghost fields and known as BRS transformations, is nilpotent, i.e. $\delta^2 = 0$. This translates into a closed cohomological algebra of which (174) is an element of the associated complex. These powerful arguments are similar to the ones used in gauge theories for solving the renormalization equations, when staging the proof of renormalizability [51]. To discuss them would go beyond the course of this work.

** Axial transformations correspond to $\gamma = \mathrm{i}(Q_\mathrm{R} - Q_\mathrm{L}) \equiv 2\mathrm{i}Q_\mathrm{R}$.

where we have used the group structure of the gauge transformations, (172), (174) and the fact that $Q_4^1(A; \Lambda)$ is linear in the gauge parameter Λ. In terms of (178), the particular solution that fulfills (177) reads

$$\Gamma_{\mathrm{WZ}}[A; U] = 2\pi N_c \int_{\mathrm{D}^5} (Q_5^0({}^U A) - Q_5^0(A)) \,, \tag{180}$$

where D_5 is a five-dimensional manifold with a spacetime boundary, i.e. $\partial \mathrm{D}_5 = \mathrm{S}^4$. Equation (180) turns out to be the properly normalized and gauged Wess–Zumino action functional that saturates Bardeen's form of the anomaly through [31],

$$\begin{aligned} \delta_\gamma \Gamma_{\mathrm{WZ}}[A; U] &= 2\pi N_c \int_{\mathrm{S}^4} Q_4^1(A; \gamma) \\ &= \int_{\mathrm{S}^4} \mathrm{TR}[\gamma D^\mu J_{5\mu}] \,. \end{aligned} \tag{181}$$

Similar equations have been derived by the Syracuse group [31] using different arguments. Equation (181) clearly shows that the non-Abelian anomaly in 4 dimensions, here in Bardeen's form, is related to the Abelian anomaly (178) in 6 dimensions via the Chern–Simons character in 5 dimensions. The latter is none but the gauged Wess–Zumino term subject to Bardeen's boundary condition. This fundamental pattern will prevail in higher-dimensional spacetime.

The explicit form of $Q_5^0(A)$ can be obtained by integrating (173). For that, define with Zumino [48] a homotopy transformation such that

$$A_t = tA \quad \text{and} \quad F_t = tF + (t^2 - t)A^2 \,, \tag{182}$$

where $t \in [0, 1]$. Injecting (182) into (173) and using the fact that $[d; \partial_t] = 0$, gives

$$\begin{aligned} \partial_t \, \mathrm{TR}[F_t^3] &= 3\mathrm{TR}[\partial_t F_t F_t^2] \\ &= 3\mathrm{TR}[(d\partial_t A_t + \partial_t A_t A_t + A_t \partial_t A_t) F_t^2] \\ &= 3\mathrm{TR}[D(\partial_t A_t) F_t^2] \\ &= 3D\, \mathrm{TR}[\partial_t A_t F_t^2] \\ &= 3d\, \mathrm{TR}[\partial_t A_t F_t^2] \,, \end{aligned} \tag{183}$$

where we have used the facts that $\partial_t A_t$ is a 1-form, the Bianchi identity and the scalar character of the trace part, respectively. Therefore

$$\mathrm{Ch}_3(F) = \frac{1}{3!} \left(\frac{\mathrm{i}}{2\pi}\right)^3 \int_0^1 \mathrm{d}t\, \partial_t \, \mathrm{TR}[F_t^3] = \frac{1}{2!} \left(\frac{\mathrm{i}}{2\pi}\right)^3 \mathrm{d} \int_0^1 \mathrm{d}t\, \mathrm{TR}[\partial_t A_t F_t^2] \,. \tag{184}$$

Comparison with (173) yields the Chern–Simons character $Q_5^0(A)$,

$$Q_5^0(A) = \left(\frac{i}{2\pi}\right)^3 \times \frac{1}{2!}\int_0^1 dt\, TR[\partial_t A_t F_t^2]$$

$$= \left(\frac{i}{2\pi}\right)^3 \times \frac{1}{2!}\int_0^1 dt\, TR[A(tF + (t^2 - t)A^2)^2]$$

$$= \left(\frac{i}{2\pi}\right)^3 \times \frac{1}{2!}\int_0^1 dt\, TR[A(F^2t^2 + A^4(t^4 - 2t^3 + t^2)) + 2A^3F(t^3 - t^2)] \tag{185a}$$

$$= \left(\frac{i}{2\pi}\right)^3 \times \frac{1}{3!} TR[AF^2 - \tfrac{1}{2}A^3F + \tfrac{1}{10}A^5]$$

$$= \left(\frac{i}{2\pi}\right)^3 \times \frac{1}{3!} TR[A(dA)^2 + \tfrac{3}{2}A^3\, dA + \tfrac{3}{5}A^5]\,,$$

i.e.

$$Q_5^0(A) = \frac{-i}{48\pi^3} TR[A(dA)^2 + \tfrac{3}{2}A^3\, dA + \tfrac{3}{5}A^5]\,. \tag{185b}$$

In the pure gauge sector G, (185) reduces to the Chern–Pontryagin index associated to $\pi_5(G)$,

$$\Gamma_{WZ}[A = 0; U] = 2\pi N_c \int_{D^5} Q_5^0(U^+ dU)$$

$$= \frac{-iN_c}{240\pi^2}\int_{D^5} TR[(U^+ dU)^5]\,, \tag{186}$$

which is the expected Wess–Zumino action (107) in Witten's form [25].

5. Vector meson dominance and the Skyrme model

5.1. Vector mesons and non-Abelian anomalies

We have shown in the previous section the fundamental relationship that exists between the Wess–Zumino term of current algebra and the non-Abelian anomalies. While the former plays an important role in our present understanding of skyrmions as baryons in the context of QCD, the latter are intimately related to anomalous meson decays at low energy, as we will discuss it in this section.

It is believed, as repeatedly emphasized throughout the course of this work, that a dynamical description of the pseudoscalar octet mesons along Skyrme's script with the proper chiral anomalies, is equivalent to QCD at low energy. While the trivial configurations should saturate current algebra requirements, their non-trivial counterparts provide through the soliton mechanism a classical playground for meson–baryon dynamics.

The purpose of this section is to extend the chiral effective description to the low-lying meson resonances such as the ω, ρ and A_1. This purpose is manifold. First, their absence in the 1 GeV scale can hardly be justified in any serious attempt towards understanding the NN interaction. Second, they provide a serious check on the large N_c expansion in which baryons are viewed as solitons within an effective theory of mesons, not only scalar but vector as well. Third, it is important to further investigate the role of vector mesons in the structure and stability of baryons and their resonances. Fourth, viewed as spin-1 gauge fields, vector mesons are likely to provide a possible realization of the non-Abelian anomaly structure of QCD. Finally, it is relevant to ask how vector meson dominance along Sakurai's precepts fits into a Skyrme-like description of baryons and mesons.

While the non-linear σ-model with the Wess–Zumino term provides an unambiguous realization of QCD at low energy based on chiral symmetry alone, the introduction of vector mesons is more ambiguous. There is however a seeming rationale for identifying these mesons with gauge bosons of a minimally broken gauge theory of pseudoscalar mesons. We are aware that while the gauging provides a systematic description of the interactions of pions with vector and axial vector mesons, it can hardly be accountable from the QCD Lagrangian. Some years ago, Weinberg [51] and later on Callan, Coleman, Wess and Zumino [52] have shown that the gauged effective chiral descriptions of which the above mentioned model is a prototype, saturates to leading order the QCD anomalous chiral Ward identities. In QCD, chiral symmetry translates into a set of Ward identities that constrain the parameters of the various correlation functions in the vacuum. At low energy, any sensible approximation to QCD must abide by these identities. A systematic procedure that keeps track of them makes use of an effective Lagrangian description involving massless pseudoscalar mesons. To leading order in the pion momentum the solution to the anomalous chiral Ward identities is the gauged non-linear σ-model with the Wess–Zumino term. The gauge fields are external sources with the quantum numbers of spin-1 mesons.

In this spirit, consider ω, ρ and A_1 as gauge particles associated with $SU(2)_L \otimes SU(2)_R \otimes U(1)_V$ and define

$$A_L^\mu = \tfrac{1}{2}(\rho_\mu + \omega_\mu + a_\mu)\,, \qquad A_R^\mu = \tfrac{1}{2}(\rho_\mu + \omega_\mu - a_\mu)\,, \tag{187}$$

where $\omega_\mu \equiv \mathrm{i}g\omega_\mu$, $\rho_\mu = \mathrm{i}g\rho_\mu{}^a T^a$ and $a_\mu = \mathrm{i}g a_\mu{}^a T^a$ are the ω, ρ and A_1 gauge fields respectively.* Their properties under local left $\times$ right transformations follow from eqs. (166)–(167). As in a local gauge theory, the coupling arises through the covariant derivative

$$D^\mu \cdot = \partial^\mu \cdot + A_L^\mu \cdot - \cdot A_R^\mu\,. \tag{188}$$

The field strengths $F_L^{\mu\nu}$ and $F_R^{\mu\nu}$ are given by

$$\begin{aligned} F_L^{\mu\nu} &= \mathrm{i}g(F_L^{\mu\nu})^a T^a = \partial^\mu A_L^\nu - \partial^\nu A_L^\mu + [A_L^\mu, A_L^\nu]\,,\\ F_R^{\mu\nu} &= \mathrm{i}g(F_R^{\mu\nu})^a T^a = \partial^\mu A_R^\nu - \partial^\nu A_R^\mu + [A_R^\mu, A_R^\nu]\,. \end{aligned} \tag{189}$$

A minimally broken $SU(2)_L \otimes SU(2)_R \otimes U(1)_V$ gauge model can be described by

$$S_1 = \int_{M_4} \mathrm{d}^4x \left\{ + \frac{1}{4g^2}\,\mathrm{TR}[F_L^{\mu\nu}F_{L\mu\nu} + F_R^{\mu\nu}F_{R\mu\nu}] - \frac{m^2}{2g^2}\,\mathrm{TR}[A_L^\mu A_{L\mu} + A_R^\mu A_{R\mu}] \right\}, \tag{190}$$

where m is the bare mass of the ρ-meson. (Introduction of this mass breaks gauge symmetry.) In the

* For convenience we have chosen the U(1) and SU(2) couplings to be identical. We can later shift the U(1) coupling relative to the SU(2) one.

presence of vector mesons the gauge invariant form of the kinetic term for the pseudoscalar fields $U(x)$ assumed to be SU(2)-valued for convenience reads

$$S_0 = \int_{M_4} d^4x \, \frac{f_\pi^2}{4} \, \mathrm{TR}[D^\mu U D_\mu U^+] , \tag{191}$$

while the gauged $SU(2)_L \otimes SU(2)_R \otimes U(1)_V$ Wess–Zumino term follows from the general analysis of the preceding section.* Since $SU(2)_L \otimes SU(2)_R$ is anomaly free, the gauged Wess–Zumino action in Bardeen's form reads

$$S_{WZ}[U, A_L, A_R] = \frac{iN_c}{24\pi^2} \int_{M_4} \omega \, \mathrm{TR}\{(dUU^+)^3 - 3\, d(A_L \, dUU^+ + A_R U^+ \, dU + A_R U^+ A_L U - A_L A_R)\} , \tag{192}$$

and clearly vanishes for $U = 1$.** Since $\pi_5(SU(2)) = 0$, there is no topological contribution to (192). The latter summarizes the non-Abelian anomaly structure of QCD in the effective language of spin-1 mesons. In short, the relevant π-ω-ρ-A_1 dynamics is described by

$$S = S_0 + S_1 + S_{WZ} . \tag{193}$$

The baryons correspond to the non-trivial configurations of spin-0 mesons. Their stability is naturally insured by the repulsive character of the vector mesons at short distances. Thus, there is no longer a need for Skyrme's fourth-order term. We will however keep calling the soliton configurations of (192) skyrmions.

5.2. ω-stabilized skyrmions

To discuss the phenomenological implications of (192), let us first specialize to the case where $\rho \equiv A \equiv 0$, i.e. to a $U(1)_V$ gauge model for π- and ω-mesons. In this case, the U(1) Lagrangian associated to (193) reads

$$\mathcal{L} = \tfrac{1}{4} f_\pi^2 \, \mathrm{TR}[\partial^\mu U \, \partial_\mu U^+] - \tfrac{1}{4} \omega^{\mu\nu} \omega_{\mu\nu} + \tfrac{1}{2} m^2 \omega_\mu^2 + g_\omega \omega^\mu B_\mu , \tag{194}$$

where $g_\omega = gN_c$ and B_μ is the topological current (32) carried by the spin-0 mesons. Since $\omega^\mu B_\mu$ scales like $[L]^{-4}$ and $g_\omega > 0$, the ω-field provides the means to stabilize the non-trivial spin-0 configurations at short distances without the additional Skyrme term. This is a well-known fact in nuclear physics, where the vector character of the ω-field accounts for the short-distance repulsion between nucleons.

* Remember that the Wess—Zumino term is non-local in (3 + 1) dimensions. Its gauging requires special care and does not follow from the simple substitution (188).

** Notice that $S_{WZ}[1, A_L, A_R] = 0$ breaks explicitly global chiral invariance. Although annoying, this breaking does not affect low-energy threshold theorems since it involves exclusively amplitudes with abnormal parity.

In the trivial sector the interaction term in (193) describes the coupling of ω to three π's to leading order in $1/f_\pi$,

$$g_\omega \omega^\mu B_\mu = -\frac{\mathrm{i}g_\omega}{\pi^2 f_\pi^3} \varepsilon^{\mu\nu\alpha\beta} \omega_\mu \partial_\nu \pi^0 \partial_a \pi^+ \partial_\beta \pi^- + \mathrm{O}\left(\frac{1}{f_\pi^3}\right). \tag{195}$$

Adkins and Nappi [53] have related g_ω to the $\omega \to \pi^0\pi^+\pi^-$ decay rate, and found $g_\omega = 25.4$. Their result constitutes an upper bound since the decay rate of $\omega \to \pi^+\pi^0\pi^-$ is dominated by the process $\omega \to \rho\pi \to \pi^+\pi^0\pi^-$ in nature. A more realistic model should include the ρ-meson. In fact, one can determine g_ω just by looking at the interaction between two nucleons, which gives $g_\omega^2/4\pi = 10$–12. Although this characterizes an effective ω-coupling that includes a large contribution from the ρ–π continuum, it is substantially smaller than that of ref. [53], indicating that $g_\omega = 25.4$ is indeed an upper bound.

To illustrate the ω-effects on the skyrmion, it is instructive to eliminate the classical ω-field from (194) using the equations of motion. From (194), we obtain the following equation of motion for ω_μ

$$\partial^\nu \omega_{\nu\mu} + m_\omega^2 \omega_\mu + g_\omega B_\mu = 0\,. \tag{196}$$

Since $\partial^\mu B_\mu = 0$ and $\omega_{\nu\mu}$ is antisymmetric, one concludes that $\partial_\mu \omega_\mu = 0$, which is a well-known property of massive vector fields. Assuming a static hedgehog configuration for the pions, we obtain ($\omega^i \equiv 0$)

$$(-\nabla^2 + m_\omega^2)\omega^0(\boldsymbol{r}) = -g_\omega B^0(\boldsymbol{r})\,. \tag{197}$$

The general solution to (197) can be constructed using the following Green function

$$(-\nabla^2 + m_\omega^2) G(\boldsymbol{r}; \boldsymbol{r}') = \delta(\boldsymbol{r} - \boldsymbol{r}')\,, \tag{198}$$

whose explicit form is

$$\begin{aligned} G(\boldsymbol{r}; \boldsymbol{r}') &= +\frac{1}{4\pi} \frac{\exp(|\boldsymbol{r} - \boldsymbol{r}'|)}{|\boldsymbol{r} - \boldsymbol{r}'|} \\ &= \mathrm{i}m_\omega \sum_{lm} (\mathrm{i}m_\omega r_<) h_l(\mathrm{i}m_\omega r_>) Y_l^m(\hat{r}) Y_l^m(\hat{r}')\,. \end{aligned} \tag{199}$$

In terms of (199), $\omega^0(r)$ reads

$$\begin{aligned} \omega^0(r) &= \int_\infty \mathrm{d}\boldsymbol{r}'\, G(\boldsymbol{r}'\, \boldsymbol{r}')(-\beta B^0(\boldsymbol{r}')) \\ &\equiv \int_0 \mathrm{d}r'\, 4\pi G_0(r, r')(-\beta r'^2 B^0(r'))\,, \end{aligned} \tag{200}$$

where $G_0(r, r')$ is the $l = 0$ component in (199). Using the explicit form of the spherical Bessel functions

$j_0(im_\omega r)$ and $h_0(im_\omega r)$, we obtain

$$4\pi G_0(r,r') = \frac{1}{2m_\omega rr'}\{\exp(-m_\omega|\boldsymbol{r}-\boldsymbol{r}'|) - \exp[-m_\omega(r+r')]\}\,. \tag{201}$$

Using the general solution (199), we can write the energy associated to (194), in the form

$$E = -\frac{f_\pi^2}{4}\int \mathrm{d}\boldsymbol{r}\,\mathrm{TR}[\partial^i U \partial^i U^+] + \frac{1}{2}\frac{g_\omega^2}{4\pi}\int \mathrm{d}\boldsymbol{r}\,\mathrm{d}\boldsymbol{r}'\,B^0(\boldsymbol{r})\,\frac{\exp(-m_\omega|\boldsymbol{r}-\boldsymbol{r}'|)}{|\boldsymbol{r}-\boldsymbol{r}'|}\,B^0(\boldsymbol{r}')\,, \tag{202}$$

thus,

$$E = +2\pi f_\pi^2\int_0^\infty\!\!\int_0^\infty r^2\,\mathrm{d}r\left[F'^2 + 2\,\frac{\sin^2 F}{r^2}\right] + 8\pi^2 g_\omega^2\int_0^\infty r^2\,\mathrm{d}r\int_0^\infty r'^2\,\mathrm{d}r'\,B^0(r)G_0(r,r')B^0(r') \tag{203}$$

for a static hedgehog pion configuration. The factor $\frac{1}{2}$ in the last term of (202) is familiar in energies from self interactions. Equation (202) illustrates clearly the ω-exchange mechanism that takes place inside the skyrmion, and underlines the repulsive character of the exchange which is at the origin of the stability of the skyrmion.

Amusingly, (202) shows the nucleon to be stabilized by ω-meson exchange between bits of nucleonic matter, analogously to how the Lorentz electron gets stabilized by the Coulomb repulsion. The finite-range interaction

$$\frac{g_\omega^2}{4\pi}\,\frac{\exp(-m_\omega|\boldsymbol{r}-\boldsymbol{r}'|)}{|\boldsymbol{r}-\boldsymbol{r}'|} \tag{204}$$

replaces Lorentz's long-range $e^2/|\boldsymbol{r}-\boldsymbol{r}'|$. For small separations, the two interactions have similar effects. Whereas Lorentz's description of the electron did not survive experiment, we know that the nucleon is an extended object with rms charge radius ~0.8 fm. A density $\rho(r)$ such that

$$\rho(r) = (m_\omega^2/8\pi)\exp(-m_\omega r) \tag{205}$$

yields a matter distribution roughly comparable to the experimental one, and is highly suggestive that the ω-stabilization determines the size of the nucleon.

The phenomenological aspects of the ω-stabilized skyrmions have been addressed by Adkins and Nappi [53], in the absence of the Skyrme term. We shall see however that the latter stems from the coupling of ρ-mesons, suggesting that both the Skyrme term and the ω-coupling enter into the problem, a prelude to vector meson dominance. In this spirit, the g_ω used by Adkins and Nappi should be decreased by a factor of more than 2.

It is clear that any term in the Lagrangian such as $g_\omega \omega^\mu \beta_\mu$, will contribute to the nucleon–nucleon force. However, neither the ρ-meson nor the ω-meson will contribute attraction in the nucleon–nucleon interaction. The latter must arise as a genuine quantum effect as we shall see later on.

5.3. The Skyrme term from ρ-mesons

Recently Iketani [54a] has argued that the Skyrme term can be understood in the context of current algebra. For that, consider (193) in the absence* of ω-mesons,

$$\mathcal{L} = \alpha \frac{f_\pi^2}{4} \mathrm{TR}[D_\mu U^+ D^\mu U] + \frac{1}{4g^2} \mathrm{TR}[F^{\mathrm{L}}_{\mu\nu} F^{\mu\nu}_{\mathrm{L}} + F^{\mathrm{R}}_{\mu\nu} F^{\mu\nu}_{\mathrm{R}}]$$

$$- \frac{m^2}{2g^2} \mathrm{TR}[A_{\mu\mathrm{L}} A^\mu_{\mathrm{L}} + A_{\mu\mathrm{R}} A^\mu_{\mathrm{R}}] . \tag{206}$$

A way to disentangle the effects of the axial particle from the pion field is through a chiral transformation to the local isospin frame. Using the field ξ defined by $U = \xi^2$ as the corresponding gauge transformation, i.e. $\xi^+ U \xi^+ = 1$, and denoting by $A'_\mu, \rho'_\mu, a'_\mu, \ldots$ the gauge transformed fields, yield (206) in the form

$$\mathcal{L} = -\alpha \frac{f_\pi^2}{4} \mathrm{TR}[a'^2_\mu] + \frac{1}{4g^2} \mathrm{TR}[F_{\mathrm{L}\mu\nu} F^{\mu\nu}_{\mathrm{L}} + F_{\mathrm{R}\mu\nu} F^{\mu\nu}_{\mathrm{R}}] - \frac{m^2}{4g^2} \mathrm{TR}[(\rho'_\mu + v_\mu)^2 + (a'_\mu + p_\mu)^2] , \tag{207}$$

where v_μ and ρ_μ are the following pure gauge fields:

$$v_\mu = -[\xi, \partial_\mu \xi^+] = -\frac{1}{4f_\pi^2} [\pi, \partial_\mu \pi] + \cdots ; \qquad p_\mu = +[\xi, \partial_\mu \xi^+]_+ = -\frac{\mathrm{i}}{f_\pi} \partial_\mu \pi + \cdots , \tag{208}$$

the dots stand for higher-order terms in $1/f_\pi$. Since

$$\mathrm{TR}[p_\mu^2] = -\mathrm{TR}[(\partial_\mu \xi^{+2})(\partial^\mu \xi^2)] , \tag{209}$$

we have

$$\mathcal{L} = -\frac{m^2}{4g^2} \left(1 - \frac{m^2}{M^2}\right) \mathrm{TR}[p_\mu^2] - \frac{m^2}{4g^2} \mathrm{TR}[(\rho'_\mu + v_\mu)^2] - \frac{m^2}{4g^2} \mathrm{TR}\left[\left(a'_\mu + \frac{m^2}{M^2} p_\mu\right)^2\right]$$

$$+ \frac{1}{4g^2} \mathrm{TR}[F'^{\mathrm{L}}_{\mu\nu} F'^{\mu\nu}_{\mathrm{L}} + F'^{\mathrm{R}}_{\mu\nu} F'^{\mu\nu}_{\mathrm{R}}] , \tag{210}$$

with $M^2 = m^2 + \alpha g^2 f_\pi^2$. The proper normalization of the pion kinetic term requires $f_\pi^2 g^2 = m^2(1 - m^2 M^{-2})$, hence,

$$M^2 = \alpha m^2 , \qquad g f_\pi = (1 - 1/\alpha)^{1/2} m . \tag{211}$$

For $\alpha = 2$, eqs. (211) yield Weinberg relation $M^2 = 2m^2$ and the KSFR relation $\sqrt{2} g f_\pi = m$. The former

* In the presence of an axial gauge particle, the pion field is "ill defined". The present Lagrangian needs proper diagonalization before one can identify the fields with their physical counterparts. Consequently, the normalization of the pion kinetic term cannot be fixed yet. To account for this, we have introduced a general parameter α.

relates the ρ-meson mass (m) to the axial mass (M), while the latter determines the ρNN-coupling ($f_{\rho NN} \sim f_{\rho\pi\pi} = g$) in terms of the pion-decay constant f_π and the ρ-meson mass m. It follows from our arguments that unless assumed, the value of α is not determined at this stage and contrary to Iketani's claim one cannot deduce the KSFR relation and Weinberg's mass formulae without further assumptions. With the conditions (211), the physical fields are

$$U = \xi^2 = \exp(i\pi/f_\pi)\,, \qquad \hat{\rho}_\mu = p_\mu + v_\mu\,, \qquad \hat{a}_\mu = a'_\mu + (1/\alpha)p_\mu\,, \tag{212}$$

and their relevant dynamics follow from (210),

$$\mathscr{L} = \frac{f_\pi^2}{4}\,\mathrm{TR}[\partial^\mu U^+ \partial_\mu U] - \frac{m^2}{4g^2}\,\mathrm{TR}[\hat{\rho}_\mu^2 + \alpha\hat{a}_\mu^2] - \frac{1}{4g^2}\,\mathrm{TR}[F'_{\mathrm{L}\mu\nu}F'^{\mu\nu}_{\mathrm{L}} + F'_{\mathrm{R}\mu\nu}F'^{\mu\nu}_{\mathrm{R}}]\,, \tag{213}$$

where the field strengths $F'_{\mathrm{L,R}}$ are complicated functions of π, $\hat{\rho}$ and $\hat{a}$,

$$\begin{aligned} F'_{\mathrm{L}\mu\nu} &= \hat{F}_{\mathrm{L}\mu\nu} + \frac{1}{4f_\pi^2}\left(1 - \frac{1}{\alpha^2}\right)[\partial_\mu\pi, \partial_\nu\pi] + \cdots\,; \\ F'_{\mathrm{R}\mu\nu} &= \hat{F}_{\mathrm{R}\mu\nu} + \frac{1}{4f_\pi^2}\left(1 - \frac{1}{\alpha^2}\right)[\partial_\mu\pi, \partial_\nu\pi] + \cdots\,. \end{aligned} \tag{214}$$

Injecting (214) into (213) gives a quartic term in the pion field of the form

$$\frac{1}{32g^2f_\pi^4} \times \left(1 - \frac{1}{\alpha^2}\right)^2 \times \mathrm{TR}[\partial_\mu\pi, \partial_\nu\pi]^2\,. \tag{215}$$

In the limit where the vector mesons ρ and A_1 decouple from the pion field, one obtains to leading order in the field gradients, $L_\mu = (i/f_\pi)\partial_\mu\pi + \cdots$,

$$\mathscr{L}_{\mathrm{SK}} = -\tfrac{1}{4}f_\pi^2\,\mathrm{TR}[L_\mu^2] + (f_\pi^2/32m^2)C(\alpha)\,\mathrm{TR}[L_\mu, L_\nu]^2 + \mathrm{O}(L^6)\,, \tag{216}$$

with $C(\alpha) = (1 - 1/\alpha^2)(1 + 1/\alpha)$. Equation (216) shows that the Skyrme term arises from the exchange of a heavy ρ-meson. This picture is consistent with vector meson dominance. Choosing $\alpha = 2$, i.e. $C(2) = \frac{9}{8}$, so as to obtain the KSFR relation and Weinberg's mass formula yields a positive Skyrme term with

$$\varepsilon^2 = (3f_\pi/8m)^2\,. \tag{217}$$

Using $f_\pi \sim 93$ MeV and $m_\rho \sim 780$ MeV give $\varepsilon^2 = 2.05 \times 10^{-3}$, which is to be compared with $\varepsilon_{\mathrm{JR}}^2 = 5.52 \times 10^{-3}$ used by Jackson and Rho.

Lately, Bando et al. [54b] have proposed to introduce the vector mesons as dynamical gauge bosons (composite of pions) of a hidden local symmetry. The idea is that the non-linear σ-model defined on the coset manifold G/H is gauge equivalent to a "linear model" with $\mathrm{G} \times \mathrm{H}_{\mathrm{local}}$ symmetry. To see this, consider the case where

$$G/H = (SU(2)_L \otimes SU(2)_R/SU(2)_V$$

and use the following parametrization for the U field,

$$U = \xi_L^+ \xi_R , \tag{218}$$

where ξ_L and ξ_R are two independent SU(2)-valued fields. It is clear, from the decomposition (218), that the non-linear σ-model as discussed so far, is also invariant under the following gauge transformation:

$$\xi'_{L,R} = h(x)\xi_{L,R} . \tag{219}$$

This is the hidden local symmetry. In the "unitary gauge", $\xi_L^+ = \xi_R = \xi$, and the action for the non-linear σ-model reads

$$\mathcal{L} = -\tfrac{1}{4} f_\pi^2 \, \mathrm{TR}[\partial_\mu \xi \cdot \xi^+ - \partial_\mu \xi^+ \xi]^2 . \tag{220}$$

At this stage the ρ-meson can be introduced as an auxiliary gauge field associated to the local symmetry (219), through the following substitution $\partial_\mu \to D_\mu = \partial_\mu - \rho_\mu$, i.e.

$$\mathcal{L}_\alpha = -\tfrac{1}{4} f_\pi^2 \, \mathrm{TR}[D_\mu \xi \cdot \xi^+ - D_\mu \xi \cdot \xi^+]^2 + \tfrac{1}{4} \alpha f_\pi^2 \, \mathrm{TR}[D_\mu \xi \cdot \xi^+ + D_\mu \xi^+ \xi]^2 , \tag{221}$$

where α is a real parameter. The second term in (221) is harmless since it reduces to

$$+ \tfrac{1}{4} \alpha f_\pi^2 \, \mathrm{TR}(2\rho_\mu - (\partial_\mu \xi \xi^+ + \partial_\mu \xi^+ \xi))^2 ,$$

and can be eliminated from the generating functional by immediate integration. Notice that ρ_μ as an auxiliary field does not propagate. To remedy this, Bando et al. [54b] added by hand a kinetic term for the ρ-field,

$$\begin{aligned} \mathcal{L}_W &= + \frac{1}{4g^2} \mathrm{TR}(F_{\mu\nu} F^{\mu\nu}) + \mathcal{L}_\alpha \\ &= \tfrac{1}{4} \mathrm{TR}(\partial_\mu \pi)^2 + \frac{1}{4g^2} \mathrm{TR}(F_{\mu\nu} F^{\mu\nu}) + \frac{af_\pi^2}{4} \mathrm{TR}(\rho_\mu^2) + \frac{\alpha}{8} \mathrm{TR}(\rho_\mu \pi, \partial_\mu \pi]) + \cdots , \end{aligned} \tag{222}$$

from which one deduces

$$m^2 = \alpha g^2 f_\pi^2 . \tag{223}$$

For $\alpha = 2$, this is the well-known KFSR relation with $g = f_{\rho\pi\pi}$.* Expression (222) agrees with Weinberg's phenomenological Lagrangian [54c]. In the heavy mass limit the ρ-field reduces to

* We will show next how to relate the gauge coupling g to Sakurai's $f_{\rho\pi\pi}$.

$$\rho_\mu = [\partial_\mu \xi^+, \xi] = \frac{1}{4f_\pi^2} [\pi, \partial_\mu \pi] + \cdots . \tag{224}$$

Thus

$$\frac{1}{4g^2} \mathrm{TR}(F_{\mu\nu}F^{\mu\nu}) = \frac{1}{32g^2} \mathrm{TR}[L_\mu, L_\nu]^2 , \tag{225}$$

which is again Skyrme's quartic term with

$$\varepsilon^2 = \left(\frac{f_\pi^2}{2m}\right)^2 . \tag{225a}$$

Using the physical parameters for f_π and m give $\varepsilon^2 = 3.3 \times 10^{-3}$, in qualitative agreement with Iketani's result.

5.4. *The skyrmion with vector mesons*

Finally, consider the skyrmion in the presence of ω, ρ and A_1 as described by (193). In terms of the anti-Hermitian fields (187), the gauge covariant kinetic term for the pions reads

$$\begin{aligned}
\mathscr{L}_0 &= \tfrac{1}{4} f_\pi^2 \,\mathrm{TR}[D^\mu U^+ D_\mu U] \\
&= -\tfrac{1}{4} f_\pi^2 \,\mathrm{TR}[L_\mu^2] + \tfrac{1}{4} f_\pi^2 \,\mathrm{TR}[\rho_\mu (R_\mu + L_\mu)] + \tfrac{1}{4} f_\pi^2 \,\mathrm{TR}[a_\mu (R_\mu - L_\mu)] \\
&\quad - \tfrac{1}{4} f_\pi^2 \,\mathrm{TR}[a_\mu^2] + \tfrac{1}{8} f_\pi^2 \,\mathrm{TR}[a_\mu U \rho_\mu U^+] - \tfrac{1}{8} f_\pi^2 \,\mathrm{TR}[\rho_\mu U a_\mu U^+] \\
&\quad + \tfrac{1}{8} f_\pi^2 \,\mathrm{TR}[\rho_\mu [U, \rho_\mu] U^+] - \tfrac{1}{8} f_\pi^2 \,\mathrm{TR}[a_\mu [U, a_\mu] U^+] .
\end{aligned} \tag{226}$$

The kinetic term for the gauge fields can be rewritten in the following form:

$$\begin{aligned}
\mathscr{L}_F &= \frac{1}{4g^2} \mathrm{TR}[F_{\mathrm{L}}^{\mu\nu} F_{\mathrm{L}\mu\nu} + F_{\mathrm{R}}^{\mu\nu} F_{R\mu\nu}] \\
&= \frac{1}{8g^2} \mathrm{TR}[\omega^{\mu\nu}\omega_{\mu\nu} + \rho^{\mu\nu}\rho_{\mu\nu} + a^{\mu\nu} a_{\mu\nu}] ,
\end{aligned} \tag{227}$$

where the chiral field strengths $\omega^{\mu\nu}$, $\rho^{\mu\nu}$ and $a^{\mu\nu}$ are given by

$$\begin{aligned}
\omega^{\mu\nu} &= \partial^\mu \omega^\nu - \partial^\nu \omega^\mu , \\
\rho^{\mu\nu} &= \partial^\mu \rho^\nu - \partial^\nu \rho^\mu + \tfrac{1}{2}[\rho^\mu, \rho^\nu] + \tfrac{1}{2}[a^\mu, a^\nu] , \\
a^{\mu\nu} &= \partial^\mu a^\nu - \partial^\nu a^\mu + \tfrac{1}{2}[\rho^\mu, a^\nu] + \tfrac{1}{2}[a^\mu, \rho^\nu] .
\end{aligned} \tag{228}$$

The gauge breaking mass term in (190) translates into

$$\mathcal{L}_{\rm M} = -\frac{m^2}{2g^2}\,{\rm TR}[A_{\rm L}^{\mu}L_{{\rm L}\mu} + A_{\rm R}^{\mu}A_{{\rm R}\mu}]$$

$$= -\frac{m^2}{4g^2}\,{\rm TR}[\omega_\mu^2 + \rho_\mu^2 + a_\mu^2] \tag{229}$$

while the Wess–Zumino term (192) becomes

$$\Gamma_{\rm WZ} = \frac{{\rm i}N_{\rm c}}{24\pi^2}\,\varepsilon^{\mu\nu\alpha\beta}\omega_\mu\,{\rm TR}[L_\nu L_\alpha L_\beta]$$

$$+ \frac{{\rm i}N_{\rm c}}{64\pi^2}\,\varepsilon^{\mu\nu\alpha\beta}\omega_{\mu\nu}\,{\rm TR}\{2\rho_\alpha(R_\beta - L_\beta) + 2a_\alpha(R_\beta + L_\beta) + 2\rho_\alpha a_\beta$$

$$-(\rho_\alpha - a_\alpha)U^{+}(\rho_\beta + a_\beta)U\}\,. \tag{230}$$

Equations (226–230) constitute the basis for discussing skyrmions with vector mesons in the context of a minimally broken $SU(2)_V \otimes SU(2)_A \otimes U(1)_V$. Due to the hedgehog character of the skyrmion and its intrinsic parity assumed positive for convenience, the tensor character of the vector mesons follows through the following time-independent ansatz

$$\rho_\mu = {\rm i}gR(r)\tau^a\varepsilon^{aik}\delta^i_\mu\hat{r}^k\,, \qquad a_\mu = {\rm i}g\tau^a(a_1(r)\delta^{ai} + a_2(r)\hat{r}^a\hat{r}^i)\delta^i_\mu\,, \qquad \omega_\mu = {\rm i}g\omega(r)\delta_{\mu 0}\,. \tag{231}$$

As mentioned earlier in the presence of the axial field, the pion field needs proper normalization. Indeed, it is not difficult to see that the coupling

$$\tfrac{1}{4}f_\pi^2\,{\rm TR}[a_\mu(R_\mu - L_\mu)] = gf_\pi a_\mu^a\partial^\mu\pi^a + \cdots \tag{232}$$

in (226), "eats up" the pion field. To avoid this, one conventionally diagonalizes the axial field. For that consider the quadratic and linear terms in a_μ, i.e.

$$-\frac{1}{4g^2}\,{\rm TR}\Big[2{\rm i}g^2 f_\pi a_\mu\partial^\mu\pi + \frac{m^2}{Z^2}\,a_\mu^2\Big]\,, \tag{233}$$

where $Z^{-2} = 1 + f_\pi^2 g^2/m^2$. To eliminate the cross term in (233), define the renormalized fields $\tilde{a}_\mu$, $\tilde{\pi}$ and the renormalized pion decay constant $\tilde{f}_\pi$ such that

$$a_\mu = \hat{a}_\mu - {\rm i}\,\frac{g^2\tilde{f}_\pi^2}{m^2}\,\partial_\mu\tilde{\pi}\,, \qquad \tilde{\pi} = Z\pi\,, \qquad \tilde{f}_\pi = Zf_\pi\,. \tag{234}$$

To leading order, the kinetic terms for the normalized pion and axial fields become

$$\mathcal{L}^{(2)} = \frac{1}{4Z^2}\Big(1 - \frac{g^2\tilde{f}_\pi^2}{m^2}\Big)\,{\rm TR}[(\partial_\mu\tilde{\pi})(\partial^\mu\tilde{\pi})] - \frac{m^2}{4Z^2g^2}\,{\rm TR}[\tilde{a}_\mu\tilde{a}^\mu]$$

$$= \tfrac{1}{4}{\rm TR}[(\partial_\mu\tilde{\pi})^2] - (m^2/4Z^2g^2)\,{\rm TR}[\tilde{a}_\mu^2]\,, \tag{235}$$

which illustrates the decoupling of $\tilde{\pi}$ from $\tilde{a}_\mu$. Although we will omit the tilde (˜) from now on, the physical fields are always defined through (234). Notice that in the presence of the axial particle, the pion decay constant gets reduced, i.e.

$$\tilde{f}_\pi = f_\pi/(1 + f_\pi^2 g^2/m^2)^{1/2} . \tag{236}$$

It is evident from (229) that the axial field acquires a mass m_A such that

$$m_A^2 - m_\rho^2 = g^2 \tilde{f}_\pi^2 (m_A^2/m_\rho^2) , \tag{237}$$

which is the mass formulae quoted by the Syracuse group. This renormalized model has recently been investigated in the $B = 1$ sector [55].

5.5. Sakurai's wisdom in the Skyrme model

Our purpose at this stage is to analyze the consistency of the gauged Skyrme model with Sakurai's precepts of vector dominance in the trivial topological sector. The corresponding analysis in the non-trivial sector involving topological baryons is given in ref. [55]. First, we will show that the gauge coupling g is Sakurai's $f_{\rho\pi\pi}$. For that consider the $\rho\pi\pi$-vertex as given by (226), i.e.

$$\begin{aligned}\mathscr{L}_{\rho\pi\pi} &= \tfrac{1}{4} f_\pi^2 \,\mathrm{TR}[\rho^\mu (L_\mu + R_\mu)] \\ &= -2\mathrm{i} g \rho_\mu^0 \pi^+ \partial^\mu \pi^- + \cdots ,\end{aligned} \tag{238}$$

where again $\pi^\pm = (1/\sqrt{2})(\pi_x \pm \mathrm{i}\pi_y)$. In the ρ center-of-mass, the decay width associated to the $\rho^0 \to \pi^+\pi^-$ process illustrated in fig. 10a, is given by

$$\Gamma(\rho^0 \to \pi^+\pi^-) = \frac{1}{32\pi^2} \times \frac{|\boldsymbol{q}_\pi|}{m_\rho^2} \int \mathrm{d}\Omega_+ \tfrac{1}{3} |M|^2 , \tag{239}$$

where $\mathrm{d}\Omega_+$ is the solid angle of the π^+ and M the transition amplitude,

$$\begin{aligned}|M|^2 &= 4g^2 \sum_a \varepsilon_\mu^a(Q)\, \varepsilon_\nu^{a*}(Q)\, q^\mu q^\nu \\ &= 4g^2 \left(\frac{Q_\mu Q_\nu}{m_\rho^2} - g_{\mu\nu} \right) q^\mu q^\nu = 4g^2 \left[\frac{(Q \cdot q)^2}{m_\rho^2} - q^2 \right] = 4g^2 |\boldsymbol{q}_\pi|^2 .\end{aligned} \tag{240}$$

Therefore, we have

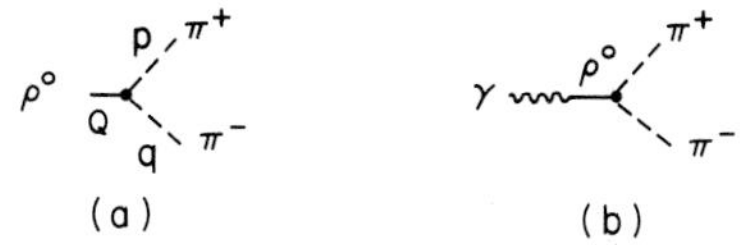

Fig. 10. (a) Illustrates the $\rho^0 \to \pi^+\pi^-$ decay process; (b) is the $(\gamma\pi^+\pi^-)$ vertex in vector meson dominance language.

$$\Gamma(\rho^0 \to \pi^+\pi^-) = \tfrac{2}{3} \times \frac{g^2}{4\pi} \times \frac{|q_\pi|^3}{m_\rho^2}\,. \tag{241}$$

Comparison with the experimental width yields

$$g^2/4\pi = f_{\rho\pi\pi}/4\pi \cong 3.0\,. \tag{242}$$

Sakurai's hypothesis of vector meson dominance stems from the fact that the low-energy electromagnetic form factors are dominated by the neutral vector mesons, namely, the ρ^0_μ and ω_μ. Electromagnetic interactions with charged vector mesons and pions occur through secondary couplings as exemplified in fig. 10b. In this spirit one can assume that the neutral vector currents are proportional to the neutral vector fields in the heavy mass limit* ($m \to \infty$). This translates into electromagnetic interactions of the form

$$\mathcal{L}_{\rm em} = e(m^2/g)A^\mu[\rho^0_\mu + \tfrac{1}{3}\omega_\mu] + O(A^2)\,, \tag{243}$$

where A_μ is the electromagnetic gauge field. The factor of $\frac{1}{3}$ is the inverse of ω to ρ SU(3) coupling constants. Equation (243) is the expression of vector meson dominance.**

To investigate the consistency of the gauged Wess–Zumino term in the context of vector meson dominance we will discuss two processes investigated recently by the Syracuse group, namely $\pi^0 \to 2\gamma$ and $\omega \to \pi^0\gamma$. The relevant trilinear vertex for the $\pi^0 \to \rho\omega \to 2\gamma$ decay as illustrated in fig. 11a, is entirely given by the Wess–Zumino term (230) i.e.

$$\mathcal{L}_{\omega\rho\pi} \equiv -(N_c g^2/8\pi^2\tilde{f}_\pi)\,\varepsilon^{\mu\nu\alpha\beta}\partial_\mu\omega_\nu\partial_\alpha\rho^a_\beta\tilde{\pi}^a\,. \tag{244}$$

For clarity, we have explicitly extracted the gauge coupling g from the vector fields. It follows that the $\omega\rho\pi$-coupling is given by

$$g_{\omega\rho\pi} = N_c g^2/8\pi^2\tilde{f}_\pi\,. \tag{245}$$

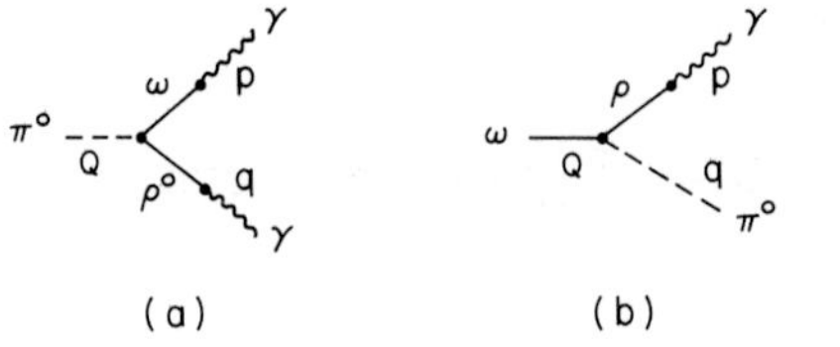

Fig. 11. (a) Chiral anomaly, $\pi^0 \to 2\gamma$, in the language of vector meson dominance; (b) shows the decay process $\omega \to \pi^0\gamma$, discussed in the text.

* This approximation is legitimate in the case where m is large compared to the inverse of the mean square radius of the nucleon. This is indeed the case for the heavy vector mesons, e.g. $m_\rho^2\langle \Gamma^2\rangle \sim 8.5$.

** In their recent arguments, Bando et al. [54b] have derived (243) by gauging the U(1) part of (221) and setting by hand $\alpha = 2$. These arguments while not unique, insure vector meson dominance only to lowest order in the photon coupling.

In the π^0 center of mass, the decay width associated to the process of fig. 11a reads

$$\Gamma(\pi^0 \to 2\gamma) = \frac{1}{32\pi^2} \times \frac{|\boldsymbol{q}|}{m_\pi^2} \times \int d\Omega_\gamma \, 2|M|^2 , \tag{246}$$

where the transition matrix element is given by ($\alpha = e^2/4\pi$)

$$|M|^2 = \left(\frac{4\pi\alpha m^4}{3g^2} g_{\omega\rho\pi}\right)^2 \varepsilon^{\mu\nu\alpha\beta} \varepsilon^{\mu'\nu'\alpha'\beta'} p_\mu q_\alpha p_{\mu'} q_{\alpha'} \times \left(\frac{1}{m^2}\right)^4 \times \left(\sum_a \varepsilon_\nu^a(p)\, \varepsilon_{\nu'}^{a*}(p)\right)\left(\sum_b \varepsilon_\beta^b(q)\, \varepsilon_{\beta'}^{b*}(q)\right).$$

Using the completeness relations for the polarizations, yields

$$|M|^2 = 8|\boldsymbol{q}|^4 \left(\frac{4\pi\alpha}{3} \times \frac{g_{\omega\rho\pi}}{g^2}\right)^2 . \tag{247}$$

Since $|\boldsymbol{q}| = m_\pi/2$ in the π^0 center of mass, we deduce

$$\begin{aligned} \Gamma(\pi^0 \to 2\gamma) &= \pi \frac{\alpha^2}{9} \times \frac{g_{\omega\rho\pi}^2}{g^4} \times m_\pi^3 \\ &= \left(\frac{N_c}{3}\right)^2 \times \frac{\alpha^2}{64\pi^3} \times \frac{m_\pi^3}{\tilde{f}_\pi^2} . \end{aligned} \tag{248}$$

For $N_c = 3$, this is just the current algebra result as noted by the Syracuse group [31]. Taking $m_\pi = 139$ MeV, $\alpha = 1/137$, $f_\pi = 93$ MeV and $m_\rho = 769$ MeV, we obtain $\Gamma(\pi^0 \to 2\gamma) \sim 7.1$ eV in good agreement with the experimental value of 7.9 eV.

The $\omega \to \rho\pi^0 \to \gamma\pi^0$ decay illustrated in fig. 11b follows through similar arguments. Indeed, the corresponding decay rate in the ω center of mass reads

$$\Gamma(\omega \to \pi^0\gamma) = \frac{1}{32\pi^2} \times \frac{|\boldsymbol{q}|}{m_\omega^2} \int d\Omega_\gamma \times \tfrac{1}{3}|M|^2 , \tag{249}$$

where the matrix element is given by

$$\begin{aligned} |M|^2 = 4\pi\alpha m^4 \left(\frac{g_{\omega\rho\pi}}{g}\right)^2 \times \left(\frac{1}{m^2}\right) \times \varepsilon^{\mu\nu\alpha\beta} \varepsilon^{\mu'\nu'\alpha'\beta'} Q_\nu P_\alpha Q_{\nu'} P_{\alpha'} \times \left(\sum_a \varepsilon_\mu^a(Q) \varepsilon_{\mu'}^{a*}(Q)\right) \\ \times \left(\sum_b \varepsilon_\beta^b(p) \varepsilon_{\beta'}^{b*}(p)\right) . \end{aligned} \tag{250}$$

Using again the completeness relations for the vector polarizations together with the on-shell conditions yield

$$|M|^2 = 8\pi\alpha \left(\frac{g_{\omega\rho\pi}}{g}\right)^2 (Q \cdot p)^2 , \tag{251}$$

whence

$$\Gamma(\omega \rightarrow \pi^0\gamma) = (\tfrac{1}{3}\alpha)(g_{\omega\rho\pi}/g)^2|\boldsymbol{q}_\pi|^3 , \tag{252}$$

using $m_\omega = 780\,\mathrm{MeV}$ and the fact that $|\boldsymbol{q}_\pi| + \sqrt{m_\pi^2 + \boldsymbol{q}_\pi^2} = m_\omega$ in the ω center of mass, one obtains a width $\Gamma(\omega \rightarrow \pi^0\gamma) \sim 0.80\,\mathrm{MeV}$, which is again in good agreement with the experimental value of $0.86 \pm 0.05\,\mathrm{MeV}$.

6. The QCD generating functional with passive gluons at low energy

In the preceding sections we have discussed the relevant framework for a low-energy analysis of strong interaction physics based solely on concepts borrowed from chiral symmetry and the large N_c expansion. The clearcut relationship between the effective degrees of freedom used and the underlying quark–gluon dynamics referred to remains yet unclear. As emphasized in this work, the low-energy version of QCD is dictated by the spontaneous breaking of chiral symmetry, where the light pions summarizing the long-wavelength properties of the intricate vacuum state play a central role.

The purpose of this section is to discuss a Skyrme-like picture within a QCD script in which the gluonic effects are reduced to a minimum, namely, a background potential for the scalar excitations, much along the recent work of Simic [56]. It is relevant to point out at this stage that there have been various attempts, namely, by Gaudin [57], Aitchison et al. [58] and Mackenzie et al. [59] to account for the stability of the skyrmion via fermion radiative corrections to 1-loop order. Although partly successful, these attempts have the repelling feature of coupling quarks to mesons everywhere in space paying no attention to confinement, nor to asymptotic freedom. Although we have nothing against hybrid models, we want to stress the fact that they cannot be used to justify the Skyrme model. We believe that they are reducible variants of the Skyrme chiral description as can be shown by the immediate integration of the quark degrees of freedom. As expected, the phenomenology that follows is similar to the Skyrme phenomenology [60].

6.1. Z_{QCD} and the low-energy collective modes

The starting point in this qualitative discussion is the QCD generating functional Z_{QCD} associated to the vacuum amplitude in the absence of sources, i.e.

$$\begin{aligned} Z_{\mathrm{QCD}} &= \langle 0_+ | 0_- \rangle \\ &= \int \mathrm{d}[A_\mu]\,\mathrm{d}[q]\,\mathrm{d}[\bar{q}]\,\mathrm{d}[c]\,\mathrm{d}[\bar{c}]\,\exp[\mathrm{i}(S_{\mathrm{QCD}} + S_{\mathrm{FP}} + S_{\mathrm{GF}})] , \end{aligned} \tag{253}$$

where S_{QCD} is the QCD action associated to the Lagrangian (1), S_{FP} is the Faddeev–Popov action involving the ghost fields c and $\bar{c}$, and S_{GF} is the gauge fixing term. As it stands, (253) carries the full vacuum properties but is unfortunately very involved. However, we already know, at least from the QCD sum rules, that the long-wavelength sector of QCD is characterized by scalar quark and gluon condensates, e.g. $\langle \bar{q}q \rangle$, $\langle GG \rangle$, etc. It is therefore relevant to divide the field configurations integrated over in (253), into classes that are characterized by a certain vacuum structure at low energy. For simplicity, we will only consider collective mesonic fields, by giving the gluons a passive role and disregarding heavier glueball excitations. We are aware that such an approximation will deprive us of the subtleties of confinement and asymptotic freedom that we know are the primers of QCD. Having said this, let us rewrite the functional form (253) by introducing unit factors

$$Z_{\mathrm{QCD}} = \int \mathrm{d}[\mathrm{QCD}]\, \mathrm{d}\mu[U]\, \mathrm{d}[\sigma]\, \mathrm{d}[\sigma^+]\, \delta(\sigma - \sigma^+)\, \delta(\bar{q}_{\mathrm{R}} q_{\mathrm{L}} - U^+\sigma)\, \delta(\bar{q}_{\mathrm{L}} q_{\mathrm{R}} - \sigma^+ U)$$

$$\times \exp \mathrm{i}(S_{\mathrm{QCD}} + S_{\mathrm{FP}} + S_{\mathrm{GF}})\,. \tag{254}$$

Now, let us define auxiliary scalar and pseudoscalar fields $S(x)$ and $P(x)$ respectively, through

$$\delta(\bar{q}_{\mathrm{L}} q_{\mathrm{R}} - \sigma^+ U) = \int \mathrm{d}[S + \mathrm{i}P] \exp\left\{-\mathrm{i} \int \mathrm{d}^4x\, \mathrm{TR}[\bar{q}_{\mathrm{L}}(S + \mathrm{i}P) q_{\mathrm{R}} - (S + \mathrm{i}P)\sigma^+ U]\right\}. \tag{255}$$

Injecting (255) into (25) and making the unitary and anomaly-free change of variable $(S + \mathrm{i}P) \to U^+(S + \mathrm{i}P)$, we obtain

$$Z_{\mathrm{QCD}} = \int \mathrm{d}[\mathrm{QCD}]\, \mathrm{d}[S]\, \mathrm{d}[P]\, \mathrm{d}\mu[U]\, \mathrm{d}[\sigma]\, \mathrm{d}[\sigma^+]\, \delta(\sigma - \sigma^+) \exp[\mathrm{i}(S_{\mathrm{QCD}} + S_{\mathrm{FP}} + S_{\mathrm{GF}})]$$

$$\times \exp\left\{-\mathrm{i} \int \mathrm{d}^4x\, \mathrm{TR}[\bar{q}_{\mathrm{L}} U^+(S + \mathrm{i}P) q_{\mathrm{R}} + \bar{q}_{\mathrm{R}}(S - \mathrm{i}P) U q_{\mathrm{L}} - 2S\sigma]\right\}. \tag{256}$$

Using the fact that $q_{\mathrm{L,R}} = \frac{1}{2}(1 \mp \gamma_5) q$, we have

$$\mathrm{TR}[\bar{q}_{\mathrm{L}} U^+(S + \mathrm{i}P) q_{\mathrm{R}} + \bar{q}_{\mathrm{R}}(\mathrm{S} - \mathrm{i}P) U q_{\mathrm{L}}] = \mathrm{TR}[\bar{q} U_5^+(S + \mathrm{i}P\gamma_5) q]\,, \tag{257}$$

where $U_5 = \exp(\mathrm{i}\gamma_5 \lambda^a \pi^a)$. In terms of (257), eq. (256) reads

$$Z_{\mathrm{QCD}} = \int \mathrm{d}[\mathrm{QCD}]\, \mathrm{d}[S]\, \mathrm{d}[P]\, \mathrm{d}[\sigma]\, \mathrm{d}[\sigma^+]\, \delta(\sigma - \sigma^+) \exp[\mathrm{i}(S_{\mathrm{QCD}} + S_{\mathrm{FP}} + S_{\mathrm{GF}})]$$

$$\times \exp\left\{-\mathrm{i} \int \mathrm{d}^4x\, \mathrm{TR}[\bar{q} U_5^+(S + \mathrm{i}P\gamma_5) q - 2S\sigma]\right\}. \tag{258}$$

Consider the gluon and ghost parts of Z_{QCD}

$$\tilde{Z}[J] = \int \mathrm{d}[A_\mu]\, \mathrm{d}[\bar{c}]\, \mathrm{d}[c] \exp\left\{\mathrm{i} \int \mathrm{d}^4x\, \{-\tfrac{1}{4}\, \mathrm{TR}[F^{\mu\nu} F_{\mu\nu}] - A_\mu^a J^{\mu,a}\}\right.$$

$$\left.\times \exp\left\{\mathrm{i} \int \mathrm{d}^4x \left\{-\frac{1}{2\alpha} (\partial^\mu A_\mu^a)^2 + \bar{c}\partial_\mu D^\mu(A) C\right\}\right\}\right. \tag{259}$$

where $J_\mu^a = g\bar{q}\gamma_\mu \lambda^a q$ is the color quark current. $\tilde{Z}[J]$ is a chiral invariant functional of J_μ^a which is not gauge invariant. It is a complicated expression. A crude approximation is to assume that the gluons, once integrated over, give a scalar potential to lowest order,

$$\tilde{Z}[J] = \exp\left\{-\mathrm{i} \int \mathrm{d}^4x\, (V_{\mathrm{G}}[\bar{q}q] + \cdots)\right\}. \tag{260}$$

A way of improving on this is to evaluate (259) using a Hartree–Fock approximation for instance, as recently discussed by Castorina and Pi [61]. Assuming (260) and rotating to Euclidean space yields (258) in the form*

$$Z_{\rm QCD} \sim \int \mathrm{d}[\bar q]\,\mathrm{d}[q]\,\mathrm{d}[S]\,\mathrm{d}[P]\,\mathrm{d}[U]\,\mathrm{d}[U^+]\,\mathrm{d}[\sigma]\,\mathrm{d}[\sigma^+]\,\delta(\sigma-\sigma^+)$$
$$\times \exp\Big[-\int_E \mathrm{d}^4x\,(\bar q(\not\partial_E + m + U_5^+(S+\mathrm{i}\gamma_5 P))q) + V_{\rm G}(\bar q q) - 2\,\mathrm{TR}([S\sigma])\Big]\,. \tag{261}$$

To a good extent, we can assume that $V_{\rm G}(\bar q q) \sim V_{\rm G}(\langle \bar q q\rangle) \sim V(\sigma)$ at low energy, where $\langle \bar q q\rangle$ is the fermion condensate in the vacuum. In this spirit, the fermion functional in (261) involves only quark bilinears and can be exactly evaluated, i.e.

$$\int \mathrm{d}[\bar q]\,\mathrm{d}[q]\,\exp\Big[-\int \mathrm{d}^4x\,\bar q(\not\partial_E + m + U_5^+(S+\mathrm{i}P\gamma_5))q\Big] = [\det(\partial_E + m + U_5^+(S+\mathrm{i}P\gamma_5))]^{N_c}$$
$$= \exp N_c\,\mathrm{TR}\ln(\not\partial_E + m + U_5^+(S+\mathrm{i}P\gamma_5))\,, \tag{262}$$

in terms of which $Z_{\rm QCD}$ reduces to a functional over the collective fields,

$$Z_{\rm QCD} \sim \int \mathrm{d}[S]\,\mathrm{d}[P]\,\mathrm{d}[U]\,\mathrm{d}[U^+]\,\mathrm{d}[\sigma]\,\mathrm{d}[\sigma^+]\,\delta(\sigma-\sigma^+)$$
$$\times \exp\Big\{N_c\,\mathrm{TR}\ln(\not\partial_E + m + U_5^+(S+\mathrm{i}P\gamma_5)) - \int \mathrm{d}^4x\,(V_{\rm G}(\sigma) - 2\,\mathrm{TR}(S\sigma))\Big\}\,. \tag{263}$$

In the large N_c limit the functional integral over the auxiliary fields is dominated by the vacuum values $\bar P$ and $\bar\sigma$ which satisfy the saddle point equations associated to (263). Because of parity $\bar P = 0$, and one obtains

$$Z_{\rm QCD} \sim \int \mathrm{d}[U]\,\mathrm{d}[U^+]\,\mathrm{d}[S]\,\exp\{N_c\,\mathrm{TR}\ln(\not\partial_E + (m + SU_5^+))\}\,\exp\Big\{\int \mathrm{d}^4x\,(2\,\mathrm{TR}(S\bar\sigma) - V_{\rm G}(\bar\alpha))\Big\}\,, \tag{264}$$

where $\bar\sigma$ is determined by the saddle-point equation in the σ direction, i.e.

$$S = \tfrac{1}{2}\delta V_{\rm G}(\bar\sigma)/\delta\bar\sigma\,. \tag{265}$$

Using now the saddle-point equation in the S-direction yields a vacuum expectation $\bar S$, solution to

$$\int \mathrm{d}^4x\,\frac{\delta}{\delta S}\,\{N_c\langle x|\,\underset{\rm I}{\mathrm{TR}}\ln(\not\partial_E + (m+S))|x\rangle + 2\,\mathrm{TR}(S\bar\sigma) - V_\sigma(\sigma)\} = 0\,,\qquad S = \bar S\,, \tag{266}$$

* In Euclidean space our conventions for the γ-matrices are $\gamma_E^4 = \gamma^0$ and $\gamma_E^k = -\mathrm{i}\gamma^k$, with $(\gamma_E^i)^+ = \gamma_E^i$ and $g^{ij} = \delta^{ij}$.

because of CPT-invariance only the real part of the TR ln term contributes to (266). With this in mind, we obtain ($\partial_E^+ = -\partial_E$)

$$\begin{aligned}\bar{\sigma} &= -N_c\langle x|\frac{\delta}{\delta S}\ln(-\not{\partial}_E^2 + m_Q^2)|x\rangle \\ &= -2N_c\langle x|(-\partial_E^2 + m_Q^2)^{-1} m_Q|x\rangle \\ &= -2N_c\int\frac{d^4k}{(2\pi)^4}\frac{m_Q}{k^2+m_Q^2}\,,\end{aligned} \tag{267}$$

where we have defined $m_Q = m + \bar{S}$. As expected, this expression is ultraviolet divergent. Using a proper time regularization with a cutoff scale Λ in the ultraviolet sector yields

$$\bar{\sigma} = -2N_c\int_{1/\Lambda^2}^{\infty} d\tau\,\exp(-\tau m_Q^2)\int\frac{d^4k}{(2\pi)^4}\exp(-\tau k^2)m_Q\,, \tag{268}$$

m_Q is suggestive of a "constituent quark mass", and eq. (268) provides a crude relationship between the vacuum condensate $\langle\bar{q}q\rangle$, m_Q and the cutoff scale Λ. In the vacuum $\bar{\sigma} = \langle\bar{q}_R q_L\rangle = \langle\bar{q}_L q_R\rangle$, and therefore

$$\langle\bar{q}q\rangle = -\frac{N_c\Lambda^3}{4\pi^2}\left(\frac{m_Q}{\Lambda}\right)\int_1^{\infty}\frac{d\tau}{\tau^2}\exp[-\tau(m_Q/\Lambda)^2]\,, \tag{269}$$

which is the relation established by Simic [56].

6.2. The Wess–Zumino term from fermions

Restricting the auxiliary field configurations to their classical vacuum expectation values yield (264) in the form

$$Z_{QCD} \sim \int d[U]\,d[U^+]\exp\{N_c\,\mathrm{TR}\ln(\not{\partial}_E + (m + \bar{S}U_5^+))\}\exp\left\{\int d^4x\,(2\,\mathrm{TR}(S\bar{\sigma}) - V_G(\bar{\sigma})\right\}. \tag{270}$$

To evaluate the TR ln term in (270), we will use the adiabatic method of Goldstone and Wilczek. This is essentially a gradient expansion in which one assumes that spatial variations are small compared to the size of the localized field configurations. Notice that the elliptic operator involved in (270) is non-Hermitian. This would translate into an action functional in the chiral field which is complex, and a priori in conflict with the requirement of CPT invariance. Fortunately the imaginary part is non-local in (3 + 1) dimensions, and consequently escapes the CPT constraint. This is nothing but the Wess–Zumino term. Indeed, consider

$$\exp(iS_1 + S_2) = \exp\{N_c\,\mathrm{TR}\ln(\not{\partial}_E + (m + \bar{S}U_5^+))\}\,, \tag{271}$$

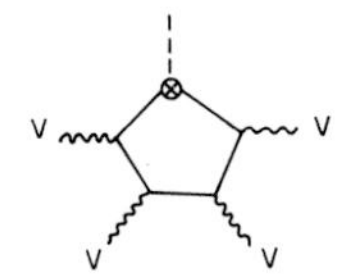

Fig. 12. Pentagonal anomaly in the fermion vacuum that accounts for the Wess–Zumino term in the adiabatic approach of Goldstone and Wilczek.

where explicitly $U_5^+ = \exp(\mathrm{i}\phi^a\lambda^a\gamma_5)$. Functional variations on both sides of (271) yield the anomaly equation for S_1,

$$\frac{S_1}{\delta\phi^a} = N_c \operatorname{Re} \mathrm{TR}\{(\not\partial_E + (m + \bar{S}U_5^+))^{-1}(\bar{S}U_5^+)\lambda^a\gamma_5\} . \tag{272}$$

Using the a posteriori fact that anomalies are mass-independent, we obtain

$$\frac{\delta S_1}{\delta\phi^a} = -N_c \operatorname{Re} \mathrm{TR}\{\lambda^a\gamma_5(-\partial_E^2 + \bar{S}^2 - \bar{S}\partial U_5^+)^{-1}(\bar{S}^2 - \bar{S}\partial U_5^+)\} . \tag{273}$$

If we denote by $\phi = \partial^\mu\phi_\mu$, $\Delta^{-1} = (-\partial_E^2 + \bar{S}^2)$, and expand (273) in field gradients, we have

$$\begin{aligned}\frac{\delta S_1}{\delta\phi^a} = &-N_c \operatorname{Re} \mathrm{TR}\{\lambda^a\gamma_5\Delta(\bar{S}^2 + \mathrm{i}\bar{S}\phi\gamma_5) - \lambda^a\gamma_5\Delta\mathrm{i}\bar{S}\phi\gamma_5\Delta(\bar{S}^2 + \mathrm{i}\bar{S}\phi\gamma_5) \\ &+ \lambda^a\gamma_5\Delta\mathrm{i}\bar{S}\phi\gamma_5\Delta\mathrm{i}\bar{S}\phi\gamma_5\Delta(\bar{S}^2 + \mathrm{i}\bar{S}\phi\gamma_5) - \lambda^a\gamma_5\Delta\mathrm{i}\bar{S}\phi\gamma_5\Delta\mathrm{i}\bar{S}\phi\gamma_5\Delta\mathrm{i}\bar{S}\phi\gamma_5\Delta(\bar{S}^2 + \mathrm{i}\bar{S}\phi\gamma_5) \\ &+ \lambda^a\gamma_5\Delta\mathrm{i}\bar{S}\phi\gamma_5\Delta\mathrm{i}\bar{S}\phi\gamma_5\Delta\mathrm{i}\bar{S}\phi\gamma_5\Delta\mathrm{i}\bar{S}\phi\gamma_5\Delta(\bar{S}^2 + \mathrm{i}\bar{S}\phi\gamma_5) - \cdots\} .\end{aligned} \tag{274}$$

The first free terms in (274) trace up to zero because of a poor isospin and spin structure, so that the leading contributions in the geometrical series (274) are

$$\frac{\delta S_1}{\delta\phi^a} = +N_c\bar{S}^4 \operatorname{Re} \mathrm{TR}[\lambda^a\gamma_5\Delta\phi\Delta\phi\Delta\phi\Delta\phi] - N_c\bar{S}^6 \operatorname{Re} \mathrm{TR}[\lambda^a\gamma_5\Delta\phi\Delta\phi\Delta\phi\Delta\phi\Delta] + \cdots , \tag{275}$$

which is just the perturbative expression of the pentagonal anomaly.

In Fourier space (275) becomes*

$$\begin{aligned}\frac{\delta S_1}{\delta\phi^a} = &+4N_c\bar{S}^4 \int \mathrm{d}^4p \,\frac{\mathrm{d}^4k_1}{(2\pi)^4}\,\frac{\mathrm{d}^4k_2}{(2\pi)^4}\,\frac{\mathrm{d}^4k_3}{(2\pi)^4}\,\frac{\mathrm{d}^4k_4}{(2\pi)^4}\,\delta\Big(\frac{k_1+k_2+k_3+k_4}{4}\Big) \\ &\times (1 - \bar{S}^2(p+k_1)^2 + \bar{S}^2)^{-1} \prod_{i=1}^{4} ((p+k_i)^2 + \bar{S}^2)^{-1} \\ &\times \epsilon^{\mu\nu\alpha\beta}\, \mathrm{TR}[\lambda^a\phi_\mu(k_2 - k_1)\phi_\nu(k_3 - k_2)\phi_\alpha(k_4 - k_3)\phi_\beta(k_1 - k_4)] + \cdots ,\end{aligned} \tag{276}$$

where the Fourier transform $\phi_\mu(k)$ is defined through

* In Euclidean space $\mathrm{TR}[\gamma_5\gamma^\mu\gamma^\nu\gamma^\alpha\gamma^\beta] = +4\varepsilon^{\mu\nu\alpha\beta}$, and consecutively (276) is real.

$$\phi_\mu(k) = \int d^4x\, e^{-ikx}\partial_\mu\phi\,. \tag{277}$$

To lowest order in the field gradients we obtain

$$\frac{\delta S_1}{\delta\phi^a} = 4N_c\bar{S}^4 \int \frac{d^4p}{(2\pi)^4}\, p^2(p^2+\bar{S})^{-5}\varepsilon^{\mu\nu\alpha\beta}\int d^4x\, \mathrm{TR}[\lambda^a\partial_\mu\phi\,\partial_\nu\phi\,\partial_\alpha\phi\,\partial_\beta\phi] + \cdots\,. \tag{278}$$

Therefore,

$$\frac{\delta S_1}{\delta\phi^a} = +\frac{N_c}{48\pi^2}\,\varepsilon^{\mu\nu\alpha\beta}\int d^4x\, \mathrm{TR}[\lambda^a\partial_\mu\phi\,\partial_\nu\phi\,\partial_\alpha\phi\,\partial_\beta\phi] + \cdots\,. \tag{279}$$

To integrate out (279) we will specialize to the set of homotopically equivalent field variations by introducing a homotopy transformation such that*

$$\phi(x,t) = t\phi(x)\,,\qquad \delta_t\phi(x,t) = \delta t\,\partial_t\phi(x,t) = \delta t\phi(x)\,,\qquad t\in[0,1]\,, \tag{280}$$

in terms of which (279) becomes

$$\delta_t S_1 = \frac{N_c}{48\pi^2}\,\varepsilon^{\mu\nu\alpha\beta}\int d^4x\, \mathrm{TR}[\delta_t\phi\,\partial_\mu\phi\,\partial_\nu\phi\,\partial_\alpha\phi\,\partial_\beta\phi] + \cdots\,, \tag{281}$$

and therefore,

$$S_1 = +\frac{N_c}{240\pi^2}\int_0^1 dt\int d^4x\,\varepsilon^{\mu\nu\alpha\beta\gamma}\,\mathrm{TR}[\partial_\mu\phi\,\partial_\nu\phi\,\partial_\alpha\phi\,\partial_\beta\phi\,\partial_\gamma\phi] + \cdots\,. \tag{282}$$

Requiring that the result (282) when performed to all orders is chiral invariant yields the Wess–Zumino term

$$S_1 = -\frac{iN_c}{240\pi^2}\int_{R^4\times[0,1]} d^5x\,\varepsilon^{\mu\nu\alpha\beta\gamma}\,\mathrm{TR}[L_\mu L_\nu L_\alpha L_\beta L_\gamma]\,, \tag{283}$$

with $U(x;0) = \mathbb{1}$ and $U(x;1) = U(x)$.

6.3. A Skyrme-like Lagrangian from fermions

To determine the real part of the action functional in (271), it is convenient to rewrite S_2 in the following form:

$$\begin{aligned} S_2 &= \tfrac{1}{2}N_c\,\mathrm{TR}\ln\{(\partial\!\!\!/_E + (m+\bar{S}U_5^+))(-\partial\!\!\!/_E + (m+\bar{S}U_5))\} \\ &= \tfrac{1}{2}N_c\,\mathrm{TR}\ln(-\partial_E^2 + M^2) + \tfrac{1}{2}N_c\,\mathrm{TR}\ln\{1 + (-\partial_E^2+M^2)^{-1}(m\bar{S}(U_5^+ + U_5) - \bar{S}\partial\!\!\!/_E U_5^+)\}\,, \end{aligned} \tag{284}$$

* This extension will be justified a posteriori by the fact that S_1 is a topological invariant.

where $M^2 = m^2 + \bar{S}^2$. The first term in (284) is the free vacuum contribution to the effective potential and will be omitted from the immediate discussion. With this in mind, we have

$$S_2 = \tfrac{1}{2}N_c\,\mathrm{TR}[(-\partial^2 + M^2)^{-1}m\bar{S}(U_5^+ + U_5)] + \tfrac{1}{2}N_c\,\mathrm{TR}\ln[1 - (-\partial^2 + M^2)^{-1}\bar{S}\partial\!\!\!/ U_5^+] + \cdots, \tag{285}$$

where the dots stand for higher-order terms in the chiral expansion. Using the proper time method with again a cutoff scale Λ, we can rewrite the mass term in (285) as follows:

$$\begin{aligned} S_{2m} &= \tfrac{1}{2}N_c\,\mathrm{TR}[(-\partial^2 + M^2)^{-1}m\bar{S}(U_5 + U_5^+)] \\ &= \tfrac{1}{2}N_c m\bar{S}\int \mathrm{d}^4x\,\mathrm{TR}[U_5 + U_5^+]\int \frac{\mathrm{d}^4k}{(2\pi)^4}(k^2 + M^2)^{-1} \\ S_{2m} &\equiv \tfrac{1}{2}N_c m\bar{S}\int \mathrm{d}^4x\,\mathrm{TR}(U_5 + U_5^+)\int \frac{\mathrm{d}^4k}{(2\pi)^4}\int_{1/\Lambda^2}^{\infty} \mathrm{d}s\,\exp[-s(k^2 + M^2)]\,. \end{aligned} \tag{286}$$

Since $\mathrm{TR}[U_5] = 4\,\mathrm{TR}[U]$, straightforward algebra yields

$$S_{2m} = \frac{N_c\Lambda^3}{4\pi^2}\left(\frac{\bar{S}}{\Lambda}\right)\int_1^{\infty}\frac{\mathrm{d}\tau}{\tau^2}\exp[-\tau(m/\Lambda)^2]\int \mathrm{d}^4x\,\mathrm{TR}\left[\frac{m}{2}(U + U^+)\right], \tag{287}$$

which stands for the chiral breaking term in conventional chiral models. From (287) one deduces a suggestive relationship between the pion mass m_π, the pion decay constant f_π, $\bar{S}$ and the cutoff scale Λ in the form $(M^2 = m^2 + \bar{S}^2)$

$$m_\pi^2 f_\pi^2 = \frac{N_c\Lambda^3}{2\pi^2}\,m\left(\frac{\bar{S}}{\Lambda}\right)\int_1^{\infty}\frac{\mathrm{d}\tau}{\tau^2}\exp[-\tau(m/\Lambda)^2]\,. \tag{288}$$

Comparison between (269) and (288) in the case of a small current quark mass m, shows that

$$m_\pi^2 f_\pi^2 = -2m\langle\bar{q}q\rangle + \mathrm{O}(m^2) \tag{289}$$

which is the expected current algebra result.

The remaining part of S_2 in (285) can be evaluated using again the adiabatic expansion of Goldstone and Wilczek. Indeed, to lowest order, the odd powers in the geometrical series vanish and one obtains

$$\begin{aligned} S_{2L} &= \tfrac{1}{2}N_c\,\mathrm{TR}\ln[1 - (-\partial^2 + \underline{M}^2)\bar{S}\partial\!\!\!/ U_5^+] \\ &= \tfrac{1}{4}N_c\,\mathrm{TR}[((-\partial^2 + M^2)^{-1}S\partial\!\!\!/ U_5^+)^2] - \tfrac{1}{8}N_c\,\mathrm{TR}[((-\partial^2 + M^2)^{-1}\partial\!\!\!/\bar{S}U_5^+)^4] + \cdots. \end{aligned} \tag{290}$$

It is straightforward to show that to fourth order in the field gradients, we have

$$-\frac{N_c}{8}\,\mathrm{TR}[((-\partial^2+M^2)^{-1}\bar{S}\not\partial U_5^+)^4]$$

$$=\frac{N_c}{192\pi^2}\left(\frac{\bar{S}}{M}\right)^4\left\{1+\left(\frac{M}{\Lambda}\right)^2\right\}\exp[-(M/\Lambda)^2]\int \mathrm{d}^4x\,\{\tfrac{1}{2}\,\mathrm{TR}[L_\mu, L_\nu]^2-\mathrm{TR}[L_\mu^2L_\nu^2]\} \tag{291a}$$

$$-\frac{N_c}{4}\,\mathrm{TR}[((-\partial^2+M^2)^{-1}\bar{S}\partial U_5^+)^2]=-\frac{N_c}{64\pi^2}\,\bar{S}^2\int_1^\infty \frac{\mathrm{d}\tau}{\tau}\exp[-\tau(M/\Lambda)^2]\int \mathrm{d}^4x\,\mathrm{TR}[\partial^\mu U^+\partial_\mu U]$$

$$+\frac{N_c}{384\pi^2}\left(\frac{\bar{S}}{\Lambda}\right)^2\int_1^\infty \mathrm{d}\tau\,\exp[-\tau(M/\Lambda)^2]\,\mathrm{d}^4x\,\mathrm{TR}[(\partial^2U)(\partial^2U^+)]\,. \tag{291b}$$

Equation (291b) gives rise to the standard kinetic term for the chiral field with the proper pion form factor to leading order in the field gradients. Indeed, let us define

$$f_\pi^2=\frac{N_c}{16\pi^2}\,\bar{S}^2\int_1^\infty \frac{\mathrm{d}\tau}{\tau}\exp\left[-\tau\left(\frac{M}{\Lambda}\right)^2\right], \tag{292a}$$

$$\hat{f}(\partial^2)=1-\frac{1}{6}\,\frac{\mathrm{d}}{\mathrm{d}M^2}\int_1^\infty \frac{\mathrm{d}\tau}{\tau}\exp\left[-\tau\left(\frac{M}{\Lambda}\right)^2\right]\partial_x^2+\mathrm{O}(\partial^4)\,, \tag{292b}$$

where $\hat{f}(\partial^2)$ stands for the inverse Fourier transform of the pion form factor.

To summarize, the effective action functional associated to (270) can be obtained by regrouping (283), (287), (291) and (292) together, i.e.

$$\mathrm{i}S_1+S_2=-\frac{\mathrm{i}N_c}{240\pi^2}\int_0^1 \mathrm{d}t\int \mathrm{d}^4x\,\varepsilon^{\mu\nu\alpha\beta\gamma}\,\mathrm{TR}[L_\mu L_\nu L_\alpha L_\beta L_\gamma]+\frac{f_\pi^2}{4}\int \mathrm{d}^4x\,\mathrm{TR}[\partial^\mu U^+(1+\hat{f}(\partial^2))\partial_\mu U]$$

$$+\frac{N_c}{192\pi^2}\left(\frac{\bar{S}}{M}\right)^4\left\{1+\left(\frac{M}{\Lambda}\right)^2\right\}\exp[-(M/\Lambda)^2]\int \mathrm{d}^4x\,\{\tfrac{1}{2}\,\mathrm{TR}[L_\mu, L_\nu]^2-\mathrm{TR}[L_\mu^2L_\nu^2]\}$$

$$+\frac{m_\pi^2f_\pi^2}{4}\int \mathrm{d}^4x\,\mathrm{TR}[U+U^+]+\cdots, \tag{293}$$

where m_π, f_π and $\hat{f}(\partial^2)$ are given by (288), (292a) and (292b) in Minkowski space, respectively. We have not included the field-independent effective potential in (293). Equation (293) is strikingly similar to the Skyrme Lagrangian introduced in the preceding sections. The procedure described above shows that at low energy, the sourceless QCD generating functional with "passive" gluons gives a Skyrme-like description provided that the gradient expansion holds. Whenever the field variations in space becomes large compared with the intrinsic size of the field, the gradient expansion breaks down. In this particular case, Simic [56] has argued that the effective description in terms of chiral degrees of freedom "disappears" below a certain critical size, leading to a free quark phase, much like in the topological chiral bag model.

7. Phenomenological aspects of SU(2) hedgehog skyrmions

In order to investigate the properties of the light baryons such as the nucleon and Δ-isobar, it is necessary to extract baryon states from the hedgehog configuration. The latter is an admixture of states with equal spin and isospin. Following the pioneering work of Skyrme [3], Adkins, Nappi and Witten [21] have quantized the spinning modes of the Skyrme Lagrangian, providing a systematic scheme that discriminates between states of different spin. In retrospect, their method is similar to well-known projective techniques used in deformed Hartree–Fock calculations [62] to extract states with good angular momentum. In brief, the idea consists in spinning adiabatically the hedgehog configuration in isospace using the relevant collective variables, generating a classical angular momentum, and then quantizing it using standard canonical rules. To within 30%, this method leads to a reasonable description of the static properties of the nucleon (N) and nucleon-isobar (Δ) such as the NΔ mass splitting, the nucleon charge radii, the nucleon magnetic moments, the πNN and πNΔ coupling. Their results were strikingly similar to the MIT bag model calculations, spurring speculations for a dual description of low-energy phenomenology along a "Cheshire cat" principle [63].

The quantization advocated by Adkins et al. does not exhaust all of the possible degrees of freedom associated with the skyrmion. It is therefore pertinent to ask about the possibility of quantizing other degrees of freedom, e.g., vibrational, rotational, . . . , and investigate their relevance to the low-lying resonances of the nucleon and nucleon isobar. It is clear that an all out theory of skyrmions would have to address both the ultraviolet and infrared problems inherent in the model. However, one can still make sensible predictions about excited states since only energy differences are involved. Various groups [64, 65, 66, 67] have investigated the weak coupling regime with the consistent conclusion that the skyrmion is too soft against deformation.

Kaulfuss and Meissner [66b] have, however, shown that when the skyrmion is stabilized by ω-meson exchange, essentially by a six-derivative term, rather than by the Skyrme four-derivative term, it is substantially stiffer against vibrations, as can easily be seen from simple scaling arguments. Quantitative predictions about vacuum fluctuations, quantum corrections to the skyrmion mass, . . . , can only be addressed within a renormalization scheme that is consistent with the $\pi\pi$-data. Recent investigations in this direction by Zahed, Wirzba and Meissner [19] show sizable attractive effects on the N and Δ-mass.

The central issue of the nucleon–nucleon interaction in the Skyrme model has been addressed by Jackson, Jackson and Pasquier [68] using the Born–Oppenheimer approximation. Their results are in qualitative agreement with the semiphenomenological Paris nucleon–nucleon potential [69] at low energy. The missing medium range attraction can be accounted for through soft pion exchange [19]. We will discuss this further in section 8.

7.1. *Nucleon and Δ-isobar masses*

To describe physical nucleon and Δ-isobar states, we need to quantize the classical skyrmion.

Skyrme's action (38) is invariant under global translations, rotations and isorotations, i.e.*

$$U(\boldsymbol{r}) \to U(\boldsymbol{r}-\boldsymbol{a})\,; \qquad U(\boldsymbol{r}) \to U(R\boldsymbol{r})\,; \qquad U(\boldsymbol{r}) \to AU(\boldsymbol{r})B^{+}\,.$$

For the hedgehog configuration (51), rotations are equivalent to isorotations. Consequently, the hedgehog wavefunctional can be thought of as a superposition of states with all possible values of A. This superposition can be understood as centrifugal effects [70] in the classical theory if we were to consider a time-dependent matrix $A(t)$.** Following Adkins et al. [21], we will choose the SU(2)-variable $A(t)$ as the pertinent collective coordinate, and substitute $U = A(t)U_0(\boldsymbol{x})A^{+}(t)$ into (38). In the adiabatic limit, we obtain

$$S = S_0 + \int \mathrm{d}t\, I\, \mathrm{TR}[\dot{A}\dot{A}^{+}] + \mathrm{O}(A^3)\,, \tag{294}$$

where S_0 is the static hedgehog action, and the soliton moment of inertia I is given by

$$I = \frac{8\pi}{3} f_\pi^2 \int_0^\infty r^2\,\mathrm{d}r\, \sin^2 F\left[1 + \frac{8\varepsilon^2}{f_\pi^2}\left(F'^2 + \frac{\sin^2 F}{r^2}\right)\right]. \tag{295}$$

The action (294) can be quantized by standard canonical methods. Since $A \in \mathrm{SU(2)}$, it can be locally parametrized by $A = a_0 + \mathrm{i}\boldsymbol{\tau}\cdot\boldsymbol{a}$ with $a_0^2 + \boldsymbol{a}^2 = 1$. Thus A in (294) is $A[a(t)]$, while the momentum conjugate to a_k is

$$\pi_k = \partial L/\partial \dot{a}_k = 4I\dot{a}_k\,. \tag{296}$$

The Hamiltonian associated to (294) reads

$$\begin{aligned} H &= \sum_{k=0}^{3} \pi_k \dot{a}_k - L \\ &= M + \frac{1}{8I}\sum_{k=0}^{3} \pi_k^2\,, \end{aligned} \tag{297}$$

using the standard canonical quantization procedure, i.e. $\pi_k = (-\mathrm{i})(\partial/\partial a_k)$, yields

$$H = M + \frac{1}{8I}\sum_{k=0}^{3}\left(\frac{1}{\mathrm{i}}\frac{\partial}{\partial a_k}\right)^2, \tag{298}$$

* For simplicity, we will disregard center of mass motion.

** A systematic approach consists in using the soliton vacuum-to-vacuum amplitude between states of good spin,

$$|Sm\rangle = \sum_{m'} \int_{\mathrm{SU(2)}} \mathrm{d}\mu(g)\, D^{s*}_{mm'}(g)g|0\rangle\,.$$

The resulting effective action will exhibit centrifugal effects that are of the type described by (294). The problem is therefore reduced to solving for the chiral fields in the presence of centrifugal effects. In general, the solution will be a squashed hedgehog [70], that will give rise to quadrupole deformations in the low-lying spin states. In the adiabatic approximation, these effects are neglected.

H involves the Laplacian on S^3 whose eigenstates are the generalized spherical harmonics or Jacobi polynomials, i.e. traceless symmetric polynomials in the a_i's. The spin ($\boldsymbol{J}$) and isospin ($\boldsymbol{I}$) operators in the space of collective coordinates can be obtained using the standard Noether construction. From appendix G, we have

$$J_k = \frac{i}{2}\left(a_k \frac{\partial}{\partial a_0} - a_0 \frac{\partial}{\partial a_k} - \varepsilon_{klm} a_l \frac{\partial}{\partial a_m}\right), \qquad I_k = \frac{i}{2}\left(a_0 \frac{\partial}{\partial a_k} - a_k \frac{\partial}{\partial a_0} - \varepsilon_{klm} a_l \frac{\partial}{\partial a_m}\right), \tag{299}$$

with

$$\hat{J}^2 \equiv \frac{1}{4} \sum_{k=0}^{3} \left(-\frac{\partial^2}{\partial a_k^2}\right). \tag{300}$$

The energy eigenstates of good spin and isospin are explicitly constructed in appendix G using standard angular momentum algebra on S^3. They are polynomials in a's of the form $(a_0 + i a_1)^l$. The latter carry $I = J = l/2$. In general, even polynomials carry integer spin and isospin, whereas odd polynomials carry half-integer spin and isospin. To quantize the SU(2) skyrmions as fermions will require odd polynomials.* Nucleons with $I = J = \frac{1}{2}$ correspond to wavefunctionals linear in the a's, i.e.

$$\begin{aligned} |p\uparrow\rangle &= \frac{1}{\pi}(a_1 + i a_2), \qquad |p_\downarrow\rangle = -\frac{i}{\pi}(a_0 - i a_3), \\ |n\uparrow\rangle &= \frac{i}{\pi}(a_0 + i a_3), \qquad |n\downarrow\rangle = -\frac{1}{\pi}(a_1 - i a_2), \end{aligned} \tag{301}$$

which are normalized on S^3.** The model discussed above generates a tower of states with $I = J = \frac{1}{2}, \frac{3}{2}, \frac{5}{2}, \ldots$. Jackson [71] has argued by using a non-relativistic quark model, that for N_c colors in a hedgehog configuration $I_{max} = J_{max} = N_0/2$, suggesting that for $N_c = 3$, only $I = J = \frac{1}{2}, \frac{3}{2}$ are relevant, the rest being spurious. It follows that (298) is diagonal in this basis,

$$\langle I = JM | H | I = JM \rangle = M + J(J+1)/2I. \tag{302}$$

Specifically, the nucleon and Δ-isobar masses are given by

$$M_N = M + 3/8I, \qquad M_\Delta = M + 15/8I, \tag{303a}$$

which shows that the NΔ mass splitting is rotational in this model,

$$M_\Delta - M_N = 3/2I. \tag{303b}$$

With the parameters advocated by Adkins et al. [21], this splitting is set up to its experimental value, while with the Jackson et al. [20] parameters $M_\Delta - M_N = 280$ MeV as shown in table 2. Notice that while $M = O(N_c)$, $M_\Delta - M_N = O(1/N_c)$ since $I = O(N_c)$. The latter follows from the facts that $f_\pi = O(\sqrt{N_c})$

* In their original work, Finkelstein and Rubinstein [5] showed that one can quantize SU(2) solitons by requiring $\psi(-A) = +\psi(A)$ for bosons and $\psi(-A) = -\psi(A)$ for fermions, as expected from the spin-statistics arguments of section 3, since $\pi_4(SU(2)) = Z_2$.

** The normalization on S^3 follows from (317).

Table 2a

The low-energy parameters and masses as predicted in the Skyrme model, together with the MIT bag model results are compared to experiment. (A.N.W.) and (J.R.) are the predictions of ref. [21] and ref. [20] respectively, for $m_\pi = 0$. (A.N.) and (M.) are the results of ref. [22a] and ref. [22b] respectively, for $m_\pi \neq 0$

	A.N.W.	A.N.	J.R.	M.	MIT	Expt
$10^3\varepsilon^2$	4.21	5.34	5.52	5.13	–	–
f_π (MeV)	64.5	54.	93*	94.3*	149	93
g_A^N	1.02	1.08	1.33*	1.23*	1.09	1.23
M_N (MeV)	939*	939*	1425	1385	939*	939
$M_\Delta - M_N$ (MeV)	293*	293*	283	310	293*	293

Table 2b

Same comparative table as in (a) for the electric and magnetic mean square radii together with the proton and neutron magnetic moments

	A.N.W.	A.W.	J.R.	M.	MIT	Expt
$\langle r_E^2\rangle_0^{1/2}$ (fm)	0.59	0.68	0.47	0.43	0.76	0.72
$\langle r_E^2\rangle_1^{1/2}$ (fm)	∞	1.04	∞		0.76	0.88
$\langle r_M^2\rangle_0^{1/2}$ (fm)	0.92	0.95			0.62	0.81
$\langle r_M^2\rangle_1^{1/2}$ (fm)	∞	1.04	∞		0	0.80
μ_p (n.m.)	1.87	1.97	2.74	2.43	1.90	2.79
μ_n (n.m.)	−1.31	−1.24	−2.24	−1.98	−1.27	−1.91

and $f_\pi/\varepsilon = O(1)$ through (295). That the splitting goes like $1/N_c$ is consistent with the large N_c limit. (Notice that vibrations are of order O(1) in this model. They will be discussed later on.) The relations of the rotational term (303b) to conventional quark models has been clarified by Brown et al. [72] and Witten [73]. Ground state baryon wavefunctions are composed of N_c quarks, in the lowest partial wave. One-gluon exchange between any pair of quarks result in the spin-dependent Fermi–Breit interaction of the form

$$\begin{aligned} U_{1G} &= \tfrac{1}{8} v \sum_{i\neq j} \boldsymbol{\sigma}_i \boldsymbol{\sigma}_j \\ &= \tfrac{1}{2} v\Big(\sum_i \tfrac{1}{2}\boldsymbol{\sigma}_i\Big)^2 - \tfrac{1}{8} v \sum_i \boldsymbol{\sigma}_i^2 \\ &= \tfrac{1}{2} v \boldsymbol{J}^2 - \tfrac{1}{8} v N_c \end{aligned}$$

where $\boldsymbol{J} = \Sigma_i \frac{1}{2}\boldsymbol{\sigma}_i$ is the total angular momentum, and v an overall constant. In this description $I = v^{-1}$.

7.2. Axial coupling g_A

The axial coupling g_A measures the spin–isospin correlation in the nucleon, and is defined as the expectation value of the axial current $A_\mu^a(x)$ in a nucleon state at zero momentum transfer. The matrix element of the axial current in a nucleon state is defined through ($q = p_2 - p_1$),

$$\langle N(p_2)|A_\mu^a|N(p_1)\rangle = \bar{U}(p_2)\frac{\tau^a}{2}[g_A(q^2)\gamma_\mu\gamma_5 + h_A(q^2)q_\mu\gamma_5]U(p_1)\,. \tag{304}$$

In the chiral limit $h_A(q^2)$ has a pion pole whose residue is $-2f_\pi g_{NN}$, i.e.

$$h_A(q^2) = d_A(q^2)/q^2 \quad \text{with} \quad d_A(0) = -2f_\pi g_{\pi NN}\,, \tag{305}$$

where $g_{\pi NN}$ is the pion–nucleon coupling. Current conservation implies that

$$2M_N g_A(q^2) + q^2 h_A(q^2) = 0\,. \tag{306}$$

Using the non-relativistic limit $(q \to 0)$ in the nucleon rest frame yields

$$\lim_{q\to 0} \langle N(p_2)|A_i^a|N(p_1)\rangle = \lim_{q\to 0} \bar{U}\, \frac{\tau^a}{2} \left[g_A(0)\sigma_i + \frac{d_A(0)}{2M_N}\, \boldsymbol{\sigma}\cdot\hat{q}\hat{q}_i \right] U$$
$$= \lim_{q\to 0} g_A(0)(\delta_{ij} - \hat{q}_i\hat{q}_j)\langle N|\sigma_i \frac{\tau^a}{2}|N\rangle\,, \tag{307}$$

where we have used the Goldberger–Treiman relation (11). As noted by Adkins et al. [21], the ambiguity in (307) can be circumvented by taking the limit symmetrically, i.e. $\hat{q}_i\hat{q}_j \leftrightarrow \frac{1}{3}\delta_{ij}$, so that

$$\lim_{q\to 0} \langle N(p_2)|A_i^a|N(p_1)\rangle = \tfrac{2}{3} g_A(0)\langle N|\sigma_i \frac{\tau^a}{2}|N\rangle\,. \tag{308}$$

For the skyrmion, the axial current A_i^a can be obtained as the pertinent $SU(2)_A$ Noether current. (For details see appendix E.) To specialize to nucleon states, we will use the collective coordinate A to project out spin. From appendix E, we have

$$\int \mathrm{d}\boldsymbol{x}\, A_i^a(\boldsymbol{x}) = -\tfrac{1}{2} g\, \mathrm{TR}(\tau_i A^+ \tau_a A)\,, \tag{309a}$$

$$g = +\tfrac{4}{3}\pi f_\pi^2 \int_0^\infty r^2\, \mathrm{d}r \left\{ \left(F' + \frac{\sin 2F}{r}\right) + 4\frac{f_\pi^2}{\varepsilon^2}\left(F'^2 \frac{\sin 2F}{r} + 2F' \frac{\sin^2 F}{r^2} + \frac{\sin^2 F}{r^3} \sin 2F\right)\right\}. \tag{309b}$$

Between the nucleon states (301), we obtain

$$\lim_{q\to 0} \int \mathrm{d}\boldsymbol{x}\, \exp(\mathrm{i}\boldsymbol{q}\cdot\boldsymbol{x})\langle N|A_i^a(\boldsymbol{x})|N\rangle = \tfrac{1}{2} g\langle N|\, \mathrm{TR}(\tau_i A^+ \tau_a A)|N\rangle$$
$$= +\tfrac{2}{3} g\langle N|\sigma_i \frac{\tau^a}{2}|N\rangle\,. \tag{310}$$

Identifying (308) with (310) yields $g_A(0) = g$ as given by (309b). The parameters of Adkins et al. yield* $g_A(0) = 0.61$, a bit too low in comparison with the experimental value of 1.33. This is expected from the asymptotic behavior of the pion field as discussed in section 2.

* See section 2 and the recent work by Karl and Paton [74] for possible N_c corrections.

7.3. Charge radii and magnetic moments

The distribution of matter in the nucleon can be characterized by the isoscalar charge radius. In terms of the normalized baryon density (58), we have

$$r_0 = \langle r^2 \rangle_{I=0}^{1/2} = \left(-\frac{2}{\pi} \int_0^\infty dr\, r^2 \sin^2 F\, F' \right)^{1/2} . \tag{311}$$

Adkins et al. [21] predict $r_0 = 0.59$ fm, while Jackson et al. [20] have $r_0 = 0.48$ fm, which are to be compared with the experimental value of 0.72 fm. It is well known that the isovector electric and magnetic charge radii are infinite for massless pions [75]. Indeed, the power fall-off of the pion field in the chiral limit causes the isovector charges to spread out at infinity, resulting in divergent quantities. In the broken phase, the pion field falls off exponentially, leading to finite electric and magnetic isovector charge radii as discussed in ref. [22].

The isoscalar and isovector magnetic moments in the nucleon rest frame are defined respectively as ($e = \hbar = c = 1$)

$$\boldsymbol{\mu}_{I=0} = \tfrac{1}{2} \int d\boldsymbol{r}\, \boldsymbol{r} \times \boldsymbol{B} , \tag{312}$$

$$\boldsymbol{\mu}_{I=1} = \tfrac{1}{2} \int d\boldsymbol{r}\, \boldsymbol{r} \times \boldsymbol{V}^3 , \tag{313}$$

where $\boldsymbol{B}$ is the baryon current, and $\boldsymbol{V}^3$ the third component of the isovector current (for more details see appendix E). For an adiabatically rotating skyrmion $\boldsymbol{B} \neq 0$, i.e.

$$B^i = \mathrm{i} \frac{\varepsilon^{ijk}}{2\pi^2} \frac{\sin^2 F}{r} F' \hat{r}^j\, \mathrm{TR}[\tau^k \dot{A}^+ A] . \tag{314}$$

Injecting (314) into (312) and using the spin-up proton state in (301), yields the isoscalar magnetic moment in the form

$$\begin{aligned} (\mu_{I=0})_3 &= -\frac{2\mathrm{i}}{3\pi} \times \int_0^\infty dr\, r^2 F' \sin^2 F \times \langle p\uparrow | \mathrm{TR}(\tau_3 \dot{A}^+ A) | p\uparrow \rangle \\ &= +\frac{\mathrm{i}}{3} \times \langle r^2 \rangle_{I=0} \times \langle p\uparrow | \mathrm{TR}(\tau_3 \dot{A}^+ A) | p\uparrow \rangle . \end{aligned} \tag{315}$$

The matrix element in (315) can be written as follows

$$\begin{aligned} \langle p\uparrow | \mathrm{TR}(\tau^3 \dot{A}^+ A) | p\uparrow \rangle &= 2\mathrm{i} \langle p\uparrow | (\dot{\boldsymbol{a}} \times \boldsymbol{a})_3 | p\uparrow \rangle \\ &\equiv -\frac{1}{2I} \varepsilon_{3kl} \langle p\uparrow | a_k \frac{\partial}{\partial a_l} | p\uparrow \rangle = -\frac{\mathrm{i}}{2I} \end{aligned} \tag{316}$$

where we have used the canonical prescription (296), and the fact that

$$\int_{S^3} d\mu(a)\, a_i a_k = \tfrac{1}{4}\delta_{ik} \int_{S^3} d\mu(a) = \tfrac{1}{2}\pi^2 \delta_{ik} . \tag{317}$$

In terms of (316) the isoscalar magnetic moment in a proton state reduces to

$$\mu_p^{I=0} = \langle r^2 \rangle_{I=0}/6I = \frac{1}{g}\,(M_\Delta - M_N)\langle r^2 \rangle_{I=0} . \tag{318}$$

The proton and neutron isoscalar magnetic moments being equal, yields

$$\mu_p^{I=0} + \mu_n^{I=0} = \left(\frac{4}{g}\,(M_\Delta - M_N) M_N \langle r^2 \rangle_{I=0}\right)\mu_B , \tag{319}$$

where $\mu_B = 1/2M_N$ is the Bohr magneton ($e = \hbar = c = 1$). In units of Bohr magnetons both Adkins et al. [21] and Jackson et al. [22] predict a value of 1.11, which is to be compared with the experimental value of 1.76.

The isovector magnetic moment can be obtained through similar arguments. Indeed, by using (313) and (E.32) we obtain

$$(\mu_{I=1})_3 = -\tfrac{4}{3}\pi f_\pi^2 \int_0^\infty dr\, r^2 \sin^2 F\left(1 + \frac{8\varepsilon^2}{f_\pi^2}\left(F'^2 + \frac{\sin^2 F}{r^2}\right)\right)\langle p\uparrow |\, \mathrm{TR}(\tau_3 A^+ \tau_3 A) | p\uparrow \rangle .$$

The matrix element in (320) reduces to

$$\begin{aligned}\langle p\uparrow |\, \mathrm{TR}(\tau_3 A^+ \tau_3 A) | p\uparrow \rangle &= 2\langle p\uparrow | (1 - 2(a_1^2 + a_2^2)) | p\uparrow \rangle \\ &= 2 - \frac{4}{\pi^2}\int_{S^3} d\mu(a)(a_1^2 + a_2^2)^2 .\end{aligned} \tag{321}$$

Using a polar parametrization of S^3, we can rewrite the remaining integral in (321) in the form

$$\int_{S^3} d\mu(a)(a_1^2 + a_2^2)^2 = \int_0^\pi d\alpha \sin^2\alpha \int_0^\pi d\beta \sin\beta \int_0^{2\pi} d\gamma \sin^4\alpha \sin^4\beta = \tfrac{2}{3}\pi^2 , \tag{322}$$

hence

$$\langle p\uparrow |\, \mathrm{TR}(\tau_3 A^+ \tau_3 A) | p\uparrow \rangle = -\tfrac{2}{3} . \tag{323}$$

In terms of (295) and (323), the isovector magnetic moment (320) in a proton state is

$$\mu_p^{I=1} = \tfrac{1}{3} I = 1/2(M_\Delta - M_N) . \tag{324}$$

Since $\mu_n^{I=1}$ has opposite sign, we deduce

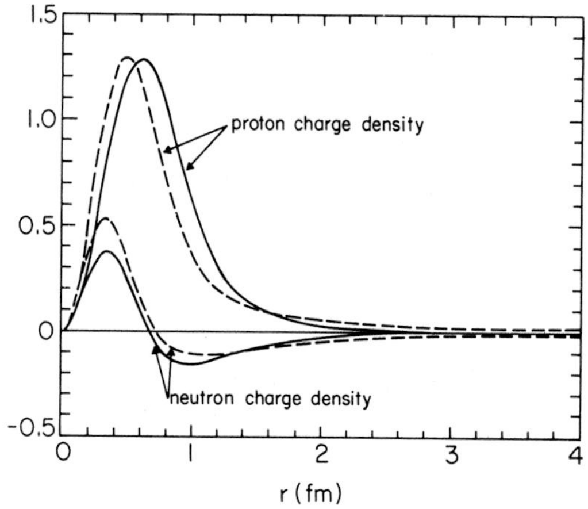

Fig. 13. Proton and neutron charge densities. The continuous curves represent the charge distributions for $m_\pi = 0$, while the dashed curves refer to the massless case $m_\pi = 0$.

$$\mu_{\rm p}^{I=1} - \mu_{\rm n}^{I=1} = \left(\frac{2M_{\rm N}}{M_\Delta - M_{\rm N}}\right)\mu_{\rm B}\,. \tag{325}$$

In units of Bohr magnetons Adkins et al. [21] predict 6.41 while Jackson et al. [20] have 10.61. These results are to be compared with the empirical value of 9.4. The magnetic moments for the proton and neutron measured in Bohr magnetons are

$$\mu_{\rm p,\,n} = \frac{1}{2\mu_{\rm B}}\,(\mu_{p,\,n}^{I=0} + \mu_{\rm p,\,n}^{I=1})\,. \tag{326}$$

Adkins et al. [21] predict $\mu_{\rm p} = 1.88$, $\mu_{\rm n} = -1.32$ and $|\mu_{\rm p}/\mu_{\rm n}| = 1.42$; Jackson et al. [20] have $\mu_{\rm p} = 2.93$, $\mu_{\rm n} = -2.38$ and $|\mu_{\rm p}/\mu_{\rm n}| = 1.23$. These results should be compared with the experimental values $\mu_{\rm p} = 2.79$, $\mu_{\rm n} = -1.91$ and $|\mu_{\rm p}/\mu_{\rm n}| = 1.46$. In the broken phase ($m_\pi \neq 0$), the predicted results are almost unaltered [22]. In order to improve the isovector magnetic moment description in this model, one needs to take into account pion rescattering corrections that yield ρ-meson-like effects [76], much in the spirit of vector meson dominance.

7.4. *Low-lying resonances in the Skyrme model*

Within 30% accuracy, the Skyrme model seems to give a reasonable description of the static properties of the nucleon and Δ-isobar. It is therefore tempting to use the present framework to investigate dynamical properties of skyrmions, and their possible relevance to the baryon spectrum. This question has been studied by a number of authors [64–67] using either the semi-classical approximation or a variational principle.

In the large N_c limit, the semi-classical treatment seems to be the relevant method for investigating chiral fluctuations about the skyrmion. Indeed, since $f_\pi \sim \sqrt{N_c}$ and $\varepsilon \sim \sqrt{N_c}$, one can factorize out N_c from the Skyrme Lagrangian (38), so that in the large N_c limit its associated vacuum-to-vacuum expectation value is dominated by the saddle point equations. Pion scattering off the skyrmion can be described by standard quadratic fluctuations about the pertinent classical configuration.

At this stage, the non-renormalizability of the model should not be an issue since we are only interested in energy differences while probing the low-lying spectrum of the skyrmion. Moreover, we will ignore the fact that the skyrmion is spinning, and neglect possible recoil corrections.

To investigate $K^\pi = 0^+$ quantum fluctuations about the skyrmion, consider the time-dependent hedgehog ansatz

$$U(\boldsymbol{r}, t) = \exp(\mathrm{i}\boldsymbol{\tau} \cdot \hat{r}(F(r) + \xi(r, t))) , \qquad (327)$$

where $F(r)$ refers to the usual skyrmion profile. The Euler–Lagrange equation associated to (327), reduces to a time-dependent equation for the spherical chiral angle $\xi(r, t)$, i.e.

$$\left(1 + \frac{16\varepsilon^2}{f_\pi^2} \frac{\sin^2 F}{r^2}\right)\ddot{\xi} - \left(1 + \frac{16\varepsilon^2}{f_\pi^2} \frac{\sin^2 F}{r^2}\right)\xi'' - \left(1 + \frac{8\varepsilon^2}{f_\pi^2} \frac{\sin 2F}{r} F'\right) \frac{2\xi'}{r}$$
$$+ \left(\frac{2\cos 2F}{r^2} - \frac{16\varepsilon^2}{f_\pi^2} \left(\frac{F''\sin 2F + F'^2\cos 2F}{r^2} - \frac{\sin^2 2F - \sin^2 F}{r^4}\right)\right)\xi = 0 , \qquad (328)$$

with $\xi(0; t) = \xi(\infty; t) = 0$ for all times, so that the entire baryon number is carried out by the skyrmion. The asymptotic behavior of (328) can be easily worked out using the harmonic ansatz $\xi(r, t) = \mathrm{e}^{\mathrm{i}\omega t}\xi(r)$. At large distances $F(r) \to K^2/r^2$, and one recovers free pions

$$\xi(r) \to \alpha(\omega)j_1(\omega r) + \beta(\omega)n_1(\omega r) , \qquad (329)$$

where $j_1(x)$ and $n_1(x)$ are the usual spherical Bessel and Neumann functions respectively. The constants α and β determine the phase shift $\delta(\omega)$ of the scattered pions. Indeed, by rewriting (329) in terms of Hankel functions, we have

$$\xi(r) \to \mathrm{const} \times (h_1(\omega r) - S_0(\omega)h_1^+(\omega r)) , \qquad (330)$$

where the S-matrix in this channel is given by

$$S_0(\omega) = \mathrm{e}^{2\mathrm{i}\delta(\omega)} = \frac{\alpha(\omega) - \mathrm{i}\beta(\omega)}{\alpha(\omega) + \mathrm{i}\beta(\omega)} , \qquad (331)$$

which lies on the unit circle. At this stage two remarks are in order. First, (331) shows a purely elastic scattering in the 0^+-channel. This feature will not hold when the skyrmion is rotated to get states of good spin and isospin. The adiabatic rotation couples $K = 0$ and $K = 1$ channels, leading to inelastic scattering in the 0^+-channel. Second, the phase shift $\delta(\omega)$ in (331) sums up both the resonant $\delta_\mathrm{R}(\omega)$ and background $\delta_\mathrm{B}(\omega)$ phase-shifts,

$$\delta(\omega) = \delta_\mathrm{R}(\omega) + \delta_\mathrm{B}(\omega) . \qquad (332)$$

To extract resonance properties in the 0^+ channel, one has to either solve the scattering equation (328) in the complex energy plane and look for the poles of (331), or equivalently fit a Breit–Wigner profile in the region where the total phase-shift (332) varies rapidly. Although for elastic scattering $\delta_R(\omega)$ has to go from 0 to π, the background phase-shift $\delta_B(\omega)$ might be strong enough to suppress it. Figure 13 displays the total phaseshift (332) up to 2 GeV excitation energies, for the parameters of Jackson et al. [20].* There is a sharp increase of $\delta(\omega)$ about $\omega \sim 300$ MeV. Using a Breit–Wigner fit, one obtains a resonance energy $E_R \sim 260$ MeV and a width $\Gamma_R \sim 180$ MeV. Although a bit too low, this resonance has been identified with the $P_{11}(1470)$ Roper resonance observed in the pion–nucleon channel.** There are no bound states in this channel, suggesting that the hedgehog configuration is locally stable against 0^+-deformations.

To investigate $K^\pi = 1^\pm$ fluctuations about the skyrmion, consider instead the time-dependent ansatz

$$U(\boldsymbol{r}, t; K=1) = \exp(\mathrm{i}\boldsymbol{\tau}\cdot\hat{r}F(r))\exp(\mathrm{i}X^{k,I}(\boldsymbol{r}, t)\tau^I)\,. \tag{333}$$

It is easy to show that (333) transforms as a vector under rotations in K-space, where $X^{K,I}(\boldsymbol{r}, t)$ is the most general tensor of rank-2 obtained by recoupling a vector in K-space ($K = 1$) to a vector in isospace ($I = 1$). Using simple tensor algebra we have***

$$[K=1]\times[I=1] = [0]+[1]+[2]\,, \tag{334}$$

hence,

$$X^{K,I}(\boldsymbol{r}, t) = \delta^{KI}X_0(r, t) + \varepsilon^{KIM}\hat{r}^M X_1(r, t) + (\hat{r}^K\hat{r}^I - \tfrac{1}{3}\delta^{IK})X_2(r, t)\,, \tag{335}$$

where $X_k(r, t)$ are regular functions of r and t, subject to the boundary conditions $X_R(0, t) = X_k(\infty, t) = 0$. Since the main role of the quartic term in the Skyrme Lagrangian is to stabilize the hedgehog configuration, its relevance to the quantum fluctuations is expected to be limited at low-excitation energies. Otherwise, the assumption of neglecting higher-order field gradients in the starting dynamics does not make sense and the baryon spectrum becomes heavily model dependent. With this in mind, the Euler–Lagrange equations for $X_k(r, t)$ decouple (remember that X_1 decouples from X_0, X_2 anyway because of parity), and one obtains

$$\left\{-4\pi f_\pi^2\left(\frac{\mathrm{d}^2}{\mathrm{d}r^2} - F'^2 - \frac{2\sin^2 F}{r^2}\right) + \omega_0^2\right\}F_0(r) = 0\,, \tag{336a}$$

$$\left\{-\tfrac{2}{3}\pi f_\pi^2\left(\frac{\mathrm{d}^2}{\mathrm{d}r^2} - \frac{6}{r^2} - F'^2 + 4\,\frac{\sin^2 F}{r^2}\right) + \omega_2^2\right\}F_2(r) = 0\,, \tag{336b}$$

$$\left\{-8\pi f_\pi^2\left(\frac{\mathrm{d}^2}{\mathrm{d}r^2} - \frac{2}{r^2} - F'^2\right) + \omega_1^2\right\}F_1(r) = 0\,,$$

* It is easy to show that for a given value of $\bar{\omega} = \varepsilon\omega/f_\pi$, the shape of the phase-shift $\delta(\bar{\omega})$ is uniquely determined. Changing ε/f_π, amounts to rescaling the energy axis in fig. 13. This explains why the results of Zahed et al. are similar to the ones obtained by Breit et al.

** As noted in the beginning of section 7, an improved description of the Roper resonance is obtained when the skyrmion is stabilized by ω-meson exchange [66b].

*** This construction can be generalized to higher K-fluctuations through

$$[K]\times[I=1] = [K-1]+[K]+[K+1]\,,$$

using vector spherical harmonics as discussed in ref. [67].

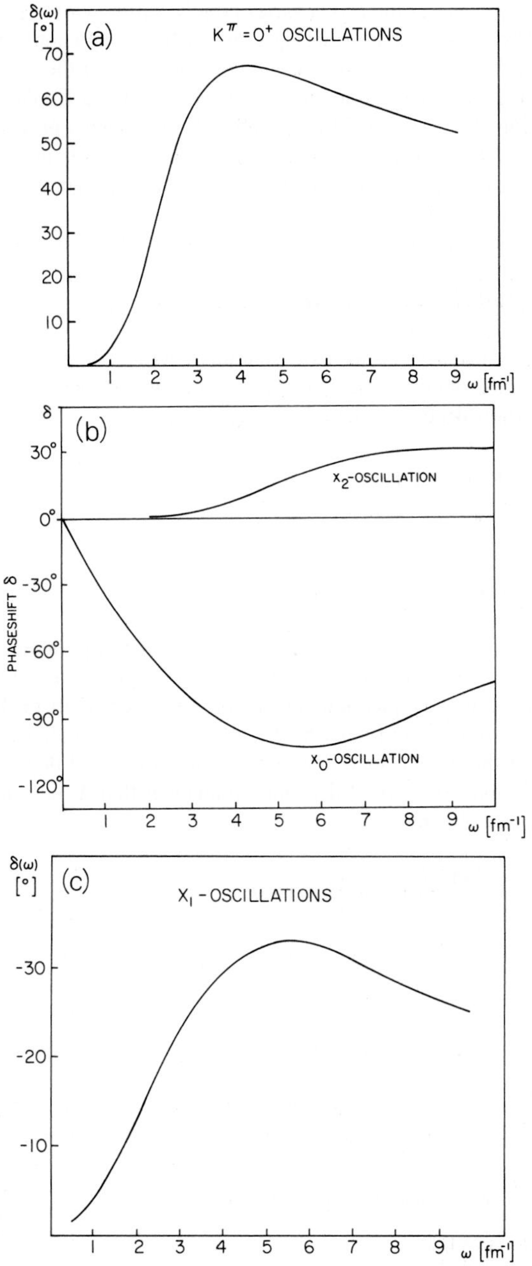

Fig. 14. (a) Phase shift of a meson scattering off a skyrmion in the $K=0^+$-channel as discussed in ref. [64]. (b) and (c) are phase shifts of a meson scattering off a skyrmion in the $K=1^-$-channel, also from ref. [64].

where we have defined $X_k(r, t) = [F_k(r)/r] \exp(i\omega_k t)$. Equation (336a) describes $K^\pi = 1^-$ fluctuations, while (336b) describes $K^\pi = 1^+$ ones. The scattering states of (336) can be analyzed in the same way as discussed above. The total phase-shifts are shown in fig. 14. In the $K^\pi = 1^-$ channel, the total phase shift associated to the X_0 fluctuations show an increase about $\omega \sim 1580$ MeV excitation energy, associated to the center of mass of the second group of resonances observed in the 1^--channel in pion–nucleon scattering. The model misses the first group of resonances which lie about $\omega \sim 700$ MeV excitation energy, also known as the $S_{11}(1650)$ and $D_{13}(1700)$ resonances. To this order, there is no resonance in the $K^\pi = 1^+$-channel.

Recently, a more thorough analysis of the baryon spectrum has been carried out independently at Siegen University [66] and SLAC [67], with similar conclusions concerning the low-lying resonances. For most of the higher partial waves, the model reproduces surprising well many of the significant features of the experimental results. Table 3 displays the SLAC results up to 2 GeV excitation energy. the bad predictions in the S- and P-channels may stem from their coupling to translational and rotational zero modes, which are expected to receive corrections at the next order in the semiclassical analysis of the scattering amplitude. It should be noted that a very natural description of

Table 3a

Skyrme model predictions for baryon resonances up to 2 GeV excitation energy from ref. [67]. Column [1] uses the parameters of (A.N.W.) with a fixed nucleon and Δ-isobar mass, while [2] uses a least square fit to the resonance spectrum

Channel	Experiment	[1]	% Error	[2]	% Error
S11	1526	1295	−15	1478	−3
S31	1610	1295	−20	1478	−8
P11	939	939	0	1190	27
P11	1723	1233	−28	1427	−17
P13	1710	1919	12	1982	16
P31	1888	1919	2	1982	5
P33	1232	1436	17	1424	16
P33	1522	1242	−18	1435	−6
P33	1868	1874	0.3	1946	4
D13	1519	1589	5	1715	13
D15	1679	1625	−3	1744	4
D33	1680	1616	−4	1737	3
D35	1901	1607	−15	1730	−9
F15	1684	1723	2	1823	8
F17	2005	1954	−3	2011	0.3
F35	1905	1856[a]	−3	1931[b]	1
F37	1913	1714	−10	1816	−5
G17	2140	2034	−5	2075	−3
G19	2268	2230	−2	2234	−2
G37	2215	2141	−3	2162	−2
G39	2468	2043	−17	2083	−16
H19	2205	2346	6	2327	6
H39	2217	2444	10	2407	9
H311	2416	2346	−3	2327	−4
I111	2577	2631	2	2558	−1
I313	2794	2658	−5	2579	−8
K113	2612	3032	16	2882	10
K315	2990	2943	−2	2810	−6

[a] Average of two peaks at 1732 and 1981 MeV.
[b] Average of two peaks at 1831 and 2032 MeV.
Fit #1 – Nucleon mass fixed.
Fit #2 – Nucleon mass allowed to vary.

Table 3b

Displays of the nucleon properties following fit [2] from ref. [67]

	A.N.W.	M.K.	Exp.
f_π (MeV)	63.5	75	93
$10^3 \varepsilon^2$	4.21	5.45	–
μ_p	1.87	2.23	2.79
μ_n	−1.31	−1.81	−1.91
$\langle r_E^2 \rangle_0^{1/2}$ (fm)	0.59	0.58	0.72
$\langle r_M^2 \rangle_0^{1/2}$ (fm)	0.92	0.90	0.81

the P-channels, essentially the odd-parity states of the nucleon, results in either the constituent quark models [77a] or in the chiral bag model [77b] in terms of promoting one quark from the $S_{1/2}$-orbit to a P-orbit.

7.5. *Soft-pion corrections to the masses*

In the context of current algebra [4], Skyrme's program will remain unsatisfactory in the absence of a consistent quantum description that would account for soft-pion skyrmion dynamics in agreement with soft-pion threshold theorems. To investigate soft-pion skyrmion dynamics we need to consider (38) beyond tree level. Unfortunately, the model as it stands is both infrared and ultraviolet divergent. What this means is that beyond tree level, the model requires counter terms of increasing complexity to cure the infinities arising from higher-loop calculations. The model is not renormalizable. However, it does yield a finite S-matrix.

Some years ago, Weinberg [16] has shown that the leading term in (38) describes uniquely $\pi\pi$-scattering to leading order in the pion momentum. Consistency with unitarity requires second to leading order terms. From ref. [19], we have*

$$\mathscr{L} = -\tfrac{1}{4}F_\pi \operatorname{TR}[L_\mu^2] + \tfrac{1}{4}\varepsilon^2 \operatorname{TR}[L_\mu, L_\nu]^2 + \tfrac{1}{4}k^2 \operatorname{TR}[L_\mu, L_\nu]_+^2 . \tag{337}$$

Since loop contributions are suppressed in the sense of a low-energy limit by powers of p^2, one needs to consider the 1-loop effect about the quadratic term in (337) along with tree-level contributions from the quartic terms to order $O(p^4)$. (For more details see appendix B.) In this spirit, it is possible to make the model meaningful to 1-loop order.

To probe the soft-pion corrections to the nucleon and Δ-masses to 1-loop order, it is sufficient to consider left or right fluctuations about the quadratic term only [19]. For that, define

$$U(x, t) = U(\boldsymbol{x})R(\phi) , \qquad R(\phi) = \exp(\mathrm{i}\phi/f_\pi) . \tag{338}$$

Injecting (338) into (337) and neglecting pion–skyrmion correlations with large momentum transfer yields

$$\mathscr{L}_1 = \mathscr{L}_0 + \tfrac{1}{4} \operatorname{TR}((\partial_\mu\phi)^2) + \tfrac{1}{4} \operatorname{TR}([\phi, \partial_\mu\phi]L^\mu) , \tag{339}$$

where we have explicitly used the fact that $U(x)$ satisfies the classical equation of motion

$$\partial^\mu\Big(L_\mu - \frac{2\varepsilon^2}{f_\pi^2}[L_\nu, [L_\mu, L_\nu]] - \frac{2k^2}{f_\pi^2}[L_\nu, [L_\mu, L_\nu]_+]_+\Big) = 0 . \tag{340}$$

In Euclidean space, the soft-pion contribution to the "vacuum-to-vacuum" amplitude follows through

$$Z \sim \exp(S_0^{\mathrm{E}}) \int_x \pi \,\mathrm{d}\phi(x) \exp\Big[-\tfrac{1}{2}\Big(\int \mathrm{d}^4x\, \phi^a(x)\, \hat{D}_{\mathrm{E}}^{ab}\phi^b(x)\Big)\Big] , \tag{341}$$

* Pion form factors are omitted in this discussion.

where the differential operator $\hat{D}_{\mathrm{E}}$ is given by

$$\hat{D}^{ab}_{\mathrm{E}} = \delta^{ab}\partial^2 + \mathrm{TR}[\tau^a L_\mu \tau^b]\partial_\mu \equiv \delta^{ab}\partial^2 + \hat{L}^{ab}_\mu \partial_\mu \,. \tag{341a}$$

Performing the Gaussian integral in (341) yields

$$\begin{aligned} S^{\mathrm{E}}_1 &= S^{\mathrm{E}}_0 - \tfrac{1}{2}\,\mathrm{TR}\ln \hat{D}_{\mathrm{E}} \\ &= S^{\mathrm{E}}_0 - \tfrac{1}{2}\,\mathrm{TR}\ln(\partial^2) - \tfrac{1}{2}\sum_{k=1}^{\infty} \frac{(-1)^{k+1}}{k}\,\mathrm{TR}((\partial^2)^{-1}\hat{L}\cdot\partial)^k \,. \end{aligned} \tag{342}$$

The general contribution to (342) is diagrammatically shown in fig. 15. The second term in (342) is the free space divergence and can be subtracted away since it does not affect the present dynamics. Consistency with the tree-level expansion (337) shows that only $k = 1, 2, 3, 4$ contribute to S^{E}_1 as shown in fig. 16. Higher-order loops yield higher-field gradients which are suppressed at low energy. Using a Pauli–Villars regulator with two mass scales m and M in the infrared and ultraviolet sector respectively, gives

$$S^{\mathrm{E}}_1 = S^{\mathrm{E}}_0 + \frac{1}{768\pi^2}\ln\left(\frac{m^2}{M^2}\right)\int \mathrm{d}^4x\,\{\mathrm{TR}[L_\mu, L_\nu]^2 - 3\,\mathrm{TR}[L_\mu, L_\nu]^2_+\} + \mathrm{O}(L^6)\,. \tag{343}$$

The nature of the divergences occurring at 1-loop order are similar to the tree-level terms retained in the chiral expansion (337), making the model renormalizable to this order. In the trivial sector, this calculation agrees with Weinberg's calculation [16] of the $\pi\pi$-scattering amplitude based on the renormalization group equations. One can set the infrared cutoff to the pion mass ($m = m_\pi \sim 140\,\mathrm{MeV}$), and adjust the ultraviolet cutoff to the η-mass ($M = m_\eta \sim 560\,\mathrm{MeV}$). At higher energies, at least in the trivial sector, the model must be extended to account for the vector meson resonances such as the ρ and ω. With this in mind, we have [19]

$$\begin{aligned} S^{\mathrm{E}}_1 = \int \mathrm{d}^4x \Big\{ &-\frac{f_\pi^2}{4}\,\mathrm{TR}[L_\mu^2] + \left(\frac{\varepsilon^2}{4} + \frac{1}{768\pi^2}\left(1 + \ln\left(\frac{m_\pi^2}{m_\eta^2}\right)\right)\mathrm{TR}[L_\mu, L_\nu]^2\right) \\ &+ \left(\frac{k^2}{4} - \frac{1}{256\pi^2}\left(1 + \ln\left(\frac{m_\pi^2}{m_\eta^2}\right)\right)\mathrm{TR}[L_\mu, L_\nu]^2_+\right)\Big\}\,, \end{aligned} \tag{344}$$

which summarizes the soft-pion corrections to (337).

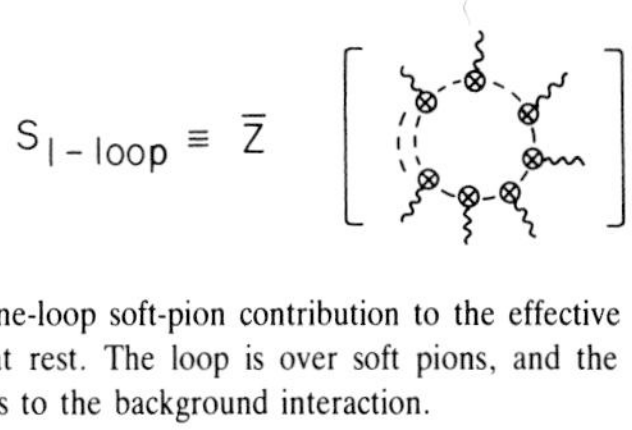

Fig. 15. One-loop soft-pion contribution to the effective action of a skyrmion at rest. The loop is over soft pions, and the wiggly line corresponds to the background interaction.

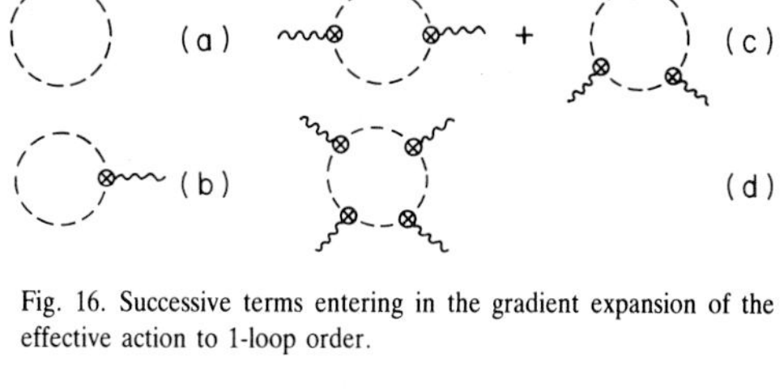

Fig. 16. Successive terms entering in the gradient expansion of the effective action to 1-loop order.

For the hedgehog configuration discussed in the preceding section ($k=0$), it is straightforward to deduce the soft-pion contribution to the mass term in the form

$$\Delta M_{\mathrm{s}} = \frac{1}{24\pi}\left(1+\ln\left(\frac{m_\pi^2}{m_\eta^2}\right)\right)\int_0^\infty \mathrm{d}r\, r^2\left\{3F'^4 + 4F'^2\,\frac{\sin 2F}{r^2} + 8\,\frac{\sin^4 F}{r^4}\right\}, \tag{345}$$

which pushes down the Skyrmion mass by 20% as quoted in table 3. The resulting corrections to the nucleon and Δ-isobar can be obtained by spinning the skyrmion. Zahed et al. [19] have shown that the decrease in the moment of inertia due to soft-pion fluctuations is given by

$$\Delta I = \frac{1}{9\pi}\left(1+\ln\left(\frac{m_\pi^2}{m_\eta^2}\right)\right)\int_0^\infty \mathrm{d}r\, r^2\sin^2 F\left\{3F'^2 + 5\,\frac{\sin^2 F}{r^2}\right\}. \tag{346}$$

This pushes down the nucleon- and Δ-isobar masses as shown in table 3.

These soft-pion corrections to the masses are attractive and substantial. As we will show next, they provide the missing medium-range attraction in the central channel of the NN-potential in the Skyrme model. This effect is important for nuclear binding.

7.6. The skyrmion–skyrmion interaction

Recently, there has been considerable interest in understanding the nucleon–nucleon interaction, and dense hadronic matter in the context of the Skyrme model. In fact, these questions were originally addressed, and qualitatively answered by Skyrme himself two decades ago. The algebraic complexity of the model at short distances together with the ambiguities to define a potential for strongly interacting skyrmions, forced Skyrme to limit his analysis of the interaction to the long-range sector, and to provide only a lower bound on the binding energy in dense hadronic matter. Most of the recent developments circumvent part of the algebraic complexity using numerical methods. Since many of the present issues are still in the process of being settled, we will be brief in our discussion.

Using the adiabatic character of the winding number, Skyrme proposed a product ansatz to investigate the asymptotic form of the potential energy of two interacting skyrmions. For two skyrmions centered around $\boldsymbol{x}_1$ and $\boldsymbol{x}_2$, and far apart from each other, the field configurations are accurately parametrized by two undistorted skyrmions with a relative spin–isospin orientation,

$$U_{\mathrm{ss}}(\boldsymbol{x};\boldsymbol{x}_1;\boldsymbol{x}_2) = U_{\mathrm{s}}(\boldsymbol{x}-\boldsymbol{x}_1)A(\boldsymbol{\alpha})U_{\mathrm{s}}(\boldsymbol{x}-\boldsymbol{x}_2)A^+(\boldsymbol{\alpha}), \tag{347}$$

where $A(\boldsymbol{\alpha}) = \exp(\mathrm{i}\boldsymbol{\tau}\cdot\boldsymbol{\alpha}/2)$, and $U_{\mathrm{s}}(\boldsymbol{x})$ is the hedgehog solution. For large separations $r=|\boldsymbol{x}_1-\boldsymbol{x}_2|$ compared to the skyrmion size, distortions of the skyrmion field can be neglected since they yield subleading corrections to the energy. As r tends to infinity, U_{ss} approaches the non-interacting configuration, and $E_{\mathrm{ss}}(r\to\infty) = 2E_{\mathrm{s}}$ independently of the isospin orientation $\boldsymbol{\alpha}$. For a static configuration the potential energy can be defined as follows:

$$V(\boldsymbol{x}_1,\boldsymbol{x}_2) = \int \mathrm{d}\boldsymbol{x}\,\{\tfrac{1}{4}f_\pi^2\,\mathrm{TR}[L_i(1,2)L_i(1,2)] + \tfrac{1}{4}\varepsilon^2\,\mathrm{TR}[L_i(1,2),\,L_j(1,2)]^2\} - E_1 - E_2\,. \tag{348}$$

Asymptotically, $L_\mu(1,2)$ splits into

$$L_\mu(1,2) = A(L_\mu(2) + U_2^+ A^+ L_\mu(1) A U_2) A^+ . \tag{349}$$

Simple algebraic manipulations give

$$\begin{aligned} V(\boldsymbol{x}_1, \boldsymbol{x}_2) = \tfrac{1}{2} f_\pi^2 \int \mathrm{d}\boldsymbol{x}\, \mathrm{TR}\{-A^+ L_i(1) A R_i(2) + A^+ L_i(1) A \hat{R}_i(2) + A^+ \hat{L}_i(1) A R_i(2)\} \\ + \tfrac{1}{2}\varepsilon^2 \int \mathrm{d}\boldsymbol{x}\, \mathrm{TR}\{[R_i(2), A^+ L_j(1) A]^2 + [R_i(2), R_j(2)] A [L_i(1), L_j(1)] A^+ \\ + [R_i(2), A^+ L_j(1) A][A^+ L_i(1) A, R_j(2)]\} , \end{aligned} \tag{350}$$

where

$$\hat{L}_i = L_i - \frac{2\varepsilon^2}{f_\pi^2} [L_j, [L_i, L_j]] , \tag{351a}$$

$$\hat{R}_i = R_i - \frac{2\varepsilon^2}{f_\pi^2} [R_j, [R_i, R_j]] , \tag{351b}$$

are conserved currents because of the equations of motion (50). Using the asymptotic form, and neglecting quartic terms in the field gradients in (350), yield

$$V(\boldsymbol{x}_1, \boldsymbol{x}_2) \sim -\tfrac{1}{2}\, \mathrm{TR}[A^+ \tau^a A \tau^b] \int \mathrm{d}\boldsymbol{x}\, \partial_i^2 \pi^a(1) \pi^a(2) . \tag{352}$$

Since $\partial_i^2 \pi^a(1)$ vanishes except around the neighborhood of $\boldsymbol{x}_1$, we can expand $\pi^b(2)$ in Taylor's series about $\boldsymbol{x}_1$. For the geometry of fig. 17, we have

$$\begin{aligned} \pi^b(2) = f_\pi (\widehat{x - r}) \sin F(|\boldsymbol{x} - \boldsymbol{r}|) \sim f_\pi K^2 \partial_r \left(\frac{1}{|\boldsymbol{x} - \boldsymbol{r}|} \right) , \\ \partial_i^2 \pi^a(1) = f_\pi \partial_i^2 (\hat{x}^a \sin F) = f_\pi \hat{x}^a \phi(x) , \end{aligned} \tag{353}$$

where $\phi(x)$ is a smooth function of x. The asymptotic contribution to (120) is

$$V(r) \sim \tfrac{1}{6} K^2 f_\pi^2\, \mathrm{TR}[A^+ \tau^a A \tau^b] \partial^b \partial^a (1/r) , \tag{354}$$

which is reminiscent of one-pion exchange.

An extensive analysis of the skyrmion–skyrmion potential has been carried out by Jackson et al. [57] at Stony Brook using the product ansatz (347) for all separations r, with undistorted hedgehogs. Their

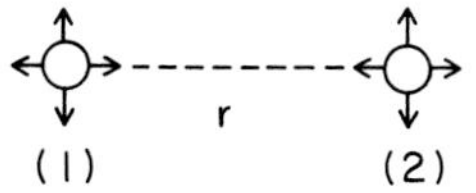

Fig. 17. Two defensive hedgehog skyrmions at very large distance r.

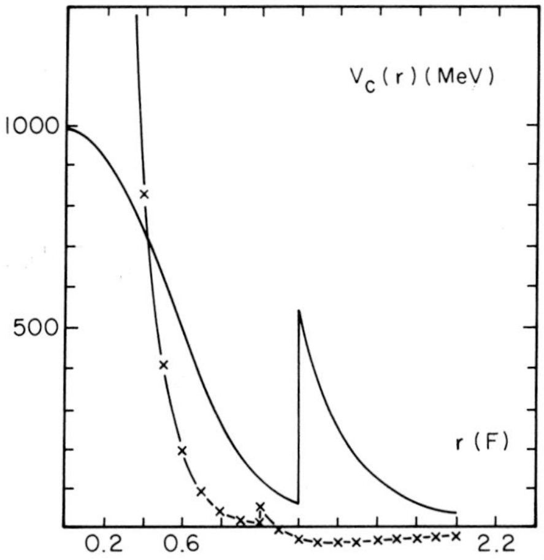

Fig. 18. Comparison of the central potential in the Skyrme model with its corresponding component of the Paris potential from ref. [68]. The break indicates a scale change by a factor of 10.

analysis shows that for a given set of Euler angles $\boldsymbol{\alpha} \equiv [\alpha\beta\gamma]$,

$$V(r, [\alpha\beta\gamma]) = V_{\mathrm{A}}(r) + V_{\mathrm{B}}(r) \cos \beta + V_{\mathrm{c}}(r) \cos(\alpha + \gamma) \cos^2(\beta/2) . \tag{355}$$

This simple parametrization is due to the hedgehog character of the solution, which allows for π- and σ-meson exchange, and restricts the ω- and ρ-meson coupling to be vectorial and tensorial respectively, in agreement with known nucleon properties. The central part of the skyrmion–skyrmion interaction ($\boldsymbol{\alpha} \equiv 0$), coincides with the central potential in the NN-interaction, as shown in fig. 18. Notice the finite-range repulsion at zero separation, which is of the order of the skyrmion mass ($\sim$1 GeV). The most intriguing feature of the central potential is the absence of the intermediate range attraction, a vital ingredient for nuclear binding. Recently, Jackson [78] and Meißner [79] have shown that by incorporating the soft-pion corrections to the skyrmion–skyrmion interaction, one can account for part of the medium range attraction in the central channel.

Given the simplicity of the calculation, many of the results of ref. [68] are qualitatively appealing. There is however major concern on the ability of the product ansatz to describe the short-distance part of the NN-interaction, where the skyrmions are known to be strongly deformed. The on-going numerical effort at Urbana [80] to overcome the limitation of the product ansatz, is in many respects important, but by no means, final. We will comment more on this point next.

7.7. *Status of the many-skyrmion problem*

Some years ago, Bogomol'ny and Fateev [81] have argued that at short distances the skyrmion–skyrmion interaction is dominated by finite two-body forces, while considering a configuration of N superimposed skyrmions. To see this, consider the variational profile

$$f_N(r) = \begin{cases} N\pi(1 - r/a_N) & r < a_N\,, \\ 0 & r > a_N\,, \end{cases} \tag{356}$$

where a_N characterizes the size of the multi-skyrmion configuration for which the energy is minimum, i.e.

$$a_N^2 = 24\,\frac{\varepsilon^2}{f_\pi^2}\left(1 + \frac{1}{N\pi}\int_0^{N\pi} \mathrm{d}x\,\frac{\sin^2 x}{x^2}\right)\left(1 + \frac{3}{\pi^2 N}\right)^{-1}. \tag{357}$$

The baryon density associated to (356) has an onion-like structure with a first minimum at a_N/N. For large N, (357) simplifies into $a_N \sim 2\sqrt{6}\varepsilon/f_\pi$, independently of N. Using (357) together with the virial theorem gives

$$E_N = 4\pi f_\pi^2 \int_0^{a_N} r^2\,\mathrm{d}r\left[F'^2 + 2\,\frac{\sin^2 F}{r^2}\right]. \tag{358}$$

For large N, this reduces to

$$E_N \sim 8\sqrt{\tfrac{2}{3}}\,\pi^3 f_\pi N^2 \sim 202\,\varepsilon f_\pi N^2\,, \tag{359}$$

which is to be compared with the asymptotic estimation of ref. [81],

$$E_\mathrm{N} \sim \tfrac{208.9}{2}\,N(N + 0.8726)\,\varepsilon f_\pi\,. \tag{360}$$

This large-N behavior is reminiscent of two-body interactions obtained by simple pairing of N skyrmions using the combinatoric factor $\frac{1}{2}N(N-1)$.

Recently, Kutschera and Pethick [82] have argued that this behavior is specific to configurations in which the characteristic lengths of the constituent skyrmions are different in different directions. Their argument is based on the fact that the energy of a skyrmion depends on the geometrical character of the volume in which it is enclosed. For a skyrmion confined to a box of dimensions $l_1 \times l_2 \times l_3$, the energy per unit volume scales like* [82]

$$\varepsilon_\mathrm{KP} \sim \varepsilon^2(\alpha_1/l_1^2 l_2^2 + \alpha_2/l_2^2 l_3^2 + \alpha_3/l_3^2 l_1^2)\,, \tag{361}$$

where the α's are constants. For a constant density $n = (l_1 l_2 l_3)^{-1}$, (361) reaches a minimum for a cubical configuration, $l_1 = l_2 = l_3$, thus**

$$\varepsilon_\mathrm{KP} \sim \varepsilon^2 n^{4/3}\,, \tag{362}$$

in agreement with the detailed calculations of ref. [83], and within Skyrme's lower bound [3]. If we

* Remember that due to the virial theorem, the quadratic and quartic term contribute equally to the ground state energy. Moreover, the currents L_i's scale like $1/l_i$.

** Notice the similarity with the energy of a free quark gas [83].

were to choose $l_1 = l_2 = Nl_3$, as suggested by the onion-shell behavior discussed above, then

$$\varepsilon_{BF} \sim N^{2/3} \varepsilon_{KP} , \tag{363}$$

which is consistent with (132). Since the energy density (134) is smaller than (135) by a factor of $N^{-2/3}$, Kutshera and Pethick concluded that in dense matter, the interaction between skyrmions is limited to nearest neighbors and behaves like r^{-1}, i.e. $F_{KP} \sim n^{1/3} \sim r^{-1}$, where r is the characteristic separation of skyrmions. They claimed that this 'screening' is due to significant many-body effects that tend to cancel out the long-range part of the two-skyrmion interaction. These results are different from the ones obtained via the product ansatz where it is found that at short distances, the interaction is finite and dominated by two-body forces.

Most of the above conclusions are tied to the scaling behavior of the fourth-order term. It is clear that by incorporating higher-derivative terms in the starting effective Lagrangian, one would end up with a different short-distance and high-density behavior, making most of the above results model dependent. Needless to recall that the Skyrme model is a low-energy approximation to QCD, that can hardly be justified at high densities. So, aside from possibly appealing scenarios, we believe that a reliable statement on central questions such as the short-distance behavior of the interaction, the high-density limit of hadronic matter and the equation of state, require a quantitative understanding of the non-perturbative character of QCD that goes beyond crude approximations. Of course, this does not include dilute phases of skyrmions where much can be learned about pion condensation and hadronic clustering at densities that are consistent with the low-energy assumptions.

7.8. Skyrmion crystal

In a recent paper, Klebanov [84] has investigated a crystalline form of neutron matter. To insure maximum attraction in the asymptotic channel, he organized the skyrmions on a cubic lattice as shown in fig. 19, and imposed twisted periodic boundary conditions on the chiral field in each cubic cell,

$$U(x, y, z) = \tau_y U(x + a, y, z)\tau_y = \tau_z U(x, y + a, z)\tau_z = \tau_x U(x, y, z + a)\tau_x . \tag{364}$$

These conditions are the analogue of the Bloch–Floquet quasi-periodicity conditions on the lattice. To eliminate the translational degeneracy due to the infinite crystal, one can fix the origin by imposing reflection symmetries both in space and isospace at a given site,

$$U(x, y, z) = \tau_x U^+(-x, y, z)\tau_x = \tau_y U^+(x, -y, z)\tau_y = \tau_z U^+(x, y, -z)\tau_z . \tag{368}$$

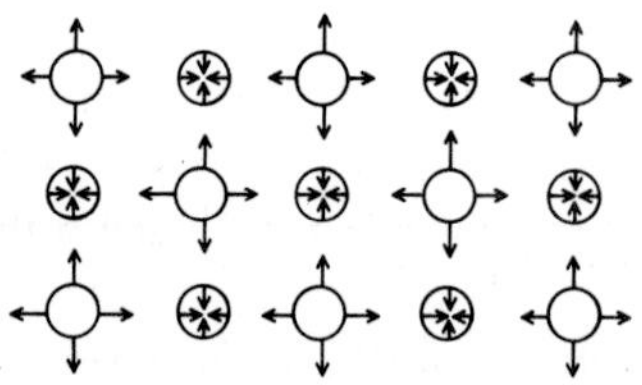

Fig. 19. Configuration before relaxation in the skyrmion crystal discussed in ref. [84].

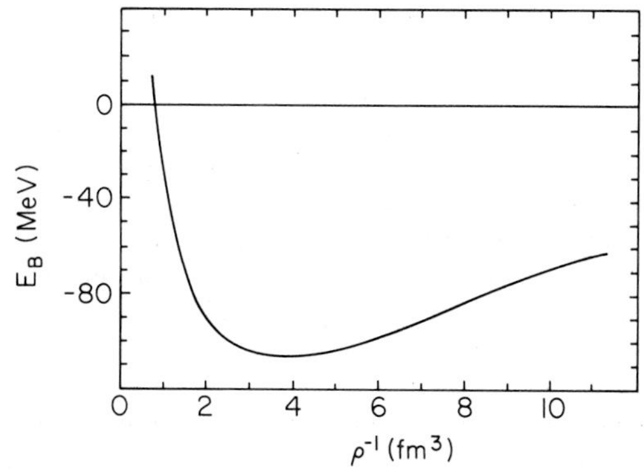

Fig. 20. Binding energy in a skyrmion crystal versus the inverse density ρ^{-1}. The minimum $E_B \sim -105$ MeV occurs at $\rho \sim 2\rho_{NM}$, with $\rho_{NM} = 0.17\,\text{fm}^{-3}$.

The initial value problem on the lattice, subject to (367) and (368) can be handled using a numerical relaxation method. Figure 20 shows the binding energy versus density for a $15 \times 15 \times 15$ lattice. (The parameters used are the ones of Adkins et al. [21]). The binding energy has a minimum at $E_B \sim -105$ MeV for $\rho \sim 2\rho_{NM}$, where $\rho_{NM} = 0.17\,\text{fm}^{-3}$ is the nuclear matter density. The binding energy is found even stronger in a neutron crystal [84]. Of course, these results are only qualitative in the semi-classical description. Because of the localization induced by the crystal, vibrational effects are important and might even unbound the system.

7.9. Exotic skyrmions

So far we have assumed that hadronic matter involves exclusively spherical skyrmions of the hedgehog type. However, it is possible that for reasonable densities, it may become energetically favorable to form new phase configurations involving clusters of skyrmions in the form of tori, cylinders, planes, etc. In ref. [11], it is argued that periodic boundary conditions on the chiral field $U(\boldsymbol{x})$ yield mappings from non-spherical group manifolds onto S^3. The specific case of periodic baryonic chains involving specific mappings of the form $(S^2 \times S^1; S^3)$ have been discussed in details in ref. [11].

8. The Skyrme model and boson exchange

8.1. The role of bosons in the Skyrme model

As noted in the Introduction, 't Hooft [1] and Witten [2] studied QCD in the large N_c limit, finding it equivalent to a local field theory of non-interacting mesons. Witten advanced arguments that the nucleon should be a soliton and the Skyrme model satisfied this requirement.

It is clear that the pion plays a dominant role in the skyrmion. The Skyrme Lagrangian can be viewed as a particularly convenient non-linear realization of chiral symmetry. We have shown above that Skyrme's ad hoc fourth-order term can be understood as a heavy ρ-exchange between bits of nucleonic matter, much as the ω-exchange we discussed in section 5. The difference is that the ρ-exchange interaction is taken to zero range. In fact, the ω-exchange can similarly be reduced [85] to an effective sixth-order term.

Whereas boson exchange has generally been considered, in nuclear physics, as providing interactions between nucleons, we see that in the generalized Skyrme model, the self same bosons are exchanged between bits of nucleonic matter within the same nucleon, holding it together at the classical level. For the many arguments we have made earlier connecting the skyrmion with the Wess–Zumino term and topology, the skyrmion in these senses represents something much deeper than simply a construction from boson exchange. However, it should be realized that the incorporation of boson exchange in a zero-range approximation is a somewhat crude way of treating it. Of course, as in Adkins and Nappi [53], one can remove this approximation.

The more serious question relates to the absence of attraction in the interaction between two nucleons in the Skyrme model, as it appears in the work of the Stony Brook group [68]. We touched on this in section 7. There we have outlined soft-pion corrections in the masses, essentially adding soft fluctuations to the Skyrme model, a related problem. We return now to the issue of the attraction in the interaction between two nucleons.

8.2. The attractive interaction between nucleons

It is well understood that in nucleon–nucleon potentials such as the Bonn [86] or Paris [69] potential, the intermediate-range attraction which leads to the binding of nucleons together into nuclei arises from the exchange of systems of two pions, coupled to $I = 0$ and $J = 0$. In this way, the interaction appears as coming from the exchange of scalars with a distributed mass spectrum. These scalars are not, however, elementary mesons, and are not present in the $N_c \to \infty$ limit, in which case interactions between the elementary mesons, pions in this case, go to zero. In other words, the attraction comes from $1/N_c$ corrections as compared with the repulsion; in the language of section 7, from loop corrections. In the sense that nuclei are bound, the attraction must predominate over the repulsion. In fact, nuclei are barely bound. The binding energy of nuclear matter is only 16 MeV, two orders of magnitude down from that of the nucleon–nucleon interaction, which is $\sim m_n c^2$. Thus, there is a very near cancellation between repulsion and attraction.

In the Paris potential the attractive scalar interaction is treated by πN and ππ dispersion relations. Pham and Truong [87] have shown that such a procedure leads to a term in the Skyrme Lagrangian additional to Skyrme's fourth-order term which was later found to summarize further effects from the ρ-meson. This non-Skyrme fourth-order term contains derivatives quartic in time, and has been recently studied by the Paris group [88]. Their numerical analysis shows that the skyrmion becomes unstable when the strength of this term increases.

These difficulties may not arise if one realizes that the effective scalar exchanges between bits of nucleonic matter should be treated in finite range, as Adkins and Nappi did [53] for ω-exchange. The non-Skyrme fourth-order term arises only in zero-range approximation in terms of the propagator of the scalar mesons. It seems, however, clear, that the attraction *between* nucleons is more naturally treated in the boson-exchange model, than through the skyrmion–skyrmion interaction.

8.3. Short-distance cutoffs in boson exchange models and the finite size of the skyrmion

When interactions between point nucleons are calculated in boson-exchange models, cutoffs must be introduced in order to make these finite as the interparticle distance goes to zero. In the Bonn potential these cutoffs are introduced via ad hoc form factors. In the Paris potential, the interaction inside of 0.8 fm is simply replaced by a phenomenological one, fited so as to reproduce nucleon–nucleon scattering data. No such troubles appear in the interaction between skyrmions [68], because they have finite sizes. It is, in fact, easy to check that by folding the ω-exchange interaction, proportional to the Green's function of eq. (199), over finite baryonic density distributions from the Skyrme model, one effectively obtains [89] form factors of the same general size as those used in the Bonn potential. This use of the solitons in regularizing boson-exchange interactions at short distances needs to be drawn much tighter. (The people who calculate boson exchange have not incorporated the soliton nature of the nucleon into their interactions yet.) It is however clear that this is a mechanism for rendering them finite and well behaved.

9. Discussion and conclusions

9.1. The Cheshire cat philosophy and the chiral bag model

The skyrmion, including effects from zero range π^-, ρ^- and ω mesons in its description seems to give a rather good description of the structure of the nucleon, much like the standard bag model description. This very strong similarity is suggestive of a Cheshire cat mechanism as discussed by Nadkarni et al. [63]. Cheshire cat models are idealized hybrid models wherein a fermionic theory inside a bag is matched to an exactly bosonic theory outside. The resulting physics is insensitive to changes in bag size or shape. In the Cheshire cat philosophy, the bag is an unphysical concept and its use is merely a matter of technical convenience, having to do with the fact that at sufficiently large momentum transfer the mesonic theory may not be known to sufficient accuracy. We believe that for a given low-energy observable, there might exist a definite range of bag radii that minimize the inaccuracy. Thus, a finite bag radius might be more efficient for describing hadronic spectra, whereas a zero bag radius like the skyrmion might be better suited for probing hadronic interactions. It should be clear however, that if enough corrections are taken into account, both descriptions should lead to the same result.

To give a complete description of the short-distance structure of the skyrmion, one would need to include heavier mesons or higher derivative terms. A way of circumventing this was offered by Vento et al. [90]. Following the perturbative work of Chodos et al. [91] and Inoue et al. [92], they built a defect containing quarks into the skyrmion. At that time, they did not know about Skyrme's work nor about the fractionation of baryon number between the quark core and the soliton cloud, although they realized that they had a soliton solution. Recently, this model has been considerably investigated [93, 94, 95]. We will not be able to give an adequate description of these developments here. A chief point is that there are possible simplifications in the description if one builds in asymptotic freedom through a two-phase model. In a boson description, a large number of bosons would be needed to achieve it.

In the interaction between two nucleons in low-energy nuclear physics it is not likely to matter much whether one builds a small quark defect into the skyrmion or not. The repulsive interaction via ω-exchange is so repulsive that two nucleons are generally kept apart and do not feel such short distances. On the other hand, the intrinsic structure and excitations of the nucleon seem to be simpler in the quark language than in the skyrmion language. As we remarked at the end of section 7, the description of the low-lying odd-parity excited states of the nucleon is simple in terms of promotion of a quark from the ground state to the first odd-parity state. In the Skyrme model, some of low-lying resonances are not well reproduced to leading order, since they tend to couple to the collective zero-mode states. So, perhaps, one can view the odd-parity states of the nucleon as an implicit manifestation of the quark substructure of the nucleon in low-energy physics.

At this stage, we should remark that a description of strange baryons in terms of the derivative expansion implied by the Skyrme model has encountered some difficulties [96, 97, 98]. Many authors have suggested that higher derivative terms play a more important role here than in the case of the nucleon. It may be that in the case of the strange baryons the size of the quark core should be taken as large as 1 fm, in line with the MIT bag model for an optimal description. In this case, the soliton cloud plays a much less important role than in the case of the nucleon, most of the strange baryons being described by an asymptotically free region. Recall that in the case of the nucleon, the strong pressure from the pion cloud compresses the quark core to a region of dimensions somewhat smaller than order-of-magnitude estimates; e.g., the latter often suggests a confinement region $\sim \Lambda^{-1} \sim 1$ fm for an optimal two-phase description. This 1 fm radius may, however, be appropriate for the strange baryons where the coupling of the π, κ, η cloud to the quarks is much weaker. This difference in confinement sizes between nucleons and strange baryons has been pursued in detail by Brown, Rho and Weise [99]. An extensive discussion of the chiral bag mode has recently been given by Brown and Rho [100].

9.2. Conclusions

We have tried in this report to sketch the relevance of the Skyrme model in the context of large N_c QCD, and clarify the many interrelationships between the model and fundamental quantities such as the Wess–Zumino term and the chiral anomalies. These connections have given much deeper insight into the model of Skyrme. Perhaps it is not surprising that essentially no one understood his model, when he formulated it more than two decades ago.

The recent developments outlined here give strong support to the belief that baryons can be described as solitons. Most importantly, they indicate that low-energy physics should be described non-perturbatively. The leading semi-classical terms in our description and the underlying geometry, etc., are thrown away once one goes over to perturbation theory. The latter can be thought of as calculating the fine structure to the leading semi-classical terms dealt with here.

In the rush of throwing away conventional nuclear physics for an exclusive quark description, interactions via vector meson exchange were generally disregarded as relating to short distances within the confinement regime. In the past, there have been many (unsuccessful) attempts to replace these interactions by descriptions in terms of quark/gluon degrees of freedom.

These vector mesons have, however, been resuscitated, and their interactions can be found by gauging the explicit or implicit chiral structure of the Skyrme model in the presence of the Wess–Zumino term. This description gives a deeper understanding to old, successful pictures such as vector dominance.

The description of the nucleon as a soliton has other extremely important implications for low-energy nuclear physics. It gives a rather definite model for regularizing boson-exchange interactions at short distances due to the finite size of the skyrmion. Because of the strong repulsion from ω-meson exchange, nucleons do not come to short distances in low-energy interactions, so modifications due to the presence of a small quark core are unlikely to matter. These statements must, however, be quantitatively tested.

For the description of the low-lying excited states of the nucleon and the properties of strange particles, we suggest that it may be more appropriate to introduce explicitly the quark structure into the baryons, essentially making a defect which contains quarks in the center of the skyrmion. This two-phase description constitutes an optimal picture in the context of the Cheshire cat philosophy.

We will end up by reporting our strong impression that followed us throughout this work. It was truly remarkable that Skyrme initiated these developments so long ago, having to invent most of the key concepts discussed in this report. It should be remembered that he published several fundamental papers, not only initiating the model, but carrying out a number of applications, with very little encouragement at the time. Once again, history should bear witness than an important work has been ignored.

Acknowledgements

We would like to thank A. Wirzba for valuable comments and a careful reading of the manuscript, along with A.D. Jackson, T.H. Hansson, U.G. Meißner, H.B. Nielsen and M. Rho for discussions. Through the course of this work, we have also benefitted from discussions with several friends and colleagues here at Stony Brook, and overseas at Nordita and the Niels Bohr Institute. In particular, we would like to thank G. Adkins, J. Ambjorn, P. Arve, A.P. Balachandran, A. Goldhaber, A. Jackson, D. Klabucar, I. Klevanov, C.J. Pethick, A. Niemi, C. Nappi, P. Simic and E. Wust.

Appendices

Appendix A: Baryon number for general textures

A remarkable property of the points where the chiral field $U = \pm\mathbb{1}$ is that the total baryon number can be expressed in terms of the topological singularities of the pion field at these points. Indeed, consider the skyrmion texture illustrated in fig. 3b, where S_β is a smooth surface surrounding the point, line or surface where $U = \pm\mathbb{1}$. All points on S_β are infinitesimally close to the points where $U = \pm\mathbb{1}$. Using current conservation, we can write

$$\frac{\mathrm{d}B}{\mathrm{d}t} = -\sum_\beta \int \mathrm{d}S_\beta(\hat{n}_\beta \cdot \boldsymbol{\beta}) , \tag{A.1}$$

where the sum runs over all the surfaces S_β, including the surface at infinity, and $B^\mu(x)$ is the covariant baryon current defined in (32). Straightforward manipulations give

$$\frac{\mathrm{d}B}{\mathrm{d}t} = -\frac{\mathrm{i}}{8\pi^2}\sum_\beta \mathrm{d}S_\beta\, \dot{\theta}\varepsilon^{ijk}n^i_\beta\{\sin^4\theta\, \mathrm{TR}[(\boldsymbol{\tau}\cdot\hat{\theta})[\boldsymbol{\tau}\cdot(\hat{\theta}\times\partial_j\hat{\theta})]\,\boldsymbol{\tau}\cdot(\hat{\theta}\times\partial_k\hat{\theta})$$
$$+\sin^2\theta\cos^2\theta\,\mathrm{TR}[(\boldsymbol{\tau}\cdot\hat{\theta})\,(\boldsymbol{\tau}\cdot\partial_j\hat{\theta})(\boldsymbol{\tau}\cdot\partial_k\hat{\theta})]\} , \tag{A.2}$$

where we have used the parametrization

$$U(\boldsymbol{x}, t) = \exp[\mathrm{i}(\boldsymbol{\tau}\cdot\hat{\theta}(\boldsymbol{x}, t)\theta(\boldsymbol{x}, t))] . \tag{A.3}$$

Now, let $\hat{e}_k(\beta)$, $k = 1, 2, 3$, be an orthogonal, direct, comoving system of coordinates on S_β with $e_3 \equiv \hat{n}_\beta$, so that ($\hat{\theta}^2 = 1$)

$$\hat{\theta} = \hat{\theta}_1 \hat{e}_1 + \hat{\theta}_2 \hat{e}_2 + \hat{\theta}_3 \hat{e}_3 \,. \tag{A.4}$$

Using the Serret–Frenet relations, and the fact that a fixed surface has zero torsion, yield

$$\begin{aligned} \partial_1 \hat{\theta} &= (\partial_1 \hat{\theta}_1 - \kappa_1 \hat{\theta}_3)\hat{e}_1 + \partial_1 \hat{\theta}_2 \hat{e}_2 + (\partial_1 \hat{\theta}_3 + \kappa_1 \hat{\theta}_1)\hat{e}_3 \,, \\ \partial_2 \hat{\theta} &= \partial_2 \hat{\theta}_1 \hat{e}_1 + (\partial_2 \hat{\theta}_2 - \kappa_2 \hat{\theta}_3)\hat{e}_2 + (\partial_2 \hat{\theta}_3 + \kappa_2 \hat{\theta}_3)\hat{e}_3 \,, \end{aligned} \tag{A.5}$$

where the κ_i's are the curvature tensions. Simple algebra using (A.5), shows that

$$\begin{aligned} &\varepsilon^{ijk} n^i_\beta \, \mathrm{TR}[(\boldsymbol{\tau} \cdot \hat{\theta})[\boldsymbol{\tau} \cdot (\hat{\theta} \times \partial_j \hat{\theta})][\boldsymbol{\tau} \cdot (\hat{\theta} \times \partial_k \hat{\theta})]] = -8\mathrm{i}\, \hat{\theta} \cdot (\partial_1 \hat{\theta} \times \partial_2 \hat{\theta}) \,, \\ &\varepsilon^{ijk} n^i_\beta \, \mathrm{TR}[(\boldsymbol{\tau} \cdot \hat{\theta})(\boldsymbol{\tau} \cdot \partial_j \hat{\theta})(\boldsymbol{\tau} \cdot \partial_k \hat{\theta})] = -8\mathrm{i} \hat{\theta} \cdot (\partial_1 \hat{\theta} \times \partial_2 \hat{\theta}) \,, \end{aligned} \tag{A.6}$$

hence

$$\frac{\mathrm{d}B}{\mathrm{d}t} = -\frac{1}{\pi^2} \sum_\beta \int \mathrm{d}S_\beta \, \dot{\theta} \sin^2 \theta \, [\hat{\theta} \cdot (\partial_1 \hat{\theta} \times \partial_2 \hat{\theta})] \,. \tag{A.7}$$

If we assume that the chiral field is turned on adiabatically from its vacuum value, then the amount of baryon number gathered around the surfaces S_β's, becomes

$$B = \frac{1}{2\pi^2} \sum_\beta \int \mathrm{d}S_\beta [\theta - \tfrac{1}{2} \sin 2\theta]_\beta \hat{\theta} \cdot (\partial_1 \hat{\theta} \times \partial_2 \hat{\theta}) \,. \tag{A.8}$$

Appendix B: Weinberg scaling argument

Some years ago Weinberg [15] has argued that loop corrections in effective chiral models were suppressed by powers of p^2 at low-energy, where p_μ is the momentum of the pseudoscalar mesons. The rationale for this argument is as follows.

Consider the general form of the S-matrix in terms of the transition amplitude M as defined through

$$S = M\delta\Big(\sum p_i\Big) \,. \tag{B.1}$$

If we denote by N_e the number of external pion wavefunctions assumed to scale like $[\mathrm{mass}]^{-1}$ according to the usual PCAC relation, then $[M] \sim 4 - N_\mathrm{e}$. In general M is a function of the total energy E flowing through the scattered amplitude, the various couplings g and the renormalization scale μ, i.e.

$$M = M(E, q, \mu) = E^\mathrm{D} f(E/\mu, g) \,, \tag{B.2}$$

with

$$D = 4 - N_e - [\alpha]\,, \tag{B.3}$$

where $[\alpha]$ is the dimension of the different couplings and propagators entering in the transition amplitude. Now, consider an effective chiral Lagrangian expressed as a series expansion in field gradients. It is easy to show that in the chiral limit, the effective coupling constants associated to interactions with d derivatives scale like ~4 – d in 4 dimensions, so that if we have N_d of such vertices in M, we obtain

$$[\alpha] = \sum_d N_d(4-d) - 2N_i - N_e\,, \tag{B.4}$$

where N_i is the number of internal propagators. Remember that the number of loops is given by

$$N_L = N_i - \sum_d N_d + 1\,. \tag{B.5}$$

This leads to an overall scaling for the transition amplitude of the form

$$D = 2 + \sum_d N_d(d-2) + 2N_L\,, \tag{B.6}$$

which shows that the dominant graphs carry the smallest value of D at low energy. In other words, the $d = 2$ term scales like p^2 at tree level, and p^4 at 1-loop, the $d = 4$ scales like p^4 at tree level, and p^6 at 1-loop, etc.

Appendix C: Asymptotic behavior of the pion field

In the broken phase, soft-pion–nucleon phenomenology can be described at tree level by a Lagrangian of the form [24]

$$\mathcal{L} = \tfrac{1}{2}(\partial_\mu \boldsymbol{\pi})(\partial^\mu \boldsymbol{\pi}) - \tfrac{1}{2}m_\pi^2 \boldsymbol{\pi}^2 + (g_{\pi NN}/2M_N)\partial_\mu \boldsymbol{\pi}\cdot(N\gamma^\mu\gamma_5 \boldsymbol{\tau} N) + \mathcal{L}_N\,, \tag{C.1}$$

where $\mathcal{L}_N$ summarizes the internal nucleon dynamics. In the static limit, the pion field in a nucleon state is described by an inhomogeneous Klein–Gordon equation of the type

$$(-\boldsymbol{\nabla}^2 + m_\pi^2)\pi^a(\boldsymbol{r}) = \frac{g_{\pi NN}}{2M_N}\boldsymbol{\nabla}\cdot(\bar{N}\boldsymbol{\gamma}\gamma_5\tau^a N)\,, \tag{C.2}$$

where the general solution is given by

$$\pi^a(r) = -\frac{g_{\pi NN}}{2M_N}\int \mathrm{d}\boldsymbol{r}'\,\boldsymbol{\nabla}_{r'}G(\boldsymbol{r}-\boldsymbol{r}')\cdot\bar{N}\boldsymbol{\gamma}\gamma_5\tau^a N\,, \tag{C.3}$$

where $G(\boldsymbol{r}-\boldsymbol{r}')$ is the Klein–Gordon propagator for massive pions, i.e.

$$G(\boldsymbol{r}-\boldsymbol{r}') = \frac{1}{4\pi}\,\frac{\exp(-m_\pi|\boldsymbol{r}-\boldsymbol{r}'|)}{|\boldsymbol{r}-\boldsymbol{r}'|}\,. \tag{C.4}$$

Using the convolution property in Fourier space, we can rewrite (C.3) in the following form:

$$\pi^a(\boldsymbol{r}) = \int \frac{\mathrm{d}^3 q}{(2\pi)^3} \exp(\mathrm{i}\boldsymbol{q}\cdot\boldsymbol{r})(\mathrm{i}\boldsymbol{q}G(\boldsymbol{q}))S^a(\boldsymbol{q}) , \tag{C.5}$$

where $G(\boldsymbol{q})$ and $S^a(\boldsymbol{q})$ are the Fourier transform of $G(\boldsymbol{r})$ and $(g_{\pi NN}/2M_N)\bar{N}\gamma\gamma_5\tau^a N$ respectively. At large distances ($r\to\infty$), the dominant contribution to (C.5) is due to the long-wavelengths ($q\to 0$) emitted by the source,

$$\begin{aligned} \pi^a(\boldsymbol{r}) &= S^a(\boldsymbol{q}\to 0) \times \int \frac{\mathrm{d}\boldsymbol{q}}{(2\pi)^3} \exp(\mathrm{i}\boldsymbol{q}\cdot\boldsymbol{r})\mathrm{i}\boldsymbol{q}\cdot G(\boldsymbol{q}) \\ &= S^a(\boldsymbol{q}\to 0)\times \nabla G(\boldsymbol{r}) . \end{aligned} \tag{C.6}$$

Straightforward algebra yields

$$\pi^a(\boldsymbol{r}) = -\left(\frac{g_{\pi NN}}{8\pi M_N}\right)(1+m_\pi r)e \frac{-m_\pi r}{r^2} \hat{r}^i \langle N|\sigma^i\tau^a|N\rangle , \tag{C.7}$$

with the nucleon matrix element given by

$$\langle N|\sigma^i\tau^a|N\rangle = \int \mathrm{d}\boldsymbol{x}\, N^+(\sigma^i \times \mathbb{1})\tau^a N . \tag{C.8}$$

Appendix D: Additivity of the baryon number

Two well separated skyrmions can be approximated by

$$U = U_1 U_2 \tag{D.1}$$

which is known as the product ansatz. This factorization is not valid at short distances where distortions in the single-particle fields are important. In terms of (D.1), we can define the 1-form on S^3

$$L_{12} = U_2^+ U_1^+\, \mathrm{d}(U_1U_2) = U_2^+(L_1 - R_2)U_2 . \tag{D.2}$$

The baryon current carried by (D.1) can be conveniently rewritten as follows:

$$B_{12} = \frac{\mathrm{i}}{24\pi^2} \mathrm{TR}(L_{12}^3) . \tag{D.3}$$

Injecting (D.2) into (D.3) and using the antisymmetric character of the 1-forms together with the Maurer–Cartan equations yields

$$B_{12} = B_1 + B_2 + \frac{\mathrm{i}}{8\pi^2}\, \mathrm{d}\,\mathrm{TR}(L_1R_2) . \tag{D.4}$$

In vector form, we have

$$B^{\mu}_{12} = B^{\mu}_{1} + B^{\mu}_{2} + \frac{\mathrm{i}}{8\pi^2}\, \varepsilon^{\mu\nu\alpha\beta} \partial_\nu\, \mathrm{TR}(L_{1\alpha} R_{2\beta})\,, \tag{D.5}$$

so up to a total divergence, the total baryon current is additive. Notice that on a closed compact manifold the total baryon charge carried by (D.1) is

$$B_{12} = \int \mathrm{d}\boldsymbol{x}\, B^{0}_{12}(\boldsymbol{x}) = B_1 + B_2\,, \tag{D.6}$$

a well known feature of product maps. This property will hold for a dilute gas of skyrmions described by

$$U = \prod_{i=1}^{N} U_i\,. \tag{D.7}$$

Appendix E: $SU(2)_{\mathrm{L}} \otimes SU(2)_{\mathrm{R}}$ *Noether currents*

The purpose of this appendix is to provide helpful details for constructing the expressions of the conserved $\mathrm{SU(2)_L} \otimes \mathrm{SU(2)_R}$ Noether currents, using the standard procedure of Gell–Mann and Levy. For that consider local and infinitesimal $\mathrm{SU(2)_{L,R}}$ transformations as defined through

$$\begin{aligned} U_{\mathrm{L}} &= (1 + \mathrm{i}Q_{\mathrm{L}})U\,, \qquad \text{and } \delta_{\mathrm{L}} U = \mathrm{i}Q_{\mathrm{L}} U\,, \\ U_{\mathrm{R}} &= U(1 + \mathrm{i}Q_{\mathrm{R}})\,, \qquad \text{and } \delta_{\mathrm{R}} U = \mathrm{i}UQ_{\mathrm{R}}\,, \end{aligned} \tag{E.1}$$

in terms of which $L_\mu = U^{+}\partial_\mu U$ transforms as

$$\delta_{\mathrm{L}} L_\mu = \mathrm{i}U\partial_\mu Q_{\mathrm{L}} U + \cdots\,, \qquad \delta_{\mathrm{R}} L_\mu = \mathrm{i}\partial_\mu Q_{\mathrm{R}} + \cdots\,, \tag{E.2}$$

$Q_{\mathrm{L,R}}$ are Lie-algebra valued. If we denote by $\mathscr{L}_Q$ the locally rotated Skyrme Lagrangian, defined for convenience through*

$$\mathscr{L}_Q = c_0\, \mathrm{TR}[L^2_\mu] + c_1\, \mathrm{TR}[L_\mu, L_\nu]^2\,, \tag{E.3}$$

then the conserved Noether currents associated to the global $\mathrm{SU(2)_L} \otimes \mathrm{SU(2)_R}$ transformations read

$$J^{a}_{\mu} = \partial\mathscr{L}_Q / \partial(\partial^{\mu}\varepsilon^{a})\,. \tag{E.4}$$

Injecting the left-transformations (E.1) into (E.3) yields

$$\begin{aligned} \delta_{\mathrm{L}}\mathscr{L}_Q &= 2c_0\, \mathrm{TR}[\delta_{\mathrm{L}} L_\mu L_\mu] + 2c_1\, \mathrm{TR}(\delta_1[L_\mu, L_\nu][L_\mu, L_\nu]) \\ &= -2c_0\, \mathrm{TR}[\mathrm{i}\partial_\mu Q_{\mathrm{L}} R_\mu] - 4c_1\, \mathrm{TR}([\mathrm{i}\partial_\mu Q_{\mathrm{L}}, R_\nu][R_\mu, R_\nu])\,, \end{aligned} \tag{E.5}$$

* Throughout we will use the Bjorken–Drell metric, paying no attention to upper or lower indices unless necessary.

in terms of which we have conserved left-handed currents $J^a_{\mu,\mathrm{L}}$,

$$J^a_{\mu,\mathrm{L}} = -2\mathrm{i}c_0\,\mathrm{TR}[Q^a_{\mathrm{L}} R_\mu] + 4\mathrm{i}c_1\,\mathrm{TR}([Q^a_{\mathrm{L}}, R_\nu][R_\mu, R_\nu])\,. \tag{E.6a}$$

Similarly, one obtains the conserved right-handed currents $J^a_{\mu,\mathrm{R}}$ in the form

$$J^a_{\mu,\mathrm{R}} = +2\mathrm{i}c_0\,\mathrm{TR}[Q^a_{\mathrm{R}} L_\mu] - 4\mathrm{i}c_1\,\mathrm{TR}([Q^a_{\mathrm{R}}, L_\nu][L_\mu, L_\nu])\,. \tag{E.6b}$$

The conserved vector and axial currents follows immediately through

$$V^a_\mu = \mathrm{i}c_0\,\mathrm{TR}(Q^a(R_\mu + L_\mu)) + 2\mathrm{i}c_1\,\mathrm{TR}(Q^a([R_\nu, [R_\mu, R_\nu]] + [L_\nu, [L_\mu, L_\nu]]))\,, \tag{E.7a}$$

$$A^a_\mu = -\mathrm{i}c_0\,\mathrm{TR}(Q^a(R_\mu - L_\mu)) - 2\mathrm{i}c_1\,\mathrm{TR}(Q^a([R_\nu, [R_\mu, R_\nu]] + [R_\nu, [R_\mu, R_\nu]]))\,. \tag{E.7b}$$

To extract the dependence of (E.7) on the collective coordinate $A(t)$ discussed in section 8, it is convenient to define*

$$A(t) \equiv \exp(\mathrm{i}\boldsymbol{\tau}\cdot\boldsymbol{\omega}t)\,, \tag{E.8}$$

so that

$$\begin{aligned}\tilde{Q}^a &= A^+ Q^a A = A^+\,\frac{\boldsymbol{\tau}}{2}\cdot\hat{a}\,A\\ &= \frac{\boldsymbol{\tau}}{2}\cdot\hat{a} - \frac{\sin 2\omega t}{2}\left[\frac{\boldsymbol{\tau}}{2}\cdot(\hat{a}\times\hat{\omega})\right] + \sin^2\omega t\left[\frac{\boldsymbol{\tau}}{2}\cdot(\hat{\omega}\times(\hat{\omega}\times\hat{a}))\right],\end{aligned} \tag{E.9}$$

$$\begin{aligned}\tilde{L}_\mu &= (AU^+A^+)\partial_\mu(AUA^+)\\ &= A[L_\mu - A_\mu + U^+A_\mu U]A^+\,,\end{aligned} \tag{E.10}$$

$$\begin{aligned}\tilde{R}_\mu &= (AUA^+)\partial_\mu(AU^+A^+)\\ &= A[R_\mu - A_\mu + UA_\mu U^+]A^+\,,\end{aligned} \tag{E.11}$$

$$\begin{aligned}U^+A_0U - A_0 &= \mathrm{i}\boldsymbol{\tau}\cdot\boldsymbol{\alpha}_-\\ &= \mathrm{i}\boldsymbol{\tau}\cdot[\sin 2F\,(\hat{r}\times\boldsymbol{\omega}) - 2\sin^2 F\,(\hat{r}\times\boldsymbol{\omega})\times\hat{r}]\,,\end{aligned} \tag{E.12}$$

$$\begin{aligned}A_0 - UA_0U^+ &= \mathrm{i}\boldsymbol{\tau}\cdot\boldsymbol{\alpha}_+\\ &= \mathrm{i}\boldsymbol{\tau}\cdot[\sin 2F\,(\hat{r}\times\boldsymbol{\omega}) + 2\sin^2 F\,(\hat{r}\times\boldsymbol{\omega})\times\hat{r}]\,,\end{aligned} \tag{E.13}$$

$$\begin{aligned}\boldsymbol{A}_i &= -\tfrac{1}{2}\mathrm{i}\,\mathrm{TR}[\boldsymbol{\tau}L_i]\\ &= \hat{r}\partial_i F + \tfrac{1}{2}\sin 2F\,\partial_i\hat{r} + (\hat{r}\times\partial_i\hat{r})\sin^2 F\,,\end{aligned} \tag{E.14}$$

* We will not concern ourselves with the $\pm\boldsymbol{\omega}$ ambiguity due to parity.

$$\begin{aligned} \boldsymbol{B}_i &= \tfrac{1}{2}\mathrm{i}\,\mathrm{TR}[\boldsymbol{\tau} R_i] \\ &= -\hat{r}\partial_i F - \tfrac{1}{2}\sin 2F\,\partial_i\hat{r} + (\hat{r}\times\partial_i\hat{r})\sin^2 F\,, \end{aligned} \tag{E.15}$$

where we have used $A_\mu = A^+\partial_\mu A = A^+\dot{A}\delta_{\mu 0}$. Under the transformation $U \rightarrow AUA^+$ the linear term in (E.6) becomes

$$f^{\mathrm{L}}_\mu = \mathrm{TR}[Q_{\mathrm{L}}\tilde{R}_\mu] = \mathrm{TR}[\tilde{Q}_{\mathrm{L}}(R_\mu - A_\mu + UA_\mu U^+)]\,, \tag{E.16}$$

since we are ultimately interested in the angle averaged currents, it is convenient to work with angle averaged quantities. With this in mind, we have

$$\begin{aligned} \bar{f}^{\mathrm{L}}_0 &= \int \mathrm{d}\hat{r}\, f^{\mathrm{L}}_0 = -\mathrm{i}\,\mathrm{TR}[\tilde{Q}_{\mathrm{L}}\tau^b] \int \mathrm{d}\hat{r}\,\alpha^b_+ \\ &= -\mathrm{i}\tfrac{16}{3}\pi\sin^2 F\,\mathrm{TR}[\tilde{Q}_{\mathrm{L}}\boldsymbol{\omega}\cdot\boldsymbol{\tau}] \\ &= -\tfrac{16}{3}\pi\sin^2 F\,\mathrm{TR}[\tilde{Q}_{\mathrm{L}}\dot{A}A^+]\,, \end{aligned} \tag{E.17}$$

$$\begin{aligned} \bar{f}^{\mathrm{L}}_i &= \int \mathrm{d}\hat{r}\, f^{\mathrm{L}}_i = \mathrm{i}\,\mathrm{TR}[\tilde{Q}_{\mathrm{L}}\tau^b] \int \mathrm{d}\hat{r}\, B^b_i \\ &= -\mathrm{i}\,\tfrac{4}{3}\pi\delta^{ib}\left(F' + \frac{\sin 2F}{r}\right)\mathrm{TR}[\tilde{Q}_{\mathrm{L}}\tau_b] \\ &= -\mathrm{i}\,\tfrac{4}{3}\pi\left(F' + \frac{\sin 2F}{r}\right)\mathrm{TR}[Q_{\mathrm{L}}A\tau_i A^+]\,. \end{aligned} \tag{E.18}$$

Similarly, the trilinear term in (E.6) becomes

$$\begin{aligned} g^{\mathrm{L}}_\mu &= \mathrm{TR}([Q_{\mathrm{L}}, \tilde{R}_\nu][\tilde{R}_\mu, \tilde{R}_\nu]) = \mathrm{TR}(Q_{\mathrm{L}}[\tilde{R}_\nu, [\tilde{R}_\mu, \tilde{R}_\nu]]) \\ &= \mathrm{TR}(\tilde{Q}_{\mathrm{L}}[R_\nu, [R_\mu, R_\nu]]) - \mathrm{TR}(Q_{\mathrm{L}}[R, [A_\mu - UA_\mu U^+, R_\nu]]) + \mathrm{O}(\dot{A}^2)\,. \end{aligned} \tag{E.19}$$

We have dropped higher-order time derivatives since they involve non-leading N_c corrections. The time component of (E.19) yields

$$\begin{aligned} g^{\mathrm{L}}_0 &= -\mathrm{TR}[\tilde{Q}_{\mathrm{L}}[R_i, [A_0 - UA_0U^+, R_i]]] \\ &= 2\,\mathrm{TR}[\tilde{Q}_{\mathrm{L}}[\boldsymbol{\tau}\cdot\boldsymbol{B}_i, \boldsymbol{\tau}\cdot(\tilde{\alpha}_+ \times \tilde{B}_i)]] \\ &= 2\,\mathrm{TR}[\tilde{Q}_{\mathrm{L}}\tau^b](\boldsymbol{B}_i \times (\boldsymbol{\alpha}_+ \times \boldsymbol{B}_i))^b\,. \end{aligned} \tag{E.20}$$

Using the following results:

$$\int \mathrm{d}\hat{r}\,\alpha^b_+ B^2_i = \tfrac{16}{3}\pi\sin^2 F\left(F'^2 + 2\frac{\sin^2 F}{r^2}\right)\omega^b\,, \qquad \int \mathrm{d}\hat{r}\, B^b_i(\boldsymbol{\alpha}_+\cdot\boldsymbol{B}_i) = \tfrac{16}{3}\pi\frac{\sin^4 F}{r^2}\,\omega^b\,, \tag{E.21}$$

we obtain

$$\bar{g}_0^{\rm L} = \int {\rm d}\hat{r}\, g_0^{\rm L} = \tfrac{64}{3}\pi \sin^2 F \left[F'^2 + \frac{\sin^2 F}{r^2}\right] {\rm TR}(Q_{\rm L}\dot{A}A^+)\,. \tag{E.22}$$

Similarly, the space component of (E.19) becomes

$$\begin{aligned} g_i^{\rm L} &= -{\rm TR}(Q_{\rm L}[R_j, [R_i, R_j]]) \\ &= -4{\rm i}\, {\rm TR}[\tilde{Q}_{\rm L}\tau^b](\boldsymbol{B}_j \times (\boldsymbol{B}_i \times \boldsymbol{B}_j))^b\,. \end{aligned} \tag{E.23}$$

Using the following results:

$$\begin{aligned} &\int {\rm d}\hat{r}\, B_i^b B_j^2 = -\tfrac{4}{3}\pi\delta^{ib}\left(F'^2 + \frac{2\sin^2 F}{r^2}\right)\left(F' + \frac{\sin 2F}{r}\right), \\ &\int {\rm d}\hat{r}\, B_j^b(\boldsymbol{B}_i \cdot \boldsymbol{B}_j) = -\tfrac{4}{3}\pi\delta^{ib}\left(F'^3 + \frac{\sin^2 F}{r^3}\sin 2F\right), \end{aligned} \tag{E.24}$$

we obtain

$$\begin{aligned} \bar{g}_i^{\rm L} &= \int {\rm d}\hat{r}\, g_i^{\rm L} \\ &= +{\rm i}\tfrac{16}{3}\pi\left\{F'^2\frac{\sin 2F}{r} + 2F'\frac{\sin^2 F}{r^2} + \frac{\sin^2 F}{r^3}\sin 2F\right\} {\rm TR}[Q_{\rm L}A\tau_i A^+]\,. \end{aligned} \tag{E.25}$$

Combining (E.17) and (E.22) gives

$$\int {\rm d}\hat{r}\, J_{0,{\rm L}}^a = -{\rm i}\tfrac{32}{3}\pi c_0 \sin^2 F\left\{1 - 8\frac{c_1}{c_0}\left(F'^2 + \frac{\sin^2 F}{r^2}\right)\right\} {\rm TR}[Q_{\rm L}^a\dot{A}A^+]\,, \tag{E.26}$$

while combining (E.18) and (E.25) yields

$$\begin{aligned} \int {\rm d}\hat{r}\, J_{i,{\rm L}}^a = &+\tfrac{8}{3}\pi c_0\left\{\left(F' + \frac{\sin 2F}{r}\right) - 8\frac{c_1}{c_0}\left(F'^2\frac{\sin 2F}{r} + 2F'\frac{\sin^2 F}{r^2} + \frac{\sin^2 F}{r^3}\sin 2F\right)\right\} \\ &\times {\rm TR}[Q_{\rm L}^a A\tau_i A^+]\,. \end{aligned} \tag{E.27}$$

To obtain the angle-averaged counterparts of the right-handed currents one can use the substitution $F \to -F$ in (E.26) and (E.27). Consequently, the angle-averaged vector and axial-vector currents become ($Q_{\rm R,L}^a \equiv \tau^a/2$)

$$\int \mathrm{d}\hat{r}\, V_0^a = -\mathrm{i}\tfrac{32}{3}\pi c_0 \sin^2 F\left\{1 - 8\,\frac{c_1}{c_0}\left(F'^2 + \frac{\sin^2 F}{r^2}\right)\right\} \mathrm{TR}[\tau_a \dot{A} A^+]\,, \tag{E.28}$$

$$\int \mathrm{d}\hat{r}\, A_i^a = +\tfrac{8}{3}\pi c_0\left\{\left(F' + \frac{\sin 2F}{r}\right) - 8\,\frac{c_1}{c_0}\left(F'^2\,\frac{\sin^2 F}{r} + 2F'\,\frac{\sin^2 F}{r^2} + \frac{\sin^2 F}{r^3}\sin 2F\right)\right\}$$
$$\times \mathrm{TR}[\tau_a A \tau_i A^+]\,. \tag{E.29}$$

For completeness, notice that

$$\begin{aligned}\int \mathrm{d}\hat{r}\, \hat{r}^c B_i^b B_j^2 &= \tfrac{4}{3}\pi\varepsilon^{ibc}\,\frac{\sin^2 F}{r}\left(F'^2 + 2\,\frac{\sin^2 F}{r^2}\right),\\ \int \mathrm{d}\hat{r}\, \hat{r}^c B_j^b (\boldsymbol{B}_i\cdot \boldsymbol{B}_j) &= \tfrac{4}{3}\pi\varepsilon^{ibc}\,\frac{\sin^4 F}{r^3}\,,\end{aligned} \tag{E.30}$$

so that

$$\int \mathrm{d}\hat{r}\, \boldsymbol{q}\cdot\hat{r} J_{i,\mathrm{L}}^a = \tfrac{8}{3}\pi c_0\,\frac{\sin^2 F}{r}\left\{1 - 8\frac{c_1}{c_0}\left(F'^2 + \frac{\sin^2 F}{r^2}\right)\right\}\varepsilon^{ibc}\,\mathrm{TR}[Q_\mathrm{L}^a A \tau^b A^+]q^c\,, \tag{E.31}$$

hence,

$$\int \mathrm{d}\hat{r}\, \boldsymbol{q}\cdot\hat{r} V_i^a = \mathrm{i}\tfrac{16}{3}\pi c_0\,\frac{\sin^2 F}{r}\left\{1 - 8\,\frac{c_1}{c_0}\left(F'^2 + \frac{\sin^2 F}{r^2}\right)\right\}\mathrm{TR}[(\boldsymbol{\tau}\cdot\boldsymbol{q})\tau^i A^+ \tau^a A]\,. \tag{E.32}$$

Appendix F: Evaluation of $\Gamma_{WZ}(2\pi)$

In discussing the spin statistics of the skyrmion in section 3, we were led to the evaluation of the Wess–Zumino term with the parameterization (132) for the chiral field, i.e.

$$V(x, t, \rho) = A^+(t, \rho)U(\boldsymbol{x})A(t, \rho)\,, \tag{F.1}$$

which is a legitimate extension to the disc D^5 since $A(t)$ is periodic and $\pi_1(\mathrm{SU}(3)) = 0$. To evaluate the Wess–Zumino term for the particular parameterization (F.1), let us denote by a, b, c the x-coordinates and by i, j the t-ρ-coordinates. With this in mind, we have

$$\varepsilon^{\alpha\beta\gamma\mu\nu}\,\mathrm{TR}[L_\alpha L_\beta L_\gamma L_\mu L_\nu] = 5\varepsilon^{abc}\varepsilon^{ij}\,\mathrm{TR}[L_i L_j L_a L_b L_c] + \cdots\,. \tag{F.2}$$

The dot terms in (F.2) do not contribute to the Wess–Zumino term, and will be dropped from now on. Injecting (F.1) into (F.2) yields

$$(\mathrm{F.2}) = 5\varepsilon^{abc}\varepsilon^{ij}\,\mathrm{TR}[L_i L_j A U^+\partial_a U U^+\partial_b U U^+\partial_c U A^+]\,. \tag{F.3}$$

Because of rotational invariance, the topological density on D^5 should be independent of the angular distribution on S^3. Therefore, any point on S^3 is equally appropriate for evaluating (F.3). In the $\hat{z}$-direction, $U(\boldsymbol{x})$ simplifies to

$$U = \cos F + \mathrm{i}\tau_3 \sin F \,, \tag{F.4}$$

with

$$\partial_{1,2} U = \mathrm{i}\,\frac{\tau_{1,2}}{r} \sin F \,, \qquad U^+ \partial_3 U = \mathrm{i}\tau_3 F' \,, \tag{F.5}$$

so that

$$\varepsilon^{abc} U^+ \partial_a U U^+ \partial_b U U^+ \partial_c U \equiv f(r) \mathbb{1}_2 \,, \tag{F.6}$$

where $\mathbb{1}_2$ is a 2×2 unit matrix and $f(r)$ a radial function of r. In terms of (F.3) and (F.6), the Wess–Zumino term in (133) reads

$$\Gamma(2\pi) = \frac{\mathrm{i}BN_\mathrm{c}}{2} \int\limits_{[0,2\pi]\times[0,1]} \mathrm{d}^2x \, \varepsilon^{ij} \, \mathrm{TR}[\partial_i A^+ \partial_j A l_2] \,, \tag{F.7}$$

where B is the winding number of the hedgehog configuration $U(\boldsymbol{x})$ in (F.1). Using a polar parameterization on $\mathrm{D}_2 = [0, 2\pi] \times [0, 1]$, namely,

$$A(t, \rho) = \begin{pmatrix} 1 & 0 & 0 \\ 0 & \rho\, \mathrm{e}^{-\mathrm{i}t} & \sqrt{1-\rho^2} \\ 0 & -\sqrt{1-\rho^2} & \rho\, \mathrm{e}^{\mathrm{i}t} \end{pmatrix} \tag{F.8}$$

we have

$$\varepsilon^{ij} \, \mathrm{TR}[\partial_i A^+ \partial_j A l_2] = -2\mathrm{i}\rho \,, \tag{F.9}$$

hence

$$(2\pi) = BN_\mathrm{c} \int_0^1 \rho \, \mathrm{d}\rho \int_0^{2\pi} \mathrm{d}t = N_\mathrm{c} B\pi \,, \tag{F.10}$$

which is the result used in the text for $B = 1$. Again, due to the topological character of Γ, this result is invariant under continuous deformations of the chiral field that leaves B unchanged.

Appendix G: Quantization of the collective coordinates

In this appendix, we will present the formal framework for discussing constrained quantization of collective coordinates. Consider the quantum mechanical Lagrangian associated to a spherical top in SU(2) as given by

$$\mathcal{L}_A = I\,\mathrm{TR}[\dot{A}^+\dot{A}]\,, \tag{G.1}$$

we will use the following parameterization of A in SU(2):

$$A = a_0[\xi] + \mathrm{i}\boldsymbol{\tau}\cdot\boldsymbol{a}[\xi] = \begin{pmatrix} a_0 + \mathrm{i}a_3 & a_2 + \mathrm{i}a_1 \\ -a_2 + \mathrm{i}a_1 & a_0 - \mathrm{i}a_3 \end{pmatrix}, \tag{G.2}$$

where the ξ's are 3 independent variables on S^3, e.g.

$$a_0 = \cos\theta\,, \qquad a_1 = \sin\theta\cos\phi\,, \qquad a_2 = \sin\theta\sin\phi\cos\psi\,,$$
$$a_3 = \sin\theta\sin\phi\sin\psi\,. \tag{G.3}$$

$\mathcal{L}_A$ is invariant under respectively "rotations" and "isorotations"

$$A \to Af \qquad \text{and} \qquad A \to fA \qquad f \in \mathrm{SU}(2)\,,$$

and also a discrete Z_2-symmetry, $A \to \pm A$. Since det $A = 1$, we have

$$A^+\,\partial A/\partial\xi^a = \mathrm{i}h_{ab}T_b\,, \tag{G.4a}$$

$$A\,\partial A^+/\partial\xi^a = \mathrm{i}k_{ab}T_b\,, \tag{G.4b}$$

with $T^a = \tau^a/2$. In terms of the independent variables ξ's, (G.1) reads

$$\mathcal{L}_A = \tfrac{1}{2}I\dot{\xi}_a h_{ab}h_{cb}\dot{\xi}_b = \tfrac{1}{2}I\dot{\xi}^{\mathrm{T}}h\,h^{\mathrm{T}}\dot{\xi}\,, \tag{G.5}$$

where the upper index T stands for transposed. If we denote by π the canonical momentum of ξ, i.e.

$$\pi^{\mathrm{T}} = \partial L/\partial\dot{\xi} = I\dot{\xi}^{\mathrm{T}}hh^{\mathrm{T}}\,, \tag{G.6}$$

then the Hamiltonian density H associated to (G.5) becomes

$$H = \pi^{\mathrm{T}}\dot{\xi} - \mathcal{L}_A = \frac{1}{2I}\,\pi^{\mathrm{T}}(hh^{\mathrm{T}})^{-1}\pi\,, \tag{G.7}$$

where ξ and π satisfy the following Poisson brackets

$$\{\xi_a, \xi_b\}_{\mathrm{P.B.}} = \{\pi_a, \pi_b\}_{\mathrm{P.B.}} = 0\,, \qquad \{\xi_a, \pi_b\}_{\mathrm{P.B.}} = \delta_{ab}\,. \tag{G.8a, b}$$

Canonical quantization consists in postulating

$$[\xi_a, \xi_b] = [\pi_a, \pi_b] = 0\,, \qquad [\xi_a, \pi_b] = \mathrm{i}\delta_{ab}\,, \tag{G.9a, b}$$

The Hilbert space attached to this problem is the space of C_∞-matrices in SU(2) valued in C, i.e. (SU(2); C). The classical spin and isospin charges associated to (G.1) follow from Noether theorem,

$$J_a = \mathrm{i}I\,\mathrm{TR}[\tau_a \dot{A}^+ A]\,, \tag{G.10a}$$

$$I_a = \mathrm{i}I\,\mathrm{TR}[\tau_a \dot{A} A^+]\,. \tag{G.10b}$$

Notice that we go from spin to isospin through the substitution $A^+ \leftrightarrows A$. In the skyrmion language this corresponds to an inversion of the spinning direction (parity). The corresponding generators in (SU(2); C) are

$$J_a = h^{-1}_{ab}\pi_b\,, \tag{G.11a}$$

$$I_a = k^{-1}_{ab}\pi_b\,, \tag{G.11b}$$

and fulfill the standard SU(2) algebra. Indeed,

$$\begin{aligned}[J_a, J_b] &= [h^{-1}_{ac}\pi_c, h^{-1}_{bd}\pi_d] \\ &= h^{-1}_{ab}[\pi_c, h^{-1}_{bd}]\pi_d - h^{-1}_{bd}[\pi_d, h^{-1}_{ac}]\pi_c \\ &= \mathrm{i}\left(h^{-1}_{bd}\,\frac{\partial h^{-1}_{ac}}{\partial \xi_d} - h^{-1}_{ad}\,\frac{\partial h^{-1}_{bc}}{\partial \xi_d}\right)\pi_c\,.\end{aligned} \tag{G.12}$$

Injecting unity, $hh^{-1} = \mathbb{1}$, and using the fact that

$$\frac{\partial h^{-1}_{ae}}{\partial \xi_d}\,h_{ef} = \tfrac{1}{2}\varepsilon_{adf} + h_{dg} + 2\mathrm{i}h^{-1}_{ae}\,\mathrm{TR}\left(A^+\,\frac{\partial^2 A}{\partial \xi_d\,\partial \xi_e}\,\tau_f\right) \tag{G.13}$$

yield

$$[J_a, J_b] = \mathrm{i}\varepsilon_{abc}J_c\,. \tag{G.14}$$

Similar relations hold for I_a's. To see how the elements of (SU(2), C) behave under (G.11), consider

$$[J_a, A] = h^{-1}_{ab}[\pi_b, A] = -\mathrm{i}h^{-1}_{ab}\,\frac{\partial A}{\partial \xi_b}\,. \tag{G.15}$$

Inserting (G.4a) gives

$$\mathrm{i}AT_b = (h^{-1})_{ba}\,\frac{\partial A}{\partial \xi_a}\,, \tag{G.16}$$

hence

$$[J_a, A] = AT_a\,. \tag{G.17}$$

Similarly, we have

Table 4

	A_{11}	A_{12}	A_{12}	A_{22}
J_3	1/2	−1/2	1/2	−1/2
I_3	−1/2	−1/2	1/2	1/2
	$\|n\uparrow\rangle$	$\|n\downarrow\rangle$	$\|p\uparrow\rangle$	$\|p\downarrow\rangle$

$$[I_a, A] = -T_a A\,. \tag{G.18}$$

In other words, J_a and I_a induce $SU(2)_R$ and $SU(2)_L$ rotations respectively. The matrix elements of A correspond to the fundamental spinor representation as summarized in the table above. Higher representations are polynomials in A. For instance,

$$[J_3, A_{11}^l] = [J_3, A_{11}]A_{11}^{l-1} + A[J_3, A_{11}^{l-1}] = \tfrac{1}{2} l A_{11}^l\,. \tag{G.19}$$

So, $A^2, A^4, \ldots$ carry integer spin, while $A, A^3, A^5, \ldots$ carry half-integer spin. The proton and neutron wavefunctions (301) follow immediately from table 4, after proper normalization on S^3. In terms of the generators (G.11), the Hamiltonian (G.7) reads

$$H = J^2/2I = I^2/2I\,, \tag{G.20}$$

which shows that the left and right Casimir operators of SU(2) are identical.

References

[1] G. 't Hooft, Nucl. Phys. B 72 (1974) 461; ibid B 75 (1974) 461.
[2] E. Witten, Nucl. Phys. B 160 (1979) 57.
[3] T.H.R. Skyrme, Proc. Roy. Soc. London 260 (1961) 127; ibid 262 (1961) 237;
T.H.R. Skyrme, Nucl. Phys. 31 (1962) 556.
[4] For a review see B.W. Lee, Chiral Dynamics (Gordon and Breach, New York, 1972).
[5] D. Finkelstein and J. Rubinstein, J. Math. Phys. 9 (1968) 1762.
[6] J.G. Williams, J. Math. Phys. 11 (1970) 2611.
[7] H.D. Politzer, Phys. Rev. Lett. 30 (1973) 1346;
D.J. Gross and F. Wilczek, Phys. Rev. Lett. 30 (1973) 1343.
[8] S. Adler, Phys. Rev. 117 (1969) 2426;
J. Bell and R. Jackiw, Nuov. Cimento 60A (1969) 47.
[9] H. Sugawara, Phys. Rev. 170 (1968) 1659.
[10] J.F. Adams, Bull. Soc. Math. (Paris) 87 (1959) 277.
[11] I. Zahad, A. Wirzba, U.-G. Meißner, C.J. Pethick and J. Ambjørn, Phys. Rev. D 31 (1985) 1114.
[12] Y.P. Rybakov and V.I. Sanyuk, NBI-HE-84 (1981) unpublished.
[13] G.H. Derrick, J. Math. Phys. 5 (1964) 1252;
R. Hobart, Proc. Roy. Soc. London 82 (1963) 201.
[14] E. Bogomolny, Sov. J. Nucl. Phys. 24 (1976) 449.
[15] S. Weinberg, Physica 96 (1979) 325.
[16] E. Cartan, La theorie des groupes finis et continus (Gauthier-Villars, Paris, 1951).
[17] M. Rho, Saclay preprint Pht 84-123 (1984).
[18] J.F. Donoghue, E. Golowich and B.R. Holstein, Phys. Rev. Lett. 53 (1984) 747.
[19] I. Zahed, A. Wirzba and U.G. Meißner, Phys. Rev. D 33 (1986) 830.
[20] A.D. Jackson and M. Rho, Phys. Rev. Lett. 51 (1983) 751.

[21] G.S. Adkins, C.R. Nappi and E. Witten, Nucl. Phys. B 228 (1983) 552.
[22] (a) G.S. Adkins and C.R. Nappi, Nucl. Phys. B 233 (1984) 109;
(b) U.G. Meißner, Ruhr Univ. Bochum preprint (1983).
[23] H.J. Schnitzer, Phys. Lett. B 139 (1984) 217.
[24] S. Weinberg, Phys. Rev. Lett. 18 (1967) 188.
[25] E. Witten, Nucl. Phys. B 223 (1983) 422; ibid 433.
[26] J. Wess and B. Zumino, Phys. Lett. B 37 (1971) 95.
[27] P.A.M. Dirac, Proc. Roy. Soc. London A 126 (1930) 360.
[28] R. Jackiw, Les Houches Summer School (1983).
[29] R. Bott, Ann. Math. (2) 60 (1954) 248; ibid 70 (1959) 179.
[30] G. De Rham, Jour. Math. (Paris) 10 (1931) 115.
[31] O. Kaymakcalan, S. Rajeev and J. Schechter, Phys. Rev. D 30 (1984) 594;
O. Kaymakcalan and J. Schechter, Phys. Rev. D 31 (1985), 1109;
see also, Y. Brihaye, N.K. Pak and P. Rossi, CERN-TH 3984 184 (1984).
[32] A.P. Balachandran, V.P. Nair, S.C. Rajeev and A. Stern, Phys. Rev. Lett. 49 (1982) 1124; Phys. Rev. D 27 (1983) 1153.
[33] J. Goldstone and F. Wilczek, Phys. Rev. Lett. 47 (1981) 986.
[34] A.P. Balachandran, F. Lizzi, V.G.J. Rodgers and A. Stern, Nucl. Phys. B 256 (1985) 525.
[35] R.L. Jaffe and C.L. Korpa, Nucl. Phys. B 258 (1985) 468.
[36] E. Guadagnini, Nucl. Phys. B 236 (1984) 35;
see also G.S. Adkins and C.R. Nappi, Nucl. Phys. B 249 (1985) 507.
[37] R.L. Jaffe, private communication.
[38] F. Wilczek and A. Zee, Phys. Rev. Lett. 51 (1983) 2250.
[39] S.L. Glashow, J. Iliopoulos and L. Maiani, Phys. Rev. D 2 (1970) 1285.
[40] C. Callan, Phys. Rev. D 25 (1982) 2141: ibid 26 (1982) 2058;
V. Rubakov, JETP Lett. 33 (1981) 644, and Nucl. Phys. B 203 (1982) 311.
[41] M. Rho, A.S. Goldhaber and G.E. Brown, Phys. Rev. Lett. 51 (1983) 747.
[42] J. Goldstone and R.L. Jaffe, Phys. Rev. Lett. 51 (1983) 1518.
[43] L. Alvarez-Gaumé and E. Witten, Nucl. Phys. B 234 (1984) 1269.
[44] W.A. Bardeen, Phys. Rev. 184 (1969) 1848.
[45] K. Fujikawa, Phys. Rev. Lett. 42 (1979) 1195; ibid 44 (1980) 1733;
Phys. Rev. D 21 (1980) 2848; ibid D 22 (1980) 1499; ibid D 23 (1980) 2262.
[46] H.B. Nielsen and M. Ninomiya, Phys. Lett. B 130 (1983) 389.
[47] R. Stora, Cargese lectures (1983), and preprint LAPP-TH-94 (1983).
[48] B. Zumino, in: Relativity, Groups and Topology II, Les Houches (1983), eds. B. de Witt and R. Stora (North-Holland, Amsterdam, 1984).
[49] M. Atiyah, V. Patodi and I. Singer, Proc. Camb. Philos. Soc. 77 (1975) 42.
[50] C. Becchi, A. Rouet and R. Stora, Ann. Phys. 98 (1976) 287.
[51] S. Weinberg, Phys. Rev. 166 (1968) 1568, and ref. 24.
[52] C. Callan, S. Coleman, J. Wess and B. Zumino, Phys. Rev. 177 (1969) 2247.
[53] G.S. Adkins and C.R. Nappi, Phys. Lett. B 137 (1984) 251.
[54] (a) K. Iketani, Kyushu Univ. 84-HE-2 (1984);
(b) M. Bando, T. Kugo, S. Uehara, K. Yamawaki and T. Yamagida, Phys. Rev. Lett. 54 (1985) 1215;
M. Bando, T. Kugo and K. Yamawaki, Prog. Theor. Phys. 73 (1985) 1541; Nucl. Phys. B 259 (1985) 493.
T. Fujiwara, T. Kugo, H. Terao, S. Uehara and K. Yamawaki, Prog. Theor. Phys. 73 (1985) 926;
(c) S. Weinberg, Phys. Rev. 166 (1968) 1568.
[55] U.-G. Meißner and I. Zahed, Phys. Rev. Lett. 56 (1986) 1035.
[56] P. Simic, Phys. Rev. Lett. 55 (1985) 40; Rockefeller Univ. preprint RU/85/124 (1985).
[57] M. Gaudin, Saclay preprint SPH-T/84-004 (1984).
[58] I.J.R. Aitchison and C.M. Fraser, Phys. Lett. B 146 (1984) 63.
[59] R. Mackenzie, Phys. Rev. D 30 (1984) 2194; 2260.
[60] S. Kahana, G. Ripka and V. Soni, Nucl. Phys. A 415 (1984) 351.
[61] P. Castorina and S.Y. Pi, Phys. Rev. D 31 (1985) 411.
[62] R.E. Peierls and J. Yoccoz, Proc. Phys. Soc. A 70 (1957) 381.
[63] S. Nadkarni, H.B. Nielsen and I. Zahed, Nucl. Phys. B 253 (1985) 308.
S. Nadkarni and I. Zahed, Nucl. Phys. B 263 (1986) 23.
[64] I. Zahed, U.-G. Meißner and U.B. Kaulfuß, Nucl. Phys. A 426 (1984) 525.
[65] J.D. Breit and C.R. Nappi, Phys. Rev. Lett. 53 (1984) 889.
[66] A. Hayashi and G. Holzwarth, Phys. Lett. B 140 (1984) 175;
the breathing mode is also discussed in

(a) C. Hadjuk and B. Schwesinger, Phys. Lett. B 140 (1984) 172;
J. Dey, J. Le Tourneux, Montreal preprint (1984);
J.D. Breit, Penn. Univ. preprint UPR-0271T (1984);
K.F. Liu, J.S. Zhang and G.R.E. Black, Phys. Rev. D 30 (1984) 2015;
A. Parmentola, Phys. Rev. D 30 (1984) 685;
(b) U.B. Kaulfuss and U.G. Meissner, Bochum Univ. preprint (1985).
[67] M.P. Mattis and M.E. Peskin, Phys. Rev. D 32 (1985) 58;
M.P. Mattis and M. Karliner, SLAC-PUB-3539 (1984);
M.E. Peskin, talk at the Niels Bohr Centennial (1985), and SLAC-Pub.-3703.
[68] A. Jackson, A.D. Jackson and V. Pasquier, Nucl. Phys. A 432 (1985) 567.
[69] M. Lacombe, B. Loiseau, J.M. Richard, R. Vin Mau, J. Cote, P. Pires and R. De Toureuil, Phys. Rev. C 21 (1980) 861.
[70] E. Braaten and J.P. Ralston, Phys. Rev. D 31 (1985) 598;
M. Bander and F. Hayot, Phys. Rev. D 30 (1984) 1837.
[71] A.D. Jackson, private communication.
[72] G.E. Brown, A.D. Jackson, M. Rho and V. Vento, unpublished notes.
[73] E. Witten, in: Solitons in Nuclear and Elementary Particle Physics, eds. A. Chodos, E. Hadjimichael and C. Tze (World Scientific, Singapore, 1984) pp. 306–312; see also references therein.
[74] G. Karl and J.E. Paton, Phys. Rev. D 30 (1984) 238.
[75] M.A. Bég and Z. Zepeda, Phys. Rev. D 6 (1972) 2912.
[76] G.E. Brown and F. Myhrer, Phys. Lett. B 128 (1983) 229.
[77] (a) N. Isgur and G. Karl, Phys. Lett. B 72 (1977) 109; Phys. Rev. D 18 (1978) 4187;
(b) H.R. Feibig and B. Schwesinger, Nucl. Phys. A 393 (1983) 349.
[78] A. Jackson, private communication.
[79] U.-G. Meißner, private communication.
[80] M. Sommerman, H.W. Wyld and C.J. Pethick, Phys. Rev. Lett. 55 (1985) 476.
[81] E.B. Bogomol'ny and V.A. Fateev, Sov. J. Nucl. Phys. 37 (1983) 134.
[82] M. Kutshera and C.J. Pethick, Nucl. Phys. A 440 (1985) 670.
[83] M. Kutshera, C.J. Pethick and D.G. Ravenhall, Phys. Rev. Lett. 53 (1984) 1041.
[84] I. Klebanov, Nucl. Phys. B 262 (1985) 133.
[85] A. Jackson, A.D. Jackson, F. Goldhaber, G.E. Brown and L.C. Castillejo, Phys. Lett. B 154 (1985) 101.
[86] K. Holinde, Phys. Rep. 68 (1981) 121.
[87] T.N. Pham and T.N. Trovona, Ecole polytechnique preprint (1985).
[88] M. Lacombe, B. Loiseau, R. Vinh Mau and W.N. Cottingham, Phys. Lett. B 169 (1986) 121.
[89] D. Klabucar, Stony Brook thesis (1986).
[90] V. Vento, M. Rho, E.B. Nyman, J.H. Jun and G.E. Brown, Nucl. Phys. A 345 (1980) 413.
[91] A. Chodos and C.B. Thorn, Phys. Rev. D 12 (1975) 2733.
[92] T. Inoue and T. Maskawa, Prog. Theor. Phys. 54 (1975) 1833.
[93] L. Vepstas, A.D. Jackson and F. Goldhaber, Phys. Lett. B 140 (1984);
G.E. Brown, A.D. Jackson, M. Rho and V. Vento, Phys. Lett. B 140 (1984) 285.
[94] I. Zahed, U.-G. Meissner and A. Wirzba, Phys. Lett. B 145 (1984) 117;
I. Zahed, A. Wirzba and U.-G. Messner, Ann. Phys. 165 (1985) 406.
[95] P.J. Mulders, Phys. Rev. D 30 (1984) 1037.
[96] M. Chemtob, CEN Saclay preprint (1984);
M. Praszalowicz, CERN-TH-4114/85.
[97] S.A. Yost and C.R. Nappi, Phys. Rev. D 32 (1985) 816.
[98] C.G. Callan and I. Klebanov, Nucl. Phys. B 262 (1985) 365.
[99] G.E. Brown, M. Rho and W. Weise, to be published.
[100] G.E. Brown and M. Rho, Comments, to be published.

Note added in proof

Some time ago, Braaten and Ralston [70] and Bander and Hayot [101] argued that if one were to include rotational effects on the soliton profile, the skyrmion mass will blow up in the chiral limit. They suggested that a rotating skyrmion will be forced to radiate massless pions, and concluded that in this case the semiclassical treatment was inappropriate. Recently, Verschelde [102] has pointed out that these arguments were not correct. He has shown that a careful analysis of the rotational and vibrational effects, using the proper canonical variables, yields variational equations in which the radiation terms cancel out exactly. Thus, there is nothing wrong with the semiclassical quantization of rotating skyrmions in the chiral limit.

Recently, Meissner and Zahed [103] have argued that the approach followed in this review to describe vector mesons as massive Yang–Mills particles is entirely similar to the seemingly alternative approach described in [54b] and based on the concept of a hidden local symmetry. The proof is based on the fact that in general, a massive Yang–Mills model is gauge equivalent to a massless Yang–Mills model gauge coupled to a non-linear σ-model. They have also shown that the bulk parameters of the nucleon and Δ-isobar are substantially improved by the addition of vector mesons.

Recently also, there has been a considerable amount of work carried out at SLAC by Karliner and Mattis [104] on the systematics of πN, KN and $\bar{\rm K}$N scattering in the context of the Skyrme model. Their results are extremely encouraging.

[101] M. Bander and F. Hayot, Phys. Rev. D 30 (1984) 1837.
[102] H. Verschelde, SUNY preprint (1986).
[103] U.G. Meissner and I. Zahed, SUNY preprint (1986).
[104] M. Karliner and M.P. Mattis, SLAC preprint 3901 (1986).